Issues in Contemporary Oil Paint

Klaas Jan van den Berg • Aviva Burnstock
Matthijs de Keijzer • Jay Krueger • Tom Learner
Alberto de Tagle • Gunnar Heydenreich
Editors

Issues in Contemporary Oil Paint

Springer

Editors
Klaas Jan van den Berg
Matthijs de Keijzer
Alberto de Tagle
Cultural Heritage Agency of the Netherlands
Amsterdam, The Netherlands

Jay Krueger
National Gallery of Art
Washington, DC, USA

Gunnar Heydenreich
Cologne Institute of Conservation Sciences
Cologne University of Applied Sciences
Köln, Germany

Aviva Burnstock
Department of Conservation and Technology
The Courtauld Institute of Art
London, UK

Tom Learner
Science Department
The Getty Conservation Institute
Los Angeles, CA, USA

ISBN 978-3-319-10099-9 ISBN 978-3-319-10100-2 (eBook)
DOI 10.1007/978-3-319-10100-2
Springer Cham Heidelberg New York Dordrecht London

Library of Congress Control Number: 2014955358

Printed on acid-free paper

Springer is part of Springer Science+Business Media (www.springer.com)

Foreword

It is a great pleasure to present the Proceedings of the Issues in Contemporary Oil Paint (ICOP) Symposium that was held in 28–29 March, 2013, in Amersfoort at the headquarters of the Cultural Heritage Agency of the Netherlands, RCE.

Our Agency is at the heart of cultural heritage in the Netherlands. Our research group is concerned with the evaluation and preservation of our heritage in the broadest sense and the research directions we need to follow in order to guarantee a sustainable heritage. Along with national and international research partners at museums, universities and archives, we conduct research, characterise materials and analyse change processes. The Agency ensures that third parties can apply the knowledge that we can provide.

The ICOP symposium was the first symposium focused on modern paints since the Modern Paints Uncovered (MPU) conference held at Tate in 2005. Whereas MPU mainly presented research on modern synthetic paints especially on acrylics, ICOP chose to focus on modern oil paints entirely. Many modern artists continue to work with oil paints, and modern oil paints increasingly become a challenge for conservators and collection keepers. Therefore it was felt by the organisers that it was time to organise a meeting which could discuss these challenges by presenting information on historical and artistic production, scientific research on degradation phenomena, and developing alternative conservation approaches.

ICOP marked the end of a 4-year Research Agenda, for our Agency.[1] In one of the programmes in the Research Agenda, *Object in Context,* the RCE research group generated knowledge on the production of and changes in heritage objects in their artistic, cultural and social contexts. Under the leadership of Klaas Jan van den Berg, the '20th century oil paint project' contributed to the outcome of the Agenda. The project brought many institutions together and was a breeding ground for

[1]Outcomes of this Research Agenda are accessible on-line: http://www.kennisvoorcollecties.nl/en/researchagenda/.

students and graduates from Dutch Universities, but also from the UK, Italy, Portugal, France and Germany, some of whom present results in these Proceedings.

I hope the Proceedings will be of interest not only to the participants of the symposium but also to a larger audience in the cultural heritage community for many years to come.

Sincerely,

Cees van 't Veen
Director, Cultural Heritage Agency of the Netherlands
Amersfoort, the Netherlands

Preface

Modern paints offer many challenges for scientists and conservators alike, and have consequently been studied in some depth over the past two decades. *Modern Paints Uncovered*, a symposium held in 2006 at Tate Modern in London, was the first professional event devoted specifically to addressing the conservation issues of modern paints. Research presented there included advances in analytical methods, improved understanding of the physical properties and surface characteristics of modern paints, and assessments of the effects of cleaning treatments on painted surfaces. However, at that time much of the profession's attention was directed towards acrylic emulsion paints, the most widely used of all the artists' synthetic polymer paints, with relatively few studies being undertaken on modern oil paints.

Artisans and artists are known to have been using oil paints for centuries[1] but particularly the industrial developments in the nineteenth and especially the twentieth centuries have resulted in significant changes to that material. The modification and application of oil paints by artists have continued to evolve in the twentieth century. In addition, many paintings are unvarnished and so atmospheric influences happen directly to the paint surface. These developments make it reasonable to say that modern oil paintings have properties which are distinctly different from traditional paintings and thus provide many unique challenges for conservators and curators.

The *Issues in Contemporary Oil Paint* (ICOP) symposium, held in 2013 in Amersfoort, the Netherlands, provided a forum to discuss these challenges and to offer new information that will help conservators and curators recognise problems and interpret visual changes on paintings, which in turn may help to give a more solid basis for decisions on the treatment of these paintings. Furthermore,

[1]Raspe RE (1781) A Critical Essay on Oil-painting; proving that the art of painting in oil was known before the pretended discovery of J. and H. van Eyck; to which are added, Theophilus de Arte Pingendi, Eraclius de Artibus Romanorum, and a review of Farinator's Lumen Animæ. London, Goldney and Cadell.

interactions with paint makers and scientists were intended to provide more information on the application and stability of paints to artists who are concerned with the longevity of their artwork.

The present proceedings represent 27 long and short presentations at the ICOP symposium. The Chapters have all undergone a peer-review process and cover a range of topics. These range from developments of paint technology, working methods of individual artists, through characterisation of paints and paint surfaces, paint degradation and the factors that influence this versus long time stability, to observations of issues in collections, cleaning and other treatment issues as well as new conservation approaches.

Research and Conservation

Chapter "Twentieth Century Oil Paint. The Interface Between Science and Conservation and the Challenges for Modern Oil Paint Research" discusses the way in which research can inform conservation by collaboration between conservators and scientists. This is done through utilisation of phenomenological and experimental methodologies to investigate the causes of optical changes in paint. Some examples of results based on case studies of paintings and the results of systematic experimental investigation of paint samples are given. This essay is corroborated by a personal account of 25 years of research, conservation treatment and display of paintings at the Stedelijk Museum in Amsterdam in chapter "Do We See What We Know or Do We Know What We See? Conservation of Oil Paintings in the Stedelijk Museum Amsterdam". The relationship between painting technique and paint quality is discussed in relation to the works of important artists in the museum's collection, including constructivists, Cobra artists and modern American painters from the 1950s to 1970s. Interestingly, while knowledge and appreciation of deterioration phenomena on these paintings has substantially increased in recent years, this in turn has often led to less conservation treatment.

Developments in Paint Technology

Many new pigments were introduced in the twentieth century, especially synthetic organic pigments. In chapter "The Delight of Modern Organic Pigment Creations", an extensive survey of their history and development in production is presented. Introduced in paints, the interaction of the organic pigments with the paint matrix can be poor which may make them vulnerable to solvents used in cleaning operations. In chapter "Sensitivity of Modern Oil Paints to Solvents. Effects on Synthetic Organic Pigments", the sensitivity of a range of these organic pigments towards dissolution and mechanical action in conservation treatment is discussed. Chapter "Distinction by Micro-Raman Spectroscopy and Chemometrical Analysis of Copper

Phthalocyanine Blue Polymorphs in Oil-Based and Acrylic Paint samples" describes one of the most important representatives of this group, PB15 copper phthalocyanine blue together with the characterisation of its polymorphous phases.

An overview of developments in oil paint manufacture is presented in chapter "Modern Oil Paints – Formulations, Organic Additives and Degradation: Some Case Studies" and highlighted with a range of analytical results from a number of tube paints and paintings. In chapter "Charting the Development of Oil-Based Enamel Paints Through the Correlation of Historical Paint Technology Manuals with Scientific Analysis", an overview of developments in technology of oil-based enamel paints in the first half of the twentieth century is given; literature published in French and English is complemented by analysis of historical oil-based Ripolin enamel paint samples produced in France.

Artists' Ideas and Working Methods: Archival Sources

Chapter "Towards Interpretation of Making, Meaning, and Change in British Twentieth Century Oil Paintings: The Relevance of an Artist's Paint Archive" describes the archive of the artist Patrick Heron (1920–1999), as well as his paints that form a great resource for study. In chapter "Hans Hofmann's Last Lesson: A Study of the Artist's Materials in the Last Decade of His Career", a technical study of a very large number of paint samples from the late-career work of the artist Hans Hofmann (1880–1966) is presented. Both studies will form a basis for a better understanding of the works of these artists and preservation of their paintings in the long term. In chapter "*Zwei Jünglinge und Zwei Mädchen*, A Tempera Painting by Otto Mueller Circa 1917. *Paint Analysis and Reconstruction*", materials used for the matte painting *Zwei Jünglinge und zwei Mädchen* (c.1917) by the *Brücke* artist Otto Mueller is discussed on the basis of archival research, analysis and paint reconstruction. The complex mixture of materials found in the painting indicates the use of animal glue/egg/oil tempera paints and highlights the later addition of materials in conservation procedures.

Case Studies of Paintings: Esthetics and Conservation Issues

Three studies are presented in these proceedings describing techniques and practical conservation procedures of the works of avant-garde artists: *Théorème de Gödel* (1957) by Georges Mathieu (chapter "*Theoreme De Godel* by Georges Mathieu, 1957. Study and Restoration: Consolidation Through Cohesive Regeneration Using a Solvent"); *Cakes* (1955) by the Croatian neo-avant-garde artist Josip Vaništa (chapter "Non-traditionally Painted Oil Painting: How to Treat It Properly? Josip Vaništa's *Cakes* (1955)") and *Fleurs Grises* (1953) by Nicolas de Staël. For all three studies, the non-traditional techniques are an issue that required careful consideration of the aesthetics to be achieved in the conservation procedure, the actual treatment of paint consolidation and surface cleaning, and climate conditions for further storage and display.

Paint Degradation and Analysis

Chemical and physical changes are occurring constantly in paintings. Degradation may be local, superficial or throughout a paint system, and constituent materials can migrate through the paint, for example to its surface, causing optical effects. These can be whitening effects and are often referred to as efflorescence, or other visual disturbances, as is the case for protrusions of metal soaps. Rationalisation of degradation phenomena in paints and paint systems is discussed throughout the Proceedings. A wide range of traditional and novel, state-of-the-art analytical techniques are used and described to analyse painting materials and their degradation products. A novel technique for localising degraded materials is described in chapter "Derivatisation Technique for Infrared Spectroscopy – Characterisation of Oxidative Ageing Products in Modern Oil Paint", employing gaseous sulfur tetrafluoride SF_4 to modify oxidised products of paints specifically, allowing for a spatially resolved localisation of these products in paint cross sections.

Zinc white pigments became increasingly popular in the twentieth century. Chapters "Conservation Issues in Several Twentieth-Century Canadian Oil Paintings: The Role of Zinc Carboxylate Reaction Products", "Zinc White and the Influence of Paint Composition for Stability in Oil Based Media" and "Its Surreal: Zinc-Oxide Degradation and Misperceptions in Salvador Dalí's *Couple with Clouds in Their Heads,* 1936" discuss the reactivity of this pigment in oil paint specifically, using a range of advanced analytical techniques. Zinc soaps are the main product of the initial reaction of zinc white with fatty acids either from the drying oil or lipidic additions. Even more pronounced is the formation of these soaps from zinc white and aluminium stearate (chapter "Zinc White and the Influence of Paint Composition for Stability in Oil Based Media"). Chapter "Conservation Issues in Several Twentieth-Century Canadian Oil Paintings: The Role of Zinc Carboxylate Reaction Products" presents more information that excessive amounts of zinc soaps may cause problems of delamination, and optically disfiguring protrusions and efflorescence. Other degradation products from zinc white, either via zinc soaps or directly, are zinc sulphide, and zinc soaps from small fatty acids, such as formate and acetate and lactate. Chapter "Its Surreal: Zinc-Oxide Degradation and Misperceptions in Salvador Dalí's *Couple with Clouds in Their Heads,* 1936" discusses the formation of such products from atmospheric products in specific reaction-diffusion processes. The formation of metal soaps from others sources is described in other Chapters. For example, in chapter "Metal Soaps and Visual Changes in a Painting by René Magritte – *The Menaced Assassin,* 1927", protrusions or pustules on the surface of the *Menaced Assassin* (1927) by René Magritte, previously thought to be mould growth, were shown to be metal soaps. Here the metal base, possibly calcium, zinc or lead, is still unidentified. The presence of visually disfiguring lead soap crusts and remineralised products such as lead carbonates on early twentieth century paintings are described in chapter "An Investigation into the Viability of Removal of Lead Soap Efflorescence from Contemporary Oil Paintings".

Modern artists have applied paints in very experimental ways, for example by painting directly from the tube, creating very thick paint films. Depending on the type and quality of the painting materials, their application, as well as environmental conditions, this occasionally leads to major problems. One of the problems is the dripping phenomenon, where apparently dry paints liquefy and start to drip years after their applications; examples are described in chapter "*Theoreme De Godel* by Georges Mathieu, 1957. Study and Restoration: Consolidation Through Cohesive Regeneration Using a Solvent", "Investigating Fluidizing Dripping Pink Commercial Paint on Van Hemert's *Seven*-Series Works from 1990–1995", "Hard Dry Paint, Softening Tacky Paint, and Exuding Drips on *Composition* (1952) by Jean-Paul Riopelle" and "Set Back the Race: Treatment Strategies for Running Oil Paint". The reason for this liquefaction process lies partly in the use of semi-drying oils with a relatively low iodine value (i.e. lack of double bonds). Hypotheses for the phenomenon are presented in chapter "Investigating Fluidizing Dripping Pink Commercial Paint on Van Hemert's *Seven*-Series Works from 1990 to 1995"; the liquefaction process and possible treatment strategies for the subsequent curing of the paint are described in chapter "Set Back the Race: Treatment Strategies for Running Oil Paint".

The problem of water sensitivity of oil paint surfaces, which offers major difficulties in the cleaning treatment of paint surfaces, is mentioned in several chapters. In chapter "Water Sensitive Oil Paints in the Twentieth Century: A Study of the Distribution of Water-Soluble Degradation Products in Modern Oil Paint Films", water sensitivity due to magnesium sulphate heptahydrate formation is explained to be largely a surface phenomenon, corroborating the evidence that this type of sensitivity is caused mainly by reaction of magnesium carbonate in the paint with atmospheric sulphurous compounds. This is further substantiated in a number of case studies of paintings from the Tate collection.

New Conservation Approaches

In the last Chapters, actual treatment strategies for surface cleaning are proposed. Chapter "Investigation into the Surface Conductivity of Water-Sensitive Modern Oil Paints" discusses how the measurement of conductivity of an oil paint surface can be a useful indicator for water sensitivity caused by water-soluble salts. In chapter "An Investigation into the Viability of Removal of Lead Soap Efflorescence from Contemporary Oil Paintings", on efflorescence of metal soap crusts, an approach is proposed which utilises EDTA chelating agents, which provides a balance between thinning down the crust layer whilst respecting the original paint surface. Chapters "Dry Cleaning: Research and Practice", "New Approaches to Surface Cleaning of Unvarnished Contemporary Oil Paintings – Moist Sponges and Cloths" and "Agar – A New Tool for the Surface Cleaning of Water Sensitive Oil Paint?" discuss approaches for removing dirt from solvent-sensitive paint surfaces: Chapter "Dry Cleaning: Research and Practice" discusses a range of dry cleaning materials, chapter "New Approaches to Surface Cleaning of Unvarnished Contemporary Oil Paintings – Moist Sponges and Cloths" reports the potential for using moistened sponges for paint surfaces, and in chapter "Agar – A New Tool for the Surface Cleaning of Water Sensitive Oil Paint?" agar gels that can be applied in liquid form and removed as a sol are discussed, limiting the amount of water released into the paint and reducing the mechanical action to a minimum.

We are grateful to the different institutions who supported the work that led to the organisation of the ICOP symposium and the compilation of articles for these Proceedings, especially the twentieth century oil paint research group at Cultural Heritage Agency of the Netherlands, *Object in Context* programme leader Tatja Scholte and the Agency's management. We furthermore thank all professionals who were involved in the research that forms the basis of these Proceedings.

Amsterdam, The Netherlands
30 June 2014

Klaas Jan van den Berg
Aviva Burnstock
Alberto de Tagle
Matthijs de Keijzer
Gunnar Heydenreich
Jay Krueger
Tom Learner

Contents

Twentieth Century Oil Paint. The Interface Between Science and Conservation and the Challenges for Modern Oil Paint Research

Aviva Burnstock and Klaas Jan van den Berg

Abstract This chapter introduces recent research related to aspects of the deterioration of modern paintings in oil media. The research was informed by collaboration between conservators and scientists and utilises phenomenological and experimental methodologies to investigate the causes of optical changes in paint, such as efflorescence and salt formation, and the sensitivity to water that occurs on surface cleaning some unvarnished oil paintings. Examples are given of results based on case studies of paintings and the results of systematic experimental investigation of paint samples, including those supplied by manufacturers and reconstructions based on paint formulations. Results of this research are given including the cause of water sensitivity related to the formation of magnesium sulphate hydrate in selected manufactured oil paints, and criteria for further investigation of the phenomena in other paints. Contributions in this volume that address other classes of material deterioration including the formation of metal soap crusts and treatment approaches are introduced.

Keywords Twentieth century oil paint • Water sensitivity • Efflorescence • Collaborative research

Introduction

The working properties and optical qualities of oil based paints continue to appeal to artists despite the availability of new paint media. A survey from the painting collection at Tate, showed that more than 70 % of its twentieth century painting

A. Burnstock (✉)
Department of Conservation and Technology, The Courtauld Institute of Art, Somerset House, the Strand, London WC2R 0RN, UK
e-mail: aviva.burnstock@courtauld.ac.uk

K.J. van den Berg
Cultural Heritage Agency of the Netherlands, Amsterdam, The Netherlands
e-mail: k.van.den.berg@cultureelerfgoed.nl

K.J. van den Berg et al. (eds.), *Issues in Contemporary Oil Paint*,
DOI 10.1007/978-3-319-10100-2_1

collection is executed in oil (Smithen 2005). Although modern oil paints contain the same oils that were used for paint media in the Middle Ages, the formulation of artists' oil paints have changed significantly (Mayer 1991). While linseed oil continues to be the most widely used medium, other oils such as safflower and sunflower have been introduced to replace the traditional, less-yellowing nut and poppy oils. Artists oil paint tubes, originally invented in the 1840s contain additives to stabilise the paints in the tube whilst controlling the rheology as well as the drying time of the paint once taken from the tube and applied. The archives of artists' paint manufacturers to some extent document these developments (Carlyle and Clarke 2005, Van den Berg et al. 2014). Manufacturers' incorporation of dryers, fillers, dispersion agents and stabilisers has developed in particular during the twentieth century (Izzo et al. 2014; Van den Berg et al. 2014). Concomitantly a range of new pigments, some of improved quality and lower cost have been introduced (de Keijzer 2011, 2014). Economic reasons have driven much of these developments, although more recently health and environmental safety have led to changes in formulations. The restrictions on the production of lead white and other toxic pigments were important changes in relation to the production of artists' paints.

These developments, however, have led to compromises in the stability of some modern oil paints. In some cases the manner in which modern paints are used by artists have influenced the material deterioration of the paint. Preferences for unvarnished surfaces from the later part of the nineteenth century prevailed in the following period of artistic production, rendering the paint surface directly vulnerable to the influences of environmental agents including dirt and pollutants. These developments present challenges for conservators. As many modern paintings are now entering conservation studios for the first time, these issues are now paramount for conservators and curators.

While experience provides the conservator with a solid framework of knowledge, the ability to question standard treatment approaches and recognize problems that require new strategies is an important part of an on-going education in conservation. In these Proceedings, several conservators as well as conservation students present aspects of their research that relates to the condition and treatment of modern oil paintings (Cooper et al. 2014; Sawicka et al. 2014; Franken et al. 2014; Blumenroth et al. 2014; Verissimo Mendes et al. 2014; Volk and van den Berg 2014; Soldano et al. 2014). This research has provided an opportunity to share experiences with professional conservators in museum and private practice, and work in well-equipped scientific laboratories. The problems investigated are important to conservators who deal with them every day in the studio, and thus there is a real desire behind the research to provide guidelines that inform future conservation practice.

The nature of conservation led research is different to scientific research in other fields. Relatively few conservation related studies are undertaken, and many have been based on cumulative evidence from individual case studies. For example, identifying common denominators or causal agents for water sensitivity observed in paintings is difficult given the range of possible materials, in particular where paintings are made using both complex manufacturers' paint formulations and artists' additions, applied heterogeneously.

More recently, collaborations have provided up to date resources for technical and scientific study, at the Netherlands Institute for Cultural Heritage (RCE), the Getty Conservation Institute (GCI), the Natural History Museum, London (NHM) and most recently the synchrotron facility in Grenoble.[1] Participating museums and galleries, notably at the Courtauld Stedelijk Museum and Tate have facilitated paintings-led research that address the condition and treatment of some important works of art made using modern oil paint from the twentieth century to the present. This first chapter in these Proceedings evaluates the approaches to research undertaken so far, gives some further research results and ideas for future study.

Fundamental Research to Inform Conservation

Research in the field of conservation, for modern oil paints and other materials is challenged by the need to design experiments that inform practice. This demands the most difficult kind of experimental work and interpretation. The notion that understanding the chemistry of materials on a molecular level can inform practice requires qualification and necessitates inferences about the macroscopic impact of practical treatment from knowledge of molecular level interactions on the surface and within the paint. This highlights an issue that is at the heart of the interface between conservation science and practice. In the context of cleaning a painting, interpreting the results of semi-standardised experiments on simplified "reconstructed" paint samples, supplemented by in-depth analyses of small and complex samples from paintings, may involve compromises and practical realities. Scientific research and cleaning paintings sometimes feel as different as making art and conserving it.

Nonetheless, with the assumption that understanding the chemical and mechanical properties of painting materials and their deterioration is integral to conservation practice, together with experience and knowledge of historical context we conserve works of art the best we can. This point can be illustrated with regard to deterioration observed in modern oil paints containing ultramarine, and replication of the phenomena in so-called reconstructions or test samples (Fig. 1). Well-prepared reconstructions offer an opportunity to perform comparative cleaning tests and analyses with a degree of confidence that the results will have relevance to cleaning painted art.

Recent research includes investigation of the causes of water sensitive oil paint; dirt removal from sensitive surfaces; causes and treatment of surface efflorescence and evaluation of other kinds of material deterioration. While some phenomena such as whitening of the surface of selected passages of paintings has long been recognized by conservators, recent studies have refined our understanding of different classes of material change that can now be addressed by more specific and

[1] See for an overview of the research undertaken in the 20th Century project: Van den Berg (2013).

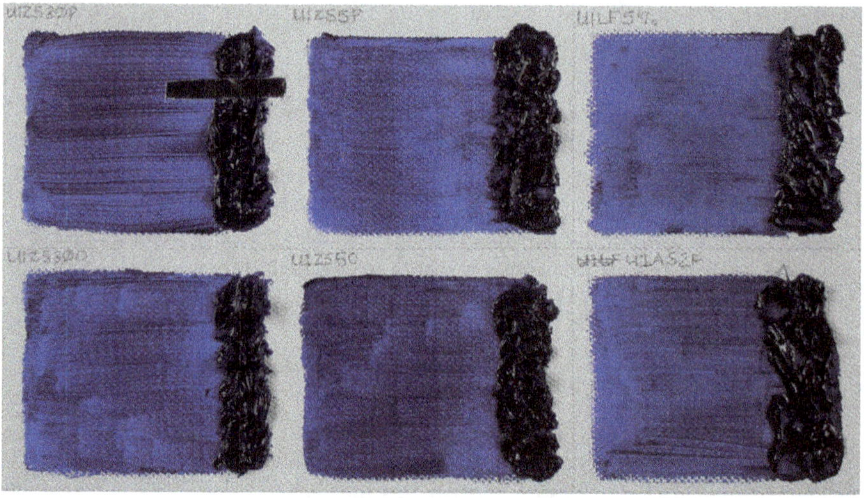

Fig. 1 Test paint samples comprising ultramarine, oil, varying concentrations of metal stearates and other additives. The hydrophilic properties of ultramarine pigment are linked with its sensitivity to swabbing with water (Tempest et al. 2013)

appropriate treatment. Surface deterioration is often evident where efflorescence, medium separation, dripping or flaking paint occur. Sometimes cases such as those of water sensitive paint offer no visual indicators. However, the problem for conservators is to remove superficial material (efflorescence, surface dirt or varnish) or to consolidate flaking paint without affecting the optical or mechanical qualities of the paint.

Classes of Surface Deterioration

Optical Whitening of the Paint Surface: Efflorescence, Salts, Metal Soap Crusts

Recent study of surface whitening in paintings has combined fundamental research and practice that has led to changes in conservation treatment methodologies. The application of new analytical techniques, including high resolution imaging, layer specific analysis and the development of methods for precise characterisation of organic materials and their components has refined understanding of different causes and classes of optical whitening, which can be of organic, inorganic and organometallic (soaps) origin (Burnstock et al. 1993; Sutherland 1995; Akerlund 2012).

Hinde's investigation of surface whitening in a series of twentieth century Spanish paintings at a National Trust property in the UK provides a good example of how research informs conservation practice (Hinde 2010; Hinde et al. 2011).

Hinde's study involved close examination of the causes of whitening, the results of which facilitated her choice of the most appropriate treatment for paintings that she then carried out. This included the removal of the white- appearing deposit of free fatty acids from the surface of the painting using non-polar solvent and or mechanical action, and removal of water soluble salt efflorescence that migrated through cracks from underlying layers. Examination of another painting identified the intentional use of a particulate white material which therefore required no treatment.

Superficially similar optical whitening identified as metal soap crusts that are insoluble in solvents requires a different approach that was subsequently developed by Sawicka and discussed in this volume (Sawicka 2013; Sawicka et al. 2014).

While whitening, or efflorescence in oil paintings may be caused by migration of fatty acids from the oil medium (Fig. 2) and the formation of metal soaps (Fig. 3) there are other examples where efflorescence is caused by reaction with airborne gases or aerosols. The light scattering surface product can consist of metal oxides, carbonates and different types of salts. Examples are sulphates (Van Loon et al. 2012), zinc formates, acetates and sulphides, (Keune and Boeve 2014), carbonates (Sawicka et al. 2014).

The Oil Paint Model

The conditions which influence the formation of efflorescence and metal soaps in oil paintings have been proposed using a model of oil paint (Van den Berg et al. 1999, 2002). The model that incorporates mobile elements, including reactive metal

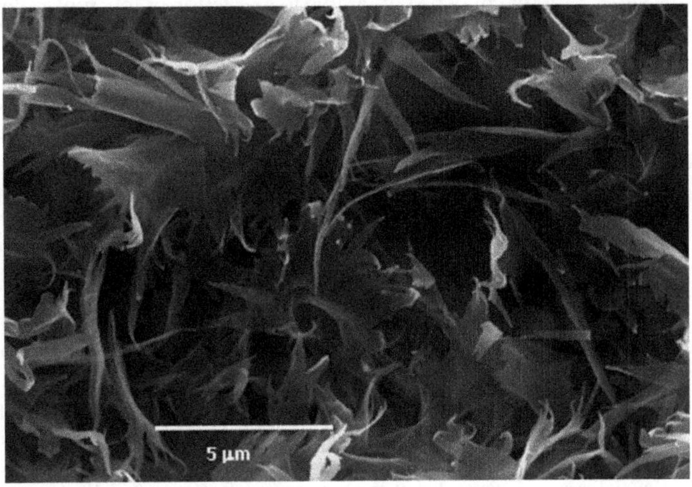

Fig. 2 SEM image of fatty acid efflorescence in a sample from an unvarnished oil painting

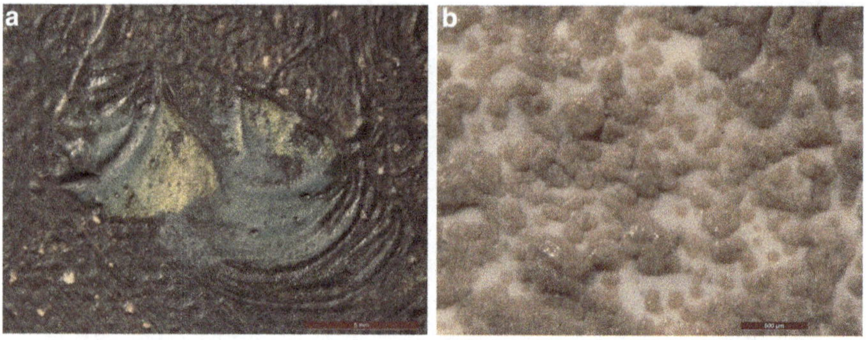

Fig. 3 *Wolmer Wood* by Philip De Lazlo, (c. 1930, private collection) (See treatment report CIA:2093 Luciana Akerlund 2012, Department of Conservation & Technology, Courtauld Institute of Art); (**a**) detail of darkened surface crust composed of zinc soaps formed from zinc ions migrated from the paint and ground reacting with free fatty acids in the oil containing varnish, (**b**) detailed photograph showing zinc soaps on an area of white paint

ions and organic molecular fragments in a cross-linked polymer matrix presents a notional dynamic system for describing the condition of oil paint, as summarised here.

In the development of oil paint drying and ageing, three processes play a role (Fig. 4). Oxidative polymerisation which causes the paint to dry creates a cross-linked network. This process occurs in competition with oxidative degradation, which leads to the formation of smaller molecules which do not participate in polymerisation. The third process is hydrolysis, which will gradually proceed from the first stages of drying of the paint and continue in the longer term.[2] Over time hydrolysis reduces the molecular weight of the binder matrix resulting in a relatively flexible paint that may be vulnerable to extraction of low molecular weight fragments by solvents. If, however, polyvalent ions are present, for example where lead or zinc is present, a more rigid, resistant paint film will be formed.

This model can be used to explain some of the mechanical and optical phenomena in modern oil paints, including aspects of efflorescence (Sawicka et al. 2014). Recent research on the reactivity of in lead and zinc white in oil media has developed a deeper understanding of the processes which involve metal soap formation, including the influence of moisture and temperature.[3]

[2]The presence of drying pigments and driers will promote the oxidative polymerisation process. The presence of alkaline pigments and fillers will promote hydrolysis.

[3]Research is currently carried out within the NWO PAinT project by A. van Loon and K. Keune (http://www.s4a-paint.uva.nl/research-topics), partly in collaboration with the authors. See A. van Loon and K. Keune et al., 'Synchrotron-based studies on the migration of lead soaps in Old Master paintings and model systems', presented at the Synchrotron Radiation and Neutrons in Art and Archeology Conference, Paris, 9–12 Sept. 2014. Manuscript in preparation for the Conference Proceedings.

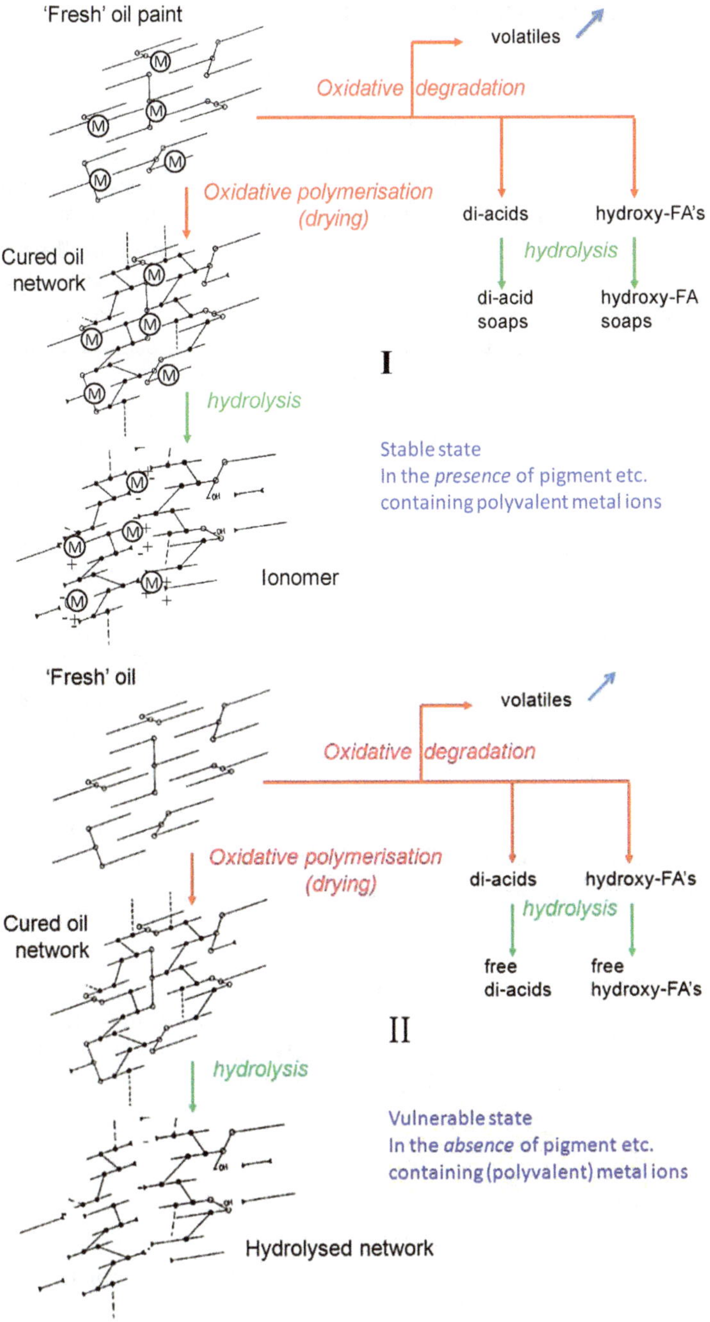

Fig. 4 The oil paint model from Van den Berg et al. (1999) and adapted to illustrate the difference between oil paint, after oxidative drying and hydrolysis, stabilised in the presence of polyvalent metal ions (I), and where the absence of polyvalent metal ions leads to a paint which is more vulnerable to external factors (II)

Water Sensitivity of Modern Oil Paints

Hypotheses based on information given by manufacturers and a review of the literature on the commercial manufacture of oil paint in the twentieth century pointed to manufacturers' addition of excess stearates to modern oil paints as the most likely cause of water sensitivity in paintings by Jasper Johns made in the 1960s (Wijnberg et al. 2007; Burnstock et al. 2006). Sets of samples were prepared following this hypothesis, based on analyses of modern oil paints from different brands with additions of aluminium and zinc stearate (Mills et al. 2008; Saltmarsh et al. 2008). Acknowledging the variable methods used by different conservators, the water sensitivity of the films was tested using a semi-standard swab rolling technique; results suggested that some paints were more sensitive than others. Ultramarine paints were consistently more sensitive to water, as were paints formulated with pre-hydrolysed linseed oil and hydrated alumina. Notable was the presence of a skin of medium on the paint surface in some samples that influenced sensitivity. Contrary to the hypothesis postulated by Burnstock et al. (2006) and Mills et al. (2008), the addition of stearate did not affect water sensitivity and samples that contained both stearate and hydrated alumina were less sensitive (Tempest et al. 2013).

Experiments carried suggested that fatty acid efflorescence in fresh paint films was dependent primarily on the presence of high concentrations of metal stearates (Tempest et al. 2013). In these samples water swabbing produced fatty acid efflorescence within days. These free fatty acids are related to technical stearates which are known to contain high proportions (4-20 %) of free fatty acids; furthermore hydrolysis of the metal stearates may form more fatty acids (Boon and Hoogland 2014; Tempest et al. 2013). Since metal stearates perform an important function both in the paint manufacture, rheology and during ageing, it is important to be able to detect the material. Fourier Transform Infrared Spectroscopy (FTIR) can be used but cannot detect concentrations of stearates below c. 5 %. More sensitive techniques have been developed to detect metal stearates using extraction methods in combination with Gas Chromatography (GC)MS (Izzo 2011) or Evolved Gas Analysis (EGA)MS (Van den Berg et al. 2011). Alternatively, the detection of fatty acids from metal stearates can be done relatively easily using Direct Electrospray Ionisation Mass Spectrometry (ESIMS) (Van den Berg et al. 2011; Van den Berg 2013).

New hypotheses about the causes of water sensitivity in modern oil paintings were developed based on evidence from wider sources. The first was a survey that identified clusters of case studies of oil paintings made in the 1950s–1960s that exhibited water sensitivity were made available for technical study (Tempest 2009). At the same time, the closure of Winsor&Newton's Wealdstone plant in 2012 led to the donation of a set of oil paint swatches painted out from every batch of

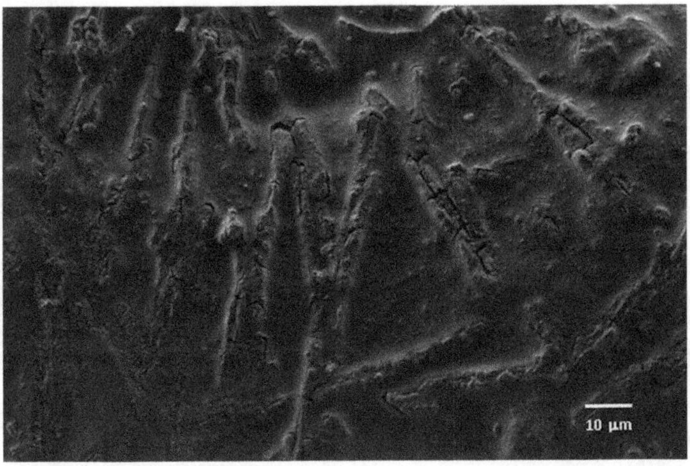

Fig. 5 SEM image of magnesium sulphate heptahydrate (epsomite) under skin of medium of a dry film of Winsor&Newton student quality oil paint (1964/1965) containing cadmium yellow pigment, barium sulphate and magnesium carbonate extenders in oil

oil paint made at the plant between the 1940s and the 1990s.[4] This provided an opportunity to relate the water sensitivity in paintings, in particular where the artists used Winsor&Newton oil paints, to naturally aged samples of the same paints from the manufacturer.

Scanning electron microscope imaging and elemental analysis (SEM-EDX) of some of the most water sensitive paints from both paintings and swatches showed consistent surface features, elongated crystals embedded in the surface of some of the paints that contained elemental Magnesium and Sulphur. The application of non-invasive X-ray diffraction (XRD) identified the crystals as magnesium sulphate hydrate that is highly water sensitive (Fig. 5). A hypothesis for the cause of water sensitivity, at least for the paints that exhibited these crystals on the surface was thus proposed, and the next phase was to try to replicate the formation of the compound by exposure to a sulphurous environment that might reflect the conditions to which paintings were exposed the before the cleaner air of the post-1980s (Silvester et al. 2014). The question of whether the water sensitive compounds were present in the bulk paint or was limited to the surface was subsequently investigated by Cooper et al. (2014).

While the formation of magnesium sulphate heptahydrate may account for water sensitivity in some oil paints (notably those that contain magnesium carbonate) there are a number of cases where water sensitivity cannot be attributed to this kind of chemical change. Friable under-bound paint may also be susceptible to

[4] Oil paint Swatches of batches of paint made by Winsor&Newton from c. 1960–1990 gifted by Ian Garrett to Tate in 2012.

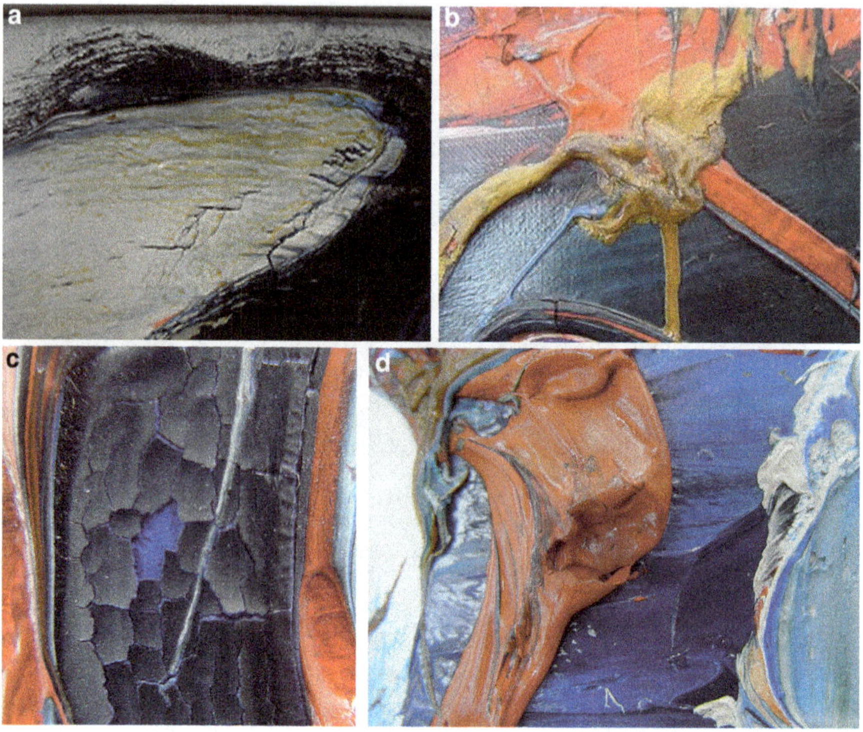

Fig. 6 Water sensitive surfaces associated with medium separation, from Karel Appel, *Les* Animaux 1961. (**a**) *Yellowed* oil medium on a *white* paint passage, (**b**) Drip of medium-rich ochre formed a few years after completion of the painting, (**c**) Flaking ultramarine blue paint with dry powdery paint visible underneath, (**d**) Organic red pigmented paint, medium has leached out into the ground and to the back of the canvas (Photographs Laura Mills 2008)

mechanical abrasion and where the paint appears to be intact there may be other causes linked to the surface deterioration of the paint medium. One example of this class of sensitivity can be seen in paintings by Karel Appel that demonstrate the deterioration of the organic medium at the surface, or separation of the medium from the binder (Fig. 6) (Burnstock et al. 2006; Mills 2008; Volk 2012).

The cause of water sensitivity in well bound paint can be further examined using paint cross sections which show the presence of a skin of medium at the paint surface. In Fig. 7a this is clearly visible in an SEM image. Figure 7b, c illustrates the medium skin in that fluoresces in a cross section photographed in UV light. UV imaging of sections taken before and after treatment has been used in some studies to monitor the effects of different methods of surface cleaning on the paint skin (Daudin et al. 2013).

Analytical results point to surface changes in the organic media in some water sensitive paints. Analysis using GCMS and ESIMS (Van den Berg et al. 2011) indicated that paint surfaces show relatively high organic diacid contents.

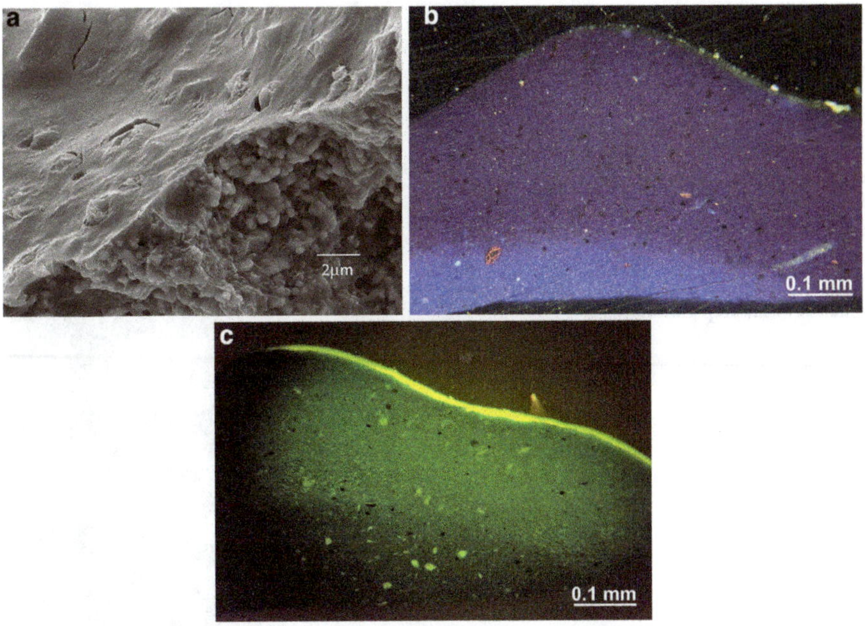

Fig. 7 (**a**) Scanning electron microscope image showing a skin of medium on a paint sample; (**b**) Microphotograph of a cross section taken from a dry film of Winsor&Newton cobalt blue tint student-quality oil paint 1964/1965, containing an organic blue pigment, magnesium carbonate and barium sulphate, (**b**) in ordinary light; (**c**) in ultraviolet light, showing strong fluorescence on the surface of the paint indicating an oxidised and medium rich surface skin (**b**, **c** from Silvester 2011)

These diacids are formed by an oxidative degradation which is in competition with the oxidative polymerisation (see Fig. 4). Following hydrolysis, these free organic diacids may render the surface relatively polar and sensitive to water, especially in medium rich paints and may increase the water sensitivity of the paint surface (Van den Berg et al. 2011; Izzo et al. 2014).

Other evidence has indicated the sensitivity of the paint medium to water increases after exposure to sulphur dioxide. Experiments carried out at the synchrotron facility at Grenoble[5] using K-Edge μ-XANES/XRF mapping of a cross section of a medium rich paint containing natural 'raw Sienna' that comprised hydrated ferric oxides and silicates, with magnesium carbonate filler, aged naturally for 50 years (Fig. 8). Sulphide was detected in a 1–3 μm thick layer on the surface that was not associated with elemental magnesium, suggesting that an organic sulphide is formed due to the reaction of the curing binding medium with atmospheric sulphur (Simendinger and Balik 1994; Silvester et al. 2014). Further research might usefully explore the influence of this reaction on the properties of the oil paint and oil paint surface, and in the bulk paint.

[5](Van den Berg, Burnstock, Pouyet and Cotte, manuscript in preparation).

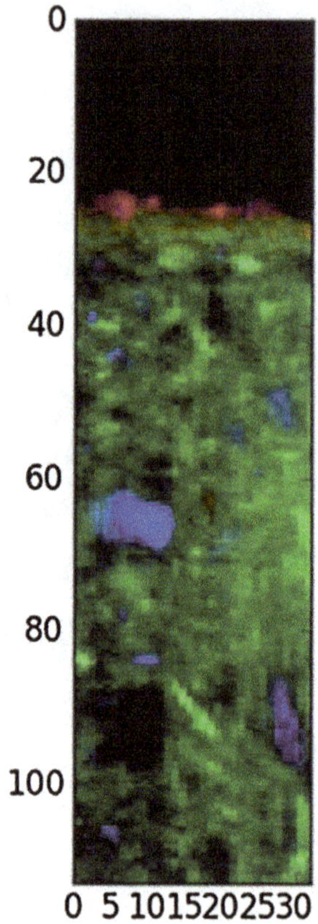

Fig. 8 K-Edge μ-XANES/XRF mapping of the cross section of Winsor&Newton raw Sienna student-quality oil paint 1964/5, indicating that organic sulphide is present on the surface. Numbers in μm. Al (*blue*); Mg (*green*); S, as sulphide (*red*) at 2.4737 keV (Van den Berg et al. manuscript in preparation)

The changes in the organic binding medium that lead to changes in the paint film, including striking examples of dripping that are only evident several years after completion of the painting are discussed by Boon and Hoogland (2014), Bronken and Boon (2014) and Franken et al. (2014).

Foreknowledge that some paints are likely to be water sensitive helps to inform conservation practice. The presence of magnesium sulphate hydrate may be predicted by simple conductivity testing of paintings (Soldano and Van den Berg 2014), a procedure that is used by conservators to adjust concentrations of aqueous cleaning solutions (Ormsby et al. 2013). It is possible, based on the results of research so far, to predict that some paints will be more or less sensitive, though the cause might be explicit where artists have used selected manufacturers' tube paints per se. Mixtures of paint and paints mixed with other materials often employed in making art will behave differently. This is a limiting factor, however some of the most water sensitive paintings comprise fields of colour that are likely to derive from a single tube of manufactured paint, and here the inferences made from reconstruction samples may have direct application.

Designing Treatments

The step that follows fundamental experimental research into the condition of paint is to design methods for treating surfaces sensitive to water and other solvents. The use of new methods that aim to evenly remove surface dirt from real paintings presents a practical challenge. The most useful experimental studies that have addressed the methods for dirt removal from unvarnished oil paint have included an evaluation of the goals for treatment. These include for instance Rachel Morrison's investigation of the use of citrate for cleaning oil paint has informed conservation practice understanding and serve as a model for future study (Burnstock and Van den Berg 2005; Morrison et al. 2007). Experimental evaluation of surface changes has employed a range of imaging methods, such as light and confocal microscopy (Van den Berg et al. 2008), SEM topographic imaging before and after treatment and other profiling techniques such as atomic absorption spectroscopy.

Naturally aged paint surfaces are the most useful for testing but are not widely available and require full characterisation of the painting materials and technique and the nature of surface dirt and its relation to the paint surface. The use of reconstruction samples aged artificially for testing practical treatments present one or more degrees of separation from the qualities of the surface of a naturally aged painting, in particular a painting with a varied physical history of conservation and display. Some experimental studies that involve comparisons between different methods of treatment have introduced artificial surface dirt, of a standardised composition. In theory this reduces the variables by providing a consistent surface for comparison of different methods for removal. Another approach is to use flat surfaces so that interstitial surface materials do not produce other anomalies for evaluation. While the advantages of comparing like with like are obvious, the compromise is that the artificial dirt is unlike the soiling that accumulates under natural conditions, in particular in relation to the interface between the dirt and the paint and the relationship between timing of both the drying of the paint and the soiling. This method is somewhat standard now for cleaning studies of unvarnished paints in different media and experiments using this approach, and the removal of naturally accumulated dirt is discussed by Volk and van den Berg (2014) in relation to selected water sensitive oil paints and by Daudin et al. (2013) for comparing methods of dry cleaning oil paint surfaces.

What has been clear from the studies of water sensitive oil paints is that the top few microns of the paint surface are critical. For paints that exhibit the formation of hygroscopic compounds the question of whether the surface or the bulk paint is likely to be vulnerable during cleaning was the focus of Cooper's research (2014).

There are differences in the nature of the surface of modern oil paints, both from paintings and reconstructed samples, some of which exhibit a skin of organic material that is vulnerable to oxidation or pollutant gasses, and there is another group of paints, perhaps those with poor paint binder interaction, that undergo changes in inorganic compulsion at the surface.

Practical and Ethical Considerations

Conservation approaches are almost always compromises that balance material preservation and aesthetic gains (that result in a more readable image) with the risks for original painting materials. The awareness of deterioration of the original paint and concomitant optical and mechanical changes in original painting materials provides important context for treatment decisions. With regard to the treatment of modern oil paint the surface, the top few micrometers of which present a critical focus, though addressing mechanical and other changes in the painting structure is also important.

Arguably the first cleaning of a painting presents the greatest challenge, in which the paint surface, that may be dirty, aged and perhaps deteriorated, is unaffected by previous treatments. This reality frames the interpretation of data from technical and experimental studies. There are many different approaches to the removal of surface dirt from unvarnished paintings and the visual difference between the results obtained using different methods for surface cleaning can be striking (Wijnberg 2014). In relation to surface cleaning of water sensitive oil paint it is important to define the questions for designing an appropriate research strategy: Is the ultimate goal to remove surface dirt completely, or is it important to do so evenly? Might a compromise be to accept the limitations of selective cleaning? In paints that include imbibed surface dirt, the goal for complete removal presupposes that dry removal of non-embedded surface dirt is unacceptable and the surface aesthetic is compromised by the presence of embedded dirt, necessitating wet or solvent based cleaning methods. Implicit in this approach are more difficult questions: How important is the integrity of the original paint? Can and should we consider cleaning methods which prioritise the clean appearance of the surface but compromise the surface skin of the paint in the pursuit of conservation goals (Fig. 9)?

Decisions about the conservation and presentation of works of art involve a wider range of criteria, may be collection based, consider the artist intention or not, and may also be influenced by the age of the work. In some cases the degradation and soiling of a paint surface might be valued as 'patina'. Conservators at the Guggenheim Museum remarked after investigation of Ad Reinhardt's monochrome paintings that we must "begin to accept signs of age on modern paintings, just as we do for historical works" even if this is at odds with the original intention of the artist (Van de Vall et al. 2011).[6]

The conservators' task to evaluate risks to the oil paint surface and desire to make the image legible is centuries old. Students in painting conservation begin by reading Feller et al. (1985), Hedley, Michalski (1990) and Phenix (1998), and all conservators navigate the practical challenges of exploiting narrow boundary

[6]http://archive-org.com/page/491723/2012-10-21/http://www.guggenheim.org/new-york/collections/conservation/conservation-projects/axa-reinhardt

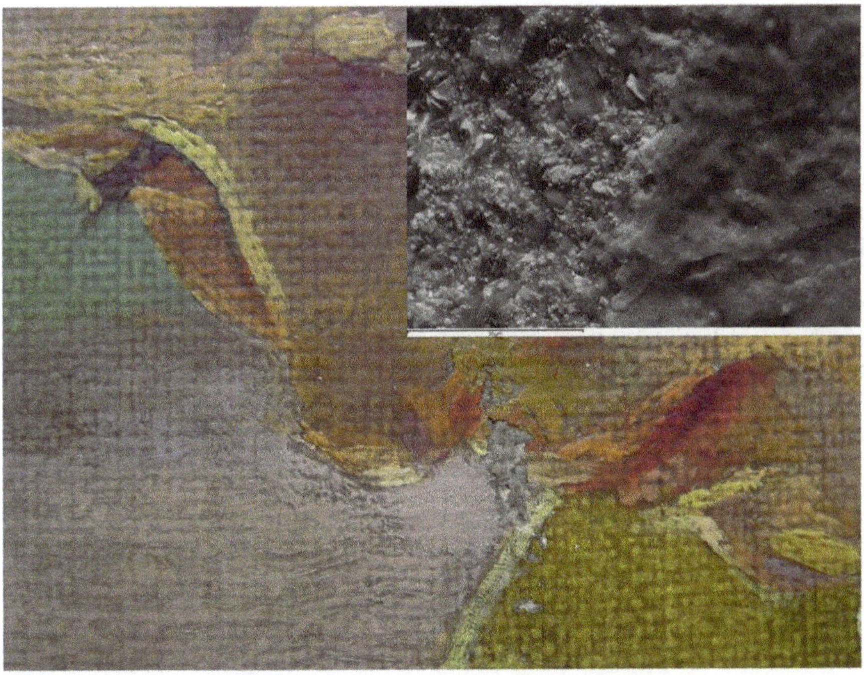

Fig. 9 Cleaning test of a detail from *Girl cleaning a Kettle* by F.G Heath, private collection (See CIA:1837 Painting treatment register, Department of Conservation & Technology, Courtauld Institute of Art.) showing the removal of dirt and medium skin by swabbing with saliva. Insert: SEM image of the surface on the edge of the test

between swelling oil paint during removal of oxidised resin and oil resin coatings, and oily repaint over original paint. The goal is to leave the original paint unaffected, and the cautious approach to cleaning is evidenced by the number of selectively cleaned paintings one encounters in practice. The use of gels that reduce capillary penetration and contain the effects of solvents and reagents to the surface can be a great advantage, and largely thanks to Richard Wolbers we now have an arsenal of methods for cleaning that can be applied to minimise the risks.

Conversely the effects of residues from cleaning reagents may not be immediately visible and require other techniques for analysis that present a whole range of interpretive challenges, and considerations for the longevity of oil paint.

Conclusion

The research described in this introduction that was summarised at the symposium and included in these proceedings serve as a starting point for further studies of the deterioration and treatment of modern oil paintings. Work so far has been based on technical investigation of case studies and experimental study of paint samples and has aimed to characterise the deterioration phenomena and its causes that include paint formulation and the influence of environmental agents. While some of the causal factors have been highlighted, there remain a number of questions that require further study, including the stability of paints in the long term, the influence of painting techniques and the development of practical approaches to treatment.

The detection of magnesium sulphate hydrate as the cause of water sensitivity in paint used by Jasper Johns' for Untitled, *1964-'65* and other paintings made in the 1960s was useful but didn't solve the problem of how to remove dirt from highly sensitive paint surfaces. The study highlighted the problem associated with ageing of unvarnished paintings that are vulnerable to dirt and other environmental agents that lead to deterioration, and where the quality of the surface is compromised. Superficial changes due to accumulation of surface accretions, of materials that include fatty acid efflorescence, soap crusts and salts may alter the surface colour saturation and gloss, resulting in an overall loss of optical balance in the painting.

Conservation treatment of paintings that include passages of water sensitive paints requires an informed assessment of the condition of the surface and consideration of treatment of the whole surface that necessitates different approaches to passages of paint with the goal of even cleaning.

Water sensitive surfaces and paintings that exhibit recurring fatty acid efflorescence present special concerns for preventive conservation and display. Protection of vulnerable and potentially reactive paint from pollutant gasses and surface dirt is paramount to retain paintings made using these materials in a displayable condition. The arguments for passive preventive framing with glazing, and provision of filtered air conditioned environments present problems, where many works are too large to be glazed and the glazing compromises the aesthetic. Air conditioning is imperfect, increasingly expensive and most paintings are not displayed in clean and controlled environments that may in some cases be critical.

Where a recommendation is made for a possible method for treatment in a Chapter based on technical research, there is an underlying assumption that the methods are deployed with clear aesthetic goals in mind. The skill and judgement of the conservator is the most critical parameter of all. An evaluation of the balance between visual gains and risks to paint are integral to every conservation intervention.

Future study of the treatment of water sensitive oil paints might include an investigation of the efficacy of removing the water sensitive compounds from the surface of paint, consideration of risks to the surface of the paint and its optical qualities, and whether this presents a practical long term solution to water sensitivity. Alternatively, methods for protecting the most vulnerable paints might be considered as a preventive measure that could include application of an aesthetically acceptable coating that offers the most sensitive paints protection from dirt. The question of whether this coating could be removed without changing the surface aesthetic is important to ascertain, and whether this approach prevent the formation of further water soluble deterioration products would be useful to know.

These options remain to be investigated, and so too the mechanical changes in modern oil paints that impact on current and future conservation practice.

Acknowledgements Modern oil Paint group at RCE: Luc Megens, Susan de Groot, Ineke Joosten, Henk van Keulen, Matthijs de Keijzer. Laura Hinde, Polly Saltmarsh, Rachel Morrison, Luciana Akerlund, Genevieve Silvester, Laura Mills, Francesca Izzo, Filipe Duarte. Tom Learner, Giacomo Chiari, Dave Carson Getty Conservation Institute. Bronwyn Ormsby, Tate. Ian Garrett Winsor&Newton, Bert Klein Ovink Talens. Katrien Keune, Annelies van Loon, University of Amsterdam. William Luckhurst, Kings College London, Alex Ball Natural History Museum London. Marine Cotte, Emeline Pouyet, ESRF Grenoble.

References

Akerlund L (2012) Efflorescence: an investigation of selected paintings from the nineteenth to the twenty-first century with an experimental study of the role of moisture gradients in development of efflorescence. Final year project dissertation, Department of Conservation & Technology, Courtauld Institute of Art 2012

Blumenroth D, Zumbühl S, Scherrer NC, Müller W (2014) Sensitivity of modern oil paints to solvents. Effects on synthetic organic pigments. These proceedings, Chapter 23

Boon J, Hoogland F (2014) Investigating fluidizing dripping pink commercial paint on Van Hemert's *Seven*-series works from 1990–1995. These proceedings, Chapter 16

Bronken IAT, Boon JJ (2014) Hard dry paint, softening tacky paint, and exuding drips on *composition* (1952) by Jean-Paul Riopelle. These proceedings, Chapter 17

Burnstock A, van den Berg KJ (2005) A pilot study of the effect of triammonium citrate solutions used for the surface cleaning of paintings. Post-prints of the American Institute for Conservation of Historic and Artistic Works; Paintings Speciality Group 17, 2005, 56–64

Burnstock A, Odlyha M, Caldwell M (1993) 'A Technical Examination of Surface Deterioration of Stanley Spencer's Paintings at Sandham Memorial Chapel', ICOM Committee for Conservation Washington, DC, USA, 231–238

Burnstock A, Van den Berg KJ, de Groot S, Wijnberg L (2006) An investigation of water sensitive oil paints in twentieth-century paintings. In: Learner T et al (eds) Modern paints uncovered: symposium proceedings, Tate Modern, London, 16–19th May 2006. Getty Conservation Institute, Los Angeles, pp 177–188

Carlyle L, Clarke M (2005) Page-image recipe databases, a new approach for accessing art technological manuscripts and rare printed sources: The Winsor&Newton archive prototype. In: I. Verger (managing ed) Preprints of the 14th Triennial Meeting of the ICOM Committee for Conservation, The Hague, The Netherlands, 12–16 Sept 2005: vol I, pp 24–29

Cooper A, Burnstock A, van den Berg KJ, Ormsby B (2014) Water sensitive oil paints in the twentieth century: a study of the distribution of water-soluble degradation products in modern oil paint films. These proceedings, Chapter 20

Daudin-Schotte M, Bisschoff M, oosten I, van Keulen H, van den Berg KJ (2013) Dry cleaning approaches for unvarnished paint surfaces. In: Proceedings from of the cleaning 2010 conference, Valencia, 26–28 May 2010, Smithsonian Institute, 2013, pp 209–219

de Keijzer M (2011) A choice of colour: modern synthetic inorganic artists' pigments, Restauratorenblätter Band 30, zum Thema: Kunst des 20. und 21. Jahrhunderts, Probleme und Perspektiven zur Erhaltung Klosterneuburg 2011, Verlag Stift Klosterneuburg, pp 33–42

de Keijzer M (2014) The delight of modern pigment creations. These proceedings, Chapter 4

Feller RL, Stolow N, Jones EH (1985) On picture varnishes and their solvents. Rev. and enlarged ed. 1985. National Gallery of Art, Washington, DC

Franken V, Heydenreich G, Jägers E, Müller W, Schulz J, Zumbühl S (2014) Set back the race: Treatment strategies for running oil paint. These proceedings, Chapter 22

Hinde L (2010) An investigation of the causes of whitening in 20th C paintings from Dudmaston Hall (NT), Shropshire, final year project. Department of Conservation & Technology, Courtauld Institute of Art

Hinde L, van den Berg KJ, de Groot S, Burnstock A (2011) Characterisation of surface whitening in 20th-century European paintings at Dudmaston Hall, UK, ICOM Committee for Conservation 16th Triennial Meeting, Lissabon 2011, Bridgland J (ed); paper nr. 1310

Izzo FC (2011) 20th century artists' oil paints: a chemical-physical survey. PhD dissertation in Chemical Sciences, Ca' Foscari University of Venice, 2011

Izzo F, van den Berg KJ, van Keulen H, Ferriani B, Zendri E (2014) Modern oil paints – formulations, organic additives and degradation: some case studies. These proceedings, Chapter 5

Keune K, Boeve G (2014) Its surreal: zinc-oxide degradation and misperceptions in Salvador Dalí's *Couple with Clouds in their Heads*, 1936. These proceedings, Chapter 19

Mayer R (1991) The artist's handbook of materials and techniques. Viking, New York

Michalski S (1990) A physical model of the cleaning of oil paint. In: Mills JS, Smith P (eds) Cleaning, retouching, coatings. International Institute for Conservation, London, pp 85–92

Mills L (2008) Water-sensitive oil paints: an experimental investigation characterising the causes of the phenomena and analysis of a case study painting Les Animaux, 1961, by Karel Appel final year project. Department of Conservation & Technology, Courtauld Institute of Art 2008

Mills L, Burnstock A, Van Keulen H, Duarte F, Van den Berg KJ (2008) Water sensitivity of modern artists' oil paints. In: ICOM Committee for Conservation (ICOM-CC) 15th Triennial Meeting New Delhi, 22–26 Sept preprints vol 2, Allied publishers 2008, pp 651–659

Morrison R, Young AB, Burnstock A, van Keulen H, van den Berg KJ (2007) The effects of triammonium citrate and variable pH for surface cleaning unvarnished oil paintings. Stud Conserv 52(4):255–270

Ormsby B, Barker R, Hellen R, Smithen P (2013) Cleaning acrylic emulsion paintings: case study treatments, evaluation and reflections. In: Mundici MC, Rava A (eds) What's changing: theories and practices in the restoration of contemporary art. Castello di Rivoli, Torino, Milano

Phenix A (1998) Solubility parameters and the cleaning of paintings: an update and review. Zeitschrift für Kunsttechnologie und Konservierung 12(2):387–409

Saltmarsh P, van den Berg KJ, Burnstock A (2008) Making paint without a recipe: modern paint reconstructions, Poster ATSR Glasgow, June 2008

Sawicka A (2013) An investigation into the formation of metal soap efflorescence on paintings with a view to developing a strategy for removal or reduction. Final year project dissertation, Department of Conservation & Technology, Courtauld Institute of Art 2013

Sawicka A, Burnstock A, Izzo F, Keune K, Boon J, Kirsch K, van den Berg KJ (2014) An investigation into the viability of removal of lead soap efflorescence from contemporary oil paintings. These proceedings, Chapter 21

Silvester G (2011) An experimental investigation of the formation of sulphates in oil paint films exposed to gaseous sulphur dioxide with particular reference to the relationship between these sulphates and water sensitivity. Final year project dissertation, Department of Conservation & Technology, Courtauld Institute of Art 2011

Silvester G, Burnstock A, Learner T, Chiari G, Megens L, Van den Berg KJ (2014) A cause of water-sensitivity in modern oil paint films: the formation of magnesium sulphate. Stud Conserv 59(1):38–51

Simendinger WH, Balik H (1994) Chemical reaction kinetics of sulfur dioxide and oxygen with unsaturated drying oils. J Coatings Tech 66:39–45

Smithen P (2005) A history of the treatment of acrylic paintings, modern paints uncovered, London, 2005. In: Learner T et al (eds) Proceedings of the conference, pp 165–174

Soldano A, Van den Berg KJ (2014) Investigation into the surface conductivity of water-sensitive modern oil paints. These proceedings, Chapter 27

Sutherland K (195) An investigation into the relationship between the organic composition of commercially primed canvases and efflorescence. Final year project dissertation, Department of Conservation & Technology, Courtauld Institute of Art 1995

Tempest H (2009) Water sensitive oil paint: an experimental investigation, with case studies, further characterising the causes of the phenomena. Final year project dissertation, Department of Conservation & Technology, Courtauld Institute of Art 2009

Tempest H, Burnstock A, Saltmarsh P, van den Berg KJ (2013) The sensitivity of modern oil paints to aqueous solvents used for cleaning. In: New insights into the cleaning of paintings (Cleaning 2010) Proceedings from the MCI Workshop Series Smithsonian Contributions to Museum Conservation, No.3 Smithsonian Institution Scholarly Press 2013, pp 107–115

van den Berg JDJ (2002) Analytical chemical studies on traditional linseed oil paints. PhD thesis, University of Amsterdam

van den Berg KJ, Daudin M, Joosten I, Wei B, Morrison R, Burnstock A (2008) A comparison of light microscopy techniques with scanning electron microscopy for imaging the surface cleaning of paintings, preprints of the art 2008 conference, 25–30 May, Jerusalem

van de Vall R, Beerkens L, van den Berg KJ (2011) Research proposal 'Patina in Modern Art', 2011

van den Berg KJ (2013) Introduction to the symposium – the 20th century oil paint project. In: van den Berg KJ et al (eds) Issues in contemporary oil paints. Extended abstracts of the symposium, 28 and 29 Mar 2013, Amersfoort, pp 17–20

van den Berg JDJ, van den Berg KJ, Boon JJ (1999) Chemical changes in curing and ageing oil paints. In: Bridgland J (ed) ICOM Committee for Conservation 12th Triennial Meeting, Lyon, France, 29 August-3 September 1999. James & James, London, pp 248–253

Van den Berg KJ, Burnstock A, Hoogland F, Izzo F, van Keulen H, Keune K, Learner T, Schilling M (2011) Novel mass spectrometric approaches for the analysis of paint media and organic additives, unpublished results presented at the Technart 2011 conference, 26–29 Apr 2011, Berlin. Article forthcoming

Van den Berg KJ, Bayliss S, van Gurp F, Burnstock A, Klein Ovink B (2014) Proceedings of the art technological sources research conference, Rijksmuseum Amsterdam, 16–17 June 2014. Manuscript submitted

Van Loon A, Noble P, Boon J (2012) The formation of complex crusts in oil paints containing lead white and smalt: dissolution, depletion, diffusion, deposition. In: Historical technology, materials and conservation: SEM and microanalysis Meeks N, Cartwright C, Meek A, Mongiatti A (eds) The British Museum/Archetype, London, pp 207–209

Veríssimo Mendes B, van den Berg KJ, Megens L, Joosten I, Daudin M (2014) New approaches to surface cleaning of unvarnished contemporary oil paintings – moist sponges and cloths. These proceedings, Chapter 25

Volk A (2012) Approaches towards the surface cleaning of water sensitive oil paint with Aqueous methods. Diploma Thesis Staatliche Akademie der Bildenden Künste Stuttgart, unpublished

Volk A, van den Berg KJ (2014) Agar – a new tool for the surface cleaning of water sensitive oil paint? These proceedings, Chapter 26

Wijnberg L (2014) Do we see what we know or do we know what we see? Conservation of oil paintings in the Stedelijk Museum Amsterdam. These proceedings, Chapter 2

Wijnberg L, Van den Berg KJ, Burnstock A, Froment E (2007) Jasper Johns's Untitled, 1964-65. ArtMatters 4:68–80

Do We See What We Know or Do We Know What We See? Conservation of Oil Paintings in the Stedelijk Museum Amsterdam

Louise Wijnberg

Abstract Some important issues in relation to the conservation, treatment and display of paintings at the Stedelijk Museum Amsterdam over the last 25 years are discussed. Examples of the treatment include key works by Kazimir Malevich, Modern American painters from the 1950s-1970s and Cobra artists. The challenges for conservators and criteria for decisions about treatment and display at the museum included reviewing former conservation treatments and the question of material authenticity. The discovery of water sensitivity in oil paint used by artists including Jasper Johns and Karel Appel in the 1950s and 1960s has had an impact in relation to cleaning these works. The relationship between painting technique and paint quality is discussed in relation to the works of Karel Appel and Asger Jorn. These studies have led to the characterisation of novel degradation phenomena and technical evidence has helped to predict water sensitivity and inform more specific cleaning methods. Despite these advances not every work of art can been treated and sometimes we have to except that '*less is more*'.

Keywords Oil painting • Water solubility • History of conservation • Stedelijk Museum • Cleaning • Malevich • Johns • Appel • Jorn • Ernst

Introduction

Conservation research begins with a close and detailed visual examination of the surface of a painting. What do we see at first sight? Does it concur with our notion of authenticity and the artistic quality of a certain painter from one period of creation? These notions are a combination of ethics, experience and research influenced by contemporary context.

During the Conference *Modern Paint Uncovered* in London in 2007, the participants were invited to take part in *The Paint Quiz* and were asked to identify

L. Wijnberg (✉)
Painting Conservation Stedelijk Museum Amsterdam, Van Baerlestraat 31,
1071 AN Amsterdam, The Netherlands
e-mail: l.wijnberg@stedelijk.nl

K.J. van den Berg et al. (eds.), *Issues in Contemporary Oil Paint*,
DOI 10.1007/978-3-319-10100-2_2

the artworks of nine different artists by looking at nine detailed pictures taken from the collection on display in the galleries of Tate Modern.[1] 'The Quiz' summarises in a way the same questions that we try to answer such as: do we recognise and understand certain artist's hands and their specific paint surfaces while we are examining a work for conservation purposes? We learn to recognise colours, brushstrokes, drips, spots or the evidence of deterioration such as wrinkling, cracking or yellowing that prompt characteristic associations. On the other hand technical studies of pictures aim to provide deeper understanding of the artistic process by ascertaining the build up, the order and quality of applied paint and the drying and deterioration processes which take place in modern oil paints.

History

In the 1990s the painting conservation lab of the Stedelijk Museum started a restoration program of 18 Suprematist paintings by Kazimir Malevich (1879–1935). Just after the acquisition of the Malevich paintings in the 1950s by the Stedelijk Museum, these paintings had been wax-lined. Subsequent display of the Stedelijk pictures with works by the artist from largely unknown Russian collections during the first Malevich exhibition in 1989 in Amsterdam prompted crucial observations of major differences in surface quality. This could be accounted for in part by the conservation history of the different collections. Surface quality of the artist's non-varnished paint surfaces were studied and a better understanding of how these paintings should look was understood by the conservators. The consequence was a new conservation treatment which involved the removal of a thick discolored 'restoration patina layer' (Fig. 1) to reveal the 'real qualities' of the paint layer.

In 1989 the Stedelijk Malevich paintings were loaned to the *Biennial of Sao Paolo* that offered another opportunity for contextual comparison with a Russian collection of the artist. The Malevich paintings from the two exhibited collections showed markedly different surface qualities as a consequence of different conservation histories. While the works from the Russian collection were unvarnished and had matte surfaces, the Dutch collection pictures on canvas that were wax lined and varnished appeared relatively glossy and saturated. This observation prompted a reconsideration of the notion of authenticity in relation to the treatment of these works that was last discussed in the 1980s (Wijnberg et al. 2007). It was understood that the return to a more authentic state of the paint layer and its presentation had not completely succeeded. With the introduction of wax and varnish, the aesthetic appearance of the paint had definitely changed in quality. We compared our paintings with ones which had also a restoration and conservation history but a different one. Were these works still authentic? In addition to the materials and techniques used by the artist, which may change in the course of time or even from

[1] Learner T, *The Paint Quiz,* unpublished.

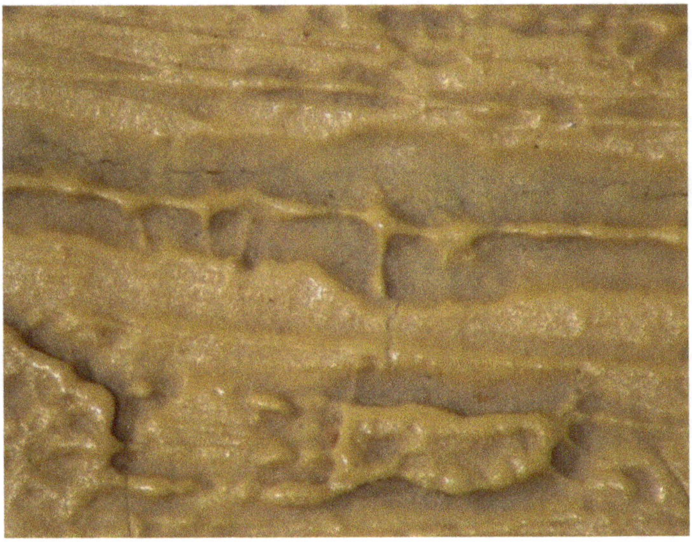

Fig. 1 Malevich, detail of paint surface with patina before cleaning in 1989

painting to painting, the final appearance of a work of art is the result of the way a collection is kept during its lifetime. Often, the works of one and the same artist are spread out over different sub-collections. Differences might become visible sooner or later and are discovered when paintings are examined and found to have aged, or former treatments have become more visible.

Along the way, our notion of authenticity has developed due to a better understanding of the consequence of the effects of physical changes in oil paintings on aging and the impact of physical history including former treatments. Ten years later, in 1999, one of the paintings had to undergo a minor treatment. Observing the painting brought us back again to the consequences of the conservation history and the idea of no real return to authenticity; we clearly noticed that the wax used for relining had become completely incorporated into the painting and some material had migrated to the surface. The migration of the wax appeared to be an ongoing process which could not be stopped. The impact of the former treatment appeared to have long term consequences for future treatment. The consequence was that it made and makes at present a conservator more and more cautious.

Wax has been used as conservation material but it was also an important component of the painting technique used by artists from ancient to the modern era. Brice Marden added beeswax to oil paint in the 1970s at the beginning of his artistic career to paint monochrome works. During the 1990s conservators noticed that his mixed media technique was problematic in spite of the carefully prepared and applied paint medium. The slow migration of the wax from the oil paint became obvious in most of his monochrome works. During the years that followed, individual paint layers showed pronounced adhesion problems which resulted in extensive flaking especially along the edges of the painting. Microscopic cross-section analyses

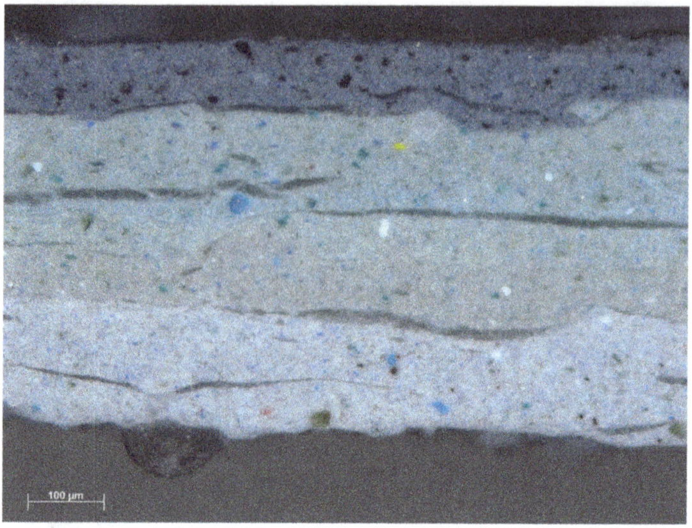

Fig. 2 B. Marden, *Horizontal Horizontal* (1972–1973), Inv. Nr. A 35202; visible transparent inter layers of beeswax (L. Megens, RCE)

(Fig. 2) illustrated the separation of wax from the other organic media and its concentration at the interfaces between the ground and individual paint layers. Since the wax has low cohesive strength the individual paint layers separate easily and flaking occurs in response to small stresses, such as those caused by changes in relative humidity.[2] The accumulation of wax between layers of paint indicated that this separation occurred shortly after the application process.[3] Droplets of wax were also present on the surface.

Marden's paintings will remain vulnerable. In addition to the difficulties in consolidating the flaking paint, monitoring of his works on exhibition and in storage is critically important for their preservation.

The Stedelijk Museum's conservation program from the late 1980s included treatment of monochrome paintings by American artists from the 1950s–1970s, including paintings by Newman, Kelly, Marden, Mangold, Tuttle, Ryman and Serra. The visual quality of the paint surfaces, mostly executed in oil, became an important consideration in relation to approaches to cleaning and retouching. New cleaning systems introduced by Richard Wolbers provided a wider choice of reagents for cleaning paintings. Some of these works that were originally unvarnished had been entirely varnished or over-painted as part of former conservation treatments that

[2]K.J. van den Berg, M. de Keijzer, L. Megens, H. van Keulen and S. de Groot, internal RCE report 2006-076, 2007.

[3]K.J. van den Berg, personal communication February 2014.

aimed at replicating a perfect surface without marks or visible damage. Removal of these superficial layers required collaboration between restorers and scientists to formulate tailor-made solutions. An example is the treatment of a painting by Ellsworth Kelly where a layer of repaint was successfully removed using tailor-made gels. Conservation and technical study highlighted the variety of media used in his works despite their visual similarities in appearance, in particular in his monochromatic paintings (Wijnberg et al. 2011). This important observation has to be taken into account in the treatment and display of works by the artist.

A recent treatment in 2012 of a monochrome work *Blanc* (1978) by Richard Serra presented particular challenges. Executed using an oil paint stick, the painting has a uniform black paint surface that can hardly be touched without changing its fragile surface texture (Chavannes et al. 2014). A visually perfect state was a pressing demand by the artist who was present at the final installation of the two canvases on two opposite walls of one cabinet of the Stedelijk Museum at its reopening in September 2012.

Water Sensitivity

The notion of non-intervention became equally relevant in relation to the treatment of a painting by Jasper Johns, *Untitled 1964–'65* ten years ago, in 2004 (Wijnberg et al. 2007). At first sight no degradation processes were visible and the painting appeared to be in perfect condition, however, attempts to remove dirt from the unvarnished paint surface revealed extreme sensitivity of the oil paint to aqueous and protic solvents. Examination using scanning electron microscopy (SEM) of the well-bound paint revealed crystals on the surface, protruding from the paint (Burnstock et al. 2006). The crystals were detected in later work as water soluble epsomite, which could explain the water sensitivity (Silvester et al. 2014).

This study again made us fully aware of these risks related to the cleaning of some modern oil paints, and deterioration that is not evident by visual examination of the painted surface.

In this case, non-intervention was preferred to a compromised risky surface treatment.

This same notion of non-intervention was followed in the treatment of a painting by William the Kooning in the Stedelijk Museum; *Rosy Fingered Dawn at Louse Point* (1963). In 2008, tests for surface cleaning demonstrated the presence of tacky water sensitive paint in particular in paints containing cadmium yellow (Fig. 3). Organic analysis of the paint identified a mixture of oils and other additives of lipid origin (Izzo et al. 2014) which were degraded. In conclusion, the conservator decided not to clean the painting and that '*less is more*' (Mills et al. 2010). In this case the results of scientific analysis helped the conservator to understand the qualities of the paint surface to avoid disruption of the painting through an intervention.

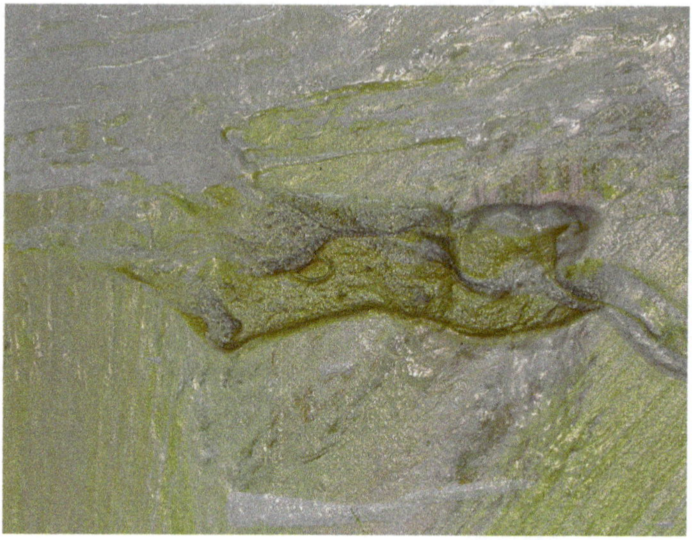

Fig. 3 W. de Kooning, *Rosy Fingered Dawn*, Inv. Nr. A 22662, detail

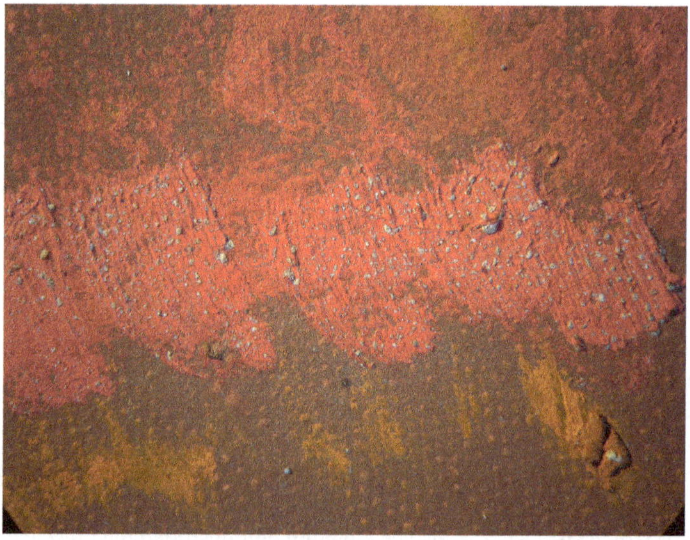

Fig. 4 M. Ernst, *La horde*, Inv. Nr. A 2500; detail

In 2010, bright white efflorescent particles were found on the paint surface of *La horde* (1927) by Max Ernst in the Stedelijk Museum collection (Fig. 4). The visible white particles were concentrated in the red passages of the unvarnished oil painting. The particles had been noted in a previous conservation campaign in 1991 but by

2010 had become more prominent and visually disturbing. The red layers were analyzed and identified as mixtures of vermillion for the darker red areas and cadmium red for the brighter parts.[4]

The white efflorescence was associated with the darker red vermillion paint, and identified as an inorganic sulphate, probably a sulphate containing zinc. The zinc sulphate was thought to be formed from zinc stearate, present in the vermillion paint as a stabiliser, and from airborne sulphur dioxide perhaps from the pigment. The water sensitivity of the paint layer was clearly related to the formation of this efflorescence.[5]

Surface deterioration phenomena have been observed in several paintings by the former Cobra artist Karel Appel, painted in the 1950s and 1960s. Visible to the naked eye, the deterioration included blooming, whitening, cracking, efflorescence and the presence of local hazes, all of which were characterised using different methods for technical analysis (Mills 2008).

In the last 8 years, several works by Appel have been restored. *Fleur de nuit* (1954) and *L'homme* (1958) from the Stedelijk collection were carefully cleaned and analysis of paint samples confirmed the presence of metal stearates and medium exudates.

New surface cleaning methods have been introduced that facilitate partial cleaning of water sensitive or degraded unvarnished paint surfaces. These include the use of an airbrush, gums, erasers, sponges, and other dry cleaning materials (Daudin and Van Keulen 2014).

The use of Agar gels with different aqueous and solvent components has also been used with a very good and convincing result (Volk and Van den Berg 2014).

An exhibition in 2008 '*karel appel, JAZZ, 1958–1962*', at the Cobra Museum for Modern Art Amstelveen, provided an opportunity to explore new perspectives of the artist's work (Sassen et al. 2008). Twenty large paintings from 1961 – 230 × 300 cm – were included in the exhibition. The making of these large paintings was the subject of a short film of nearly 15 min by Jan Vrijman (1925–1997).[6] The film documents the artist in full action in the creation of the works in 1961. Appel commented on his working style: '*I paint a barbaric art, in a barbaric Age*', sloshing paint around freely with a palette knife, squeezing tubes onto the canvas and manipulating the paint with his gloved hands or big flat brushes. One of the paintings in the film is *Archaic Life* (1961).[7] This painting was loaned by the Stedelijk for the *Jazz* exhibition and was installed together with the others made in the same year. In the film the work can be viewed at the moment of creation, when

[4]Van den Berg, Megens, de Groot (RCE), and Burnstock (CIA), unpublished results 2009.

[5]Bracht, E, van den Berg, KJ, de groot, S, Megens, L (FTRI, XRF), RCE 2010, Project in progress.

[6]Dutch filmmaker who became well known with its presentation. It is called *The Reality of Karel Appel. Musique Barbare by Karel Appel* (1962).

[7]K. Appel donated the painting to the Stedelijk Museum end 1962 after the exhibition '*Art since 1950*', Seattle, as a part of the collection Sandberg. The painting was then titled '*From the beginning*'.

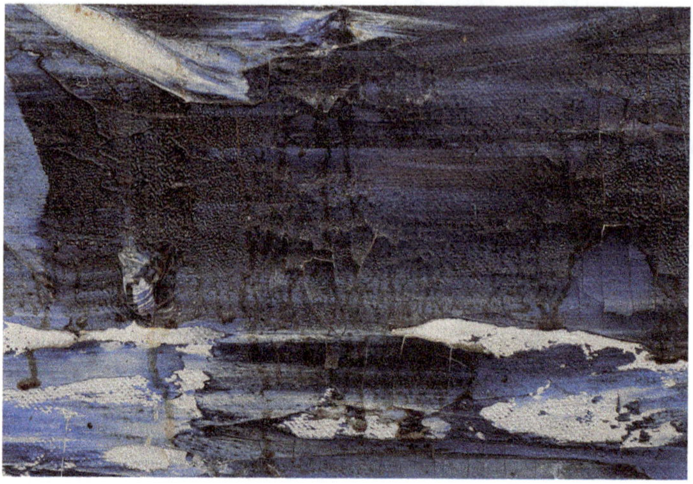

Fig. 5 K. Appel, *Archaic Life*, Inv. nr. A 21604 detail ultramarine pigment and dripping binding medium

the paint was freshly applied without dust and drying phenomena that is apparent today. The work is large and the heavily impastoed paint contributes to its significant weight. Some of the oil has been absorbed into the canvas, and resulting staining is visible at the back.[8]

Before lending the work to the exhibition the painting underwent conservation treatment.[9] Brittle blue paint, that was water sensitive was tenting and flaking. The paint also exhibited pigment-medium separation, and brown vertical drips of oil medium were visible at the paint surface (Fig. 5).

The blue paint shows characteristics typical of ultramarine pigment in oil media, known for its difficult drying properties. Research has provided explanations for this phenomena including water sensitivity and phase separation.

In the same painting, the white painted areas were in good condition, apart from the presence of disfiguring surface dust on the tops of the paint strokes, that had become embedded at early drying phase (Fig. 6). Appel's paintings are characterised by a great variety of colours each with its characteristic drying cracks, wrinkles or other deterioration.

This was clear in the exhibition that showed that all the works made in the same year shared similar painting phenomena. These included traces of dripping binding media now dried to a yellow or brownish hue, and a range of other surface phenomena. Appel stated '*The paint is applied to the canvas in all manner of expressive ways, because emotion has to be beaten into paint*' (Vrijman 1962).

[8]The film was shot on location in a build studio for the film in 'Huize Groeneveld', Baarn, The Netherlands at the time that Appel was still living in Paris.

[9]2008, conservator M. Engel and 2010 V. Blok.

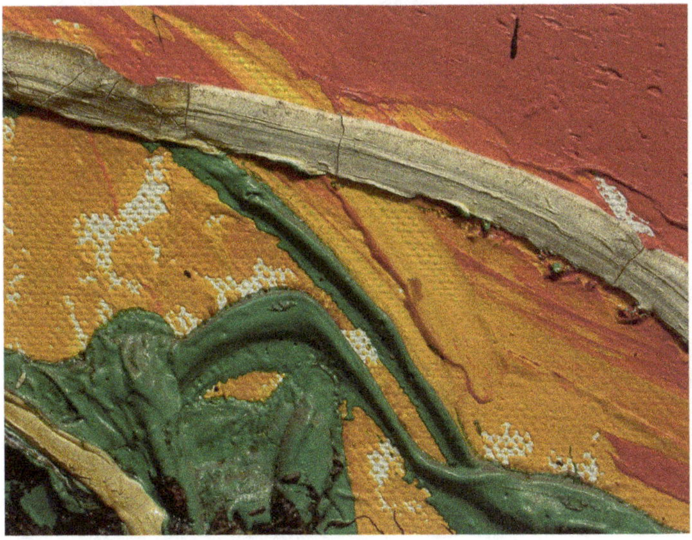

Fig. 6 K. Appel, *Archaic Life*, nr. A 21604; detail dirty white paint

Comparison of the works in the exhibition prompted the question of whether the paint quality is the result of the particular materials employed or the painting methods used by Appel, or both.[10]

With regard to treatment, the water sensitivity observed in Appel's paintings was linked to the presence of water soluble crystals which formed on the paint by reaction with atmospheric pollutants. In 2010, conductivity measurements were proposed as a low-tech means to identify water sensitivity caused by this crystal formation on oil paints characterised by high conductivity at the paint surface (Soldano and van den Berg 2014). Tests were carried out on several paintings including a passage of ultramarine blue paint from *Fleur de nuit,* that registered high conductivity before cleaning. The conductivity decreased by 50 % after surface cleaning (Fig. 7), suggesting that ionic salts were removed in the process.

The most recent treatment of a painting by Karel Appel in the Stedelijk Museum conservation studio in 2013 was *Michel Tapié de Celeyran* (1956). It proved typical in its patterns of deterioration exemplified by flaking of blue passages of paint that required consolidation, together with pronounced cracking and cupping paint. The characteristic separation of blue paint medium from pigment showed as the thinner underlayers of leanly bound bright blue were covered by more medium rich ultramarine paint. Flaking of the top layer revealed a homogenous thin layer of bright blue on the canvas. Other paints exhibited different characteristic surfaces:

[10]Appel used a table as a palette. Among the metal jars we can distinguish a pot with ETA paint (emulsion of casein).

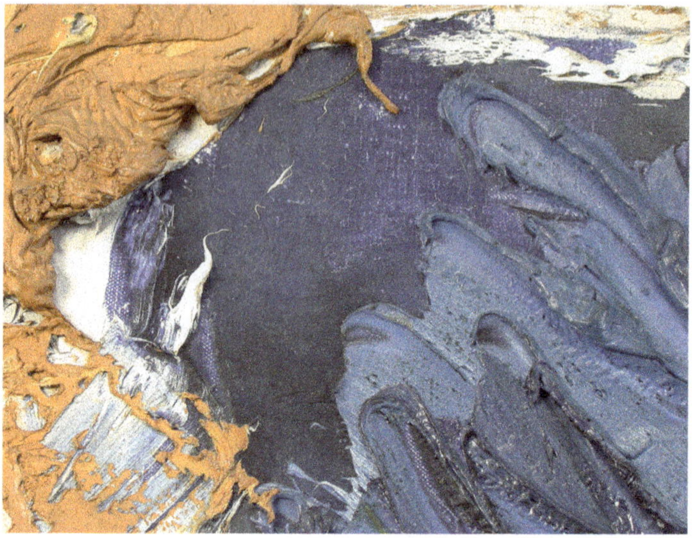

Fig. 7 K. Appel, *Fleur de nuit*, Inv. nr. A 29186; detail cleaned (*left*) and non-cleaned (*right upper part*) ultramarine oil paint

the grey layers were severely wrinkled while thickly painted white areas formed deep linear cracks.

In conclusion, while the knowledge gained from advanced scientific investigations has undoubtedly informed the eye of the conservator, the particular painterly qualities of an artist's oeuvre and changes that occur on ageing of the works can largely be catalogued by a discerning eye.

Technique or Paint Quality?

The changes in oil paint discussed in relation to Appel can also be found in works by other artists. The works of Asger Jorn in the Stedelijk collection present problems of delaminating paint probably related to the artist's use of gouache mixed with oil paint, exemplified in the case study of *Abstract Minded Figure* (1962) (Witlox 2003). The result of the investigation also highlighted delamination of cadmium yellow pigmented paint.

Another painting by the artist in the Stedelijk Museum, *Guillaume Apollinaire* (1956) had been wax relined in a former treatment that aimed to consolidate the flaking yellow and blue paint. Jacqueline de Jong (1939) explained that when Jorn was living in Paris in the 1950s, he used good quality oil paints from Lefèvre – Foinet and that the artist was mixing oil paint with gouache was not known to de Jong.[11] In 1961 she was for a short time the assistant of Appel in his studio where she was asked to mount canvases on stretchers and to clean the large and numerous brushes he normally stored standing straight in tins.[12] De Jong remembered that Appel bought his paint in the '*Quincaillerie*' in the rue Brézin just in front of the entrance of his studio. She commented that 'He was not interested in the quality of paint but only in colour!'.

We can conclude that the conservation problems with the oil paintings of Appel and Jorn are the consequence of the quality of the paint they bought but linked to their very personal paint technique which is predominant for the conservation of their works.

Conclusion

Do we see what we know or do we know what we see? The '*20th Century Oil Paint Project*' RCE, could give many answers to both questions. We have partly cleared the haze. With the years we entered in new levels of science where phenomena detected by the restorer were little by little explained. Surfaces are visually well known by us but even so oil paintings still hide their interior qualities from us. It is a quest without an end but one that becomes more and more detailed, a continuing search for the truth and reality that these paintings have in store for us.

References

Burnstock A, van den Berg KJ, de Groot S, Wijnberg L (2006) An investigation of water-sensitive oil paints in twentieth century paintings. In: Modern paints uncovered, proceedings symposium Tate Modern London, The Getty Conservation Institute, Los Angeles, 2006, pp 177–188

Chavannes M, Leeuwenstein M, Bongaarts M, de froot S, Diependaal H, van Keulen H, Wijnberg L (2014) I try to keep surfaces an anymous as possible. Conservation issues in Richard Serra's monumental paintstick drawings on canvas, 14–19 September 2014

Daudin M, van Keulen H (2014) Dry cleaning: research and practice. These proceedings, Chapter 24

[11]Interview with the author on the 25th Mars 2013. Jacqueline de Jong, Dutch artist, had a relationship with Asger Jorn from 1959 to 1970. Paint material was bought at Sennelier, 3 Quai Voltaire, Paris (Since 1887).

[12]17 rue Brézin, XIV Arrondissement, Paris, where he lived and worked from 1958 on. The place is shown in the very beginning of the film *The reality of Karel Appel* (Vrijman 1962): Two man with a wooden chariot working for '*Quincaillerie Couleurs de Montparnasse*' are delivering quite large canvases packed in brown paper and big jars probably containing paint.

Izzo FC, van den Berg KJ, van Keulen H, Zendri E, Ferriani B (2014) Modern oil paints – formulations, organic additives and degradation: some case studies. In: van den Berg KJ et al (eds) Issues in contemporary oil paint. Springer, Dordrecht

Mills L (2008) Water-sensitive oil paints: an experimental investigation characterizing the causes of the phenomena and analysis of a case study painting 'Les Animaux', 1961, by Karel Appel, Courtauld Institute, January 2008

Mills L, Bracht E, Wijnberg L (2010) Less is more – the rationale behind the decision not to treat a painting. Poster presentation, CAWC symposium, 9–11 June 2010, Amsterdam

Sassen JH, Vrieze J et al (2008) 'karel appel, JAZZ, 1958–1962', 09.02 – 01.06.2008. Exhibition catalogue, Cobra Museum of Modern Art, Amstelveen 2008:23

Silvester A, Burnstock A, Megens L, Learner T, Chiari G, van den Berg KJ (2014) A cause of water-sensitivity in modern oil paint films: the formation of magnesium sulphate. Stud Conserv 59(1):38–51

Soldano A, van den Berg KJ (2014) Investigation into the surface conductivity of water-sensitive modern oil paints. These proceedings, Chapter 27

Volk A, van den Berg KJ (2014) Agar – a new tool for the surface cleaning of water sensitive oil paint? These proceedings, Chapter 26

Vrijman J (1962) Film the reality of Karel Appel (1962), Jan Vrijman Cineproductie

Witlox M (2003) Asger Jorn's abstract minded figure (1962) Opstaande verf door schildertechniek en consolidatie. In: Cr 1, voorjaar 2003, pp 26–32

Wijnberg L, van den Berg KJ, Burnstock A, Froment E (2007) Jasper Johns 'Untitled 1964–'65. In: ArtMatters (4), Zwolle, 2007, pp 68–80

Wijnberg L, Bracht E, van den Berg KJ, de Keijzer M, van Keulen H (2011) A study of the grounds used by three post-war American Artists (1954–1974): Barnett Newman, Ellsworth Kelly and Brice Marden, ICOM Committee for Conservation 16th Triennial Meeting, Lissabon 2011, Bridgland J (ed); paper nr. 1324

Towards Interpretation of Making, Meaning, and Change in British Twentieth Century Oil Paintings: The Relevance of an Artist's Paint Archive

Mary Bustin

Abstract A rare opportunity to document materials in an artist's studio brought unexpected insight into his material preferences and working patterns. It revealed practical solutions to picture-making long-since absorbed by the artist, Patrick Heron (1920–1999), into the instinctive act of painting and left un-articulated in interviews or commentaries. The large quantity of oil paint in 155 different colours led to questions as to why so much variety, which brands and qualities were favoured, which paints had been in current use and in what combination, and which had been rejected. A selection of this paint was donated by the Heron Estate to Tate, London, for research purposes. On its own, the Heron Paint Archive is a simple collection of tubes. In the context of documentation and oral histories, a more subtle and complex reality emerges with implications for the interpretation of physical properties, visual phenomena and analysis of British colourfield paintings of the mid to late twentieth century. Heron's oil paintings, purposefully unvarnished, are water-sensitive.

Keywords Patrick Heron • St Ives • Roberson • Winsor & Newton Ltd • Paint archive • Colourfield • Water-soluble oil paint • Paint swatches • Painting conservation

Introduction

In the summer of 1953, Patrick Heron (1920–1999) in his essay for *Space in Colour*, an exhibition of modern abstraction at the Hanover Gallery, London, declared that colour is 'the sole means of bestowing that physical vibrancy and resonance without which no picture is alive.' (Heron 1953)

M. Bustin (✉)
Mary Bustin Conservation, London, UK
e-mail: conservation@marybustin.co.uk

K.J. van den Berg et al. (eds.), *Issues in Contemporary Oil Paint*,
DOI 10.1007/978-3-319-10100-2_3

> One reels at the colour possibilities now: the varied and contrasting intensities, opacities, transparencies, the seeming density and weight, warmth, coolness, vibrancy or the superbly inert "dull" colours – such as the marvellously uneventful expanse of the surface of an old green door in the sunlight. (Heron 1973).

When his materials were documented in 2000, it was perhaps not surprising that his studios were replete with an extraordinary range of oil and gouache colours. His oil paint range encompassed 155 colours with 790 tubes in stock (not counting empty tubes).[1]

The concept of an artist's paint archive is to capture a moment in art history when the artist has made his selection from the commercially available range of materials and has done so for a particular purpose. Studio contents represent a snapshot of a collection of possibilities. They are in no way comprehensive and by default represent works in progress, relics of part-used materials, or works in planning but not yet realised. Realistically their contextual information will balance investigations and analyses that rely on general reference collections for standards: raw materials, for example, being actively collected by the Getty Conservation Institute in the GCI Reference Collection[2] and the commercial products collected by the National Gallery of Art in the Art Materials Research and Study Center.[3] The contents of artists' studios, the ephemera as well as materials, help to refine the range of possibilities and provide context to their decision-making.[4] Patrick Heron's studio as a well-ordered, clean space affords a reading of his process that is diametrically opposed to the archaeological analogies raised by Francis Bacon's accumulation of clutter (Dawson and Cappock 2001; Russell et al. 2012).

Heron's stock of oil paint afforded an opportunity to re-evaluate preconceptions about artists' use of materials. For instance the assumption that date of paint manufacture will be the same as that of the execution of a painting; or that an artist will only use one brand; or that if they prefer artists' quality paint they would never use student quality. Heron's fascination with colour and simple rule of never mixing colours (except with white) led to a reminder that handling alters the optical properties of paint: a single tube can produce a variety of shades, hues and tints.

---

[1]Documented by Mary Bustin and Jo Crook on the invitation of the Artist's Estate, Heron Paint Survey, Tate Conservation, May 2000, Tate Gallery Records. This paper draws on further work carried out on the master copy of the archive in a private collection by Mary Bustin Conservation (MBC) in 2011–2012.

[2]www.getty.edu/conservation/publications_resources/reference_collection/index.html. Accessed 1 Oct 2013.

[3]www.nga.gov/content/ngaweb/conservation/materials-study-center.html. Accessed 1 Oct 2013.

[4]For example Harvard Art Museums' Center for the Technical Study of Modern Art which actively collects studio documentary materials including 'detritus and discarded trials'. www.harvardartmuseums.org/study-research/research-ceters/cfsma. Accessed 3 Oct 2013.

The Artist and St Ives, Cornwall

Patrick Heron (1920–1999) was a British painter and art critic who worked in London, UK, and St Ives, Cornwall. Born in Leeds, Heron studied painting part-time at the Slade School of Fine Art while designing for Cresta Silks, his father's company. By 1945 he also supported his painting by writing for the New English Weekly, the start of a lifelong influence on contemporary art issues.

Heron was at the forefront of painterly abstraction in the UK.[5] The artists who gathered in St Ives after the Second World War – Peter Lanyon, Terry Frost, Bryan Wynter, William Scott and others – exhibited internationally. They forged close dialogues with European and American abstract painters including Pierre Soulages, Mark Rothko, Robert Motherwell and Sam Francis. Heron as art critic championed artistic interchange, reporting on exhibitions of American painters in the UK and British painters in New York.

Heron's fascination with pure colour in juxtaposed rhythmic accents evolved through vast un-modulated fields of single colours into organic shapes redolent of the Cornish landscape jostling in his 'wobbly hard edge' works of the 1960s and 1970s. In late works of the 1980s and 1990s his colour travelled in painterly notations on white primed canvas, on occasion softened with a touch of white. He rarely mixed colours, preferring the subtle variations of hue created by the manufacturers in paint of ostensibly the same identity, there being no standardisation of colour in respect of nomenclature: a boon to the artist wishing for subtle variation between brands; a nightmare for matching if the artist ran out of paint and had to purchase from another batch.[6]

No. 5 Porthmeor Studios

Porthmeor Studios was originally an old pilchard shed that overlooked Porthmeor beach and the gasworks (replaced by Tate St Ives in 1992). Converted for artists' use at the end of the nineteenth century, the studios have been recently renovated to correct the leaky roof and bitter winter cold of Heron's day. (Borlase Smart John Wells Trust 2013) Heron took up the lease of No. 5 in the summer of 1958. It had triple the height of his previous studio, the dining room at Eagles Nest where 'the light ... is purely horizontal'. No. 5 was to be 'a godsend both for

[5]For biography and list of writings see Sylvester (1998), Gooding (1994).

[6]The American colourman, Sam Golden, reflected on the problems of achieving a precise match from one batch to the next. When Barnett Newman asked for some more cadmium red to finish a painting, Golden matched a new batch to the previous one using a colour lab's spectrophotometer to check for accuracy. Nonetheless, on receiving the new paint, Barnett Newman rang Golden to complain that the colour was not the same Mancusi-Ungaro (2004).

its overhead neutral light and its superb, quiet space!'[7] North light reflected off the sea has 'a brilliance that is unique in England'. Heron blocked all side light from the window, which overlooked the inner courtyard, with a discarded panel by the studio's previous occupant, Ben Nicholson.[8] The studio was a neutral white space (Fig. 1).

Recording the Artist's Materials

In May 2000, when the contents were recorded, the studio was found to be well-organised and clean. There was a stock of paint piled high on one table behind the sofa. Clusters of tubes were laid ready for use on three painting tables. Paintings in progress were hanging around the walls. The approach was to record in a simple Excel spread-sheet the manufacturer, colour nomenclature, and other data from the tube plus quantity, location and groupings. The aim was to capture context as well as identity. Subsequently, a selection of tubes was donated to Tate for research purposes.[9] This gift proved to be felicitous. Following a fire at Momart warehouse in 2002, in which key paintings by Heron were destroyed, swatches of each colour were made on the request of Susanna Heron to act as 'colour memories' to aid accurate rendering of lost works in printed reproductions (Fig. 2).

Paint Swatches

Eighty-nine colours from six manufacturers (Art Spectrum, Lefranc, Lefranc & Bourgeois, C. Roberson & Co., George Rowney & Co. and Winsor & Newton Ltd.) were painted out onto two substrates in July 2004[10]:

1. pH neutral conservation card primed with Lascaux Acrylic Primer.
2. Oil-primed linen canvas: Brodie & Middleton oil primed linen canvas.

Each colour was applied in four different ways to mimic the application methods used by Heron and to capture colour shifts that occur when paint is thinned:

---

[7]Patrick Heron to Philip James, 19 June 1958, Arts Council Archive quoted in Tufnell (2006).

[8]Ben Nicholson's board bearing Heron's drawing. Personal communication, the artist, 1997.

[9]Tate Conservation Archive Q04033.1–2.

[10]Paint swatches made by Susanna Heron, Robert Holyhead and 10 members of Tate conservation staff: Mary Bustin, Jo Crook, Maureen Cross, Patti Favero, Rica Jones, Julia Jonsson, Adrian Moore, Sarah Morgan, Carole Towers, and Natasha Walker, 22 July 2004. The swatches on primed canvas now in Tate Conservation Archive Q04033.3–9.

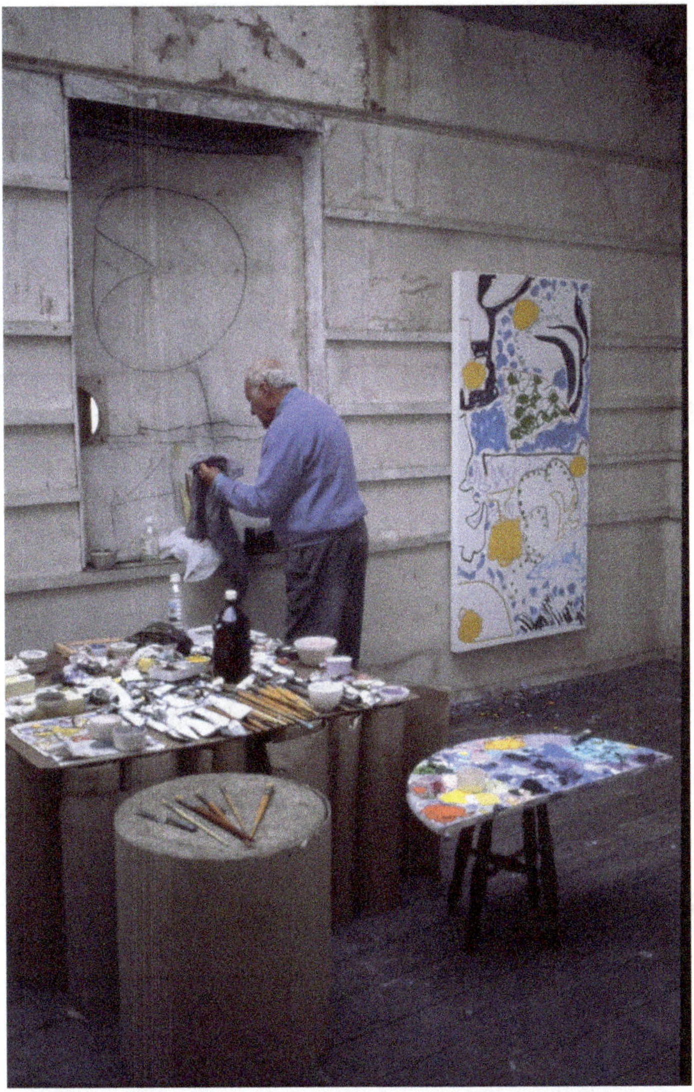

Fig. 1 Patrick Heron at 5 Porthmeor Studios St Ives, 1997 (© Susanna Heron)

1. Flat colour without brushstrokes colour pulled across the surface with a new glass microscope slide using 3M low-tack masking tape as a mask and guide.
2. Impasto paint brushed without dilution.
3. Paint half-cut with turpentine to a creamy consistency.
4. Thin wash of turpentine with a small dab of paint.

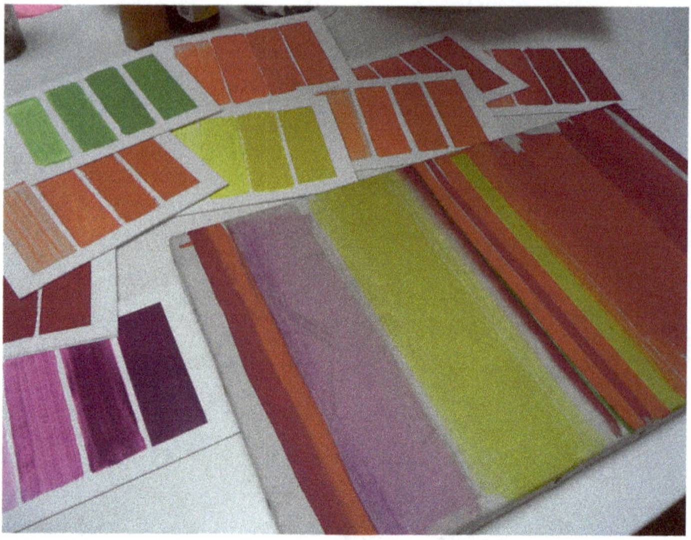

Fig. 2 'Sketch for Mural (Violet, Red, Lemon): February 1958' with colour swatches (Photograph Mary Bustin. © Estate of Patrick Heron)

Changing the texture and dilution of each swatch revealed colour shifts in both opaque and transparent paints; cadmium green, for example, appears more yellow in thin films. Heron's Garden Paintings employ similar shifts to broaden his palette, e.g. *Summer Painting: August 1956.*

It is serendipitous certainly that the desire to make a colour record in 2004 resulted in a body of paint samples that will be of increasing help to preserve Heron's paintings the older they get. They also form a resource for the study of mid to late twentieth century oil paint.

Paint

Heron selected paint by British, French, and Australian manufacturers (Fig. 3) Out of 155 colours, the majority was artist's quality paint from the Selected List of C. Roberson & Co. Ltd.[11]: 61 colours in 627 tubes out of the 790 tubes present; followed by Winsor & Newton Ltd[12] (23 colours), Art Spectrum[13] (19), Lefranc (19), and its later incarnation as Lefranc & Bourgeois[14] (12). Winsor & Newton's

[11]Charles Roberson & Co. Ltd. UK 1819–1987 (Woodcock 1995).

[12]Winsor & Newton Ltd UK 1832 – present (NPG Artists Suppliers Database).

[13]Art Spectrum, Australia 1960s – present (Tate Archive).

[14]Lefranc 1720.

Manufacturer	Quantity of oil paint tubes (all sizes) both part-used and unused stock in 1999	Number of different colours by brand	black	blue	brown	green	orange	red	crimson	vermilion	Cadmium red	violet	yellow	Cadmium yellow	white
Art Spectrum	25	19	1	2		3	2	3	2			4	1		1
Lefranc	32	19		1	1	3		4	2	1		3	3	1	
Lefranc & Bourgeois	41	12				3	2	2				4	1		
C.Roberson & Co. Ltd.	627	61	3	8	4	7	3	12		2	4	4	10	3	1
Geo.Rowney & Co.	11	6	1					2		1			2		
Talens	2	1							1						
Winsor & Newton Ltd. (artists' oil colours)	38	23	2	6	2	1	1	1	3			2	2	1	2
Winsor & Newton Ltd. (Winton Oil Colours)	14	14		2	2	1		1	3	1			2	1	1
Totals	**790 tubes**	**155 colours**	**7**	**19**	**9**	**18**	**8**	**25**	**11**	**5**	**4**	**17**	**21**	**6**	**5**

Fig. 3 Table of Heron's brand preferences and quantity of useable paint. Red is sorted into opaque reds (*red*) and transparent red (*crimson*). Heron further sub-divided 'red' into vermilion and cadmium. Yellows are subdivided here into cadmium yellow and other yellows for comparison

Winton student colours (14) including a box set of 6, while Rowney[15] (6) and Talens (1) represent less commonly-used paints. Talens Van Gogh Permanent Rose oil paint had been a gift to Heron by a studio neighbour.[16]

Of the 45 different reds which Heron classified into vermilion, cadmium, red, and crimson, he preferred Roberson for cadmium reds, and Winsor & Newton for transparent crimsons. For violets, his preference was for Lefranc, and Lefranc & Bourgeois, Roberson and Art Spectrum brands, purchasing cobalt violet from both Roberson and Lefranc & Bourgeois at the same time.[17]

---

[15]George Rowney & Co Ltd 1848–1985 when incorporated with Daler to form Daler-Rowney. (NPG Artists Suppliers Database).

[16]Personal Communication, Janet Axten, March 2013.

[17]Studio shopping list mid 1990s, transcribed MBC 18 February 2013.

Heron played with the subtle differences in shade produced by different manufacturing processes and source materials used by competing brands. Unmixed colour was essential to his work because, by 1962:

> I had already raised those colours to an undiluted maximum chromatic intensity… [they] were as bright and strong in themselves as it was physically possible to make them. (Heron 1973)

He chose to manipulate the shade and hue by altering his brush technique:

> You can manipulate space in dozens of different ways by the varied ways in which you paint-out an identical area-shape with an identical colour-mixture. The mere brush-work of an area endows it with differing kinds of space-creating power. You can weave it thick and tight, dense and opaque, in mutually self-obliterating heavy loops for instance… Or the pigment can be semi-transparent, more lightly and rapidly applied, full of nervous flicking movements. (Heron 1970)

Critically, because intensity and purity of colour is important to the reading of Heron's paintings, any accumulation of surface dirt which mutes colour and tonality will visually-compromise his work. Cleaning is a major challenge because Heron's paintings are unvarnished. Current conservation dilemmas include how to remove dirt without disturbing a 'patina', a waxy bloom, evenly from fields of single colour, e.g. *Horizontal Stripe Painting: 1957–58* (Tate) (Tempest 2009; Cooper et al. 2014). This is complicated by the water-sensitivity of some, but not all, of the oil paint colours that he used. An initial survey of the 8 year-old painted swatches in 2012 revealed that the majority of his paints are sensitive to saliva, at worst swelling and releasing colour instantly on the swab.[18] Apart from Art Spectrum, all of whose colours are unaffected by moisture, each manufacturer produced water-sensitive oil paint.

Date of Manufacture vs. Date of Execution

Dating the paint tubes revealed a far broader age-range than anticipated. Roberson tubes were dated by comparing label design and price tags to the manufacturer's marketing literature.[19] Winsor & Newton paints bear a notch and die-stamp identifying the manufacturing date.[20] Dates of Lefranc and Lefranc & Bourgeois

[18]Heron paint swatches water-sensitivity test report, 24 July 2012, MBC. These young paint films were tested for sensitivity to saliva and to de-ionized water assessed on a scale of 1–5. 1 = no response, 5 instantly soluble.

[19]Colourmen and manufacturer's trade catalogues and marketing literature assembled by Jo Crook, Tate Gallery Archive. Primarily twentieth century British.

[20]Personal communication Winsor & Newton tube dating conventions of notches, stamps on the fin, printed numbers, and bar codes, Ian Garrett, 2 April 2013.

paint were estimated from brand and label design upgrades. In addition, the artist's assistant wrote the date of purchase (1991) on the Art Spectrum paints.[21]

The 30 year span can be explained in part by the accumulation of paint left over from earlier projects when continuity of batch was critical, particularly in his 'wobbly-hard edge' colourfields of the 1960s, but it was mainly due to bulk-buying of prime quality paint after Roberson went into liquidation in 1987 prompting concerns about potential dwindling stocks.[22] A shopping list pinned to the studio wall revealed that a large purchase of colours had been made in the mid-1990s not in the mid-1960s as the paint labels might suggest: quantities, brand and colour correlated with stock levels.

Significance

How significant though is paint left in an artist's studio on his death to our understanding of his paintings? Knowing it has direct relevance to his last paintings, what assumptions can be drawn to inform the interpretation of his earlier works?

Contextual information such as the location of the tubes in Heron's studio, clusters in boxes, and which tubes had been used, clarified that the majority of the paint was not only in active use, the artist had stockpiled particular colours. Old used tubes, squashed beyond recognition were piled in a cardboard box.[23] Whether the disposition of paint around the studio reflected a single day or week of painting is probably unlikely given that there was little change between the photograph of Heron in the studio taken in 1997 (Fig. 1) and our visit in 2000. The purple, blue and the shock of red on his marble palette hints at the portraits of A.S. Byatt painted in 1997: *A.S. Byatt: Red, Yellow, Green and Blue: 24 September 1997* (NPG London). The painting on the wall in Fig. 2 is *16 September 1996.*

The archive reveals another aspect of the artist's process and rationale that is not often articulated by them in interviews. Richard Mulholland captured this in his study of the American sculptor, David Smith (1906–1965), who:

> understood as most artists do, that technical procedures must be so well absorbed that control is subconscious, so innate, that process proceeds fully and becomes 'second nature' and that the mind's occupation in developing concept of vision is not hindered by simple technical concerns. (Mulholland 2009)

Heron's statements in 1957 and 1969 reveal a deepening intuition:

> By the time the picture is actually being painted, it is too late to think... thought is there all the time but the thought that goes into the painting is already something that has been done before the act of painting began. (Heron 1957)

[21] Note that this is date of purchase, not date of manufacture.

[22] For the history of C. Roberson Co. Ltd, see Woodcock (1995).

[23] Crumpled used paint tubes and stock were sifted later by the artist's family.

> The only rule I follow while painting is this: I always allow my hand to surprise me. I always follow impulse – for instance in the choice of colours; deliberation is fruitless. (Heron 1969)

It is tempting to read too much into the groups of tubes on the three painting tables. The main painting table, the active one, held an accumulation of rags, bowls, turpentine, brushes, and a similar group of tubes to those in the photograph.

Paint clusters on all three painting tables revealed Heron mixing brands and qualities. The smallest table bore two shades of French Ultramarine blue: – artist's quality by Roberson and student-grade Winton Oil Colour by Winsor & Newton; three variations of violet from Art Spectrum (Permanent Mauve), Lefranc (Cobalt Violet Deep), Lefranc & Bourgeois (Cobalt Violet Light); one transparent crimson, Rowney Rose, two vermilions: Vermilion hue by Lefranc, Orange Vermilion by Rowney, and the remainder by Roberson: two opaque reds (Indian Red and Venetian Red), three yellows (Cadmium Deep and Pale, with Yellow Ochre) and one green (Cobalt Green). Their dates of manufacture ranged over a 30 year period from the mid-1960s (Roberson) to the 1980s (Art Spectrum, Lefranc & Bourgeois, Rowney) and 1990s (Winsor & Newton).

Heron's late garden series, and his portraits of A.S. Byatt painted between 1996 and 1997, might be expected therefore to follow a similar pattern and be painted in oil paint made by several different manufacturers including the possibility that an ultramarine blue, for example, may be two different products and may therefore have different physical properties.

Each group of tubes therefore probably represents an idea for a painting but not the full palette. Especially as the artist's practice was to take up a bowl of oil colour cut with turpentine to brush-able state, and go around the room working on different paintings at the same time, including any earlier paintings left unfinished from previous projects that happened to be in the studio.[24] It is possible therefore that early canvases were completed with paint purchased much later. Art Spectrum paint purchased during his Australian residency (1989) could, potentially, be present on pre-1989 canvases.

Because he could move fluidly from one painting to another with paint from any of the tables the significance of the groups lies in demonstrating Heron's brand preferences for opaque or transparent colours in the mid 1990s. They confirm that he combined old and new paint including old paint newly purchased when very old stock was released onto the market by shops or businesses closing down.

Conclusion

The repetition of colours and brands and bulk-purchases reveal Heron's preferred materials in the 1990s. Anomalous tubes are readily identifiable by their sparse quantity and uncommon manufacturer. An archive made of paint collected from the

[24]Personal communication Susanna Heron 30 September 2013.

artist's studio is particularly relevant to the study of his paintings when contextual data is also captured. It is in the quantity, location and groupings of materials that a deeper level of understanding of their use by the artist in his last works may be gained. It is logical to select a palette on the basis of colour not by brand or by age of the paint; the studio contents confirm this notion and reveal a less-obvious reality that has other implications for the interpretation of analysis of paintings. For if ultramarine blue, for example, on a single painting appears to be slightly different in character from one place to another, it may be not one but two different paints made by different manufacturers from raw materials of different origins purchased many years apart, and separated by technological innovation as much as recipe.

Acknowledgments With many thanks to the artist, Patrick Heron, for his warm kindliness and generosity in discussions about his working practices, Susanna Heron and the Estate of Patrick Heron for inspiration and permission to reproduce images. Jo Crook for her expertise in documenting artists' archives, Robert Holyhead, and Tate staff past and present: Maureen Cross, Patti Favero, Rica Jones, Julia Jonsson, Susan Lamb, Adrian Moore, Sarah Morgan, Carole Towers, Joyce Townsend and Natasha Walker. I am grateful to the Clothworkers Foundation for their bursary to attend ICOP.

References

Borlase Smart John Wells Trust (2013) www.bsjwtrust.co.uk/porthmeorrenvaa.html. Accessed 3 Oct 2013

British Artist Suppliers 1650–1950 – R, National Portrait Gallery. ww.npg.org.uk/research/programmes/directory-of-suppliers/r.php. Accessed 30 Sept 2013

Cooper A et al (2014) Water sensitive oil paints in the twentieth century: a study of the distribution of water-soluble degradation products in modern oil plant films. These proceedings, Chapter 20

Dawson B, Cappock M (2001) Francis Bacon's Studio at the Hugh Lane Municipal Gallery of Modern Art, Dublin

Gooding M (1994) Patrick Heron. Phaidon, London

Heron P (1953) Space in Colour. Hanover Galleries, London

Heron P (1957) Arts (New York), Sept 12

Heron P (1969) Colour in my paintings. Studio International, London, December p 205

Heron P (1970) Two Cultures. Studio International, London, December, p 247

Heron P (1973) Notes on my painting 1953–1972. In: Heron P (ed) Perception: the shape of colour II. Midland Arts Centre, Birmingham

Mancusi-Ungaro C (2004) The paintings of Barnett Newman involved intuition on the highest level in Barnett Newman: a catalogue Raisonée, Yale, p 130

Mulholland (2009) Technique and process in the papers of David Smith. In: Hermans E, Townsend JH (eds) Sources and serendipity: testimonies of artists' practice 3rd symposium of ATSR Working Groups, London, p 133

Russell J, Singer B, Perry J, Bacon A (2012) Investigation of the materials found in the studio of Francis Bacon (1909–1992). Stud Conserv 57(4)

Sylvester D (ed) (1998) Patrick Heron. Tate Gallery Publishing, London

Tempest H (2009) Water-sensitive oil paint, Unpublished student project, Courtauld Institute of Art

Tufnell B (2006) On the very edge of the ocean. Tate Publishing, London/St Ives, p 63

Woodcock S (1995) The Roberson archive, content and significance. In: Wallert A et al (eds) Historical painting techniques, materials, and studio practice. IIC, Leiden, pp 30–37

The Delight of Modern Organic Pigment Creations

Matthijs de Keijzer

Abstract This article presents a survey of the history and development of synthetic organic artists' pigments discovered, produced and used in the twentieth century. The following organic groups and pigments will be described: the β-naphthol pigment lakes, BON pigment lakes, β-naphthols, Hansa Yellows, pyrazolones, diarylides, Naphthol Reds, Pigment Green B, phthalocyanine blues and greens, Nickel Azo Yellow, Dioxazine Purple, perinones, perylenes, quinacridones, isoindolinones, benzimidazolones and diketopyrrolo-pyrroles. Their discovery, industrial production and introduction as artists' pigments were compiled from the patent literature and from pigment lists of paint manufacturers. Chronologies of the relevant pigments are tabulated, grouped by colour and by chemical composition. Notable occurrences of pigments are presented both from the literature and from identification of works of art.

Keywords Modern organic artists' pigments • Twentieth century • History • Patents • Chronology

Introduction

Modern pigments have been studied in a framework of several projects at the Cultural Heritage Agency of the Netherlands (RCE). The published outcomes provide an overview of pigments used by well-known European artists' paint manufacturers, such as Royal Talens B.V., Winsor & Newton Ltd., Daler-Rowney Ltd., H. Schmincke & Co., Lukas Künstlerfarben- und Maltuchfabrik Dr. Fr. Schoenfeld GmbH & Co. and Lefranc & Bourgeois. At present, half of the artists' paints of A quality contain organic and inorganic pigments discovered in the last century.

M. de Keijzer (✉)
Cultural Heritage Agency of the Netherlands, Amsterdam, The Netherlands

Department Archaeometry, University of Applied Arts Vienna, Salzgries 14, A-1013 Vienna, Austria
e-mail: m.de.keijzer@cultureelerfgoed.nl

K.J. van den Berg et al. (eds.), *Issues in Contemporary Oil Paint*,
DOI 10.1007/978-3-319-10100-2_4

For the quality B this proportion may be up to 70 %. The history and development of twentieth century pigments have been studied from literature (Gaertner 1963; Lewis 1988; Herbst et al. 2004), especially the international patent literature, because the invention and start of the industrial production can be ascertained from patents, and from artists' pigment lists. The modern synthetic inorganic artists' pigments have been published recently (de Keijzer 2011a, b). This article presents a survey of developments which have led to the creation of high-grade organic artists' pigments. The data are represented in Table 1. They were provided by the manufacturers and suppliers of artists' materials concerning the type of paints in which a certain pigment was first used and the date it was introduced. The notable occurrences were found in the literature (Berrie and Lomax 1996/1997; Lake and Lomax 2007; Lutzenberger 2009) or established in the RCE and are to be found in the institute's documentation files. This has been a great help in establishing the pigment chronology.

Modern Synthetic Organic Pigments Before 1950

In a short period of time, at the beginning of the twentieth century, the following organic groups: the β-naphthol pigment lakes, BON pigment lakes, β-naphthols, Hansa Yellows, pyrazolones, diarylides and Naphthol Reds were synthesized based on the diazo-reaction invented by the German chemist Johann Peter Griess (1829–1888) in 1858.

β-Naphthol Pigment Lakes

The development of the azo pigment lakes started with the invention of the β-naphthol pigment lake Lithol Red (PR 49) discovered by Paul Julius of the Badische Anilin- & Soda-Fabrik (BASF) in 1899. Lithol Red, which is synthesized from β-naphthol with 2-naphthylamine-1-sulfonic acid as a diazonium compound, was used in the form of sodium, barium, calcium and strontium salts, which were precipitated onto inorganic carrier materials. The pigment was used in its pure form after it became apparent that the carriers contribute little to the application properties of the product. The breakthrough of Lithol Red led to the preparation of a whole family of hard-to-dissolve red and orange azo compounds of which Red Lake C (PR 53), diazotized 2-amino-5-chloro-4-methyl-benzenesulfonic acid coupled with alkaline β-naphthol, became an important member. This one was, discovered in 1902 by Karl Schirmacher at Farbwerke vorm. Meister Lucius & Brüning. The ethyl homologue of Red Lake C is called Clarion Red (PO 46) and was invented by Frederic Henry Adams of the American Cyanamid Company in 1949. Other red and orange pigments of this class are PR 50, PR 51, PR 68 and PO 17.

BON Pigment Lakes

This group of pigments derives its name from 2-hydroxy-3-naphthioc acid (BON, BONA, BONS, β-oxynaphthoic acid) used as the coupling component. Lithol Rubine (PR 57) was produced by coupling 2-amino-5-methylbenzenesulfonic acid with BON, synthesized by the chemists Richard Gley and Otto Siebert at Aktien-Gesellschaft für Anilin-Fabrikation (AGFA) in 1903. Another important pigment lake is Permanent Red 2B (PR 48), prepared by 1-amino-3-chloro-4-methylbenzenesulfonic acid (2B acid) with BON, and was first synthesized in the late 1920s. Alfred Siegel at E.I. du Pont de Nemours & Company (DuPont) solved the bleed problem by precipitating the dye with calcium, barium or manganese in the late 1920s/beginning of the 1930s. Other red members are Red 2G (PR 52), also known as BON Lake Red C, Rubine G Pigment (PR 58), Pigment Scarlet (PR 60), BON Maroon (PR 63) and PR 64.

β-Naphthols

The group of the β-naphthols possesses several famous red pigments synthesized and produced by different German factories. The product β-naphthol was prepared by Schaeffer in 1869. Para Red (PR 1) was obtained by Moritz von Gallois and Eduard Ullrich in 1885 at Farbwerke, vormals Meister, Lucius & Brüning by diazotisation of 4-nitroaniline coupled with β-naphthol and this red pigment was manufactured in 1889. Fifteen years later, in 1904, Toluidine Red (PR 3), produced by diazotized 2-nitro-4-toluidine coupled with alkaline β-naphthol, was invented by the same firm. In 1906 Wilhelm Herzberg and Oscar Spengler discovered Chlorinated Para Red (PR 4) and AGFA patented the manufacture by diazotizing 2-chloro-4-nitroaniline combined with β-naphthol. In the same year the chemists Conrad Schraube and Erhart Schleicher at BASF revealed that the isomer compound of 4-chloro-2-nitroaniline gave a brilliant orange-red pigment with good properties; it was called Parachloro Nitroaniline Red (PR 6). This class also obtained one well-known orange pigment, named Dinitroaniline Orange (PO 5), synthesized by coupling diazotized 2,4-dinitroaniline with β-naphthol, prepared in 1907 by Richard Lauch and the preparation as Permaton Orange was disclosed by AGFA in 1909 (Fig. 1).

Pyrazolones

In 1883 the German chemist Ludwig Knorr (1859–1921) prepared methylphenyl-pyrazolone and 1 year later the first yellow pigment of this group Tartrazine Yellow (PY 100) was produced. In 1909 other yellow pyrazolones like Hansa

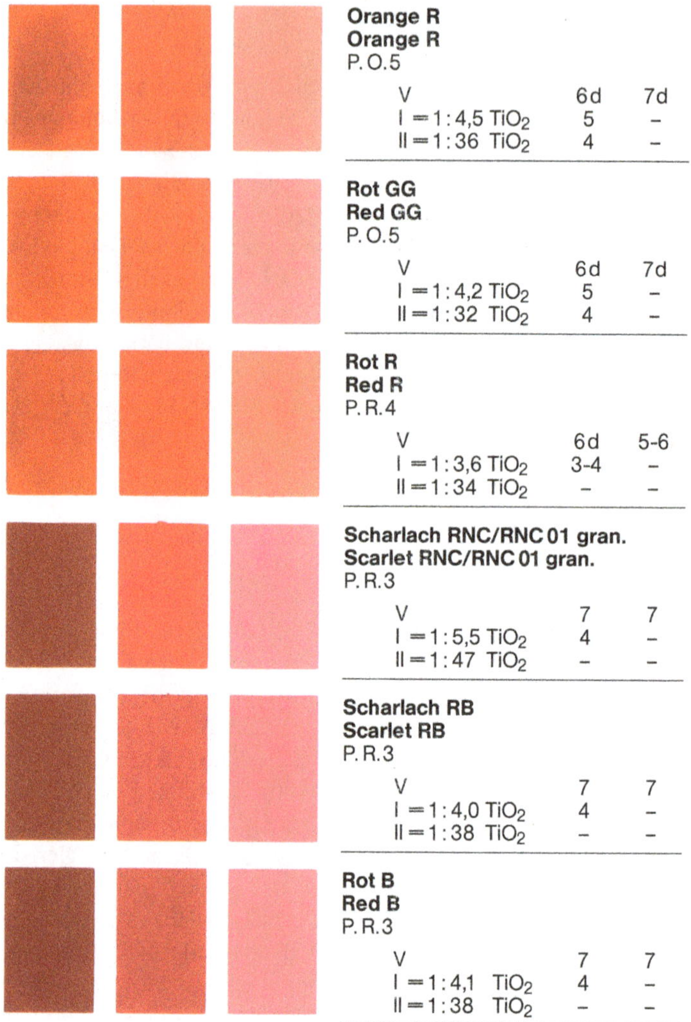

Fig. 1 Light-fastness of the β-naphthols. Leaflet: Organic pigments for the paint industry (DP 7017 DE; Hoechst Aktiengesellschaft)

Yellow R (PY 10) was prepared by 2,5-dichloroaniline and 3-methyl-1-phenyl-5-pyrazolone. Hansa Yellow 4R (PY 60), followed by Benzidine Orange (PO 13), also known as Pyrazolone Orange was obtained from 3,3′-dichlorobenzidine and 3-methyl-1-phenyl-5-pyrazolone. These were developed in 1910 by the Chemische Fabrik Griesheim-Elektron, but the commercial application starts 20 years later in the 1930s. In 1934 Pyrazolone Red B (PR 38), prepared by 3,3′-dichlorobenzidine and 3-carbethoxy-1-phenyl-5-pyrazolone was patented. Dianisidine Red (PR 41) produced by o-dianisidine and 3-methyl-1-phenyl-5-

pyrazolone, was patented in the same year, followed by Tolyl Orange (PO 34), synthesized by 3,3′-dichlorobenzidine and 3-methyl-1-p-tolyl-5-pyrazolone, 1 year later (in 1935).

Hansa Yellows

With β-naphthol, BON and the arylide of BON as coupling components the range of colours was restricted for red and orange pigments, but it was not possible to manufacture yellow shades. The breakthrough came with the use of acetoacetarylide reacting with diazo components as coupling compounds. This led eventually to the fundamental German patent 257.488 that deals with the production of azo pigments, made by coupling diazotized nitro- or chloro-aromatic amines with acetoacetarylides. The first pigment, Hansa Yellow G (PY 1), was prepared by diazotized 2-nitro-p-toluidine coupled with acetoacetarylide, prepared by Hermann Wagner at Farbwerke vorm. Meister Lucius & Brüning in 1909. This pigment group led to a search for other possible combinations using different derivatives. The second pigment in this range, Hansa Yellow 10G (PY 3), synthesized by diazotized 4-chloro-2-nitroaniline coupled with o-chloroacetoacetarylide, was also prepared by Farbwerke vorm. Meister Lucius & Brüning in 1910 and was produced 2 years later, in 1912. By 1928, more Hansa Yellows were marketed by this German factory, such as Hansa Yellow GR (PY 2), Hansa Yellow 13G (PY 4), Hansa Yellow 5G (PY 5) and Hansa Yellow 3G (PY 6). Hansa Yellow RN (PY 65) was patented by the chemists A. Funke, A. Sieglitz and R. Huβ of IG. Farbenindustrie AG. (IG. Farben) in 1936. One orange pigment, named Hansa Yellow 3R (PO 1), was patented by the same factory in 1926.

In 1957 improved Hansa Yellow pigments were synthesized and entered the pigment market, such as Hansa Brilliant Yellow 4GX (PY 73), diazotized 4-chloro-2-nitroaniline with o-acetoacetanisidide, patented by Hercules Inc., Hansa Brilliant Yellow 5GX (PY 74), prepared by diazotized 4-nitro-o-anisidine with o-acetoacetanisidide, was protected by DuPont in 1958. In 1960 Farbwerke Hoechst AG. introduced Hansa Yellow FGL (PY 97), produced by diazotized 4-amino-2,5-dimethoxybenzene-sulfonanilide coupled with 4′-chloro-2′,5′-dimethoxyacetoacetarylide, and Hansa Brilliant Yellow 10GX (PY 98), synthesized by diazotized 4-chloro-2-nitroaniline coupled with 4-chloro-o-acetoacetotolide (Fig. 2).

Diarylides

In the same period that the Hansa Yellows were invented the diarylides became known. In 1911 the Chemische Fabrik Griesheim-Elektron patented (fundamental German patent 251.479) this class of pigments, but this invention was not utilized,

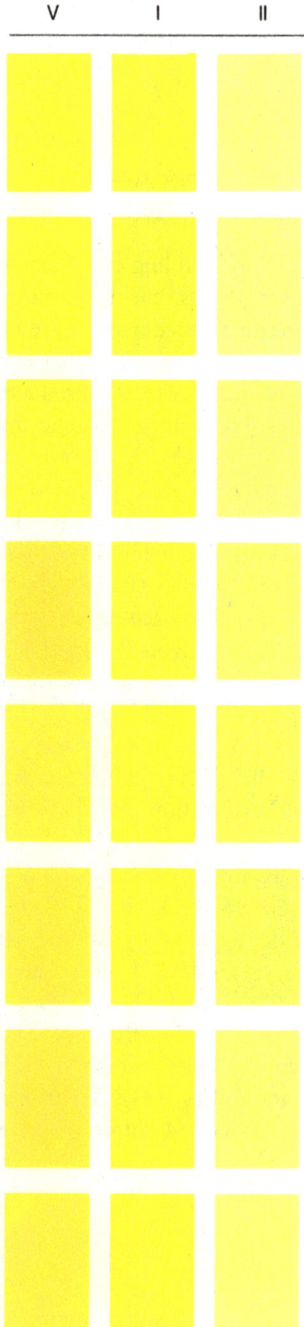

Hansa-	V / I / II	LL	EL
Gelb 10G 41 gran. **Yellow 10G 41 gran.** P.Y.3	V	7-8	7-8
	I = 1 : 3,3 TiO_2	7	–
	II = 1 : 42 TiO_2	6	–
Gelb 10G **Yellow 10G** P.Y.3	V	7-8	8
	I = 1 : 2,6 TiO_2	7-8	–
	II = 1 : 31 TiO_2	6-7	–
Brillantgelb 10GX 01 **Brilliant Yellow 10GX 01** P.Y.98	V	7-8	8
	I = 1 : 4,5 TiO_2	7-8	7-8
	II = 1 : 52 TiO_2	7	–
Brillantgelb 5GX 02 **Brilliant Yellow 5GX 02** P.Y.74	V	7-8	7-8
	I = 1 : 14,0 TiO_2	5-6	–
	II = 1 : 132 TiO_2	4-5	–
Brillantgelb 5GX **Brilliant Yellow 5GX** P.Y.74	V	7-8	8
	I = 1 : 12,0 TiO_2	6	–
	II = 1 : 118 TiO_2	5	–
Brillantgelb 4GX **Brilliant Yellow 4GX** P.Y.73	V	7-8	7-8
	I = 1 : 6,9 TiO_2	6	–
	II = 1 : 74 TiO_2	5	–
Gelb G 41 gran. **Yellow G 41 gran.** P.Y.1	V	7-8	7-8
	I = 1 : 7,5 TiO_2	5	–
	II = 1 : 78 TiO_2	4	–
Gelb G 02 **Yellow G 02** P.Y.1	V	7-8	7-8
	I = 1 : 7,4 TiO_2	5	–
	II = 1 : 76 TiO_2	4	–

Fig. 2 Light-fastness of the Hansa Yellows. Leaflet: Organic pigments for the paint industry (DP 7017 DE; Hoechst Aktiengesellschaft)

because the Hansa Yellows had entered the pigment market and were more lightfast than the diarylides. Three diarylide yellows were known: Benzidine Yellow AAA (PY 12), Permanent Yellow GR (PY 13) and Benzidine Yellow AAOT (PY 14). In 1935 these pigments were sold for the use in rubber by IG. Farben as Vulcan Fast Yellows and Oranges. The diarylides are based on 3,3′-dichlorobenzidine. They are prepared by coupling of tetraazotized benzidines with acetoacetarylides. The general weak point is their bad light-fastness. After 1945 better diarylides were introduced, like Benzidine Yellow AAOA (PY 17) in 1951, Permanent Yellow HR (PY 83) in 1953 and Permanent Yellow H10G (PY 81) in 1961.

Naphthol Reds

The last group based on the diazo-reaction are the Naphthol Reds and derive from the coupling component arylide of BON (Naphtol AS). The invention by the chemists Adolf Winther, August Leopold Laska and Arthur Zitscher at Teerfarbenwerk Oehler, part of the Chemische Fabrik Griesheim-Elektron, led to the fundamental German patent 256.999 in 1911 (Fig. 3).

Between 1920 and 1930 the development of the Naphthol AS technology in Germany was initiated by IG. Farben and in the 1940s this pigment group began to be developed in the USA. Today there are nearly 80 known Naphthol AS pigments. Eight of them are of importance for the use as artists' pigments: Permanent Red FRR (PR 2), Permanent Red F4RH (PR 7), Permanent Red F4R (PR 8), Permanent Red FRLL (PR 9), Permanent Bordeaux FRR (PR 12) and Permanent Red FGR (PR 112) (Fig. 4).

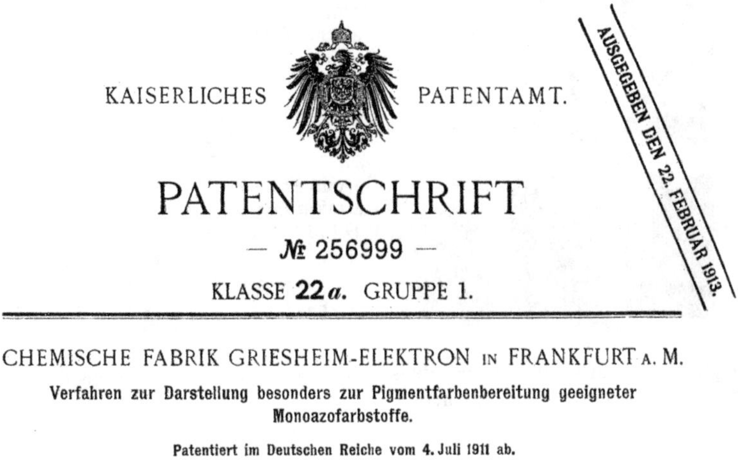

KAISERLICHES PATENTAMT.

AUSGEGEBEN DEN 22. FEBRUAR 1913.

PATENTSCHRIFT

— № 256999 —

KLASSE **22*a*.** GRUPPE 1.

CHEMISCHE FABRIK GRIESHEIM-ELEKTRON IN FRANKFURT A. M.

Verfahren zur Darstellung besonders zur Pigmentfarbenbereitung geeigneter Monoazofarbstoffe.

Patentiert im Deutschen Reiche vom 4. Juli 1911 ab.

Fig. 3 Fundamental German patent for the Naphthol Reds

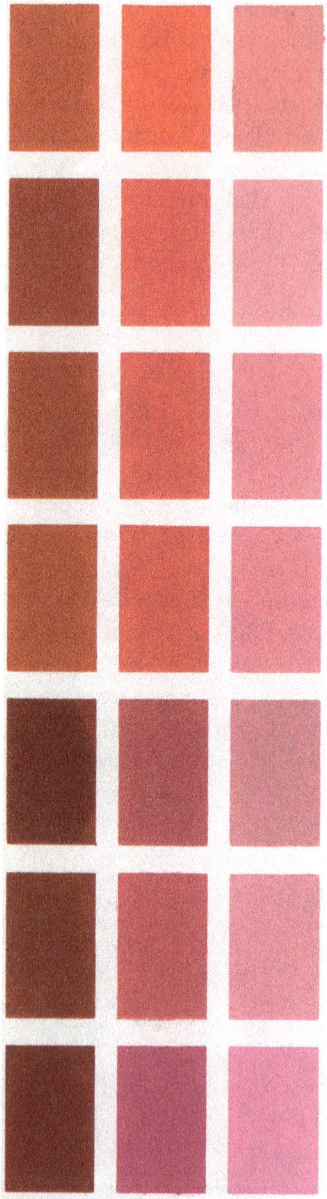

Pigment			
Rot FRLL **Red FRLL** P.R.9	V	7	-
	I = 1 : 4,2 TiO_2	6-7	-
	II = 1 : 45 TiO_2	5-6	-
***Rot FRR** ***Red FRR** P.R.2	V	6	-
	I = 1 : 4,1 TiO_2	3	-
	II = 1 : 66 TiO_2	1-2	-
Rot FGR **Red FGR** P.R.112	V	7-8	8
	I = 1 : 5,5 TiO_2	6-7	7-8
	II = 1 : 70 TiO_2	5-6	6
Rot FGR 70 **Red FGR 70** P.R.112	V	7-8	8
	I = 1 : 4,7 TiO_2	6-7	7-8
	II = 1 : 59 TiO_2	5	6
Rot F4RH **Red F4RH** P.R.7	V	7-8	7-8
	I = 1 : 8,4 TiO_2	6-7	6-7
	II = 1 : 84 TiO_2	5-6	-
Carmin FB 01 **Carmine FB 01** P.R.5	V	7d	7
	I = 1 : 6,6 TiO_2	6	6-7
	II = 1 : 73 TiO_2	4-5	5
Carmin FBB 02 **Carmine FBB 02** P.R.146	V	5-6d	6d
	I = 1 : 7,3 TiO_2	4-5	5
	II = 1 : 102 TiO_2	3-4	3-4

Fig. 4 Light-fastness of some Naphthol Reds. Leaflet: Organic pigments for the paint industry (DP 7017 DE; Hoechst Aktiengesellschaft)

Better pigments were invented using new substituents like the sulfoamide and carbonamide groups, such as Permanent Carmine FBB (PR 146) and Permanent Red F5RK (PR 170).

Pigment Green B

Pigment Green B (PG 8), the iron complex of 1-nitroso-2-naphthol, was first reported in an article by Otto Hoffmann in 1885. Since he foresaw no possibility for its technical utilization he did not apply for a patent to cover its manufacture. Much later, in 1921, BASF developed a procedure which yielded the compound in a form suitable for pigment use and exploited it under their trade name Pigmentgrün B. The first large-scale application of Pigment Green B was for wallpaper. It also found early use in rubber and research in the 1930s was directed toward texture improvement, particularly for rubber colouration. Significant quantities of Pigment Green B began to be employed in the textile printing industry in the 1940s. It reached a peak in the early 1950s before copper phthalocyanine green (PG 7) began to make inroads into its markets.

The pigments of these groups have their weak points as well as their very good ones, from inferior via medium to good light and weather fastness and from poor to good solvent resistance and migration fastness. For instance: the ß-naphthol pigment lakes and the BON pigment lakes are generally resistant to solvents, but show a poor light-fastness and are sensitive to alkalis and acids. The neutral metal free compounds like the Hansa Yellows, pyrazolones, diarylides and Naphthol Reds generally exhibit good resistance to alkalis and acids, but are often too easily soluble in solvents and plasticisers. These pigments cannot be considered as 'all-round pigments', employable for any kind of application, but rather as products well suited to specific fields of use such as inks, coatings and plastics. Nowadays the pigments of these classes have mainly been replaced by later discovered organics with better fastness properties.

Phthalocyanines

The discovery of copper phthalocyanine blue (PB 15), in 1927, was a turning point in the pigment chemistry and a start for the 'all-round' and 'high-grade' pigments. The shade was brilliant, its chemical resistance outstanding and its application possibilities all-embracing that its appearance on the market in 1935 set up entirely new standards for high-grade pigments.

In 1907, A. V. Braun and J. Tcherniac at the South Metropolitan Gas Company in London, found a trace of a blue metal free substance. They noted that when o-cyanobenzamide was heated above its melting point for a protracted period a blue compound was formed in yields of less than 1 %. They did not characterize the compound, nor investigated it further at the time. In 1927 Henri de Diesbach and Emmanuel von der Weid in Fribourg, Switzerland, heated o-dibromobenzene and cuprous cyanide in pyridine at 200 °C in a sealed tube, attempting to obtain phthalonitrile, but instead they got a deep blue product containing copper, their product was undoubtedly a copper phthalocyanine. They observed the remarkable stability to alkalis, concentrated sulphuric acid and heat. A year later, in

1928, independently, the chemists Arthur Gilbert Dandridge, Hugh Albert Edvard Drescher and John Thomas at the Grangemouth Works of Scottish Dyes Ltd. observed the formation of a blue-green, highly coloured impurity, when ammonia was passed through molten phthalic anhydride in an iron vessel. The compound was iron phthalocyanine. In 1929 the first patents were issued by these chemists (British patent 322.169 and German patent 586.906). In 1932 Imperial Chemical Industries Ltd. (ICI. Ltd.) patented the manufacture of phthalocyanines (U.S. patents 2.000.051 and 2.000.052, followed by the British patents 410.814, 464.126 and 476.243). Reginald Patrick Linstead and his students at the Imperial Institute of Science and Technology of London, supported by grants from ICI. Ltd. during 1932/34, elucidated the structure and the synthetic pathway of the phthalocyanines. The structure of phthalocyanine was published in 1934.

For producing phthalocyanine blues many industrial processes were developed, but from technical point of view three methods are important. The first process uses phthalonitrile with finely divided copper. The second method is based on the reaction of phthalonitrile with cuprous chloride. The last process uses phthalic anhydride, urea and cuprous chloride. This process is known as the urea-process. It was discovered by Max Wyler at ICI. Ltd. and improved by Albert Riley (ICI. Ltd., British patent 464.126 and U.S. patent 2.214.477). The urea-process has the advantage over the phthalonitrile method that it involves the use of the readily accessible phthalic anhydride. The yields are much improved by the addition of boric acid and ammonium molybdate (British patents 476.243, 633.713 and U.S. patent 2.469.663).

The first pigments appeared in 1935 by ICI. Ltd. under the trade name Monastral Fast Blue BS. As an industrial pigment copper phthalocyanine blue was first showed on an exhibition of ICI. Ltd. in London in 1935. In 1934 research at IG. Farben in Ludwigshafen and Leverkusen, led to patents (German patents 658.019 and 663.552). At the end of 1935 this firm produced the pigment under their trade name Heliogenblau B Pulver. In 1936 DuPont started to produce this blue pigment in the USA (U.S. patents 2.173.699 and 2.247.752).

Since the development of the first phthalocyanine blue, there have been many improvements. One high point is the β-modification of phthalocyanine blue (U.S. patents 2.556.728 and 2.556.730). DuPont introduced the first stable crystalline β-type of copper phthalocyanine (Peacock shade) in 1949, followed by the BASF AG. in 1953 as Heliogenblau BG. Another one is the ε-crystal form of phthalocyanine blue patented by the American Cyanamid Company in 1962 (U.S. patent 3.051.721).

Chemists tried to produce pigments of equal fastness, but in varying shades, by subjecting this molecule to every substitution and variation. This led to products exhibiting lower fastness and/or less attractive shades, with one exception: copper phthalocyanine green (PG 7). In August 1935, soon after the determination of the structure of phthalocyanine, Reginald Patrick Linstead and Charles Enrique Dent discovered copper phthalocyanine green, while the chemists Georg Niemann, Willi Schmidt, Fritz Mühlbauer and Georg Wiest at IG. Farben did the same in December 1935. Phthalocyanine green is a polychloro copper phthalocyanine obtained by direct chlorination of copper phthalocyanine suspended in molten phthalic anhydride or sodium aluminium chloride. The commercial product contains 14–15 atoms of chlorine. ICI. Ltd. started in 1936 with the production under the trade name Monastral Fast Green GS, followed 2 years later by IG. Farben as Heliogengrün B and in 1940 by DuPont in the USA. In 1957 the development of phthalocyanine green received a boost when Harmon Colors discovered a copper phthalocyanine green which contains both chlorine and bromine and the shade becomes yellower by increasing the bromine content. These yellow-green phthalocyanines (PG 36) came into use as Viridine green in the USA in 1959. Unfortunately, the phthalocyanine shades remained limited to blues and greens (Fig. 5).

Metal free phthalocyanine blue (PB 16) was already discovered by Jocelyn Field Thorpe, Reginald Patrick Linstead and John Thomas in 1931 (Scottish Dyes Ltd., British patent 389.842), but is of less importance because its rather complex preparative techniques led to dull shades and low strengths.

Nickel Azo Yellow

The chemist Ernst Fussenegger at BASF, discovered it as an unmetalised dye and was patented in 1904 and 1905. In the beginning of the 1940s Nickel Azo Yellow (PG 10) was developed by Donovan E. Kvalnes and Harold E. Woodward at DuPont. It is a metallised monoazo pigment synthesized by 4-chloroaniline and 2,4-dihydroxyquinoline. In 1946 it was covered by U.S. patents and introduced to the trade by DuPont under the name Green-Gold YT-562-D in 1947.

Modern Synthetic Organic Pigments After 1950

There was a growing demand for high-grade pigments in all shades primarily in the fields of automobile finishes, organic coatings, plastics, printing inks and man-made fibres. The chemists investigated other chemical classes for pigments attaining the standard of the phthalocyanines. The first thing they did was to turn back to the fastest of the textile dyes, the vats, to find out whether any of them could be used as high-grade pigments.

Although this appeared to be a simple proposition, it presented a large problem. The usual production methods for vats yield a form which is suitable for the dyeing of textiles, but in no way takes into account the special requirements of pigment chemistry in respect of chemical purity, particle size and crystal modification. For a long time the vats in general had the reputation of being deficient in both brilliance and tinctorial strength when used as pigments.

In the beginning of the 1950s chemists have been successful in modifying the manufacturing processes for vats in such a manner that they have become acceptable for the manufacture of pigments. Practically all textile vats placed on

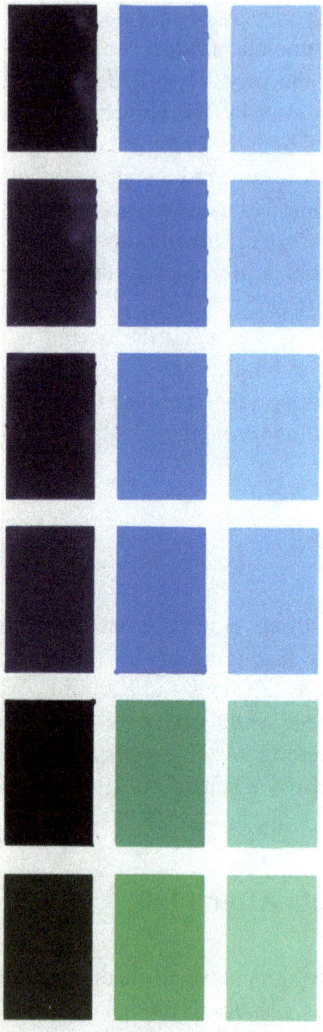

Blau A2R
Blue A2R
P.B.15:1

V	8	8
I = 1:12 TiO_2	8	8
II = 1:164 TiO_2	8	8

Blau AFL
Blue AFL
P.B.15:2

V	8	8
I = 1:11,5 TiO_2	8	8
II = 1:153 TiO_2	8	8

Blau B2G
Blue B2G
P.B.15:3

V	8	8
I = 1:9,7 TiO_2	8	8
II = 1:145 TiO_2	8	8

Blau BFL
Blue BFL
P.B.15:4

V	8	8
I = 1:11,2 TiO_2	8	8
II = 1:164 TiO_2	8	8

Grün GG 01
Green GG 01
P.G.7

V	8	8
I = 1:5,4 TiO_2	8	8
II = 1:88 TiO_2	8	8

Grün 8G
Green 8G
P.G.36

V	8	8
I = 1:3,8 TiO_2	8	8
II = 1:56 TiO_2	8	8

Fig. 5 Light-fastness of the phthalocyanine blues and greens. Leaflet: Organic pigments for the paint industry (DP 7017 DE; Hoechst Aktiengesellschaft)

the market have been examined for their use as pigments. However, the yield was not overwhelming, only about a dozen made their mark as pigments, and only a fraction of these approached the fastness standard of the phthalocyanines.

Therefore, the pigment chemists began to devise new compounds, but in other shades, which led to new pigment groups specially tailored to pigment use which could carry the names high-grade and all-round.

These classes are: the dioxazines (1953), perinones (1953), condensation pigments (1954), perylenes (1955), quinacridones (1955), benzimidazolones (1960) and diketopyrrolo-pyrroles (DPPs) (1986). By virtue of continued research the pigment industry expanded the spectrum of the quinacridones, benzimidazolones and DPPs.

Perinones

The perinone group has one orange (PO 43) and one red pigment (PR 194), found as a mixture of isomers by Wilhelm Eckert and Heinrich Greune at Farbwerke vorm. Meister Lucius & Brüning in 1924. The trans-form of Perinone Orange is by far the more important of the two. In the beginning of the 1950s one was able to separate the isomers and the use of Perinone Orange as a pigment started. The pigment was put by Farbwerke Hoechst AG. as Hostapermorange GR on the pigment market in 1953.

Perylenes

They are diimides of perylene-3,4,9,10-tetracarboxylic acid. In 1913 up to the middle of the 1950s the perylenetetracarboxylic acid dyes were used as vats. The first introduction of these compounds as pigments came in 1950 when Harmon Colors did research on the vats to bring them to a suitable pigment form and further research of the perylenes was continued by Farbwerke Hoechst AG. Perylene Vermilion (PR 123) was prepared by Harmon Colors in 1952 and Perylene Scarlet (PR 149) was patented by Farbwerke Hoechst AG. in 1956. Harmon Colors developed Perylene Red (PR 190) and BASF AG. protected Perylene Maroon (PR 179) in 1959 and followed with Pigment Red 178 in 1966.

Dioxazine Purple

The most important dioxazine pigment is Dioxazine Purple (PV 23). In 1928 this group was discovered by the chemists Georg Kränzlein, Heinrich Greune and Max Thiele at Farbwerke vorm. Meister Lucius & Brüning, when they were exploring the chemistry of oxazine dyes. In 1952 Farbwerke Hoechst AG. found that by special conditioning processes a good dioxazine pigment was obtained and 1 year later, in 1953, it was introduced as Permanentviolett RL. In 1958 DuPont started to produce dioxazine pigments.

Quinacridones

The name quinacridone was coined by Niementowski in 1896. Between 1915 and 1932 chemists synthesized different quinacridone systems. Unsubstituted linear-trans-quinacridone, a red-violet powder, was first synthesized by H. Liebermann in 1935. The quinacridones attracted no particular attention until 1955, when chemists of DuPont discovered its excellent lightfastness and methods for the preparation of its different crystal forms. The differences in the X-ray patterns are mentioned in the U.S. patent 2.844.485 by William S. Struve in 1958. The first quinacridones were introduced by DuPont in 1958 under the trade name Monastral. Three colours were available: Monastral Red B, Monastral Red Y and Monastral Violet R (PV 19). William S. Struve described in the fundamental U.S. patent 2.821.529 and complementary patents a method to produce linear-trans-quinacridone. Farbwerke Hoechst AG. was the first company in Europe which produced linear-trans-quinacridone soon after DuPont did this in the USA. The pigment industry expanded the spectrum of the quinacridones. These have been marketed since the late 1960s in colours ranging from orange, gold, red and maroon to magenta and violet (Fig. 6).

Examples are Sandorin Brilliant Red (PR 192), Monastral Magenta (PR 202), Monastral Maroon (PR 206), Monastral Scarlet (PR 207), Hostaperm Red E (PR 209), Monastral Gold (PO 48) and Monastral Deep Gold (PO 49).

Isoindolinones

At the end of the 1950s the isoindolinone pigments, derived from 4,5,6,7,-tetrachloroisoindolinone, were discovered. In 1946 William Owen Jones at ICI. Ltd. described in two patents dyes and pigments wherein the benzene rings of the isoindolinone radical contain a maximum of two substituents. In 1952 J. A. Elvidge and R. P. Linstead and 1 year later P. F. Clarck, J. A. Elvidge and R. P. Linstead published the synthesis of condensed m-phenylenediamine, 2,6- and 3,5-diaminopyridine, 2,7-diaminonaphthalene, and 2,4-diaminopyrimidine with 3-aminoisoindolinin-1-one. The final step for the development of the isoindolinones was made by the firm J. R. Geigy AG. and led to Swiss patents in 1956. The isoindolinones are registered under the trade names of Irgazin and Cromophtal. The most important member is Irgazin Yellow 3RLT (PY 110). Other pigments of this class are Irgazin Yellow 2GLT (PY 109), PY 139, PY 173, PY 185, PO 61, PO 66 and PBr 38.

Benzimidazolones

Systematic research of diazonium compounds in the second half of the twentieth century opened the door for other azo-groups, like the benzimidazolones, which possesses benzimidazolone as the coupling component. New methods of syntheses have made it possible in build up high molecular azo pigments resulted in high molecular azo pigments reaching a higher standard of fastness than the earlier ones.

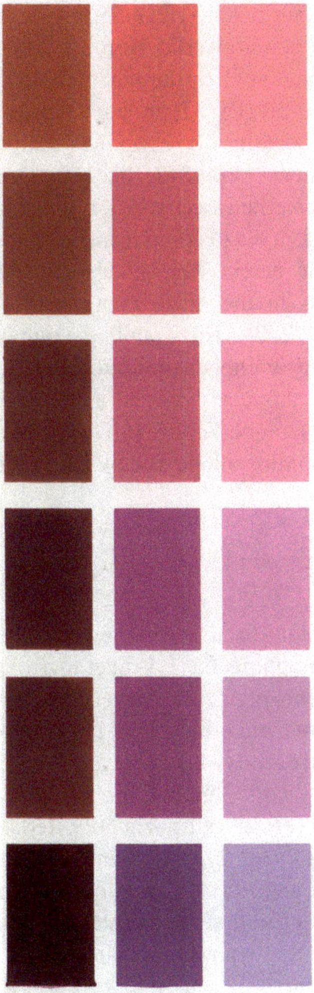

Rot EG transp.
Red EG transp.
P.R.209

V	7d	7-8d
I = 1 : 1,9 TiO_2	7-8d	8
II = 1 : 32 TiO_2	7-8	7-8

Rot E3B
Red E3B
P.V.19

V	7d	7d
I = 1 : 3,4 TiO_2	7-8d	7-8d
II = 1 : 42 TiO_2	7-8	8

Rot E5B 02
Red E5B 02
P.V.19

V	7d	7d
I = 1 : 4,2 TiO_2	7-8d	7-8d
II = 1 : 61 TiO_2	7-8	8

Rosa E
Pink E
P.R.122

V	7-8d	7-8d
I = 1 : 4,9 TiO_2	8	8
II = 1 : 71 TiO_2	7-8	8

Rosa E transp.
Pink E transp.
P.R.122

V	7-8d	7-8d
I = 1 : 5,3 TiO_2	8	8
II = 1 : 80 TiO_2	7-8	7-8

Rotviolett ER 02
Red Violet ER 02
P.V.19

V	7-8d	7-8d
I = 1 : 6,7 TiO_2	8	8
II = 1 : 79 TiO_2	7-8	7-8

Fig. 6 Light-fastness of the quinacridones. Leaflet: Organic pigments for the paint industry (DP 7017 DE; Hoechst Aktiengesellschaft)

Further research on this class provided a variety of shades from greenish yellow to brown and bluish red. The development of the benzimidazolones started in the early 1960s by the chemists Karl Schilling and Erich Dietz at Farbwerke Hoechst AG. The first patent (U.S. patent 3.124.565) was applied by this firm in 1960 to protect red and brown pigments obtained by diazotising simple amines such as nitroanilines and chloronitroanilines coupling with BON derivatives of 5-aminobenzimidazolone. In the middle of the 1960s PV Fast Maroon HFM (PR 171), PV Fast Red HFT (PR 175), PV Carmine HF3C (PR 176) and PV Fast Brown HFR (PBr 25) entered the pigment market.

The U.S. patent 3.137.686 and the German patent 1.213.552 by Erich Dietz, Manfred Förderreuther and Karl Schilling at Farbwerke Hoechst AG. disclose benzimidazolones with aminobenzene sulfonic acid as the diazo component. PV Carmine HF4C (PR 185) and PV Bordeaux HF3R (PV 32) are good representatives.

Yellow and orange benzimidazolones contain the coupling component 5-acetoacetylaminobenzimidazolone (U.S. patent 3.109.842) like PV Orange HL (PO 36). The Belgian patent 723.012 by Farbwerke Hoechst AG. protects pigments with 3-aminoisophthalic acid dimethyl ester as a diazo component, such as PV Fast Yellow H2G (PY 120) and another Belgian patent 723.711 also by Farbwerke Hoechst AG. mentions that ester functions in the diazo component yield red benzimidazolones like PV Red HF2B (PR 208). The coupling component for this red pigment is 5-(2-hydroxy-3-naphthoyl)-aminobenzimidazolone.

In the 1970s further systematic study of this class opened the door for more yellow benzimidazolones, such as Benzimidazolone Yellow H4G (PY 151), Benzimidazolone Yellow H3G (PY 154) and Benzimidazolone Yellow H6G (PY 175).

Diketopyrrolo-Pyrroles (DPPs)

In the early 1980s the Ciba Geigy Corporation discovered a new type of heterocyclic pigments, based on a systematic chromophore, the 1,4-diketopyrrolo-pyrrole system. This group is the latest important organic pigment discovery of the twentieth century. In the middle of the 1980s very fast red and orange DPP pigments came on the market. These pigments turned out to be extremely insoluble and have excellent fastness properties. Five reds and three orange DPP pigments are available: Irgazin DPP Red BO (PR 254), Irgazin DPP Red 5G (PR 255), Irgazin DPP Red 4013 (PR 264), PR 270, Cromophtal DPP Flame Red FP (PR 272), Irgazin DPP Orange TR (PO 71), Irgazin DPP Orange 16A (PO 73) and PO 81. Their only disadvantage is the price. Irgazin DPP Red BO (PR 254), the first representative, is the most important one of this group. However, after the expiration of the patent in 2004, several products of European and Asian competitors caused a drop in price. Their applications are in paints, plastics, printing inks and in the automobile industry, which require outstanding application properties.

Acknowledgement The manufacturer of colour materials Royal Talens B.V. has provided the author with data regarding the production of pigments and marketing of their artists' materials. Apart from this the author acknowledges gratefully the use of the Royal Talens' archives and library.

Appendix

Table 1 Survey of modern synthetic organic pigments with year of discovery, manufacture and use as an artists' pigment

Red Pigments	Discoverers and patents		Industrial introduction and *occurrences as an artists' paint*	
β-naphthol pigments	1880	Read Holliday & Sons Ltd., England, British Pat. 2.757	1889	Farbwerke vorm. Meister Lucius & Brüning, Germany
Para Red, PR 1, 12070	1885	M. von Gallois and E. Ullrich, Farbwerke, vormals Meister, Lucius & Brüning, Germany		
Toluidine Red PR 3, 12120	1905	Farbenfabriken vorm. Friedr. Bayer & Co., Germany, German Pat. F. 20.265, Verfahren zur Darstellung roten Pigmentfarben	1915	*Olga Rozanova, Simultane Komposition: Spielkarten*
			1924	*Vilmos Huszár, Miss Blanche wall advertisement*
	1905	Société Farbenfabriken vorm. Friedr. Bayer et Cie., Germany, French Pat. 357.858, Production de laques	1927	*Talens N.V., Netherlands, list of artists' paints, Rembrandt series*
	1905	British Pat. 19.100	1931	*Talens N.V., Netherlands, list of artists' paints, Rembrandt series, Helioechtrot RL*
			1964	*Raoul Dubuffet, Chain de Mémoire III*
			1965	*Raoul Dubuffet, Le Train de Pendules*
			m. 1980	*Winsor & Newton, England, Artists' Oil Colour and Winton Oil Colour: Bright Red and Vermilion Hue*
			1988	*Lukas, Germany, A and B quality artists' oil colours, Lukasrot, Karmin, Krapplack hell, Kadmiumrot dunkel and Zinnober dunkel*
Chlorinated Para Red PR 4, 12085	1906	Act.-Ges. für Anilin-Fabrikation, Germany, German Pat. 180.301, Verfahren zur Darstellung eines orangefarbenen Monoazofarbstoffes	c. 1918	*Ernst Ludwig Kirchner, Plastik: Mann und Frau*
			1929	Missiemuseum, Steyl-Tegelen, Netherlands, pigment collection
	1906	Actien-Gesellschaft für Anilin-Fabrikation, Germany, French Pat. 368.259, Procédé de production d'un nouveau colorant monoazoïque orangé	1988	*Lukas, Germany, A quality artists' oil colours, Helio-Echtrot and Helio-Echtorange*
	1906	W. Herzberg and O. Spengler, Actien Gesellschaft für Anilin Fabrikation, Germany, U.S. Pat. 865.587, Red-Orange Monoazo Dye and Process of Making Same	1988	*Lukas, Germany, B quality artists' oil colours, Kadmiumorange, Echtrot and Zinnoberrot hell*
Parachloro Nitroaniline Red, PR 6, 12090	1906	Badische Anilin-& Soda-Fabrik, Germany, German Pat. 200.263, Verfahren zur Darstellung eines wasserunlöslichen Monoazofarbstoffs		-
	1907	C. Schraube and E. Schleicher, Badische Anilin & Soda Fabrik, Germany, U.S. Pat. 860.575, Azo Dye		
	1907	British Pat. 6.237		
Dinitroaniline Orange PO 5, 12075	1907	Actien-Gesellschaft für Anilin-Fabrikation, Germany, German Pat. 217.266, Verfahren zur Darstellung eines orangefarbenen Lackfarbstoffs	1909	Aktien-Gesellschaft für Anilin-Fabrikation, Germany, Permaton Orange
			1931	*Talens N.V., Netherlands, list of artists' paints, Rembrandt series*
	1908	R. Lauch, Germany, French Pat. 394.754, Nouveau colorant monoazoïque pour laques et ses application industrielles	1933/34	*Wassily Kandinsky's Paris palette*
	1908	AGFA, Germany, British Pat. 18.736	1936	*Max Beckmann, Stilleben mit Ausblick auf den Wannsee*
	1909	R. Lauch, Actien Gesellschaft für Anilin Fabrikation, Germany, U.S. Pat. 912.138, Orange Lake Dye and Process of Making Same		
β-naphthol pigment lakes **Lithol Red** PR 49, 15630	1899	P. Julius, Badische Anilin- und Soda-Fabrik, Germany, German Pat. 112.833, Verfahren zur Herstellung rother Farblacke mittels des durch Combination von β₁-Naphtylamin-α₁-sulfosäure und β-Naphtol erhaltenen Azofarbstoffen	1961	*Mark Rothko, Untitled, Harvard Murals*
	1899	British Pat. 25.511		
	1900	P. Julius, Badische Anilin und Soda Fabrik, Germany, U.S. Pat. 650.757, Azo Dye for Lakes		
	1900	French Pat. 297.330		
	1900	Badische Anilin- und Soda-Fabrik, Germany, German Pat. 120.322, Verfahren zur Darstellung rother Farblacke aus β-Naphtylaminmonosulfosäuren und β-Naphtol (additional to German patent 112.833)		
	1905	Badische Anilin- & Soda-Fabrik, Germany, German Pat. 175.630, Verfahren zur Darstellung roter Farblacke (additional to German patent 112.833)		

(continued)

Table 1 (continued)

β-naphthol pigment lakes				
Red Lake C PR 53, 15585	1902	Farbwerke vorm. Meister Lucius & Brüning, Germany, German Pat. 145.908, Verfahren zur Darstellung eines roten, besonders zur Bereitung von Farblacken geeigneten Monoazofarbstoffe aus o-Chlor-m-toluidin-p-sulfosäure und β-Naphthol	1993	*Markus Lüpertz, Parsifal oder Männer ohne Frauen*
	1902	Cie Parisienne de Couleurs, France, French Pat. 328.131, Procédé pour la préparation d'une matière colorante azoïque rouge		
	1902	British Pat. 23.831		
	1903	K. Schirmacher, Farbwerke vorm. Meister Lucius & Brüning, Germany, U.S. Pat. 733.280, Red Azo Dye and Process of Making Same		
Clarion Red PO 46, 15602	1949	F.H. Adams, American Cyanamid Company, USA, U.S. Pat.2.598.483, Pigments from 2-Chloro-5-Amino-Ethylbenzene-4-Sulfonic Acid and Beta Naphthol		-
BON pigment lakes				
Lithol Rubine PR 57, 15850	1903	Aktien-Gesellschaft für Anilin-Fabrikation, Germany, German Pat. 151.205, Verfahren von be-sonders für die Herstellung von Farblacken geeigneten Monoazofarbstoffen aus 2-3-Oxynaphtoësäure	1958	*Fritz Winter, Teile*
			1958	*Fritz Winter, Später Garten*
			1963	*Fritz Winter, Winterliches Geäst*
	1903	Actien-Gesellschaft für Anilin-Fabrikation, Germany, French Pat. 332.145, Procédé de fabrication de colorants monoazoïques destinés spécialement à la préparation de laques rouges	1991	*Royal Talens B.V., Netherlands, list of artists' paints, Rembrandt Oil Colours, Geraniumlak and Talensrood purper*
	1903	R. Gley and O. Siebert, Actien-Gesellschaft für Anilin Fabrikation, Germany, U.S. Pat. 741.029, Red Azo Lake	1991	*Royal Talens B.V., Netherlands, list of artists' paints, Van Gogh Oil Colours, Cadmiumrood azo, Vermiljoen and Karmijn*
	1903	R. Gley and O. Siebert, Actien Gesellschaft für Anilin Fabrikation, Germany, U.S. Pat. 743.071, Monoazo Dye and Process of Making Same	1991	*Royal Talens B.V., Netherlands, list of artists' paints, Amsterdam Oil Colours, Donkerrose*
	1903	British Pat. 11.004	m. 1990	*Royal Talens B.V., Netherlands, replaced*
Permanent red 2B PR 48, 15865	1904	Badische Anilin- und Soda-Fabrik, Germany, German Pat. 175.378, Verfahren zur Darstellung eines besonders für die Herstellung von Farblacken geeigneten roten Monoazofarbstoffs	l. 1920s	A.Siegel, E.I. du Pont de Nemours & Company, USA, trade name: Watchung Red
	1904	P. Julius, Badische Anilin und Soda Fabrik, Germany, U.S. Pat. 759.716, Red Azo Dye	1991	*Royal Talens B.V., Netherlands, list of artists' paints, Van Gogh Oil Colours, Cadmiumrood donker azo*
	1926	A. Siegel, Mesne Assignments, to the Grasselli Chemical Company, USA, U.S. Pat. 1.803.657, Azo Dyestuffs and Method for their Preparation	1991	*Royal Talens B.V., Netherlands, list of artists' paints, Amsterdam Oil Colours, Cadmiumrood donker azo*
	1936	A. Siegel, E.I. du Pont de Nemours & Company, USA, U.S. Pat. 2.117.860, Colored Rubber Products		
	1939	A. Siegel, E.I. du Pont de Nemours & Company, USA, U.S. Pat. 2.225.665, Azo Pigment and Coating Composition Containing the Same		
BON Red PR 52, 15860	1910	O. Ernst and H. Eichweide, Farbwerke vorm. Meister Lucius & Brüning, Germany, U.S. Pat. 983.486, Monoazo Dyestuff and Process of Making Same		-
BON Maroon PR 63, 15880	1906	Farbwerke vorm. Meister Lucius & Brüning, Germany, German Pat. 205.080, Verfahren zur Darstellung eines roten, besonders für die Herstellung blauroter Farblacke geeigneten Monoazofarbstoffe		-
	1907	O. Ernst and G. Gullbranson, Farbwerke, vorm. Meister, Lucius & Brüning, Germany, U.S. Pat. 858.065, Monoazo Dye for Lakes and Process of Making Same		
	1907	Farbwerke vorm. Meister Lucius et Brüning, Germany, French Pat. 385.570, Procédé de fabrication d'une matière colorante appropriée à la préparation de laques		
	1907	British Pat. 7.351		

Pyrazolones **Benzidine Orange G** PO 13, 21110	1910 1911 1910 1911	Chemische Fabrik Griesheim-Elektron, Germany, German Pat. 236.856, Verfahren zur Darstellung besonders zur Pigmentfarbenbereitung geeigneter Disazofarbstoffe A.L. Laska, Chemische Fabrik Griesheim-Elektron, Germany, U.S. Pat. 1.001.286, Diazo Dyes and Process of Making Same French Pat. 425.429 British Pat. 1.730		-
Pyrazolone Red PR 38, 21120	1934 1934	W. Scheurer and G. Niemann, I.G. Farbenindustrie Akt.-Ges., Germany, German Pat. 596.755, Verfahren zur Herstellung von Disazofarbstoffen I.G. Farbenindustrie Akt.-Ges., Germany, German Pat. 618.342, Verfahren zum Färben von Kautschuk	1934	I.G. Farbenindustrie Akt.-Ges., Germany, on the market for colouring rubber
Tolyl Orange PO 34, 21115 **Dianisidine Red** PR 41, 21200	1934 1935	H. Eichweide, General Aniline Works, USA, U.S. Pat. 1.977.936, Water-Insoluble Disazo-Dyestuffs W. Scheurer and G. Niemann, General Aniline Works, USA, U.S. Pat. 1.995.902, Disazo Dyestuffs and Their Production	1934	I.G. Farbenindustrie Akt.-Ges., Germany, on the market for colouring rubber
Naphthol Reds	1911 1912 1912 1912 1913	Chemische Fabrik Griesheim-Elektron, Germany, German Pat. 256.999, Verfahren zur Darstellung besonders zur Pigmentfarbenbereitung geeigneter Monoazofarbstoffe Chemische Fabrik Griesheim-Elektron, Germany, German Pat. 258.654, Verfahren zur Darstellung besonders zur Pigmentfarbenbereitung geeigneter Monoazofarbstoffe Chemische Fabrik Griesheim-Elektron, Germany, German Pat. 261.594, Verfahren zur Darstellung von Monoazofarbstoffen Chemische Fabrik Griesheim-Elektron, Germany, German Pat. 264.527, Verfahren zur Darstellung von Nitroarylamiden der 2.3-Oxynaphthoesäure Chemische Fabrik Griesheim-Elektron, Germany, German Pat. 293.897, Verfahren zur Darstellung von Arylamiden der 2.3-Oxynaphthoesäure		-
Permanent Red FRR PR 2, 12310 **Permanent Red F4R** PR 8, 12335	1911 1912 1912 1912	Chemische Fabrik Griesheim-Elektron, Germany, German Pat. 256.999, Verfahren zur Darstellung besonders zur Pigmentfarbenbereitung geeigneter Monoazofarbstoffe Chemische Fabrik Griesheim-Elektron, Germany, French Pat. 441.333, Procédé de préparation de matières colorantes mono-azoïques se prêtant particulièrement à la fabrication de laques A. Winther, A.L. Laska and A. Zitscher, Chemische Fabrik Griesheim-Elektron, Germany, U.S. Pat. 1.034.853, Azo Dyes for Lakes British Pat. 6.379	1967 1988 1988	*Georg Baselitz, Drei Feldarbeiter,* Permanent Red FRR (PR 2) *Lukas, Germany, A quality artists' oil colours, Indischgelb, Geraniumlack and Grüner Lack hell* (PR 8) *Lukas, Germany, B quality artists' oil colours, Geraniumlack and Kobaltviolett dunkel* (PR 8)
Permanent Carmine FB PR 5, 12490	1931 1931 1932	W. Neelmeier and W. Lamberg, I.G. Farbenindustrie Akt.-Ges., Germany, German Pat. 575.216, Verfahren zur Herstellung eines Monoazofarbstoffs I.G. Farbenindustrie Aktiengesellschaft, Germany, British Pat. 391.468, Process for the Manufacture of an Azo-dyestuff and an Intermediate Product therefor I.G. Farbenindustrie Aktiengesellschaft, Germany, French Pat. 742.326, Nouvelles matières colorantes azoïques	1982	*Georg Baselit, Orangenesser IV*
Permanent Red F4RH PR 7, 12420 **Permanent Bordeaux FRR,** PR 12, 12385	1921 1922 1922	H. Wagner, Firma Chemische Fabrik Griesheim-Elektron, Germany, German Pat. 421.205, Verfahren zur Herstellung von Azofarbstoffen A.G. Bloxham, Chemische Fabrik Griesheim Elektron, Germany, British Pat. 199.771, Manufacture of Azo-Dyestuffs (revoked) Chemische Fabrik Griesheim Elektron, Germany, French Pat. 549.020, Procédé pour la préparation de colorants azoïques		-

(continued)

Table 1 (continued)

Permanent Red FRLL PR 9, 12460	1922 1923	Chemische Fabrik Griesheim-Elektron, Germany, German Pat. 390.627, Verfahren zur Darstellung von Azofarbstoffen A.L. Laska and A. Zitscher, Corporation Chemische Fabrik Griesheim-Elektron, Germany, U.S. Pat. 1.457.114, Azo Dyestuffs and Process of Making Same	1970 1970/71 1993	*Bernard Rancillac, La Suite Américaine* *David Hockney, Mr. And Mrs. Persey* *Georg Baselitz, Bildsechzehn*
Permanent Red FGR PR 112, 12370	1939	IG. Farbenindustrie AG., Germany	c. 1975 1991	*A.R. Penck, Young Generation* *Royal Talens B.V., Netherlands, list of artists' paints, Rembrandt Oil Colours, Vermiljoen chin., Talensrood donker and Permanentrood*
Permanent Carmine FBB, PR 146, 12485	1953	Farbwerke Hoechst Aktiengesellschaft vormals Meister Lucius & Brüning, Germany		-
Permanent Red F5RK PR 170, 12475	1963	Farbwerke Hoechst Aktiengesellschaft vormals Meister Lucius & Brüning, Germany, British Pat. 1.033.917, Water-Insoluble Monoazo Dyestuffs and processes for their manufacture and use	1983 1985/86	*J. Dubuffet, Mire G 137 Kowloon* *Patrick Caulfield, Interior with a Picture*
Perylenes **Perylene Vermilion** PR 123, 71145	1948	M.O. Shrader, General Aniline & Film Corporation, USA, U.S. Pat. 2.543.747, Preparation of N,N'-Diaryl Derivatives of Perylene Diimid		-
Perylene Scarlet PR 149, 71137	1956	W. Eckert and H. Remy, Farbwerke Hoechst Aktiengesellschaft vormals Meister Lucius & Brüning, Germany, German Pat. 1.067.157, Verfahren zur Herstellung eines Farbstoffes	1957 1961 1996	Farbwerke Hoechst AG., launched the product in the market *Société Lefranc, France* *Royal Talens B.V., Netherlands, list of artists' paints, Scarlet*
Perylene Maroon PR 179, 71130	1913 1913 1919 1922 1960	M. Kardos, Germany, German Pat. 276.956, Verfahren zur Darstellung von Küpenfarbstoffen der Naphtalinreihe Badische & Co., British Pat. 26.690 Kalle & Co., Akt.-Ges., Germany, German Pat. 386.057, Verfahren zur Darstellung von Küpenfarbstoffen Kalle & Co., Aktiengesellschaft, Germany, British Pat. 201.786, Process for producing Vat Colouring-matters E. Daubach, B. Hirsekorn, H. Schroeder, H. Sixtus and H. Wuest, Badische Anilin- & Soda-Fabrik AG., Germany, British Pat. 923.721, Pigment Coloured Compositions and Mixtures		-
Perinones **Perinone Orange** PO 43, 71105	1924 1932	W. Eckert and H. Greune, Farbwerke vorm. Meister Lucius & Brüning, Germany H. Vollmann and M. Corell, I.G. Farbenindustrie Akt.-Ges., Germany, German Pat. 602.445, Verfahren zur Herstellung von Naphthalin 1.4.5.8-tetracarbonsäure	1953 1978 1978 1991	Farbwerke Hoechst AG., Germany, trade name: Indanthrenbrillantorange GR *Royal Talens B.V., list of artists' paints, Rembrandt Oil Colours* *Sam Francis, Untitled* *Royal Talens B.V., Netherlands, list of artists' paints, Rembrandt Oil Colours, Geraniumlak, Talensrood purper and Cadmiumrood donker Van Gogh Oil Colours, Cadmiumrood donker azo, Vermiljoen and Karmijn, Amsterdam Oil Colours, Cadmiumrood donker azo*
Benzimidazolones **PV Fast Maroon HFM** PR 171, 12512 **PV Fast Red HFT** PR 175, 12519 **PV Carmine HF3C** PR 176, 12515 **PV Fast Brown HFR** PBr 25, 12510	1960	K. Schilling and E. Dietz, Farbwerke Hoechst Aktiengesellschaft vormals Meister Lucius & Brüning, Germany, U.S. Pat. 3.124.565, Water-Insoluble Benzimidazolone Mono-Azo-Dyestuffs	1964 1965 1965 1966 1988 1988	Farbwerke Hoechst AG., Germany, commercial production of PR 171 Farbwerke Hoechst AG., Germany, commercial production of PR 175 Farbwerke Hoechst AG., Germany, commercial production of PR 176 Farbwerke Hoechst AG., Germany, commercial production of PBr 25 *Lukas, Germany, list of artists' pigments, A quality Karmin, Krapplack hell dunkel and dunkelst, Casslerbraun, Sepia, Vandyckbraun and Indigo* (PR 176) *Lukas, Germany, list of artists' pigments, B-quality, Karminrot, Krapplack hell and dunkel, Casslerbraun and Vandyckbraun* (PR 176)

Pigment	Year	Patent	Year	Production / use
PV Carmine HF4C PR 185, 12516 **PV Bordeaux HF3R** PV 32, 12517	1962 1963	E. Dietz, M. Förderreuther and K. Schilling, Farbwerke Hoechst Aktiengesellschaft vormals Meister Lucius & Brüning., Germany, German Pat. 1.213.552, Verfahren zur Herstellung von wasserunlöslichen Monoazofarbstoffen E. Dietz, M. Förderreuther and K. Schilling, Farbwerke Hoechst Aktiengesellschaft vormals Meister Lucius & Brüning, Germany, U.S. Pat. 3.137.686, Water-Insoluble Benzimidazolone Monoazo-Dyestuffs	1965 1967	Farbwerke Hoechst AG., Germany, commercial production of PV 32 Farbwerke Hoechst AG., Germany, commercial production of PR 185
PV Red HF2B PR 208, 12514		Farbwerke Hoechst Aktiengesellschaft vormals Meister Lucius & Brüning, Germany, Belgian Pat. 723.711	1970	Farbwerke Hoechst AG., Germany, commercial production
PV Orange HL PO 36, 11780	1961	K. Schilling and E. Dietz, Farbwerke Hoechst Aktiengesellschaft vormals Meister Lucius & Brüning Farbwerke, Germany, U.S. Pat. 3.109.842, Water-Insoluble Benzeneazo-5-Aceto-Acetylaminobenzimidazolone Dyestuffs	1964	Farbwerke Hoechst AG., Germany, commercial production
PV Fast Yellow H2G PY 120, 11783		Farbwerke Hoechst Aktiengesellschaft vormals Meister Lucius & Brüning, Germany, Belgian Pat. 723.012	1969	Farbwerke Hoechst AG., Germany, commercial production
Quinacridones **Quinacridone Magenta** PR 122, 73915 **Sandorin Brilliant Red BL,** PR 192, - **Monastral Magenta** PR 202, 73907 **Monastral Maroon** PR 206, 73900 + 73920 **Monastral Scarlet** PR 207, 73900 + 73906 **Hostaperm Red E** PR 209, 73905	1958 1958 1958 1958 1958 1958	W.S. Struve, E.I. du Pont de Nemours & Company, USA, U.S. Pat. 2.821.530, Tetrahalogen Substituted Quinacridones W.S. Struve, E.I. du Pont de Nemours & Company, USA, U.S. Pat. 2.821.541, Production of Dialkyl 2,5-Diarylamino-3,6-Dihydroterephthalates W.S. Struve, E.I. du Pont de Nemours & Company, USA, U.S. Pat. 2.830.990, Organic Pigment A.D. Reidinger and W.S. Struve, E.I. du Pont de Nemours & Company, USA, U.S. Pat. 2.844.484, Organic Pigments W.S. Struve, E.I. du Pont de Nemours & Company, USA, U.S. Pat. 2.844.485, Organic Pigments C.W. Manger and W.S. Struve, E.I. du Pont de Nemours & Company, USA, U.S. Pat. 2.844.581, Organic Pigments	1935 1955 1958 1961 1974 1978 1980 1982 1991 1991 1991	H. Liebermann, Germany, Article: Über die Bildung von Chinakridonen aus p-Di-arylamino-terephtalsäuren E.I. du Pont de Nemours & Company, USA Introduction trade name: Cinquasia, Monastral Red B *Société Lefranc, France* *Royal Talens B.V., Netherlands, list of artists' paints, Rembrandt Oil Colours* (PR 122) *Sam Francis, Untitled* (PR 122) *Schmincke, Germany* *Georg Baselitz, Orangenesser IV* (PR 202) *Royal Talens B.V., Netherlands, list of artists' paints, Rembrandt Oil Colours, Rembrandtrose, Permanentviolet and Permanen troodviolet* (PR 122) *Royal Talens B.V., Netherlands, list of artists' paints, Van Gogh Oil Colours, Permanenrose, Permanent roodviolet and Permanent blauwviolet* (PR 122) *Royal Talens B.V., Netherlands, list of artists' paints, Amsterdam Oil Colours, Violet* (PR 122)
Diketopyrrolo-pyrroles **Irgazin DPP Red BO** PR 254, 56110 **Irgazin DPP Red 5G** PR 255, 561050 **Irgazin DPP Red 4013** PR 264, 561300 **Cromophtal DPP Orange TR** PO 71, 561200 **Irgazin DPP Orange 16A** PO 73, 561170	1982 1983 1984 1986 1989 1994	A. Iqbal and L. Cassar, Ciba-Geigy Corporation, Switzerland, U.S. Pat. 4.415.685, Process for Dyeing High-Molecular Organic Material, and Novel Polycyclic Pigments A. Iqbal and L. Cassar, Ciba-Geigy Corporation, Switzerland, U.S. Pat. 4.490.542, 1,4-Diketo- Pyrrolo [3,4-C] Pyrroles M. Jost, A. Iqbal and A.C. Rochat, Ciba-Geigy Corporation, Switzerland, U.S.Pat. 4.585.878, N-Substituted 1,4-Diketopyrrolo [3,4-C] Pyrroles M. Jost, A. Iqbal and A.C. Rochat, Ciba-Geigy Corporation, Switzerland, U.S.Pat. 4.791.204, 1,4-Diketopyrrolo [3,4-C] Pyrrole Pigments O. Wallquist and A. Iqbal, Ciba-Geigy AG., Switzerland, German Pat. 4.037.556, Neue Stoffzusammensetzung auf Basis von Diketopyrrolopyrrolen Z. Hao, I. Schlöder and A. Iqbal, Ciba-Geigy AG., Switzerland, European Pat. 690.058, Neue Kristallmodifikation eines Diketopyrrolopyrrolpigments (PR 254, β-modification)	1974 1974 1983 1986 1994 1996 2000 2003 2004	Discovery of the DPP group by Ciba Geigy AG., Switzerland Discovery of Irgazin DPP Red BO (PR 254) Discovery of Irgazin DPP Red 5G (PR 255) First representative of the DPPs on the market Discovery of Irgazin DPP Red 4013 (PR 264) *Royal Talens B.V., Netherlands, list of artists' paints, Rembrandt Oil Colours* (PR 254, PR 255, PR 264 and PO 73) *Schmincke, Germany, MUSSINI* (PR 254, PR 255, PR 264 and PO 71) *Schmincke, Germany, PRIMAcryl* (PR 254, PR 255 and PO 71) *Lukas, Germany* (PR 254, PR 255 and PR 264)

(continued)

Table 1 (continued)

Yellow Pigments		Discoverers and patents		Industrial introduction and *occurrences as an artists' paint*
Hansa Yellows	1897	Farbwerke vorm. Meister Lucius & Brüning, Germany, German Pat. 99.381, Verfahren zur Darstellung wasch- und lichtechter Azofarbstoffe aus ß-Diketonen	1912	*Schmincke, Germany*
Hansa Yellow G PY 1, 11680	1909	Farbwerke vorm. Meister Lucius & Brüning, Germany, German Pat. 257.488, Verfahren zur Darstellung wasserunlöslicher gelber Monoazofarbstoffe	1918	*Lithographische Tinte, Poster Mallinson's*
	1910	H. Wagner, Farbwerke vorm. Meister Lucius & Brüning, Germany, U.S. Pat. 1.082.719, Monoazo Dyestuffs and Process of Making Same	1919	*Alexander Rodchenk, Konstruktivistische Komposition*
	1910	Farbwerke vorm. Meister Lucius & Brüning, Germany, British Pat. 9.633	1919	*Ludwig Meidner, Der Krieg*
	1910	V. Villiger and E. Fussenegger, Badische Anilin & Soda Fabrik, Germany, U.S. Pat. 990.173, Yellow Azo Dyes	1924	*Daler-Rowney, England*
	1910	Farbwerke Meister Lucius & Brüning, Germany, start of commercial production	1925	*Marianne Ullmann, Komposition mit zwei Akten*
	1914	Farbenfabriken vorm. Friedr. Bayer & Co., Germany, German Pat. ap. F. 33.190	1926	British Dyestuffs Corporation, England, tradename: Monolite Yellow G
	1914	Farbenfabriken vorm. Friedr. Bayer & Co., Germany, German Pat. 287.569, Verfahren zur Darstellung von Azofarbstoffen	1926	*J.W. Leitch and Co., England, trade name: Pigment Fast Yellow HGL*
	1914	Farbenfabriken vorm. Friedr. Bayer & Co., Germany, German Pat. 293.429, Verfahren zur Darstellung von Azofarbstoffen	1931	*Talens N.V., Netherlands, list of artists' paints, Rembrandt series*
	1914	British Pat. 9.252	1935	*Schmincke, Germany, Gouache paint-outs: Sunproof Yellow I*
			1948	*Victor Pasmore, Abstract Design for a Poster*
			1948	*Clyfford Still, 1948-C*
			1958	*Alfred Pellan, Jardin Vert*
			1958	*Helen Frankenthaler, Basque Beach*
			1982	*Sigmar Polke, Stairwell (Treppenhaus)*
Hansa-Yellow 10G PY 3, 11710	1909	Farbwerke vorm. Meister Lucius & Brüning, Germany, German Pat. 257.488, Verfahren zur Darstellung wasserunlöslicher gelber Monoazofarbstoffe	1927	*Talens N.V., list of artists' paints, Rembrandt Tempera Paints, Talens Groen Licht*
	1910	H. Wagner, Farbwerke vorm. Meister Lucius & Brüning, Germany, U.S. Pat. 1.082.719, Monoazo Dyestuffs and Process of Making Same	1931	*Talens N.V., Netherlands, list of artists' paints, Rembrandt Extra Fijne Olieverven, Talens Geel Citroen, Talens Geel Licht, Talens Geel Middel and Talens Groen Licht*
	1911	Farbenfabriken vorm. Friedr. Bayer Co., Germany, German Pat. ap. F. 33.190, Verfahren zur Darstellung eines grünstichig gelben Lackfarbstoffes	1935	*Schmincke, Germany, Gouache paint-outs: Sunproof Yellow II*
	1911	Farbenfabriken vorm. Friedr, Bayer Co., Germany, British Pat. 23.791	1935	*Vaclav Vytlacil, Construction No. 3*
	1912	Farbwerke Meister Lucius & Brüning, Germany, start commercial production	1936	*Alexander Calder, Form Against Yellow*
	1912	K. Desamari, Farbenfabriken vorm. Friedr. Bayer & Co., Germany, U.S. Pat. 1.051.565, Azo Color	1936	*Max Beckmann, Stilleben mit Ausblick auf den Wannsee*
	1912	K. Desamari, Farbenfabriken vorm. Friedr. Bayer & Co., Germany, U.S. Pat. 1.059.599, Azo Dye	1938	*Pablo Picasso, Portrait of Dora Marr*
	1912	Farbenfabriken vorm. Friedr. Bayer & Cie., Germany, French Pat. 448.958, Procédé de production d'une couleur jaune-verdrate appropriée à la préparation de laques	1939	*Vaclav Vytlacil, Construction*
	1914	British. Pat. 9.252	c. 1945	*Jackson Pollock, There Were Seven in Eight*
	1914	Farbenfabriken vorm. Friedr. Bayer & Co., Germany, German Pat. 287.569, Verfahren zur Darstellung von Azofarbstoffen	1948	*Jackson Pollock, Summertime: Number 9A*
	1914	Farbenfabriken vorm. Friedr. Bayer & Co., Germany, German Pat. 293.429, Verfahren zur Darstellung von Azofarbstoffen	1949	*Barnett Newman, Dionysius*
	1925	H. Wagner and A. Funke, Grasselli Dyestuff Corporation, USA, U.S. Pat. 1.644.003, Greenish-Yellow Azo Dyestuff and Process of Making Same	1953	*Ernst Briggs, Untitled*
			1957	*Yves Klein, Monochrome Vert*
			1958	*Alfred Pellan, S'abstenir*
			1965	*Tom Wesselmann, Study for 'First Illuminated Nude'*
			1968	*Lucebert, Baby Elephant*
			1970	*Philip King, Dunstable Reel*
			1973	*Bridget Riley, Cantus Firmus*
			1982	*Terry Setch, Once Upon a Time There Was Oil III*
			1985	*Patrick Caulfield, Interior with Picture*
			1989	*Katharina Fritsch, Display Stand with Madonnas*

Hansa Yellow 3RN PY 65, 11740	1936 1936	A. Funke, A. Sieglitz and R. Huβ, IG. Farbenindustrie AG., Germany, German Pat. 679.481, Verfahren zur Herstellung von wasserunlöslichen Azofarbstoffen I.G. Farbenindustrie Aktiengesellschaft, Germany, British Pat. 481.290, Manufacture of Azo-dyestuffs		IG. Farbenindustrie AG., Germany, before World War II
Hansa-Brilliant Yellow 4GX, PY 73, 11738	1961 1961 1962	J.W. Conley and W.W. Williams, Hercules Powder Company, USA, British. Pat. 938.047, Improvements in or Relating to Yellow Monoazo Pigments J.W. Conley and W.W. Williams, Hercules Incorporated, USA, German Pat. 1.231.367, Verfahren zur Herstellung von Hansagelb-Pigmentfarbstoffen Hercules Powder Company, USA, French Pat. 1.309.211, Procédé de preparation d'un pigment jaune Hansa		-
Hansa-Brilliant Yellow 5GX, PY 74, 11741	1959 1961	R.A. Johnson, E.I. du Pont de Nemours and Comp. Inc., USA, U.S. Pat. 3.032.546, Yellow Azo Pigment Canadian Pat. 612.395 E.I. du Pont de Nemours & Comp. Inc., USA, trade name: Dalamar Yellow	1958 1961 1961 1973 1994	E.I. du Pont de Nemours & Company, USA, trade name Dalamar Yellow Sherwin Williams Comp., USA, trade name: Permansa Yellow GY Farbwerke Hoechst AG., West Germany, trade name: Hansa-Brillant 5GX *Alfred Jensen, The Sun Rises Twice (Per I, Per II, Per III, Per IV)* *Gillian Ayres, Sundark Blues*
Hansa Yellow FGL PY 97, 11767	1950 1950 1951	E. Fisher, Naphthol-Chemie Offenbach, Germany, German Pat. 845.374, Verfahren zur Herstellung von wasserunlöslchen Monoazofarbstoffen Farbwerke Hoechst Aktiengesellschaft Vormals Meister Lucius & Brüning, Germany British Pat. 705.590, Manufacture of Water-Insoluble Monoazo Dyestuffs E. Fischer, Farbwerke Hoechst A.G. vormals Meister Lucius und Brüning, Germany, U.S. Pat. 2.644.814, Water-Insoluble Monoazo-Dyestuffs	1960	Farbwerke Hoechst AG., Germany, introduction of Hansa Yellow FGL
Hansa Brilliant Yellow 10GX, PY 98, 11727	1960 1961	O. Braun, G. Bandel and E-A. Theiling, Farbwerke Hoechst Aktiengesellschaft vormals Meister Lucius & Brüning, Germany, German. Pat. 1.212.237, Verfahren zur Herstellung eines wasserunlöslichen gelben Monoazofarbstoffes O. Braun, G. Bandel and E-A. Theiling, Farbwerke Hoechst Aktiengesellschaft vormals Meister Lucius & Brüning, Germany, U.S. Pat. 3.165.507, Water-Insoluble Yellow Monoazo Dyestuff	1960	Farbwerke Hoechst AG., Germany, introduction of Hansa Brilliant Yellow 10GX
Hansa Yellow 3R PO 1, 11725	1925 1925 1926 1926	H. Wagner and A. Funke, Grasselli Dyestuff Corporation, USA, U.S. Pat. 1.595.269, Yellowish-Red Azo Dyestuffs and Process of Making Same H. Wagner and A. Funke, Grasselli Dyestuff Corporation, USA, U.S. Pat. 1.644.003, Greenish-Yellow Azo Dyestuff and Process of Making Same H. Wagner and A. Funke, IG. Farbenindustrie Akt.-Ges., Germany, German Pat. 461.499, Verfahren zur Darstellung von gelbroten Azofarbstoffen British Pat. 272.580		-
Diarylides **Benzidine Yellow AAA** PY 12, 21090 **Permanent Yellow GR** PY 13, 21100 **Benzidine Yellow AAOT** PY 14, 21095 **Benzidine Yellow AAOA** PY 17, 21105	1911	Chemische Fabrik Griesheim-Elektron, Germany, German Pat. 251.479, Verfahren zur Darstellung besonders zur Pigmentfarbenbereitung geeigneter Disazofarbstoffe	1988 1988	*Lukas, Germany, A quality artists' oil colours, Indischgelb, Geraniumlack and Grüner Lack hell* (PY 12) *Lukas, Germany, B quality artists' oil colours, Geraniumlack* (PY 12)

(continued)

Table 1 (continued)

Pigment	Year	Patent	Year	Introduction
Permanent Yellow H10G PY 81, 21127	1958	J. Ribka, Farbwerke Hoechst Aktiengesellschaft vormals Meister Lucius & Brüning, Germany, German Pat. 1.088.634, Verfahren zur Herstellung eines wasserunlöslichen Disazofarbstoffes		-
Permanent Yellow HR PY 83, 21108	1949 1952	R. Huss and R. Weigele, Farbwerke Hoechst Aktiengesellschaft vormals Meister Lucius & Brüning, Germany, German Pat. 899.696, Verfahren zur Herstellung von wasserunlöslichen Disazofarbstoffen R. Weigele and R. Huss, Farbwerke Hoechst Aktiengesellschaft vormals Meister Lucius & Brüning, Germany, German Pat. 921.404, Verfahren zur Herstellung von wasserunlöslichen Disazofarbstoffen	1958	Farbwerke Hoechst AG., inroduction of Permanent Yellow HR (PY 83)
Permanent Yellow GRL PY 127, 21102	1964	J. Ribka, Farbwerke Hoechst Aktiengesellschaft vormals Meister Lucius & Brüning, Germany, U.S. Pat. 3.361.736, Mixtures of Water-Insoluble Disazo Dyestuffs		-
Isoindolinones	1946 1946 1952 1956 1957 1957	W.O. Jones, Imperial Chemical Industries Ltd., England, British Pat. 615.697, New Colouring Matters of the Azamethine Series W.O. Jones, Imperial Chemical Industries Ltd., England, U.S. Pat. 2.537.352, Dyestuffs from Phthalimide Derivatives J.B. Campbell, E.I. du Pont de Nemours and Comp. Inc., USA, U.S. Pat. 2.692.267, Phthalimide Derivatives of 2,2'- Di-amino-4,4'-bithiazole J.R. Geigy AG., Switzerland, Swiss Pat. 346.218 Ciba Geigy AG., Switzerland, Swiss Patents. 348.496, 363.979 and 363.980 A. Pugin, J.R. Geigy AG, Switzerland, U.S. Pat. 2.973.358, 4,5,6.7-Tetrachloro-3-Imino-Isoindolinone-1-One-Derivatives (production patent, trade name: Irgazin)	1952 1953 1965 1966	J.A. Elvidge and R.P. Linstead, England, Journal of the Chemical Society, Article: Heterocyclic Imines, Part I. Imino-derivatives of isoIndoline and their Reaction with Primary Bases P.F. Clark, J.A. Elvidge and R.P. Linstead, England, Journal of the Chemical Society, Article: Hetero Imines and Amines, Part II. Derivatives of isoIndoline and isoIndolenine Introduction of isoindolinones in the USA A. Pugin and J. von der Crone, Switzerland, Farbe und Lack, Article: Chemie und Eigenschaften von neuen Pigmenten der 4, 5, 6, 7-Tetrachlorisoindolin-1-on.Reihe
Isoindolinone Yellow G PY 109, 56284 **Isoindolinone Yellow R** PY 110, 56280	1956 1977	A. Pugin, J.R. Geigy A.G., Switzerland, German Pat. 1.098.126, Verfahren zur Herstellung von farbigen Kondensationsprodukten F. Bäbler, Ciba-Geigy AG., Switzerland, German Pat. 2.804.062, Neue stabile Modifikation eines Isoindolinon-Pigments, Verfahren zu deren Herstellung und deren Verwendung	1965 1978 1980	USA, trade name: Irgazin Yellow 3RLT (PY 110) *Royal Talens B.V., Netherlands, list of artists' paints, Rembrandt Oil Colours, Indian Yellow* (PY 110) *Royal Talens B.V., Netherlands, list of artists' paints, Rembrandt Oil Colours* (PY 109)
Lithol Fast Yellow 1840 PY 139, 56298	1980	M. Lorenz, Bayer AG., Germany, German Pat. 3.022.839, Isoindoleninderivate, Verfahren zu ihrer Herstellung sowie ihre Verwendung als Zwischenprodukte zur Herstellung von Farbstoffen		
Benzimidazolones **Benzimidazolone Yellow H4G,** PY 151, 13980		-	1971	Farbwerke Hoechst AG., Germany, commercial production
Benzimidazolone Yellow H3G, PY 154, 11781		-	c. 1975 1996	Farbwerke Hoechst AG., Germany, commercial production *Royal Talens B.V., Netherlands, list of artists' paints, Rembrandt Oil Colours, Permanent Yellow, light, medium and dark, Permanent Green light, medium and dark, Permanent Yellowish Green, Cinnabar Green light and medium*
Benzimidazolone Yellow 6G, PY 175, 11784		-	a. 1980	Farbwerke Hoechst AG., Germany, commercial production
Other yellow pigments **Irgazin Yellow 5GT** **PY 129, 48042**	1970	K-H. Schündehütte, Bayer AG., Germany, German Pat. 2.064.093, Azofarbstoffsalze und -komplexe mehrwertiger Metalle sowie deren Verwendung als Pigmentfarbstoffe	2008 2008	*Lukas, Germany* *Royal Talens B.V., Netherlands, list of artists' paints, Rembrandt Oil Colours, Transparent Yellow Green*
Helio Fast Yellow 4G **PY 150, 12764**	1970	K-H. Schündehütte, Bayer AG., Germany, German Pat. 2.064.093, Azofarbstoffsalze und -komplexe mehrwertiger Metalle sowie deren Verwendung als Pigmentfarbstoffe	2008	*Royal Talens B.V., Netherlands, list of artists' paints, Golden Acrylic*

Blue Pigments		Discoverers and patents		Industrial introduction and *occurrences as an artists' paint*
Phthalocyanine Blues	1928	A.G. Dandridge, H.A.E. Drescher and J. Thomas, Scottish Dyes Ltd., Scotland, British Pat. 322.169, Improvements in and relating to the manufacture and use of Colouring Matters	1927	H. de Diesbach and E. von der Weid, Switzerland, Helvetica Chimica Acta, Article: Quelques sels complexes des o-dinitriles avec le cuivre et la pyridine
Copper Phthalocyanine Blue, PB 15, 74160	1929	Scottish Dyes Ltd., Scotland, German Pat. 586.906, Verfahren zur Herstellung von Farbstoffen	1934	R.P. Linstead, England, Journal of the Chemical Society, Article: Phthalocyanines. Part I. A New Type of Synthetic Colouring Matters
	1931	J.F. Thorpe, R.P. Linstead and J. Thomas, Scottish Dyes Ltd., Scotland, British Pat. 389.842, Improvements in and relating to the Production of Coloured Products	1932/34	R.P. Linstead, England, revelation of chemical structure, name given: phthalocyanine
	1931	J.F. Thorpe and R.P. Linstead, Scottish Dyes Ltd., Scotland, Imperial Chemical Industries Ltd., England, British Pat. 390.149, Improvements in and relating to the Manufacture and Use of Colouring Matters	1934	G.T. Byrne, R.P. Linstead and A.R. Lowe, England, Journal of the Chemical Society, Article: Phthalocyanines. Part II. The Preparation of Phthalocyanine and some Metallic Derivatives from o-Cyanobenzamide and Phthalimide
	1932	J.F. Thorpe, R.P. Linstead and J. Thomas, Scottish Dyes Ltd., Scotland, U.S .Pat. 2.000.051, Metal Compound Containing an Isoindole Nucleus and Process of Making the Same	1934	R.P. Linstead and A.R. Lowe, England, Journal of the Chemical Society, Article: Phthalocyanines. Part III. Preliminary Experiments on the Preparation of Phthalocyanines from Phthalonitrile
	1932	I.M. Heilbron, F. Irving, R.P. Linstead and J.F. Thorpe, Imperial Chemical Industries Ltd., England, British Pat. 410.814, Manufacture of Colouring Matters from o-arylene Dicyanides (processes with phthalonitrile and precipitated copper; phthalonitrile and cuprous chloride)	1934	C.E. Dent and R.P. Linstead, England, Journal of the Chemical Society, Article: Phthalocyanines. Part IV. Copper Phthalocyanines
	1932	J.F. Thorpe and R.P. Linstead, Imperial Chemical Industries Ltd., Scotland, U.S. Pat. 2.000.052, Process for the Manufacture of Colored Compounds and the Product thereof	1934	R.P. Linstead and A.R. Lowe, England, Journal of the Chemical Society, Article: Phthalocyanines. Part V. The Molecular Weight of Magnesium Phthalocyanine
	1934	K. Holzach and G, Niemann, IG, Farbenindustrie AG., Germany, German Pat. 658.019, Verfahren zur Herstellung von Farbstoffen	1934	C.E. Dent, R.P. Linstead and A.R. Lowe, England, Journal of the Chemical Society, Article: Phthalocyanines. Part VI. The Structure of the Phthalocyanines
	1935	M. Wyler, Imperial Chemical Industries Ltd., England, British Pat. 464.126, Manufacture of Colouring Matters (urea-process)	1935	Imperial Chemical Industries Ltd., England, trade name: Monastral Fast Blue BS
	1935	G. Niemann, F. Mühlbauer, W. Schmidt and G. Wiest, IG. Farbenindustrie AG., German Pat. 717.164, Verfahren zur Herstellung von Farbstoffen der Phthalocyaninreihe	1935/36	I.G. Farbenindustrie AG., Germany, trade name: Heliogenblau B Pulver
	1936	M. Wyler, Imperial Chemical Industries Ltd., England, British Pat. 464.673, Manufacture of Colouring Matters (urea-process)	1937	E.I. du Pont de Nemours & Company,, USA, start of producing copper phthalocyanine blue
	1936	A. Riley, Imperial Chemical Industries Ltd., England, British Pat. 476.243, Improved Manufacture of Phthalocyanines	1937	*Winsor & Newton, England, artists' paints*
	1936	A. Siegel, E.I. du Pont de Nemours & Comp., USA, U.S. Pat. 2.173.699, Liquid Coating Compositions and Lakes therefor	1940	*Talens N.V., Netherlands, list of artists' paints, Rembrandtblauw*
	1937	A. Riley, Imperial Chemical Industries Limited, England, U.S. Pat. 2.214.477 Manufacture of Phthalocyanines (process of making phthalocyanine)	1949	E.I. du Pont de Nemours & Company, USA, introduction of the β-modification of phthalocyanine (Peacock shade)
	1940	M. Wyler, Imperial Chemical Industries Limited, England, U.S. Pat. 2.197.458 Process of Producing Coloring Matters of the Phthalocyanine Series	1950	*Talens N.V., Netherlands, list of artists' paints, Rembrandtblauw*
	1943	G.M. Neal, USA, British Pat. 633.713, Manufacture of Copper Phthalocyanine Dyestuffs	1953	BASF AG., West Germany, ß-modification phthalocyanine blue, trade name: Heliogenblau BG
	1946	F.H. Moser, The Standard Ultramarine Company, USA, U.S. Pat. 2.469.663, Process for Producing Metal Phthalocyanines and their Derivatives	1953/55	*Hans Schmithals, Gletscherbach*
	1948	R.H. Wiswall, American Cyanamid Company, USA, U.S. Pat. 2.486.351, Solvent Stable Metal Phthalocyanine Pigments and Methods of Making the Same	1958	*Fritz Winter, Federblüten (alpha-form)*
	1949	D.P. Graham, E.I. du Pont de Nemours & Company, USA, U.S. Pat. 2.556.728, Process of Producing Tinctorially Stable Phthalocyanine Coloring Matters (β-modification)	1967	*Barnett Newman, Who's afraid of Red, Yellow and Blue III*
	1949	D.P. Graham, E.I. du Pont de Nemours & Company, USA, U.S. Pat. 2.556.730, Process of Producing Tinctorially Stable Phthalocyanine Coloring Matters (β-modification)	c. 1975	*A.R. Penck, Young Generation (PB 15:4)*
	1959	F.L. Pfeiffer, American Cyanamid Company, USA, U.S. Pat. 3.051.721, Pigmentary Copper Phthalocyanine in "R" Form and its Preparation (patented in 1962)		

(continued)

Table 1 (continued)

Metal-Free Phthalocyanine Blue PB 16, 74100	1931	J.F. Thorpe, R.P. Linstead and J. Thomas, Scottish Dyes Ltd., Scotland, British Pat. 389.842, Improvements in and relating to the Production of Coloured Products	1907	A.V. Braun and J. Tcherniac, Germany, Berichte der Deutschen Chemischen Gesellschaft, Article: Über die Produkte der Einwirkung von Acetanhydrid auf Phthalamid
	1932	I.M. Heilbron, F. Irving, R.P. Linstead and J.F. Thorpe, Imperial Chemical Industries Ltd., England, British Pat. 410.814, Manufacture of Colouring Matters from o-arylene Dicyanides	1939	Imperial Chemical Industries Ltd., England, trade name: Monastral Fast Blue G
	1935	J.Y. Johnson, I.G. Farbenindustrie Aktiengesellschaft, Germany, British Pat. 457.526, Improvements in the Manufacture and Production of Valuable Pigment Dyestuffs	1939	I.G. Farbenindustrie Aktiengesellschaft, Germany, trade name: Heliogenblau G
	1936	M. Wyler, Imperial Chemical Industries Ltd., England, British Pat. 486.782, Improved Manufacture of Phthalocyanine Compounds	1962	*Georg Baselitz, Der Zwerg*
	1936	H. Dimroth, I.G. Farbenindustrie Akt.-Ges., Germany, German Pat. 696.334, Herstellung von metallfreien Phthalocyaninen	1964	*Ellworth Kelly, Blue Green Red* I
	1937	C.E. Dent, Imperial Chemical Industries Ltd., England, U.S. Pat. 2.214.454, Manufacture of Finely Divided Phthalocyanine Pigments		
	1948	W.L. Rintelman, E.I. du Pont de Nemours & Company, USA, U.S. Pat. 2.485.167 and 2.485.168, Manufacture of Metal-Free Phthalocyanine (Pigmentary β-form)		
	1949	O. Stallmann, E.I. du Pont de Nemours & Company, USA, U.S. Pat. 2.556.729, Metal-Free Phthalocyanine of Novel Color Characteristics (Pigmentary β-form)		

Green pigments	Discoverers and patents		Industrial introduction and *occurrences as an artists' paint*	
Phthalocyanine Greens				
Polychloro Copper Phthalocyanine Green PG 7, 74260	1935	R.P. Linstead and C.E. Dent, Imperial Chemical Industries Ltd., England, British Pat. 461.268, Manufacture of Halogenated Compounds	1936	Imperial Chemical Industries Ltd., England, trade name: Monastral Fast Green GS
	1935	G. Niemann, F. Mühlbauer, W. Schmidt and G. Wiest, IG. Farbenindustrie AG., German Pat. 717.164, Verfahren zur Herstellung von Farbstoffen der Phthalocyaninreihe	1938	I.G. Farbenindustrie AG., Germany, trade name: Heliogengrün G
	1936	G. Niemann, W. Schmidt, F. Muehbauer and G. Wiest, General Aniline & Film Corporation, USA, U.S. Pat. 2.276.860, Dyestuff of the Phthalocyanine Series	1939	*Winsor & Newton, England, artist's paints*
	1936	C.E. Dent and W.A. Silvester, Imperial Chemical Industries Ltd., British Pat. 478.256, Manufacture of Halogenated Phthalocyanine Useful as Pigments	1940	E.I. du Pont de Nemours & Comp. Inc., USA
	1936	R.P. Linstead and C.E. Dent, Imperial Chemical Industries Ltd., England, U.S. Pat. 2.214.469, Process for Manufacturing Halogenated Phthalocyanines	1940	*Talens N.V., Netherlands, list of artists' paints, Rembrandtgroen*
	1936	C.E. Dent and W.A. Silvester, Imperial Chemical Industries Ltd., England, U.S. Pat. 2.195.984, Halogenated Phthalocyanines	1950	*Talens N.V., Netherlands, list of artists' paints, Rembrandtgroen*
	1936	G. Wiest, W. Schmidt, G. Niemann and F. Mühlbauer, I.G. Farbenindustrie AG., Germany, German Pat. 740.053, Verfahren zur Darstellung von Farbstoffen der Phthalocyaninreihe	1950	*Max Beckmann, Frau mit Mandoline in Gelb und Rot*
	1936	I.G. Farbenindustrie Aktiengesellschaft, Germany, French Pat. 815.088, Procédé pour préparer des colorants	1961	*Morris Louis, Alpha Phi*
	1937	G.W. Johnson, IG. Farbenindustrie AG., Germany, British Pat. 474.740, Improvements in the Manufacture and Production of Dyestuffs		
	1937	A.L. Fox, E.I. du Pont de Nemours & Company, USA, U.S. Pat. 2.247.752, Process for Halogenation of Phthalocyanines		
	1943	A.L. Fox and K.C. Johnson, E.I. du Pont de Nemours & Company, USA, U.S. Pat. 2.377.685, Production of Chlorinated Phthalocyanines		

Chlorobromo Copper Phthalocyanine Green PG 36, 74265	1935	R.P. Linstead and C.E. Dent, Imperial Chemical Industries Ltd., England, British Pat. 461.268, Manufacture of Halogenated Compounds	1957	Harmon Colors, USA, trade name: Viridine-green
	1958	A. Kershaw and R. Fazackerley, Imperial Chemical Industries Ltd., England, British Pat. 850.237, Phthalocyanine Pigments	1959	E.I. du Pont de Nemours & Comp. Inc., USA
	1960	CIBA Société Anonyme, Switzerland, French Pat. 1.273.802, Procédé de préparation de phtalocyanines halogénées mixtes	1959	Introduction of Viridine green in the USA
			1960	BASF AG., Germany, trade name: Heliogrün 6G
			1961	ICI Ltd., England, trade name: Monastral Fast Green 6 YS
			1964	*Ellsworth Kelly, Blue Green Red I*
			1967	BASF AG., Germany, trade name: Heliogrün 8 GA
			1971-72	*Georg Baselitz, Seeschwalbe*
Other green pigments				
Pigment Green B PG 8, 10006	1884	Frankfurter Anilinfarbenfabrik, Gans & Co., Germany, German Pat. 28.065, Verfahren zur Darstellung grüner, brauner und gelber Farbstoffe durch Einwirkung von Metallen auf Nitrosonaphtolsulfosäuren	1885	O. Hoffmann, Germany, Berichte der Deutschen Chemischen Gesellschaft, Article: Ueber die Verbindungen der Nitroso-naphtole und Nitroso-naphtolsulfosäuren mit Eisen und Kobalt
	1921	Badische Anilin- & Soda-Fabrik, Germany, German Pat. 356.973, Verfahren zur Darstellung grüner Pigmentfarben		
	1925	C. Immerheiser and B. Zachimmer, Badische Anilin- & Soda-Fabrik, Germany, U.S. Pat. 1.529.891, Production of Pigment Colors		
	1935	D.E. Mac Queen, Krebs Pigment & Color Corporation, USA, U.S. Pat. 1.993.971, Quinone-Oxime Pigment		
	1937	A.A. Brizzolara and A. Siegel, E.I. du Pont de Nemours & Company, USA, U.S. Pat. 2.092.750, Process of Making Green Pigment		
	1938	E.R. Allen and G. Lutz, E.I. du Pont de Nemours & Company, USA, U.S. Pat. 2.127.374, Process of Making an Iron Compound of Nitroso-Beta-Naphthol		
	1941	S.C. Horning, E.I. du Pont de Nemours & Company, USA, U.S. Pat. 2.259.641, Nitroso-Beta-Naphthol Iron Pigment and a Process of Making It		
	1948	W.G. Huey and W.G. Drautz, General Aniline & Film Corporation, USA, U.S. Pat. 2.454.453, Pigment Green and Process for Producing the Same		
Nickel Azo Yellow PG 10, 12775	1904	Badische Anilin- und Soda-Fabrik, Germany, German Pat. 165.327, Verfahren zur Darstellung von Azofarbstoffen der Chinolinreihe (unmetalised dye)	1947	E.I. du Pont de Nemours & Comp. Inc., USA, trade name: Greengold-YT-562-D
	1905	E. Fussenegger, Badische Anilin & Soda Fabrik, Germany, U.S. Pat. 806.077, Azo-Coloring-Matter (unmetalised dye)	1963	Siegle & Co. GmbH., Germany, trade name: Grüngold Z.
	1905	British Pat.11.205 (unmetalised dye)	1963	BASF AG., Germany, trade name: Litholechtgelb 0830
	1941	D.E. Kvalnes and H.E. Woodward, E.I. du Pont de Nemours & Company, USA, U.S. Pat. 2.396.327, Metallized Azo Pigment (patented in 1946)		
	1942	D.E. Kvalnes and H.E. Woodward, E.I. du Pont de Nemours & Company, USA, U.S. Pat. 2.396.328, Monoazo Pigments (patented in 1946)		

(continued)

Table 1 (continued)

Violet pigments		Discoverers and patents		Industrial introduction and *occurrences as an artists' paint*
Quinacridones				
Quinacridone Violet PV 19, 46500	1955	W.S. Struve, E.I. du Pont de Nemours & Company, USA, U.S. Pat. 2.821.529, Process for the Preparation of Linear Quinacridones	1935	H. Liebermann, Germany, Justus Liebigs Annalen der Chemie, Article: Über die Bildung von Chinakridonen aus p-Di-arylamino terephtalsäuren
	1956	E.I. Du Pont de Nemours & Company, USA, British Pat. 828.052, Linear Quinacridone Pigments	1955	E.I. du Pont de Nemours & Company, USA
	1958	W.S. Struve, E.I. du Pont de Nemours & Company, USA, U.S. Pat. 2.821.530, Tetrahalogen Substituted Quinacridones	1965-70	*Royal Talens B.V., Netherlands, list of artists' paints, Rembrandt Oil Colours*
	1958	W.S. Struve, E.I. du Pont de Nemours & Company, USA, U.S. Pat. 2.821.541, Production of Dialkyl 2,5-Diarylamino-3,6-Dihydroterephthalates	1967	S.S. Labana and L.L. Labana, Chemical Reviews, Article: Quinacridones
	1958	W.S. Struve, E.I. du Pont de Nemours & Company, USA, U.S. Pat. 2.830.990, Organic Pigment	1974	*Royal Talens B.V., Netherlands, list of artists' paints, Rembrandt Oil Colours*
	1958	A.D. Reidinger and W.S. Struve, E.I. du Pont de Nemours & Company, USA, U.S. Pat. 2.844.484, Organic Pigments	1980	*Schmincke, Germany*
	1958	W.S. Struve, E.I. du Pont de Nemours & Company, USA, U.S. Pat. 2.844.485, Organic Pigments	1983	*Gerhard Richter, Bild 525, Prag*
	1958	C.W. Manger and W.S. Struve, E.I. du Pont de Nemours & Company, USA, U.S. Pat. 2.844.581, Organic Pigments	1991	*Schoenfeld, Germany*
	1958	E.I. du Pont de Nemours & Company, USA, British Pat. 851.976, Linear Quinacridone Pigments	1991	*Royal Talens B.V., Netherlands, list of artists' paints, Rembrandt Oil Colours, Rembrandtrose*
	1958	W.S. Struve, E.I. du Pont de Nemours & Company, USA, U.S. Pat. 3.009.916, Alpha-Phase Dihydroquinacridone	1991	*Royal Talens B.V., Netherlands, list of artists' paints, Van Gogh Oil Colours, Permanentrose*
	1960	H. Vollmann, W. Burneleit and W. Hohmann, Farbenfabriken Bayer Aktiengesellschaft, Germany, French Pat. 1.254.084, m-N,N'-quinacridones linéaires		
	1972	E. Greth, Lonza AG., Switzerland, German Pat. 2.313.329, Verfahren zur Herstellung von Succinylobernsteinsäurediester		
Dioxasines				
Dioxazine Purple PV 23, 51319	1928	G. Kränzlein, H. Greune and M. Thiele, I.G Farbenindustrie Akt.-Ges., Germany, German Pat. 517.194, Verfahren zur Darstellung von Farbstoffen	1928	G. Kränzlein, H. Greune and M. Thiele, Farbwerke Hoechst AG., Germany
	1928	I.G. Farbenindustrie Aktiengesellschaft, Germany, British pat. 387.565, Manufacture of Dyestuffs	1953	Farbwerke Hoechst AG., Germany, trade name: Permanent Violet RL
	1930	G. Kränzlein, H. Greune and M. Thiele, General Aniline Works, USA, U.S. Pat. 2.016.013, Dyestuffs of Oxazine Series and Their Production	1967	*Schmincke, Germany*
	1932	K. Thieβ and F.Maennchen, I.G Farbenindustrie Akt.-Ges., Germany, German Pat. 600.102, Verfahren zur Darstellung von Farbstoffen	1973	*Georg Baselitz, Straßenbauarbeiter*
	1932	G. Kränzlein, H. Greune and M. Thiele, I.G Farbenindustrie Akt.-Ges., Germany, German Pat. 606.672, Verfahren zur Darstellung von Farbstoffen	1974	*Royal Talens B.V., Netherlands, list of artists' paints, Rembrandt Oil Colours*
	1933	G. Kränzlein, H. Greune, W. Schultheiss and G. Langbein, I.G Farbenindustrie Akt.-Ges., Germany, German Pat. 637.020, Verfahren zur Darstellung von Farbstoffen	1991	*Royal Talens B.V., Netherlands, list of artists' paints, Rembrandt Oil Colours, Permanentviolet and Permanen troodviolet*
	1933	I.G. Farbenindustrie Aktiengesellschaft, Germany, British Pat. 387.565, Manufacture of Dyestuffs	1991	*Royal Talens B.V., Netherlands, list of artists' paints, Van Gogh Oil Colours, Permanent roodviolet and Permanent blauwviolet*
	1952	R. Sedlmayr, Farbwerke Hoechst AG., Germany, German Pat. 946.560, Verfahren zur Herstellung feinverteilter Pigmentfarbstoffe der Dioxazinreihe	1991	*Royal Talens B.V., Netherlands, list of artists' paints, Amsterdam Oil Colours, Violet*
	1958	J.H. Cooper, E.I. du Pont de Nemours & Comp. Inc., USA, U.S. Pat. 2.857.400, Pigment Production		
	1980	Th. Hufnagel and M. Hetschko, Cassella AG., Germany, German Pat. 3.010.949, Verbessertes Verfahren zur Herstellung eines violetten Pigmentfarbstoffs		

References

Berrie BH, Lomax SQ (1996/1997) Azo pigments: their history, synthesis, properties, and use in artists' materials. Studies in the history of art 57, Monograph series ll, Conservation Research 1996/1997, National Gallery of Art, Washington, University Press of New England, Hannover and London, pp 9–33

de Keijzer M (2011a) A choice of colour: modern synthetic inorganic artists' pigments, Restauratorenblätter Band 30, zum Thema: Kunst des 20. und 21. Jahrhunderts, Probleme und Perspektiven zur Erhaltung Klosterneuburg 2011, Verlag Stift Klosterneuburg, pp 33–42

de Keijzer M (2011b) Survey of modern synthetic inorganic pigments, discovered, manufactured and used as artists' pigments (Table l) Restauratorenblätter Band 30, zum Thema: Kunst des 20. und 21. Jahrhunderts, Probleme und Perspektiven zur Erhaltung Klosterneuburg 2011, Verlag Stift Klosterneuburg, pp 184–192

Gaertner H (1963) Modern chemistry of organic pigments. J Oil Colour Chem Assoc 46:14–45

Herbst W, Hunger K, Wilker G (2004) Industrial organic pigments: production, properties, applications, 3rd edn. WILEY-VCH Verlag GmbH & Co. KGaA., Weinheim

Lake S, Lomax SQ (2007) Arylide (Hansa) Yellow Pigments. In: Berrie BH (eds) Artists' Pigments: a handbook of their history and characteristics, vol 4, , National Gallery of Art, Washington, Archetype Publications, London, pp 179–222

Lewis PA (ed) (1988) Pigment handbook, vol l, 2nd edn, Properties and economics. Wiley, New York

Lutzenberger K (2009) Künstlerfarben im Wandel, Synthetische organische Pigmente des 20. Jahrhunderts und Möglichkeiten ihrer zerstörungsarmen analytischen Identifizierung, Herbert Utz Verlag, München

Modern Oil Paints – Formulations, Organic Additives and Degradation: Some Case Studies

Francesca Caterina Izzo, Klaas Jan van den Berg, Henk van Keulen, Barbara Ferriani, and Elisabetta Zendri

Abstract This study examines some of the degradation phenomena exhibited by modern oil paintings and twentieth century oil paint formulations and possible correlations with the various organic additives typically introduced by manufacturers. The research takes into consideration historical and modern tube oil paints produced by different European manufacturers and selected twentieth century oil paintings which showed degradation problems such as efflorescence, softening, cracks and sensitivity towards water and polar solvents. The composition of oil paints and paintings samples was studied with XRF, SEM-EDX, ATR-FTIR and GCMS. The results showed that both artists' oil paints and paintings samples contain a complex mixture of additives to the oil binders and pigments. The presence of aluminium and zinc stearates, added as dispersion agent gelling agents, was detected in most paint formulations of HKS, W&N, Talens and Maimeri (c. 1940-present). Several paint films showed an unusually high content of fatty diacids, suggesting that, in drying, an oxidative reaction was favoured above polymerisation. This occurrence, together with the diversity of lipidic media, additives, pigments and driers present in the industrial formulations could play a crucial role in paint failure and instability, as showed by the analysed case studies. The obtained results may help understanding the behaviour of modern oil paints and painting surfaces and contribute to improved conservation methods.

F.C. Izzo (✉)
Ca' Foscari University of Venice, Venice, Italy

Cultural Heritage Agency of the Netherlands, Amsterdam, The Netherlands
e-mail: fra.izzo@unive.it

K.J. van den Berg (✉) • H. van Keulen
Cultural Heritage Agency of the Netherlands, Amsterdam, The Netherlands
e-mail: k.van.den.berg@cultureelerfgoed.nl

B. Ferriani
Ca' Foscari University of Venice, Venice, Italy

Barbara Ferriani Srl, Milan, Italy

E. Zendri
Ca' Foscari University of Venice, Venice, Italy

K.J. van den Berg et al. (eds.), *Issues in Contemporary Oil Paint*,
DOI 10.1007/978-3-319-10100-2_5

Keywords Oil paints • Metal stearates • Dispersion agents • GCMS • Castor wax • Paint binder oxidation • Willem de Kooning • Karel Appel

Introduction

From the nineteenth century onwards, in relation to scientific and technological developments, artists' oil paints have increasingly been produced on a commercial scale. The formulation and quality control of these products shifted from the artist's studio to the realm of industry (Patton 1964; Mayer 1991; Carlyle 2001). Manufacturing oil paints changed significantly with the invention of metal tubes, which were introduced in the art market for the first time in 1841 by the British company Winsor & Newton. Paint manufacturers developed formulations that could be adapted to large-scale production processes (Patton 1964; Myers and Long 1968). A variety of vegetable oils were used in these new paint formulations; in addition to traditional drying oils (linseed, poppy seed and walnut oil), other drying, semi- and non-drying lipidic binders became employed, probably as inexpensive and readily available substitutes for walnut and poppy oil. Examples include safflower, sunflower, castor and cotton-seed oils (Mayer 1991; Doerner 1984). In contrast to the well known properties of traditional lipidic binding media (Sabin 1911; Wexler 1964; van den Berg et al. 1999; Erhardt et al. 2005; Rudnik et al. 2001), they present different properties and behaviour in drying processes and in the formation of pictorial films (Myers and Long 1968; Korus et al. 1984; Schönemann et al. 2006).

Commercial paints may also contain fatty acids of animal origin such as tallow; sometimes free fatty acids – not directly involved in the curing processes – were added as stabilisers. Other materials used as stabilisers are alumina and waxes or wax-like materials such as hydrogenated castor oil known also as castor wax (Mayer 1991; Doerner 1984; Gettens and Stout 1966). Free fatty acids or metal stearates, particularly aluminium, zinc and magnesium stearates were added for fast wetting and dispersion of the pigments, but also used as a gelling agent (Tumosa 2001). Driers, fillers and extenders were introduced, as well as natural resins, often as adulterants in order to produce cheaper paints (Mayer 1991; Doerner 1984). The additives are chosen to provide good stability by e.g. keeping the pigments in suspension and the right consistency and uniform handling properties of the paints.

However, modern oil paints are known to exhibit a higher occurrence of stability problems than traditional oil paints, such as phase separation in the paint, efflorescence, insufficient drying, appearance of drying cracks and the presence of efflorescence and water sensitivity.[1] (Burnstock et al. 2007; Mills et al. 2008; Tempest et al. 2013; Izzo et al. 2013). The presence of the various additives in modern formulations is unlikely to be completed unrelated to these degradation

[1]These proceedings.

phenomena, although other factors – such as the modification by artists of their paints – can also be a cause.

The present study is part of a larger research project dedicated to the nature of twentieth century artists' oil paints (Van den Berg and Burnstock 2014). The aims of the current paper are:

- to provide an overview of materials detected in a range of manufactured artists' oil paints;
- to present potential correlations between the formulations of twentieth century manufactured oil paints and the vulnerability observed in several well-bound oil paints from unvarnished artworks.

Commercial oil paints from different European manufacture brands dating c.1920s–2000s and selected paintings created between 1919 and 1963 were investigated using several analytical techniques. The identification of binding media and additives in paint samples was found to be an essential tool to an understanding of painting techniques and their degradation phenomena, as both of these aspects can play a role in the planning of preventive and sustainable conservation procedures.

Experimental

Analysed Paints

Artists' Oil Paints

Paints produced by Haagsche Kunstschilderverven Fabriek (HKS), Royal Talens and Old Holland, the Netherlands, Winsor & Newton (W&N), United Kingdom, Maimeri, Italy and Lefranc, France were studied. Taking into account the different availability and drying conditions of the paints, the samples ranged from liquid paint, fresh from the tube, to completely dried and cured from the paint-outs. A detailed description of the commercial oil paints is listed in Table 1.

Paints samples were taken from:

- liquid paints from the interior or dry paint leaked from the "historical" collapsible tubes from HKS, Royal Talens and Old Holland dating 1920–1960 collected from Dutch collections of oil paints[2] (Fig. 1a–c);
- paint-outs on canvas produced in 1964–1965 by W&N as part of set of paint swatches prepared by the manufacturer to ensure the quality of the paints;

[2]As described in S. Litjens, *Nederlandse Olieverftubes uit de Twintigste Eeuw*, dissertation MA, University of Amsterdam, 2010, the "historical" oil tubes from Talens, HKS and Old Holland come from the Historisch Museum Rotterdam, Chabot Museum Rotterdam, Mauritshius Den Haag and Collectie Old Holland were dated and catalogued.

Table 1 Name, manufactures, description and composition of analysed oil paint samples from "historical" collapsible tubes, paint-outs on canvas and cured paint-outs by HKS, Royal Talens, Old Holland, W&N, Maimeri, Le Franc

Brand	Paint labels	Production Period	Curing state/age/ application year	EDX/XRF	ATR-FTIR	GC-MS[a]	Conclusions
	Series						
HKS, The Netherlands	Goud Oker (Gold ochre)	c.1940	Dry paint leaked from the tube	Al, Si, S, K, Ca, Ba, Fe	Oil, barium sulphate, calcium carbonate, kaolin, iron oxide?, stearates?	Drying oil, P/S = 1.3, A/P = 5, stearates	Iron oxide in linseed oil with barium sulphate, calcium carbonate, kaolin and aluminium stearate
	Kobalt Violet (Cobalt Violet)	c.1940	Dry paint leaked from the tube	Al, P, Ca, Co	Oil, calcium carbonate, phosphates	Drying oil, P/S = 1.2, A/P = 8.1, stearates	Cobalt violet phosphate in linseed oil with calcium carbonate and aluminium stearate
	Monastraal Blauw (Monastral Blue)	c.1940	Dry paint leaked from the tube	Al, Si, S, Ca, (Ti), (Cr), (Fe)	Oil, gypsum, iron oxide?, stearates?	Drying oil, P/S = 1.5, A/P = 7.1, stearates	Phataloblue mixed with titanium white, chromium oxide, iron oxide and gypsum in linseed oil with aluminium stearates
	Chroom Groen (Chrome Green)	c.1940	Dry paint leaked from the tube	Al, S, Ca, Ba, Cr	Oil, barium sulphate, calcium carbonate	Drying oil, P/S = 1.7, A/P = 1.9, stearates	Chromium oxide in linseed oil with barium sulphate, calcium carbonate and aluminium stearate

	Smaragdgroen (Emerald Green)	c.1940	Dry paint leaked from the tube	Al, Ca, S	Oil, gypsum, alumina, beeswax?	Drying oil, P/S > 5, A/P = 3, stearates, odd numbered alkanes, 15-hydroxy-palmitic acid	Phtaloblue pigment mixed with gypsum and alumina in drying oil with beeswax and aluminium stearate
Royal Talens, The Netherlands	Karmijnlak (Carmine lacquer) Talens&Zoon	c.1920	Fresh paint from tube	/	/	Drying oil, P/S = 1.8, stearates	Linseed oil and stearates
	Permanentgroen (Permanent green) Talens&Zoon	c.1920	Fresh paint from tube	/	/	Drying oil, P/S = 1.5, stearates	Linseed oil and stearates
	Ultramarijn Donker (Ultramarine deep) Rembrandt Artists' Quality	c.1950	Dry paint leaked from the tube	Na, Al, Si, Zn, Pb	Oil, lead carbonate, silicates (ultramarine)	Drying oil, P/S = 1.4, A/P = 5.5, stearates	Ultramarine pigment mixed with zinc white and lead white in linseed oil and stearates
	Koningsblauw (Queen's Blue) Rembrandt Artists' Quality	c.1950	Fresh paint tube	Cl, K, Ti, Fe, Cu, Zn, Pb	Oil, Prussian blue pigment, lead carbonate, kaolin	Drying oil, P/S = 3.5	Prussian blue (potassium ferric ferrocyanide) mixed with zinc white, lead white and kaolin in poppyseed oil, unidentified chlorine compound

(continued)

Table 1 (continued)

Brand	Paint labels Series	Production Period	Curing state/age/ application year	EDX/XRF	ATR-FTIR	GC-MS[a]	Conclusions
	Cobaltblauw licht (Cobalt blue light) Rembrandt Artists' Quality	c.1950	Fresh paint from tube	Al, Ca, Fe, Co, Zn, Ba, Pb	Oil, lead carbonate, calcium carbonate, barium sulphate, iron oxide?	Drying oil, P/S = 3.4	Cobalt blue pigment mixed with iron oxide, zinc white and lead white in poppyseed oil with calcium carbonate and barium sulphate
	Zinkwit (Zinc White) Rembrandt Artists' Quality	c.1950	Dry paint leaked from the tube	Ca, Zn	Oil, calcium carbonate, zinc stearates?	Drying oil, P/S = 2.4, A/P = 1.4, stearates	Zinc white pigment in poppyseed oil and stearates with calcium carbonate
	Mangaanblauw Phtl (Manganese Blue) Rembrandt Artists' Quality	c.1950	Dry paint leaked from the tube	Al, Si, Ca, Mn, Ba	Oil, calcium carbonate, kaolin, stearates?	Drying oil, P/S = 1.8, A/P = 4.2, stearates	Organic blue pigment mixed with barium manganese oxide in linseed oil with calcium carbonate and stearates
	Groene Aarde (Green Earth) Rembrandt Artists' Quality	c.1950	Dry paint leaked from the tube	Ca, Ti, Cr, Fe	Oil, calcium carbonate, kaolin, stearates?	Drying oil, P/S = 2.5, A/P = 5.2, stearates	Iron oxide mixed with chromium oxide, titanium dioxide, calcium carbonate and kaolin in safflower oil[d] with stearates

	Cadmiumgeel Licht (Cadmium yellow light) Talens-Van Gogh	c.1950	Fresh paint from tube	Al, S, Si, Zn, Cd	Oil, kaolin, stearates?	Drying oil, P/S = 1.6, stearates	Cadmium sulphide mixed with zinc white and kaolin in linseed oil and stearates
	Lampenzwart (Lamp Black) Talens-Van Gogh	c.1950	Fresh paint from tube	Ca, P	Oil, calcium carbonate, Bone black?, stearate?	Drying oil, P/S = 1.3, stearates	Carbon black in linseed oil with stearates and calcium carbonate
	Ultramarijn Blauw (Ultramarine blue)[b] Rembrandt Artists' Quality	2000s	Cured paint-outs/dry brittle paint	Na, Al, Si, Ca, S, Mg	Oil, silicates (ultramarine), magnesium carbonate, gypsum, stearates?	Drying oil, P/S = 1.6, A/P = 2.1, stearates	Ultramarine blue pigment in linseed oil with magnesium carbonate, gypsum and aluminium stearate
	Cadmiumgeel (Cadmium yellow)[b] Rembrandt Artists' Quality		Applied in 2008, artificially aged ± 45 years	Al, S, Cd, Zn, Ti	Oil, aluminium stearate	Drying oil, P/S = 1.6, A/P = 3.4, stearates	Cadmium sulphide pigment mixed with titanium -and zinc white in linseed oil with aluminium stearate
	Chroomgroen (Chrome green)[b] Rembrandt Artists' Quality			Al, Ca, Cr	Oil, calcium carbonate, aluminium stearate, chromium oxide (?)	Drying oil, P/S = 1.6, A/P = 3.2, stearates	Chromium oxide green in linseed oil with calcium carbonate and aluminium stearate.

(continued)

Table 1 (continued)

Brand	Paint labels / Series	Production Period	Curing state/age/ application year	EDX/XRF	ATR-FTIR	GC-MS[a]	Conclusions
Old Holland, The Netherlands	Cadmiumrood (Cadmium red) Artists' oil colours	c.1960	Fresh paint from tube	Cd, Se, Zn	/	Drying oil, P/S = 1.6	Cadmium selenide pigment mixed with zinc oxide in linseed oil
	Rauwe Omber (Raw Umber) Artists' oil colours	c.1960	Fresh paint from tube	Mn, Fe	/	Drying oil, P/S = 1.5	Iron and manganese oxide in linseed oil
	Vermilijoen Extra (Vemilion Extra) Artists' oil colours	c.1960	Fresh paint from tube	Al, S, Hg	/	Drying oil, P/S = 1.4	Vermilion red pigment in linseed oil; unidentified aluminium compound
	Gebrande Omber (Burnt Umber) Artists' oil colours	c.1960	Fresh paint from tube	Mn, Fe	/	Drying oil, P/S = 1.5	Iron and manganese oxide in linseed oil
	Ultramarijn Blauw (Ultramarine blue)[b] Classic Oil Colour	2000s	Cured paint-outs/dry brittle paint Applied in 2008, artificially aged ± 45 years	Na, Al, Si, Ca	Oil, silicates (ultramarine), calcium carbonate	Drying oil, P/S = 1.3, A/P = 2.4, 12-hydroxy stearic acid	Ultramarine blue in linseed oil with calcium carbonate and castor wax

	Cadmiumgeel (Cadmium yellow)[b] Classic Oil Colour			Al, S, Cd, Zn	Oil, kaolin	Drying oil, P/S = 1.4, A/P = 2.6, 12-hydroxy stearic acid	Cadmium sulphide mixed with zinc white and kaolin in linseed oil with castor wax
	Chroomgroen (Chrome green)[b] Classic Oil Colour			Ca, Cr	Oil, calcium carbonate	Drying oil, P/S = 1.6, A/P = 2.8,12-hydroxy stearic acid	Chromium green in linseed oil calcium carbonate and with castor wax
Winsor & Newton, UK	Titanium White[c]	1964–1965	Paint-outs on canvas by W&N for the quality control/dry brittle paint	S, Ti, Zn, Ba	Oil, barium sulphate and zinc stearate?	Drying oil, P/S = 3.9 A/P = 11.5, stearates	Titanium dioxide mixed with zinc oxide and/or sulphide in poppy seed/safflower oil with barium sulphate and zinc stearates
	Student quality paints		Applied in 1965	(Co, Zr)			Possibly zirconium and cobalt salts added as driers.
	Burnt Sienna[c] Student quality paints			Mg, Si, Fe (Co, Zr)	Oil, magnesium carbonate, silica	Drying oil, P/S = 3.4, A/P = 14.3	Iron oxide in poppy seed/safflower oil with magnesium carbonate and silica. Possibly zirconium and cobalt salts added as driers.

(continued)

Table 1 (continued)

Brand	Paint labels / Series	Production Period	Curing state/age/ application year	EDX/XRF	ATR-FTIR	GC-MS[a]	Conclusions
	Raw Sienna[c] Student quality paints			Mg, Si, Fe (Co, Zr)	Oil, magnesium carbonate, silica	Drying oil, P/S = 1.8, A/P = 11.2	Iron oxide in linseed oil with magnesium carbonate and silica. Possibly zirconium and cobalt salts added as driers.
	Flake White[c] Student quality paints			S, Zn, Ba, Pb (Co, Zr)	Oil, barium sulphate	Drying oil, P/S = 3.4, A/P = 3.4	Lead white mixed with zinc oxide in poppy seed/safflower oil with barium sulphate
	Ultramarine Blue[c] Student quality paints			Na, Mg, Al, Si, S, K (Co, Zr)	Oil, silicates (ultramarine), magnesium carbonate	Drying oil, P/S = 5.9, A/P = 6.4	Synthetic ultramarine in poppy seed oil with magnesium carbonate and kaolin. Possibly zirconium and cobalt salts added as driers.
	Cobalt Blue[c] Student quality paints			Na, Mg, Al, Si, S, Ca, Ti, Zn, Ba (Co, Zr)	Oil, barium sulphate, magnesium carbonate, silicates	Drying oil, P/S > 5, A/P = 15.2	An organic pigment mixed with lead sulphate in cottonseed oil (?) with barium sulphate, magnesium carbonate and kaolin. Possibly zirconium and cobalt salts added as driers.

	Prussian Blue[c]			Na, Mg, Si, S, K, Ca, Zn, Fe	Oil, Prussian blue pigment, magnesium carbonate, calcium carbonate, silicates (ultramarine)	Drying oil, P/S = 2.4, A/P = 3.9	Prussian blue (potassium ferric ferrocyanide) mixed with artificial ultramarine in safflower oil with magnesium and calcium carbonate. Possibly zirconium and cobalt salts added as driers.
	Student quality paints			(Co, Zr)			
	1285 L (yellow)	1969	Cured paint-outs/dry brittle paint	Mg, S, Ca, Zn, Cd	Oil, magnesium and/or calcium carbonate	Drying oil, P/S = 1.3, A/P = 4.2	Cadmium sulphide mixed with zinc sulphide/oxide in linseed oil with magnesium and/or calcium carbonate
			Applied in 2004				
	1425 L (red)	1969	Cured paint-outs/dry brittle paint	Al, Mg, Se, Cd	Oil, magnesium carbonate, beeswax, stearates?	Drying oil, P/S = 1.8, A/P = 2.9, odd numbered alkanes, 15-hydroxy-palmitic acid	Cadmium selenide in linseed oil with magnesium carbonate, aluminium stearate and beeswax
			Applied in 2004				
	Cobalt blue	1969	Cured paint-outs/dry brittle paint	Al, Ca, Mg, Co	Oil, magnesium carbonate, aluminium stearate	Drying oil, P/S = 3.3, A/P = 2.4, steatates	Cobalt blue in safflower oil with magnesium carbonate and aluminium stearate
			Applied in 2004				

(continued)

Table 1 (continued)

Brand	Paint labels Series	Production Period	Curing state/age/ application year	EDX/XRF	ATR-FTIR	GC-MS[a]	Conclusions
	Permanent Blue	1969	Cured paint-outs/dry brittle paint	Na, Al, Si, S, Ca	Oil, silicates (ultramarine), stearate	Drying oil, P/S = 3.4, A/P = 2.7, stearates	Ultramarine blue in safflower oil with magnesium carbonate, aluminium stearates and beeswax
			Applied in 2004				
	Ultramarine blue (263)[b]	2000s	Cured paint-outs/dry brittle paint	Na, Al, Si, S, Mg	Oil, silicates (ultramarine), magnesium carbonate	Drying oil, P/S = 2.5, A/P = 1.5	Ultramarine blue in safflower oil/sunflower oil with magnesium carbonate
	Artists' Oil Colour		Applied in 2008, artificially aged ± 45 years				
	Cadmium yellow (108)[b]	2000s	Cured paint-outs/dry brittle paint	Al, S, Cd, Zn,	Oil, magnesium carbonate, barium sulphate, stearates?	Drying oil, P/S = 1.3, A/P = 4.1, stearates	Cadmium yellow in linseed oil with barium sulphate and magnesium carbonate and (aluminium) stearate
	Artists' Oil Colour		Applied in 2008, artificially aged ± 45 years				
	Chrome green (459)[b]	2000s	Cured paint-outs/dry brittle paint	Al, Mg, Cr	Oil, magnesium carbonate	Drying oil, P/S = 1.8, A/P = 2.3, stearates	Chromium oxide green in linseed oil with magnesium carbonate and (aluminium) stearates
	Artists' Oil Colour		Applied in 2008, artificially aged ± 45 years				

Maimeri, Italy	Titan White	2000s	Cured paint-outs/dry brittle paint	Si, S, Ti, Zn	Oil, silica, stearates?	Drying oil, P/S = 1.2, A/P = 3.9, stearates	Titanium dioxide pigment with zinc oxide/sulphide in linseed oil with silica and (zinc) stearate
	Extra fine artists' oil		Applied in 2008				
	Cadmium Red	2000s	Cured paint-outs/dry brittle paint	Si, Cd, Se	Oil, kaolin	Drying oil, P/S = 1.6, A/P = 3.5, stearates	Cadmium selenide in linseed oil with silicates and stearates
	Extra fine artists' oil		Applied in 2008				
	Raw Umber	2000s	Cured paint-outs/dry brittle paint	Al, Si, Fe, Mn	Oil, kaolin	Drying oil,/S = 1.6, A/P = 5.1	Iron and manganese oxide in linseed with kaolin
	Extra fine artists' oil		Applied in 2008				
	Mars Red	2000s	Cured paint-outs/dry brittle paint	Al, Si, Ca, Fe	Oil, calcium carbonate, kaolin	Drying oil, P/S = 1.8, A/P = 3.4, stearates	Iron oxide in linseed oil with kaolin, calcium carbonate and stearates (?)
	Fine artists' oil		Applied in 2008				
	Cadmium yellow	2000s	Cured paint-outs/dry brittle paint	Si, Ca, S, Cd	Oil, calcium carbonate, kaolin	Drying oil, P/S = 1.3, A/P = 3,5, stearates	Cadmium yellow pigment in linseed oil with kaolin, calcium carbonate and stearates
	Fine artists' oil		Applied in 2008				

(continued)

Table 1 (continued)

Brand	Paint labels	Production	Curing state/age/				
	Series	Period	application year	EDX/XRF	ATR-FTIR	GC-MS[a]	Conclusions
	Chrome Oxide Green	2000s	Cured paint-outs/dry brittle paint	Al, Si, Ca, Cr	Oil, calcium carbonate, kaolin	Drying oil, P/S = 1.1, A/P = 4.1, stearates	Chromium oxide green in linseed oil with kaolin, calcium carbonate and stearates
	Fine artists' oil		Applied in 2008				
	Prussian Blue	2000s	Cured paint-outs/dry brittle paint	Al, Si, Ca, Fe	Oil, Prussian Blue, calcium carbonate, kaolin, stearates?	Drying oil, P/S = 1.4, A/P = 2.3, stearates	Ferric ferrocyanide pigment in linseed oil with kaolin, calcium carbonate and stearates
	Fine artists' oil		Applied in 2008				
Le Franc, France	Sap Green	2000s	Cured paint-outs/dry soft paint	/	/	Drying oil, P/S = 1.6, A/P = 1.5, 12-hydroxy stearic acid, Phtalocyanine PG7?	Mixture of organic pigments (Copper Phtalocyanine, PG7?) and inorganic pigments (iron oxide?) in linseed oil
	Louvre oil colour		Applied in 2008				

	Orange	2000s	Cured paint-outs/dry soft paint	/	/	Drying oil, P/S = 3.2, A/P = 1.8, 12-hydroxy stearic acid, arylide yellow pigment (PY?), diarylide orange/red pigment?	Mixture of organic pigments in safflower and/or sunflower oil with castor wax
	Louvre oil colour		Applied in 2008				
	Yellow	2000s	Cured paint-outs/dry soft paint	/	/	Drying oil, P/S = 1.7, A/P = 1.5, 12-hydroxy stearic acid, arylide yellow pigment PY74	Organic yellow pigment (PY74) in linseed oil with castor wax
	Louvre oil colour		Applied in 2008				

[a]GC-MS results: the P/S ratios of drying oils were calculated on the extracted solution after the extraction method described in the Experimental; the presence of stearates was detected on the extracts when the P/S ratios of the residues were around 0.5–0.8; the A/P ratios were not calculated for fresh paints from the tubes

[b]Cured paint-outs s prepared in 2008 by Polly Saltmarsh and Klaas Jan Van den Berg at the RCE, Amsterdam; Artificial ageing of test paint films was carried out at Stichting Restauratie Atelier Limburg (SRAL), Maastricht. The total ageing time of 2,800 h is calculated to be equivalent to circa 45 years of exhibition in recommended museum conditions, 200 lx, 8 h a day, assuming reciprocity, 36 W Philips colour 96.5 fluorescent lamps with UV filtering (transmission 15 W/lm) (Mills L., 2009)

[c]Inorganic analysis on paint-outs on canvas by W&N were performed by G. Silvester, A. Burnstock and coworkers at CIA, London

[d]P/S ratio indicates a modern oil such as safflower; however the Talens archives show that around 1950, linseed oil was used for, most colours, while only poppyseed oil was used for white and blue pigments

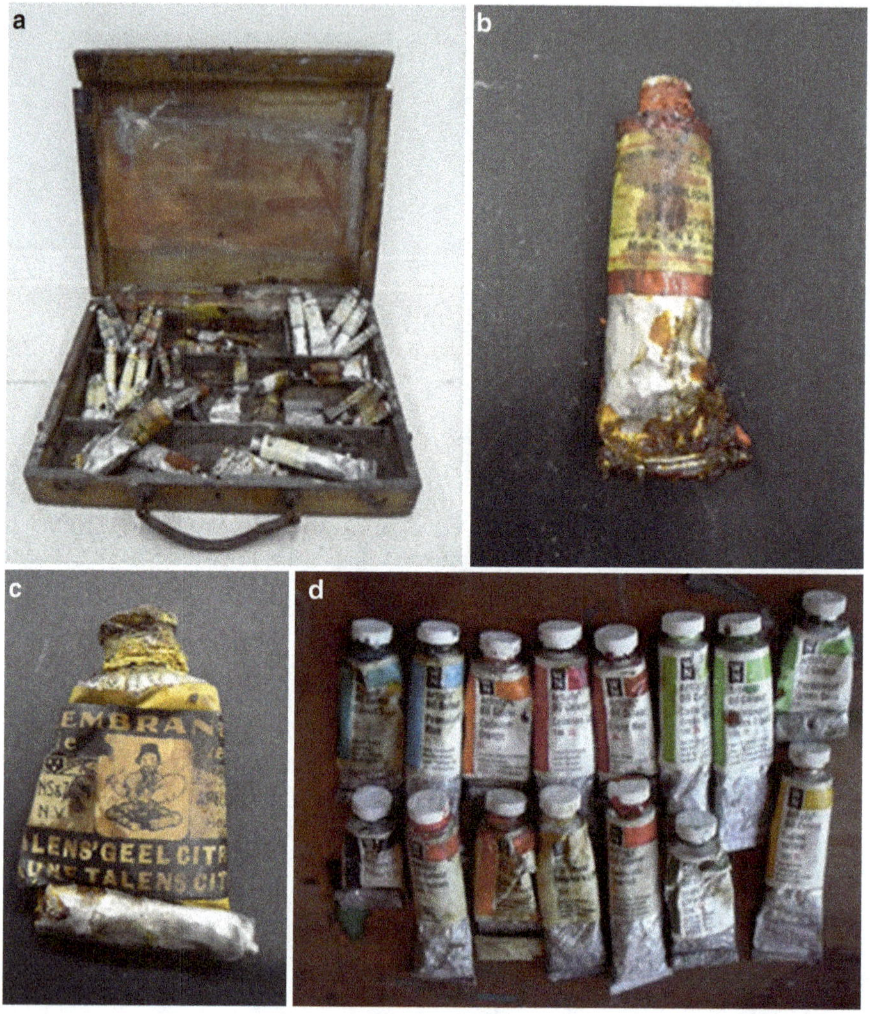

Fig. 1 (**a**) The art box of the Dutch conservator Roy Hesterman, containing various historical tube paints by Dutch manufactures (courtesy of RCE, Amsterdam); (**b**) Old Holland oil paint, 1960s; (**c**) Talens-Rembrandt oil paint, 1950s; (**d**) W&N oil paints, 1969

- paint-outs on Melinex® sheets of W&N student quality paints dating 1969, cured for 5 years (a picture of the tubes is reported in Fig. 1d);
- paint-outs on Melinex® sheets of modern (after 2000) Talens-Rembrandt, Old Holland, W&N, Maimeri and Lefranc paints, naturally or artificially cured (see descriptions and notes to Table 1).

Case Studies

Table 2 reports the works of art examined in this study. All are unvarnished oil paintings dating from 1919 to 1963 which were presented for investigation as part of conservation treatment. The selection of paintings showed several degradation phenomena, which will be discussed in the Results section and reported in the Table 2. Micro-fragments of paint were taken for analysis.

Instrumental Details and Experimental

Optical and Electronic Microscopy

Samples were investigated with an Olympus SZX16 optical microscope equipped for microphotography both in visible and UV light. Polished cross-sections of selected painting samples were prepared and observed with a JEOL JSM 5600 LV scanning electronic microscope coupled with an OXFORD-Link Isis series 300 microanalysis system (SEM-EDX).

X-Ray Fluorescence Spectroscopy

XRF analyses were carried out using a Philip Minipal spectrometer equipped with a low power Rh-tube operating at operating at 40–120 kV and 2.2 uA.

Fourier-Transform Infrared Spectrometry

Fourier-transform infrared spectrometry (FTIR) was performed on a double beam Nexus Nicolet IR spectrometer combined with an Endurance Golden Gate Single Reflection Diamond ATR (Attenuated Total Reflectance) accessory. Transmittance (%) was recorded in the range of 4,000–400 cm^{-1} in 64 scans with 4 cm^{-1} resolution. The elaboration of IR spectra was done with Omnic® 10 software.

Gas Chromatography-Mass Spectrometry

Before running GCMS analysis, paint samples were subjected to an extraction procedure especially designed for twentieth century oil paints and unvarnished painted surfaces, with the aim to separate oil components from metal stearates[3] (Izzo 2011). A mix of chloroform/methanol (2:8, v:v) was used to allow the separation of fatty acids from the oil binders and stearates and palmitates present

[3]The metal stearates are composed of stearates and palmitates and this complicates their distinction from stearic and palmitic acids present in drying oils.

Table 2 A short description of the analysed paintings and the observations about their state of conservation before restoration

Author	Title and collection (inventory nr.)	Image/detail	Year	Provenance (courtesy of)	Conservation problems
Emma Ciardi (1879–1933)	Ritorno del Colleoni (2139/4527)	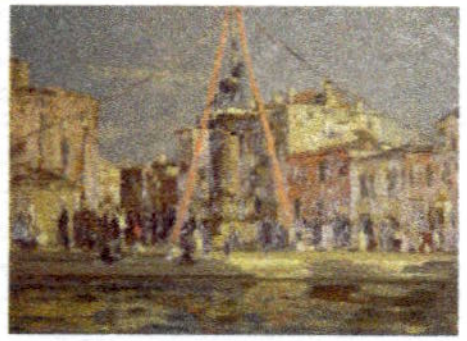	1919	International Gallery of Modern Art Ca' Pesaro, Venice	Widespread craquelures; whitish efflorescences; dirt deposition; water sensitivity
Karel Appel (1921–2006)	La grande Fleur de la nuit (24222)	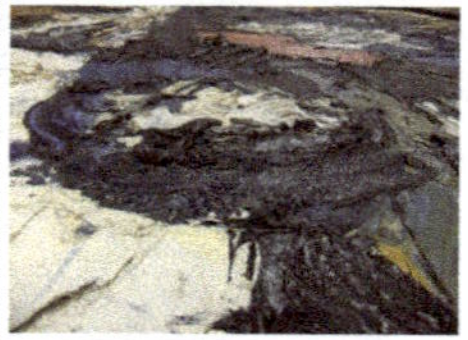	1954	Stedelijk Museum Amsterdam	Water sensitivity; dirt deposition; softness of painted layers
Lucio Fontana (1899–1968)	Concetto spaziale (60 O 81)	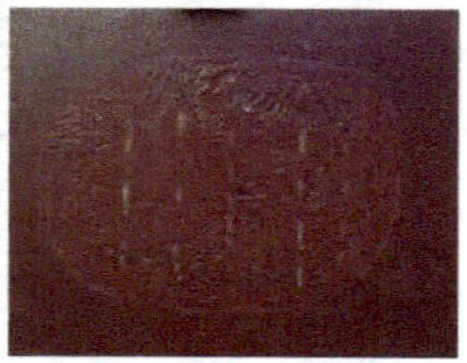	1960	Fondazione Lucio Fontana, Milan	Widespread craquelure, detachments and wrinkles on the pictorial layer; in some areas the paint is extremely brittle and fragile, while in others, it is still sticky and soft
Willem De Kooning (1904–1997)	Rosy-fingered dawn at Louse point (A22662)	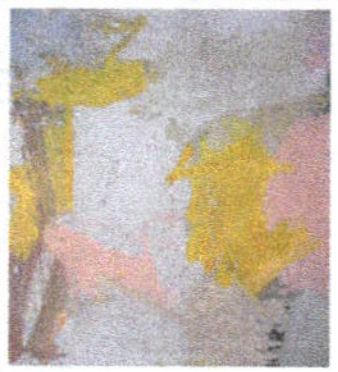	1963	Stedelijk Museum Amsterdam	Disfiguring layer of dirt; water sensitivity (especially in the yellow areas); softness and stickiness of painted layers

as added metal soaps. Extracts and solutions were treated with m-(trifluoromethylphenyl)trimethylammonium hydroxide, overnight reaction. This procedure enabled the simultaneous identification of lipidic materials, natural waxes and resins (Pitthard 2005; Sutherland 2007). Fatty acids from lipidic sources were detected as their corresponding methyl ester, while several peaks related to glycerol derivatives were identified as (trifluoromethyl)phenyl ethers (Sutherland 2007; Izzo 2011). Gas chromatography-mass spectrometry (GC-MS) analyses were performed on a Thermo Quest GC-8000 instrument with a Supelco Equity®-5 column, 30 m, i.d. 0.25 mm, ft 0.5 μm interfaced with MS MD-800. The inlet temperature was 300 °C; the temperature program was 120 °C to 300 °C with a ramp of 10 °C/min, then held at 300 °C for 3 min. The MS was run in Full Scan mode (m/z 40–600), 1.9 scans/s. Solvent delay was set at 4.5 min. The MS interface was at 240 °C and the source temperature was 220 °C. Electron Ionisation (EI) energy was 70 eV. Quantitative GC-MS analysis was performed using nonadecanoic acid as internal standard and calculating molar ratios among the most significant fatty acids (palmitic, stearic, azelaic acids).

Results and Discussion

Artists' Oil Paints

Table 1 reports the composition of the manufactured oil paints from analysis with EDX, XRF, ATR-FTIR and GCMS.

Drying Oils

Analysis of the artists' paints showed that all the examined paints were bound in lipidic media. The typical composition of a drying oil-based paint – in terms of saturated, unsaturated and difatty acids (Wexler 1964; Schilling et al. 1997; Van den Berg 2002) is displayed in Fig. 2. Table 1 reports, for every analysed paint, the indication of the different kind of oils used in the manufactured formulations. This identification was based on GCMS results, not only on the palmitate/stearate (P/S) ratios[4] but also considering the occurrence of any special fatty acids[5] (Colombini and Modugno 2009; Mills 1966; Mills and white 1996; Izzo 2011).

[4]P/S ratios alone cannot provide absolute proof of the kind of oil in modern formulations, since there are different contributions which may interfere (e.g. metal stearates, castor wax, semi drying oils, non-drying oils, etc).

[5]The presence of ricinoleic acid is indicative of the use of castor oil, while 11-eicosenoic (gondoic) acid and 13-docosenoic (erucic) acid are the biomarkers for rapeseed oil. In regard to sunflower oil, it is important to say that arachidic and behenic acids are not specific markers; they might be present in linseed oil for example. Nevertheless, in sunflower oil their weight percentage is higher

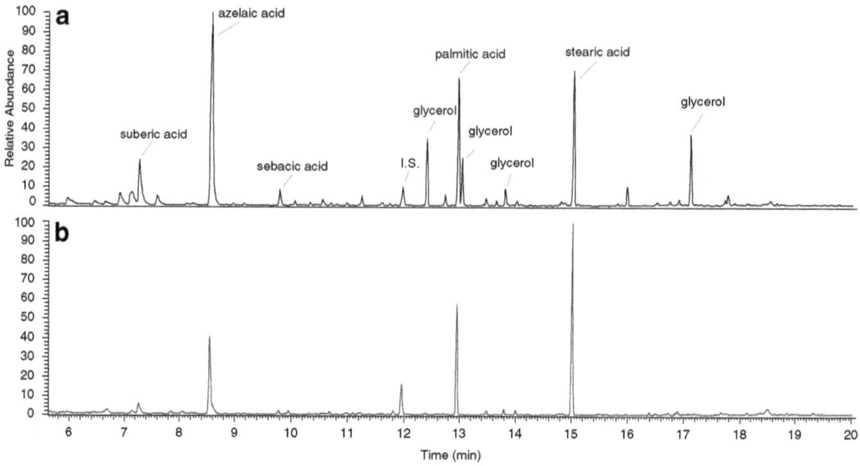

Fig. 2 Total ion current chromatogram of Ultramarijn Blauw (Ultramarine Blue) Rembrandt Artists' Quality, Talens, 2000: (**a**) TIC of the extracted solution; (**b**) TIC of the residue after extraction (see extraction method described in the Experimental section), in which a higher content of stearic acid compared to palmitic acid (P/S = 0.7) could indicate the presence of stearates added as dispersion agent gelling agents in paint manufacturing

Most of the paints analysed indicate the use of linseed oil, which is in accordance with paint manufacturers' catalogues. This oil was found in all colours from Old Holland and Maimeri. In general, white and blue pigments are bound in poppyseed, safflower and sunflower oils, most probably to avoid yellowing occurring in linseed oil during curing (Carlyle 2001; Mayer 1991). This is the case, for instance, of Queen's Blue, Cobalt Blue and Zinc White paints from the 1950s by Royal Talens and Titanium White, Flake White, Ultramarine and Prussian Blue paints dating 1964–1965 by W&N. The use of poppyseed oil is confirmed by recent research in the Talens' paint manufacturing archives. Safflower oil only replaced poppyseed oil in the late 1960s (Van den Berg et al. 2014).

Organic Additives

All the analysed commercial oils contain organic additives, except for "historical" Old Holland paints from the 1960s in which no evidence was found, (Table 1). The presence of the stabilising agent castor wax was detected in all Lefranc and Old Holland paints produced in the 2000s, on the basis of the presence of a specific marker, namely 12-hydroxy-octadecanoic acid.

---

than in linseed oil. The same consideration can be done for safflower oil, in which arachidic acid is present in a higher amount (Dubois et al. 2007).

Beeswax is a more traditional stabiliser. It was identified by GCMS due to the presence of 15-hydroxy-palmitic acid and odd-numbered alkanes in the Emerald Green paint by HKS and in the red 1425 L and Permanent Blue paints by W&N dating 1969.

Interestingly, the presence of aluminium and zinc stearates, added as dispersion gelling agents, was detected in most paint formulations of HKS, W&N, Talens and Maimeri (see Table 1 and Fig. 2), while they were not detected in Old Holland and LeFranc. These findings correspond to archival information from Talens and W&N as well as verbal information from Old Holland.[6] Their presence was detected in the residues after extraction (see Experimental) on the basis of their P/S ratios which were consistently around 0.5–0.8. These values are comparable to the P/S ratio of metal stearates which is generally between 0.4 and 0.7.

The detectability of metal stearates in old dry paints such as W&N is interesting as the hydrolysis of metal stearates seems to occur in oil paints and has been proposed as a potential explanation for water sensitivity exhibited by modern oil paintings (Burnstock et al. 2007; Mills et al. 2008). The fatty acids from the hydrolysed metal stearates would be detected as part of the extract rather than in the residue and lower the P/S ratio of the residue – this may explain the relatively low P/S ratios for poppyseed oil containing Old Holland and W&N paints.

Aged Tube Paints

Although the commercial paints were available in different states of drying and ageing, some suggestions can be made about ageing products on the paint-outs that were cured and aged under the same conditions. Noteworthy is the general presence in the W&N sketches dating 1964–1965 of extremely high contents of fatty diacids, in particular azelaic acid, compared to traditional well-bound oil paintings (Schilling and Khanjian 1996). In addition, the presence of oxidised fatty acids and diacids, such as 4-oxo-nonanoic, 4-oxo-decanoic and 4-oxo-undecanedioc acids is peculiar. Figure 3 reports the example of Raw Sienna paint from this W&N series. The detection of these oxidised diacids is reported here for the first time; unfortunately we cannot explain how these molecules were formed or whether they were present in the original paint already.

Table 1 reports the azelaic to palmitic acid (A/P) ratios which are all higher than 3.4: Cobalt blue paint has an A/P of 15.2. Interestingly, in addition to the sketches produced by W&N in 1964–1965, also other paints exhibit rather high content of oxidation products: Cobalt Violet and Monastral Blue by HKS, Manganese Blue and Green Earth by Royal Talens and Raw Umber and Chrome Oxide Green by Maimeri, all have A/P ratios higher than 5.

At this point the explanation is unknown but we could speculate that this might be linked to the presence of zirconium and cobalt salts added as driers, which

[6]Talens: see (Van den Berg et al. 2014); W&N: verbal communication Ian Garrett, W&N to Klaas Jan van den Berg, Jan. 2010; Old Holland: verbal communication W. van der Zwan to Klaas Jan van den Berg, April 2014.

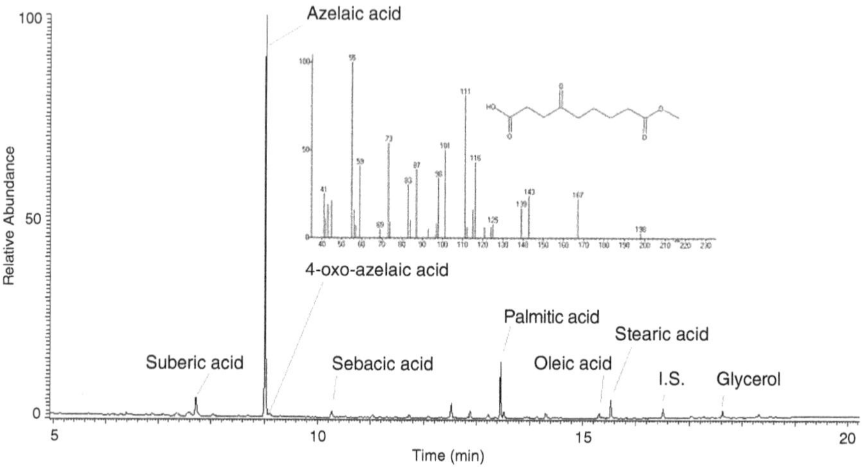

Fig. 3 Total ion current chromatogram of Raw Sienna Student quality paints, W&N, 1964–1965. The typical drying oil fatty acids (palmitic, stearic and oleic acids and glycerol in particular) are visible, together with the typical oxidation products (azelaic, sebacic and suberic fatty diacids). Interesting is the extremely high content of azelaic compared to palmitic acid (A/P = 11) and the presence of oxidised fatty acids and diacids, such as 4-oxo-azelaic acid whose corresponding mass spectrum (m/z) is reported in the Figure

were detected in the paint-outs (Silvester 2011), or the exposure to atmospheric sulphur dioxide (Silvester et al. 2014). Carboxylic diacids are the typical oxidation products of drying oils. Drying oils contain high portions of polyunsaturated fatty acids, which oxidise into hydroperoxides at the first stages of the drying process. The following decomposition of these hydroperoxides can arise by two different mechanisms; one will lead into polymerisation of the oil, the other in the formation of diacids (Mallégol et al. 2000; Šimůnková et al. 1985).

The excessive presence of the diacids in the analysed samples show that, in drying, an oxidative reaction was favoured above polymerisation.[7] Ageing products appear to influence the behaviour of painted layers in terms of water sensitivity, since fatty diacids are more polar than saturated and unsaturated fatty acids present in oils. There is likely a relation between the high content of diacids and the water sensitivity shown by these W&N paint-outs while cleaned and tested for sensitivity (Van den Berg and Burnstock 2014).

Pigments, Fillers and Extenders

XRF, EDX and ATR-FTIR identified the pigments used in oil paints such as the modern cadmium, chrome, zinc, titanium, chrome and cobalt-based pigments,

[7]Izzo, Van Keulen, Van den Berg, manuscript in preparation 2014.

commercially produced since the end of the nineteenth century (West Fitzhugh 1988). In addition, GCMS helped in detecting the presence of organic synthetic pigments, as for example the Phthalocyanine blue present in the Emerald green by HKF and the Hansa yellow in Lefranc oil paints (Table 1).

Although the analysed paints were not applied and aged the same way, some considerations about the influence of metals in pigments and driers on oils can be made. As seen in the previous paragraph, when lead, cobalt and manganese had been introduced, they have a catalytic effect on the polymerisation of the oil media (Tumosa and Mecklenburg 2005). All manufacturers adjust the drying rates of the paints by adding driers. The most representative case, as already mentioned, is the addition of zirconium and cobalt salts in W&N paints in the 1960s. The majority of the paints contain fillers and extenders: gypsum, aluminium-potassium silicates, kaolin, magnesium and calcium carbonate and barium sulphate were detected, especially in student quality oil paints (for example W&N student quality paints or Maimeri Fine artists' oils).

Zinc Oxide and Oleic Acid

Analysis of zinc white containing paint indicated a notably high oleic acid content in the zinc white paints by Talens-Rembrandt dating 1950s, and Titanium White by W&N dating 1964–1965. This oleic acid concentration was consistently higher than in paints of comparable age or drying time with other pigments. These findings are in line with literature (Rogala et al. 2010; Osmond 2011), in which the phenomenon is explained. In an oil-based medium zinc white forms a packed structure, which seems to be able to trap oleic acid present in the binder.

The presence of zinc white in modern paints is an important issue, since this pigment could influence the mechanical behaviour and stability of the paint layer (Mecklenburg 2007) and can react with fatty acids to form zinc carboxylates after hydrolysis of the fatty acid moieties (Robinet and Corbeil 2003; van der Weerd et al. 2005; Osmond 2014).

Case Studies

Emma Ciardi, *Ritorno del Colleoni*, 1919

Ritorno del Colleoni is a work of art by the Venetian artist Emma Ciardi characterised by bodied brushstrokes of pure and mixed colours, applied directly on a board. Being the daughter of a famous artist, Emma started using manufactured oil paints already in the 1910s (Zerbi and Esposito 2003). Although the general state of conservation was good, the painting exhibited widespread white efflorescence and whitish protrusions, particularly in the blue painted areas from the sky.

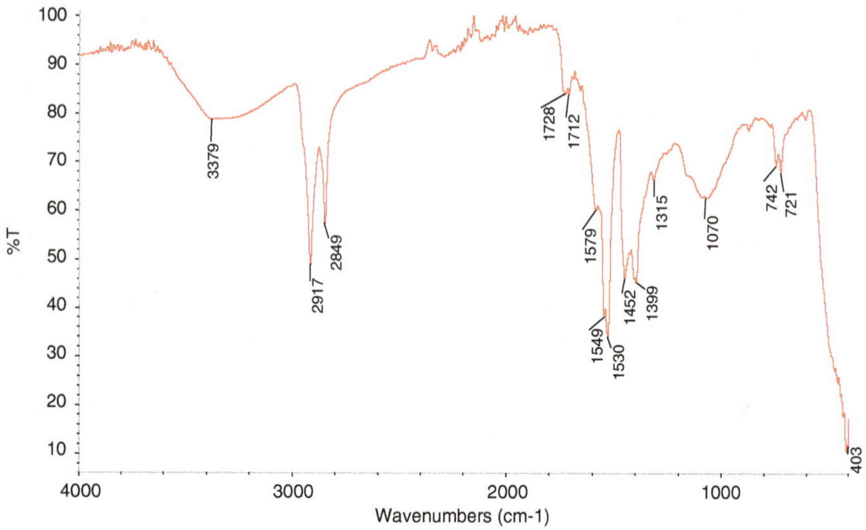

Fig. 4 ATR- FTIR spectrum of the whitish scraping from the sky area of Emma Ciardi, *Ritorno del Colleoni*, 1919. The typical IR zinc stearate absorption peaks at 1,549, 1,452 and 1,399 cm^{-1} are visible

XRF analysis indicated that zinc, likely zinc oxide, is present in the white ground layer, while SEM-EDX mapping could identify the presence of this pigment in the painted layers also in mixture with other inorganic pigments, such as green chrome oxide and yellow iron oxide. The sky was painted by mixing zinc oxide, chrome green and yellow ochre pigments. ATR-FTIR and GCMS analyses performed on the white paint from the ground showed the presence of drying oil. Most evident in the IR spectra of the whitish efflorescence scraped from the painting surface are the carboxylate absorption peaks at 1,549, 1,452 and 1,399 cm^{-1} (visible in Fig. 4). These peaks are possibly the result of the influence of zinc white in the drying and ageing of the oil medium as they could refer to the formation of metal soaps with the carboxylic acids present (Robinet and Corbeil 2003; van der Weerd et al. 2005; Osmond 2011).

In the gas chromatogram, among the characteristic compounds of linseed oil (P/S = 1.7), two uncommon fatty acids were detected: erucic (13-docosenoic) and gondoic (11-eicosenoic) acids and their oxidation products (13,14-dihydroxydocosanoic acid and 11,12-dihydroxyeicosanoic acid respectively). These unsaturated fatty acids are the bio-markers for *Brassicaceae* oils, such as rapeseed oil. The latter is a slow-drying oil introduced in oil paint manufacturing in the nineteenth century as a lubricant for paint production (Doerner 1984) or as a cheap adulteration (Van Keulen 2014). An unusually high content of oleic acid for an aged painting was registered: this could be linked to the presence of zinc oxide as previously seen for the manufactured tube paints (see 5.3.1 and Rogala et al. 2010).

The degradation phenomena observed on the painting surface might be the result of the presence of a slow-drying oil (rapeseed oil) containing a large amount of unsaturated acids (oleic, erucic and gondoic) which reacted with zinc white from the painted layers to form metal soaps, visible as efflorescence and causing protrusions in the painted layers (Keune 2005).

Karel Appel, *La Grande Fleur de la Nuit*, 1954

Karel Appel's painting *La Grande Fleur de la Nuit* is a complex overlapping of thick painted layers, characterised by both flat and impasto areas. Some areas showed water sensitivity[8]; two samples of paint were taken from these red and yellow areas. During sampling, a layer of dirt was observed and a certain softness of the painted layers was noticed.

The red paint was found to be composed of oil paint, containing a mixture of cadmium selenide sulphide pigment and Thioindigo Red B (also called Antinolo Red B, PR 88), lightened by the addition of lead white and zinc oxide. The high proportion of dicarboxylic acids (pimelic, suberic, azelaic, sebacic, etc.) and the occurrence of their oxidation by-products suggest that the lipidic binder used is a pre-treated oil, likely linseed oil since the P/S ratio is 1.8. In the paint formulation aluminium stearate dispersion agent as well as barium sulphate and calcium carbonate as extenders were detected.

The light yellow paint layer consists of an iron oxide earth pigment mixed with zinc white, calcium carbonate and barium sulphate extenders together with aluminium stearates. The oil binder is possibly linseed oil (P/S ratio of 1.5). Interestingly, he medium contains a high amount of oleic and linoleic acids suggesting that it was relatively unoxidised even after 60 years. This could be explained by the extreme thickness of the painted layers. This thickness is typical of Appel's style, which involves application of dense brushstrokes of colours, sometimes directly from the tubes. This creates many areas of thick impasto with a combination of numerous overlapping paints: thick layers have lower oxygen availability. The significant amounts of oleic acid may again partly be explained by the presence of zinc oxide (see above).

The stability problems of the painting seem to be related both to the paints formulations (including the presence of zinc oxide) and the artist's technique that led to the presence of a rather fresh organic layer could be the explanation of the extreme softness and dirt absorption of the painted surface.

[8]Wijnberg L., Soldano A., personal communication during conservation treatments, 2010.

Lucio Fontana, *Concetto spaziale*, 1960

After experimenting different traditional and synthetic binding materials, by the end of the 1950s Fontana started his own research on oil, the classical paint medium. As recently outlined (Gottschaller et al. 2012; Chiantore et al. 2012; Izzo et al. 2013), several oil paintings by Fontana revealed discrepancies in terms of both the physical-mechanical behaviour of the pictorial film and in alterations resulting from degradation processes. An example of these anomalies is presented by the red *Concetto spaziale,* dating 1960. This artwork presented many wrinkled areas, which were partially soft to the touch, fragile and very deformed, while in some areas the paint was still sticky. The red paint contains alizarin (1,2-dihydroxyanthraquinone) mixed with zinc white, calcium carbonate and barium sulphate, the latter two probably present as extenders. The red paint is bound in a mixture of linseed oil (P/S = 1.6) and castor oil as its analytical marker ricinoleic acid (12-hydroxy-9-cis-octadecenoic), was detected by GCMS analysis. The presence of this non-drying additive in a thick, bodied, layer of oil may be the explanation of the problematic drying and degradation phenomena evident in the painting (Erhardt et al. 2005; Izzo et al. 2013).

It may not be clear at first whether Fontana added castor oil by himself or that he bought a ready-made, linseed-castor oil mixed paint. Unfortunately, very little is known about the brands of oil paints he used, except the evidence that he used to buy commercial paints in local hardware suppliers (Ferriani et al 2009). Fontana, however, wanted his oil paintings to dry fast to be exposed in art galleries all around the world: this seems quite in contradiction with a conscious addition of a castor oil into the oil paint formulation and therefore it seems likely that the artist was not aware of the presence of this oil in the paint.

Willem de Kooning, *Rosy-Fingered Dawn at Louse Point*, 1963

Rosy-Fingered Dawn at Louse Point was created in 1963 by Willem de Kooning. The painting showed a significant, disfiguring layer of surface dirt and in some areas it was soft and sticky. During the conservation treatments in 2008 it was furthermore found that is was rather water-sensitive, especially in the yellow tacky passages (Tempest 2009).

A fragment of the yellow paint, containing cadmium zinc sulphide paint with barium sulphate, was analysed by GCMS and the corresponding chromatogram is reported in Fig. 5. The paint binder is a drying oil whose P/S ratio is 2.9, this value may refer to the use of safflower oil, the slow-drying oil used by De Kooning to prolong the workability of the painted surfaces (see below). Among the typical drying oil fatty acids, 12-hydroxy stearic acid was detected, likely associated with the use of castor wax, a stabilisation agent found also in other paint samples from the painting (Tempest 2009). The A/P ratio is 1.6 and suggests that oxidation into diacids was favoured in the drying process of the oil film, but the content in terms of fatty diacids is not so high to explain the sensitivity in terms of oxidised surface (as previously seen for the W&N sketches).

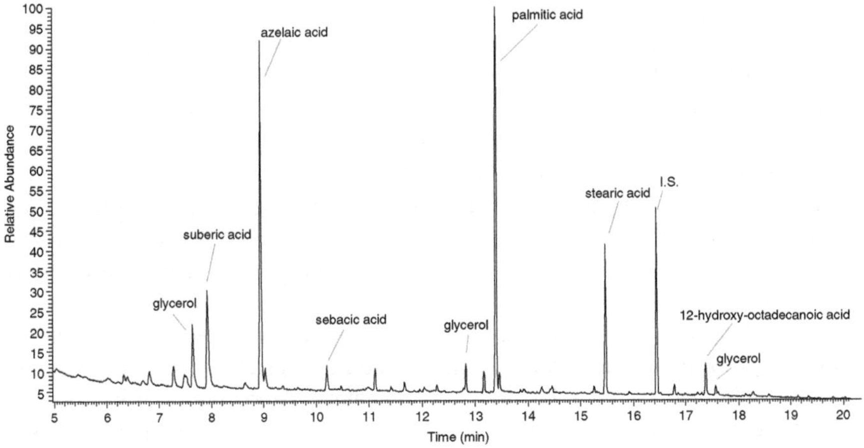

Fig. 5 Total ion current chromatogram of a yellow oil paint sample from Willem de Kooning, *Rosy-Fingered Dawn at Louse Point*, 1963. Among the typical drying oil fatty acids, the presence of castor wax, a stabilisation agent used by paints manufacturers, was detected thanks to the identification of 12-hydroxy-octadecanoic acid

For the blue paint, which contained synthetic ultramarine with calcium carbonate bound in oil (Tempest 2009), GCMS analysis outlined a P/S value of 4: this ratio is traditionally referred to poppy seed oil. However, the co-presence of safflower and/or sunflower cannot be excluded. In fact, significant amounts of behenic and arachidic acid were found in the oil paints, which might refer to these slow-drying oils commonly used by modern manufacturers. Furthermore, safflower oil, as previously seen, was used by De Kooning to modify commercial paints (Lake 1999; Lake et al. 1999). In this period, the artist was known to use linseed oil and castor oil, although he frequently mixed poppy and/or safflower oil and water into his paints; furthermore, from the 1964 until the middle of the 1970s safflower became his medium of choice (Lake 1999).

De Kooning was, indeed, an experimentalist: he wanted to achieve different effects by using radical paint formulations with extensive re-working of the painted canvases. Nevertheless, the addition of the slow-drying safflower oil could be related to the negative long-term stability of *Rosy-Fingered Dawn at Louse Point*, in which many passages remained sticky and soft even 50 years after its artistic production, absorbing dirt and atmospheric gases.

Conclusions

The results obtained by this study have provided much information regarding the formulations of commercial twentieth century oil paints by several oil paints manufacturers. These paints generally consist of rather complex formulations.

A variety of lipidic binders, dispersion agents, stabilisers, extenders, fillers and driers were detected: these data are indeed according with the specific literature and here confirmed for the first time on a wide set of manufactured paints.

Among the analysed paint-outs, the sketches produced in the 1964–1965 by W&N were found to have an unusually high content of fatty diacids, suggesting that, in drying, an oxidative reaction was favoured above polymerisation.

This occurrence, together with the use of specific lipidic media (in particular the semi- and non drying oils), pigments (zinc white) and driers present in the industrial formulations could play a crucial role in the vulnerability of the painted surfaces, as underlined by the selected case studies. The results have provided useful elements that facilitate a better understanding of the paintings materials used by artists studied in this paper– Emma Ciardi, Karel Appel, Lucio Fontana and Willem de Kooning. All four artists used commercial oil paints containing different lipidic binders, pure or in mixture: non-drying oils such as castor oil, semi-drying oils such as safflower and rapeseed oils and the traditional siccative binder, linseed oil.

Stability and durability of the painted surfaces is indeed also linked to the artists' interest and consciousness in selecting pictorial materials. Except for de Kooning, who declared to deliberately add safflower oil to commercial paints, it is not clear other artists wished to experiment with different kinds of oils in order to achieve different effects or if they simply used readymade paints without knowing their composition.

Acknowledgements The authors thank Dr. Louise Wijnberg from the Stedelijk Museum, Amsterdam, Dr. Matteo Piccolo from the International Gallery of Modern Art Ca' Pesaro, Venice and Fondazione Lucio Fontana, Milan for providing the interesting case studies for this research. Special thanks go to Dr. Luc Megens and BSc. Suzan de Groot of RCE for their assistance in performing XRF and FTIR-ATR analyses.

Prof. Dr. Aviva Burnstock from the Courtauld Institute of Art, London is acknowledged for providing analysis results of inorganic materials present in W&N paints from 1964 to 1965.

References

Burnstock A, van den Berg KJ, de Groot S et al (2007) An investigation of water-sensitive oil paints in the 20th century paintings. In: Learner TJS, Smithen P, Krueger J et al (eds) Modern paints uncovered. The Getty Conservation Institute, Los Angeles, pp 177–188

Carlyle L (2001) The artist's assistant: oil painting instruction manuals and handbooks in Britain, 1800–1900, with reference to selected eighteenth-century sources. Archetype Publications, London

Chiantore O, Ploeger R, Poli T, Ferriani B (2012) Materials and techniques in the pictorial oeuvre of Lucio Fontana. Stud Conserv 57:92–105

Colombini MP, Modugno F (2009) Organic mass spectrometry in art and archaeology. Wiley, New York

Doerner M (1984) The materials of artists and their use in painting. Harvest, San Diego/New York/London

Dubois V, Breton S, Linder M, Fanni J, Parmentier M (2007) Fatty acid profiles of 80 vegetable oils with regard to their nutritional potential. Eur J Lipid Sci Technol 109(7):710–732

Erhardt D, Tumosa SC, Mecklenburg MF (2005) Long-term chemical and physical processes in oil paint films. Stud Conserv 50:143–150

Ferriani B, Ploeger R, Poli T et al (2009) Lucio Fontana. Between tradition and innovation. In: Art d'aujourd'hui. Patrimoine de demain. Conservation et restauration des œuvres contemporaines, Paris

Gettens RJ, Stout GL (1966) Painting materials. A short encyclopedia. Dover Publications, New York

Gottschaller P, Khandekar N, Lee LF, Kirby D (2012) The evolution of Lucio Fontana's painting materials. Stud Conserv 57:76–91

Izzo FC (2011) 20th century artists' oil paints: a chemical-physical survey. PhD dissertation in Chemical Sciences, Ca' Foscari University of Venice

Izzo FC, Ferriani B, Van den Berg KJ, Van Keulen H, Zendri E (2013) 20th century artists' oil paints: the case of the Olii by Lucio Fontana. J Cult Herit. doi:10.1016/j.culher.2013.11.003

Keune K (2005) Binding medium, pigments and metal soaps characterised and localised in paint cross-sections. PhD dissertation, Amsterdam

Korus RA, Mousetis TL (1984) Polymerisation of safflower and rapeseed oils. J Am Oil Chem Soc 61(3):537–540

Lake S (1999) The relationship between style and technical procedure: Willem de Kooning's paintings of the late 1940s and 1960s. PhD dissertation thesis, University of Delaware

Lake S, Lomax SQ, Schilling MR (1999) A technical investigation of Willem de Kooning's paintings from the 1960s and 1970s. In: ICOM Committee for conservation preprints, vol 1. James, Lyon, pp 381–385

Mallégol J, Lemaire J, Gardette J (2000) Drier influence on the curing of linseed oil. Prog Org Coat 39:107–108

Mayer R (1991) The artist's handbook of materials and techniques, Revisited Edition. Viking, New York

Mecklenburg MF (2007) Meccanismi di cedimento dei dipinti su tela: approcci per lo sviluppo di protocolli di consolidamento. Il Prato, Padova

Mills JS (1966) The gas chromatographic examination of paint media. Part I. Fatty acid composition and identification of dried oil film. Stud Conserv 11:92–108

Mills SJ, White R (1996) The organic chemistry of museum objects, 2nd edn. Butterworth, London

Mills L, Burnstock A, Duarte F et al (2008) Water sensitivity of modern artists' oil paint. In: ICOM Committee for conservation, vol 2. Allied publishers, India, pp 651–659

Myers RR, Long JS (1968) Treatise on coatings, vol I. Marcel Dekker, New York

Osmond G (2011) Zinc white: a review of zinc oxide pigment properties and implications for stability in oil-based paintings. AICCM Bull 32:20–29

Osmond G (2014) Zinc white and the influence of paint composition for stability in oil based media. These proceedings, Chapter 18

Patton TC (1964) Paint flow and pigment dispersion. Wiley, New York

Pitthard V, Stanek S, Griesser M et al (2005) Gas chromatography–mass spectrometry of binding media from early 20th century paint samples from Arnold Schönberg's palette. Chromatographia 62:175–182

Robinet L, Corbeil MC (2003) The characterization of metal soaps. Stud Conserv 48:23–40

Rogala D, Lake S, Maines C, Mecklenburg M (2010) Condition problems related to zinc oxide underlayers: examination of selected abstract expressionist paintings from the collection of the Hirshhorn Museum and sculpture garden, Smithsonian Institution. J Am Inst Conserv 46: 96–113

Rudnik E, Szczucinska A, Gwardiak H et al (2001) Comparative *studies of oxidative stability of linseed oil*. Thermochim Acta 370:135–140

Sabin AH (1911) Linseed oil. J Ind Eng Chem 3:84–86

Schilling MR, Khanjian HP (1996) Gas chromatographic determination of the fatty acid and glycerol content of lipids. I. The effect of pigments and aging on the composition of oil paints. In: ICOM Committee for conservation preprints, vol 1. James & James, London, pp 220–227

Schilling MR, Khanjian HP, Carson DM (1997) Fatty acid and glycerol content of lipids; effect of ageing and solvent extraction on the composition of oil paints. Techne 5:71–78

Schönemann A, Frenzel W, Unger A et al (2006) An investigation of the fatty acid composition of new and aged tung oil. Stud Conserv 51:99–110

Silvester G (2011) An experimental investigation of the formation of sulphates in oil paint films exposed to gaseous sulphur dioxide with particular reference to the relationship between these sulphates and water sensitivity. Postgraduate Dissertation, Coutrauld Institute of Art, London

Silvester G, Burnstock A, Megens L, Learner T, Chiari G, van den Berg KJ (2014) A cause of water- sensitivity in modern oil paint films: the formation of magnesium sulphate. Stud Conserv 59(1):38–51

Šimůnková E, Brothánková-Bucifalová J, Zelinger J (1985) The influence of cobalt blue pigments on the drying of linseed oil. Stud Conserv 30:161–166

Sutherland KR (2007) Derivatisation using m-(trifluoromethyl)phenyltrimethylammonium hydroxide of organic materials in artworks for analysis by gas chromatography–mass spectrometry: unusual reaction products with alcohols. J Chromatogr A 1149:30–37

Tempest H (2009) Water-sensitive oil paint: an experimental investigation, with case studies, further characterizing the causes of the phenomenon. Post graduate diploma in the Conservation of easel paintings, Courtauld Institute of Art

Tempest H, Burnstock A, Saltmarsh P, van den Berg KJ (2013) Progress in the water sensitive oil project. In: Proceedings from of the cleaning 2010 conference, Valencia, 26–28 May 2010, Smithsonian Institute, pp 107–117

Tumosa CS (2001) A brief history of aluminum stearate as a component of paint. Waac Newslett 23:10–11

Tumosa CS, Mecklenburg MF (2005) The influence of lead ions on the drying of oils. Rev Conserv 6:39–47

Van den Berg JDJ (2002) Analytical chemical studies on traditional linseed oil paints. PhD dissertation, University of Amsterdam, Amsterdam

Van den Berg KJ, Burnstock A (2014) Twentieth century oil paint. The interface between science and conservation and the challenges for modern oil paint research. These proceedings, Chapter 1

van den Berg JDJ, van den Berg KJ, Boon JJ (1999) Chemical changes in curing and ageing oil paints. In: ICOM Committee for conservation preprints, vol 1. James, Lyon, pp 248–253

Van den Berg KJ, Bayliss S, Burnstock A, Van Gurp F, Klein Ovink B (2014) The Talens Archive. Presented at ATSR, Amsterdam; manuscript in preparation

van der Weerd J, van Loon A, Boon JJ (2005) FTIR studies of the effects of pigments on the ageing of oil. Stud Conserv 50:3–22

Van Keulen H (2014) Slow-drying oil additives in modern oil paints, and application in conservation treatments. ICOMCC (in press)

West Fitzhugh E (ed) (1988) Artists' pigments, a handbook of their history and characteristics. Archetype publications, London

Wexler H (1964) Polymerization of drying oils. Chem Rev 64:591–611

Zerbi M, Esposito M (2003) Emma Ciardi pittrice veneziana tra Ottocento e Novecento, Tra ombra e sole. Canova, Treviso

Distinction by Micro-Raman Spectroscopy and Chemometrical Analysis of Copper Phthalocyanine Blue Polymorphs in Oil-Based and Acrylic Paint Samples

Catherine Defeyt, Jolien Van Pevenage, Peter Vandenabeele, Tom Learner, and David Strivay

Abstract Copper phthalocyanine (CuPc) blue, commonly named phthalo blue is the most important synthetic organic blue pigment in the twentieth and twenty-first century artists paints. Phthalo blue, which is adopted by artists since 1936, is a polymorphous pigment. Currently, the α, β and ε-CuPc polymorphs are used in artists paint formulations. The identification of the CuPc crystal form provides technical and chronological information relevant for studying artworks. Raman Spectroscopy (RS) is a very valuable technique for the detection of phthalo blue in paint layers. However, the spectral interpretation is not straightforward concerning the CuPc polymorph distinction. To overcome the problem we have previously developed a procedure combining RS and chemometrical analysis. The experimental results that we obtained have demonstrated its efficiency for predicting the CuPc crystal form in unknown paint samples. In the present work, this procedure was applied on oil-based and acrylic paints from Sam Francis' studio and the Getty Conservation Institute (GCI) Reference Collection.

Keywords Copper phthalocyanine polymorphs • Raman spectroscopy • Chemometrical analysis • Modern paints

---

C. Defeyt (✉) • D. Strivay
Centre Européen d'Archéométrie, Institut de Physique Nucléaire, Atomique et de Spectroscopie, Université de Liège, B-4000 Liège, Belgium
e-mail: Defeyt.Catherine@hotmail.com

J. Van Pevenage
Raman Spectroscopy Research Group, Department of Analytical Chemistry, Ghent University, Krijgslaan 281, B-9000 Ghent, Belgium

P. Vandenabeele
Archaeometry Research Group, Department of Archaeology, Ghent University, Sint-Pietersnieuwstraat 35, B-9000 Ghent, Belgium

T. Learner
Science Department, Getty Conservation Institute, 1200, Getty Center Drive, Los Angeles, CA-90049, USA

K.J. van den Berg et al. (eds.), *Issues in Contemporary Oil Paint*,
DOI 10.1007/978-3-319-10100-2_6

Introduction

Copper phthalocyanine (CuPc) blue, commonly named phthalo blue is the most important synthetic organic blue pigment used in contemporary artists' paint formulation. Almost half of the blue oil-based paints contains the actual blue (Defeyt 2013). Commercially available since 1935 (New York Times 1935), phthalo blue has been adopted by worldwide famous twentieth century artists, such as Barnett Newman (De Keijzer 2002), Vassily Kandinsky (Dagron 1998) and Jackson Pollock (Khandekar et al. 2010). The phthalo blue identification in artworks contributes to the knowledge of artists' techniques and, at the same time, provides chronological information for their dating, which may reveal anachronisms, and therefore, unveil forgeries. For example, because they were supposedly made before the pigment appearance, fake paintings attributed to James Ensor (De Keijzer 2002), Max Ernst and Fernand Léger (Rohrs et al. 2012) were unmasked thanks to CuPc blue detection.

CuPc blue ($C_{32}H_{16}CuN_8$) is an organometallic compound belonging to the class of polycyclic pigments and exhibiting polymorphism. The α-, β- and ε-CuPc polymorphs are industrially produced. The Colour Index Generic Name of the CuPc blue polymorphs is Pigment Blue 15 (PB15) (Colour Index Dyes and Pigments 1971). Currently, PB15:0, the unstabilized α-type, PB15:1, the non-crystallizing α-type, PB15:3, the unstabilized β-type, PB15:4, the non-floculating β-type, and PB15:6, the unstabilized ε-type, are widely used by artists' color makers (Defeyt 2010).

These five pigments are characterized by different properties such as hue and solubility (Bernstein 2002). The α-, β- and ε- crystal forms were patented respectively in 1928 (Moser and Arthur 1963), 1949 (Wiswall 1949) and 1962 (Pfeiffer 1962). Consequently, the identification of the CuPc crystal form may provide supplementary technical and chronological information useful for getting a better knowledge of an artist's technique as well as for dating and authenticating works of art. However, for most of the cases of CuPc identification from works of art, artists' studio-materials and forgeries, which are reported in literature, the crystal form remains unspecified. And when it is specified authors did not give any explanation about how they achieved to the polymorphic identification (Defeyt 2013).

Recently, it has been shown (Defeyt et al. 2012) that the combination of X-Ray Diffraction (XRD), Raman spectroscopy (RS) and Fourier-Transform Infra-Red spectroscopy (FTIR) allow us to discriminate between the α-, β- and ε-CuPc in paint layers. However, in some specific cases the identification of the CuPc crystal form may not be achieved in this way. Indeed, in the case of oil-based paints FTIR results cannot be taken into account because the IR bands of the CuPc and the binder are overlapping (van der Weerd et al. 2005). Also, in case of very complex mixtures or paint mixtures very poor in phthalo blue, the detection of the blue pigment by the XRD and FTIR methods may become very complicated because the CuPc signals are too weak or absent (Brostoff et al. 2009). Inversely, RS remains efficient for the phthalo blue identification in paint systems, irrespective to the CuPc concentration or the nature of the binding medium (Defeyt et al. 2012).

Unfortunately, the spectral interpretation is not straightforward, and the Raman data treatment requires some improvements for discriminating the CuPc polymorphs. That is why we have recently developed a procedure combining RS and chemometrical analysis (Defeyt et al. 2013), namely Linear Discriminant Analysis (LDA) (Duda et al. 2000). While RS is a molecular analytical technique, LDA is a supervised classification method that requires a training set, e.g. a set with spectra of well-known nature. The aim of supervised classification methods is to find a function for allocating a new object of unknown nature to the correct group (Miller and Miller 2005).

In the procedure developed for the CuPc polymorphs distinction, a training set of 150 Raman spectra recorded on well-known α-, β- and ε-CuPc samples and, 12 intensity ratios of selected Raman bands used as variables, are employed for allocating new Raman spectra recorded for an unknown phthalo blue sample to the correct crystal form.

From our experimental results, the percentage of correct classification obtained in this way is 94 % (Defeyt et al. 2013).

In the present work this procedure was applied to characterize the phthalo blue pigments favored by the worldwide famous American twentieth century painter, Sam Francis (1923–1994). Three blue paints from Sam Francis' studio have been analyzed. Additionally, 14 phthalo blue-based paints from the Getty Conservation Institute (GCI) Reference Collection have been characterized by the RS and LDA procedure, XRD and FTIR. All the samples investigated in this study have been kindly supplied by the GCI. Information on the studied samples is given in Table 1.

Case Studies

Paints from Sam Francis' Studio

Sam Francis' painting exhibits a very sensitive approach of the colors. The brightness and the transparency of the blues achieved by Francis remain an original part of his work (Ruhrberg et al. 2005). In 2008, Bouchard et al. (2009) analyzed the paints from Sam Francis's studio in Santa Monica by ATR-FTIR, RS and SEM-EDX in order to identify the binders and the pigments present in the mixtures.

Among the paint pots left in his studio after his death, the authors found out by means of RS that the blue paints of GCI codes SF18, SF42 and SF60 (Fig. 1) are based on phthalo blue. For each of them a pEA/MMA (ethacrylate/methylmethacrylate) binder has been identified (Bouchard et al. 2009). These three paints are described in Table 1. Here, to obtain further information on the phthalo blue pigments preferred by Francis, XRD and new Raman measurements

Table 1 Information on the 17 paint samples investigated in this work. All the studied paints have been kindly supplied by the Getty Conservation Institute (GCI), Los Angeles. GS stands for Green Shade and RS for Red Shade

Paint reference	Range and brand	Binding media	Acquisition date
Paints from Sam Francis' studio			
SF18	Studio-made paint	Acrylic	1997
SF42	Studio-made paint	Acrylic	1997
SF60	Studio-made paint	Acrylic	1997
Paints from the GCI Reference Collection			
Phthalo Blue GS1	Golden Acrylic	Acrylic	2003
Phthalo Blue RS1	Golden Acrylic	Acrylic	2003
Phthalo Blue1	Perfect Acrylic, Creative Product Corp.	Acrylic	1976
Phthalo Blue2	Acryl. Art. Col., Liquitex	Acrylic	1990
Phthalo Blue3	Acryl. Art. Col. Prof., Liquitex	Acrylic	2003
Phthalo Blue4	Art. Oil Col., Acrylex Tinta Espec.	Oil-based	1992
Phthalo Blue5	Marie's Oil Colours	Oil-based	Unspecified
Phthalo Blue GS2	Orig. Oil Col., Daniel Smith	Alkali refined linseed oil	2006
Phthalo Blue RS2	Orig. Oil Col., Daniel Smith	Alkali refined linseed oil	2006
Phthalo Blue6	Oil Col. for Artists, Dilo	Oil-based	1992
Phthalo Blue7	Oil-based Water Miscible, Royal Talens	Oil-based	2005
Phthalo Blue RS3	Artisan Water Mixable Oil Col., Winsor & Newton	Modified linseed oil	2003
Phthalo Turquoise	Orig. Oil Col., Daniel Smith	Alkali refined linseed oil	2006
Iridescent Blue	Luminescent Oil Col., Daniel Smith	Alkali refined linseed oil	2006

Fig. 1 Paint films of SF60 (*left*), SF18 (*center*) and SF42 (*right*), from Sam Francis' studio. The three paints were applied on glass slides for further XRD and Raman measurements

have been performed on SF18, SF42 and SF60 paints (applied on standard microscope glass slides). LDA has been performed only on the Raman spectra acquired from phthalo blue pigment grains.

Paints from the GCI Reference Collection

Started in the early 1990s, the GCI Reference Collection currently contains more than 13,000 artists' reference materials that provide a valuable resource for conservation research. In the present work, the phthalo blue-based paints from this major collection have been investigated in order to identify the CuPc crystal form used in their formulation.

The artists' paints used in the present work, dating between 1976 and 2006, include five acrylic paints and nine oil-based paints. The GCI artists' paint samples are described in Table 1. For 10 paints out of 14, the type of the CuPc pigment used in the formulation is not specified by the producer. Here, the 14 paints have been applied on microscope glass slides directly from the tubes. All of them have been analyzed by XRD, μ-Raman spectroscopy (μ-RS) in combination with LDA and attenuated total reflectance Fourier transform infrared spectroscopy (ATR-FTIR).

Experimental

All the samples were analyzed by μ-RS. Raman analyses were performed with a Bruker Senterra spectrometer by using a diode laser with a wavelength of 785 nm. The laser power was reduced at 10 % ($\sim$4.22 mW at the sample). Three spectra from different pigment grains were recorded in the spectral range of 80–2,660 cm^{-1} by using a 50X objective. Raman spectra were recorded with an accumulation time of 30 s. Every Raman spectrum has been manually baseline corrected using the Grams software. Accordingly to the procedure described in Defeyt et al. (2013) LDA using the leave-one-out classification method (each spectrum included in the analysis is classified according to the functions derived from all spectra other than that spectrum) and intensity ratios as variables, was performed with the SPSS software (George and Mallery 2001).

The CuPc crystal form of the unknown paint spectra was predicted by including 150 well-known spectra in the analysis (Defeyt et al. 2013).

XRD measurements were performed with a Bruker D8 Focus diffractometer equipped with a Cu tube (K_α 0.154056 nm). A step time of 1s and an angular step of 0.04° were used, the range of the 2θ was between 3° and 50°.

The IR spectra were recorded in the 4,000–650 cm^{-1} range. The spectrometer is a Bruker ALPHA provided with an ATR Ge crystal. The spectral resolution is 4 cm^{-1}.

The FTIR spectra were recorded with the automatic suppression of CO_2 and H_2O vapor bands.

Results and Discussion

Paints from Sam Francis' Studio

After performing LDA on the Raman spectra recorded for the SF60 and SF18 paint samples we obtained the following results: SF60 contains a β-CuPc blue pigment and SF18 is based on the α-CuPc polymorph. As shown in Fig. 1, the SF42 dried paint film exhibits a light turquoise blue shade and is slightly transparent. For this paint sample our Raman measurements did not allow us to detect the presence of a phthalo blue pigment. Actually, the examination of the three Raman spectra recorded for SF42 leaded to the identification of a mixture combining a polychloro-CuPc green (PG7), Cobalt blue (PB28) and titanium dioxide. The PG7 identification is based on Raman data reported in literature (Lutzenberger 2009; Scherrer et al. 2009). Because phthalo blue has not been detected from the SF42 paint film, the related spectra were not investigated by LDA. In addition, the XRD measurements performed on the SF42 sample have confirmed the cobalt blue and titanium dioxide identification but did not indicate the presence of any blue or green CuPc pigment. The XRD, RS and LDA results obtained for the Francis' paints are given in Table 2.

The SF60 dried paint film is completely opaque and shows a dark green shade (Fig. 1). Two out the three Raman spectra recorded for SF60, leaded to the phthalo blue identification. For both of them the LDA classification is β. LDA was not performed on the third spectrum because it pointed out the presence of only PG7.

As shown in the Fig. 2, in the SF60 diffraction pattern it is quite easy to recognize the two strongest peaks induced by a β-CuPc blue pigment. The related experimental interplanar distances, $d = 12.91$ and 9.8, indicate a β-form (Brown 1968; Robertson 1935). XRD results then confirm the LDA classification. Additionally the XRD measurements allowed the identification of chalk ($CaCO_3$) and baryte ($BaSO_4$) in the mixture. The presence of PG7 is supported by the diffraction peaks observed at $d = 14.94$ and 3.34. The relative intensities of these peaks are respectively 27 and 52 %. The diffraction peak at $d = 3.34$ could also be attributed to $BaSO_4$ but its relative intensity is much more in accordance with the strongest peak reported for PG7 (Lomax 2010). Although phthalo blue in the β-form is the greenest in shade (Herbst et al. 2004), the SF60 greenish blue shade mainly results from the presence of PG7 in the mixture.

Table 2 The Colour Index Generic Names of the pigments identified by RS in SF18, SF60 and SF42. For SF18 and SF60, the CuPc crystal form identified by XRD confirms the LDA classification

Paint reference	Pigment(s) identified by RS	PB15 type identified by LDA	PB15 type identified by XRD
SF18	PB15	α	α
SF42	PG7, PB28, PW6	–	–
SF60	PB15, PG7	β	β

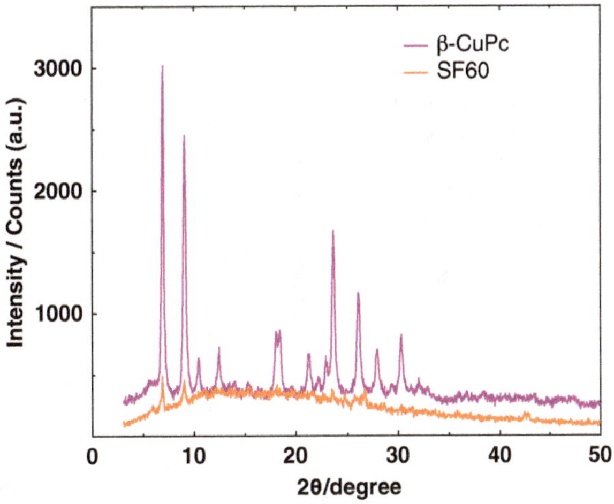

Fig. 2 SF60 diffractogram and diffraction pattern obtained for a dry pigment of β-CuPc (Heliogen 23060 from Kremer). The position of the CuPc diffraction peaks appearing in the SF60 diffractogram totally fits with the β-crystal form

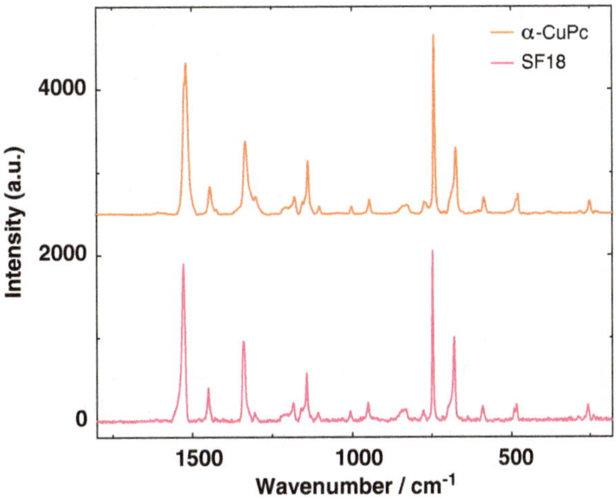

Fig. 3 Raman spectra (1,800–100 cm^{-1}) with baseline correction obtained for the SF18 paint sample and for a phthalo blue dry pigment sample supplied by Kremer (Heliogen 23050). All the Raman bands displayed in the dry pigment spectrum arise in the SF18 spectrum

The blue hue of the SF18 paint (Fig. 1) looks like the bright and transparent blue hue characterizing various Sam Francis' paintings from the early 1960s, such as *Mantis* and *Blue Balls VIII*, both from 1961 to 1962. Phthalo blue has been successfully identified from the three Raman spectra recorded on the SF18 sample. Actually, in the three SF18 spectra, all the Raman bands reported for dry pigments of PB15 can be observed (Fig. 3). For each of them the LDA classification indicates the α-form. The experimental interplanar distances recorded at 13.38 and 12.43 for the two strongest PB15 diffraction peaks displayed in the SF18 diffractogram (Fig. 4) completely fit with the XRD data reported for the α-form (Ashida 1966; Robinson and Klein 1952). It is well established that the α-form provides the brightest and the most transparent phthalo blue pigments (Herbst et al. 2004). Therefore, it appears

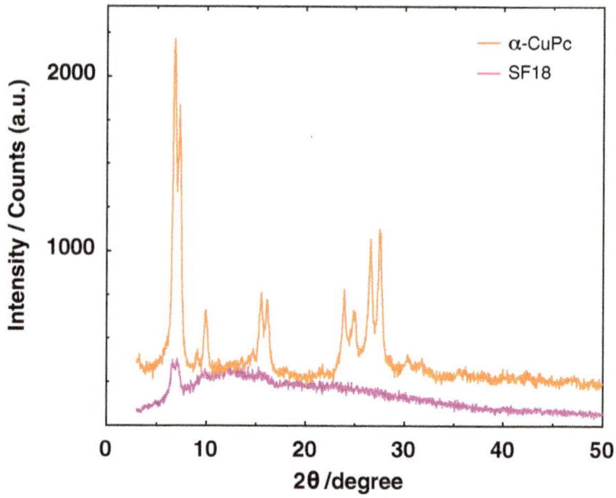

Fig. 4 SF18 diffractogram and diffraction pattern obtained for a dry pigments of α-CuPc (Heliogen 23050 from Kremer). The position of the two strongest CuPc diffraction peaks appearing in the SF18 diffractogram totally fit with the α-crystal form

that the sharpness of the blue hue and the transparency characterizing the SF18 paint film, partly result from some of the specific pigmentary properties of the CuPc blue pigments of the α-form.

Paints from the GCI Reference Collection

Oil-Based Paints

The results obtained for the oil-based paints and the acrylics are presented separately in Table 3. On the basis of the LDA classification, six out of nine oil-based paints contain a phthalo blue of the β-form. The α-form has been identified only in the Daniel Smith's Phthalo Blue RS2 and in the Winsor&Newton's Phthalo Blue RS3.

The LDA results obtained for the Iridescent Blue from Daniel Smith are not in accordance as two Raman spectra have been classified in the α group while the third one has been classified in the ε group. Therefore, for this paint the procedure combining RS and LDA did not work.

Iridescent Blue belongs to the Luminescent Oil Colours range. The Color Index generic names mentioned on the paint tube are PB15:1 that indicates a non-crystallizing α-CuPc, aluminium (PM1) and Mica (PW20). In the diffractogram recorded for this paint only one small diffraction peak at $d = 13.12$ is induced by the phthalo blue pigment. This spacing only matches with the α-form but it is not sufficient to assert that this CuPc crystal form is present.

Concerning the other paints, all the LDA classifications, with the exception of the Marie's Oil Colours and the Dilo paint samples, have been confirmed thanks to the XRD analysis (Table 3). However it is important to note that the number of CuPc diffraction peaks strongly differs according to the sample as it varies between two to ten. Surprisingly, in the diffractograms obtained from various paint films that

Table 3 Comparison of the LDA classification, XRD and FTIR results obtained for the paint samples from the GCI Reference Collection

	PB15 type			
Paint reference	LDA	XRD	FTIR	Supplier's declaration
Acrylic paints				
Phthalo Blue GS1	β	β	(β)	β
Phthalo Blue RS1	α	α	(α)	α
Phthalo Blue1	β	β	(β)	Unavailable
Phthalo Blue2	β	β	(β)	Unavailable
Phthalo Blue3	β	β	β	Unavailable
Oil-based paints				
Phthalo Blue4	β	β	?	Unavailable
Phthalo Blue5	β	?	?	Unavailable
Phthalo Blue GS2	β	β	?	Unavailable
Phthalo Blue RS2	α	α	?	α
Phthalo Blue6	β	?	?	Unavailable
Phthalo Blue7	β	β	?	Unavailable
Phthalo Blue RS3	α	α	?	Unavailable
Phthalo Turquoise	β	β	?	Unavailable
Iridescent Blue	?	?	?	α

have in common the darkness of their blue shade, the CuPc diffraction pattern may be very marked or almost absent depending on the nature of the binding medium. This difference could be explained by the paint formulation and/or the CuPc crystals orientation in the paint film. In accordance with the results of the study carried out by Lomax (2010), in the present study the XRD method has appeared significantly less efficient for oil-based paints than for acrylic paints.

For the paint from Marie's Oil Colours, the presence of the CuPc blue has not been detected by means of the XRD method. All the recorded diffraction peaks are induced by the extender ($BaSO_4$). For the oil paint from Dilo the experimental d values related to the CuPc diffraction peaks did not allow the crystal form identification.

The interpretation of the FTIR results is much more complicated due to the IR band overlapping of the other components present in the paint system. It has been reported that the positions of the IR bands arising at 715–735, 765–785 and 865–885 cm^{-1}, allow identifying the CuPc crystal form (Defeyt et al. 2012). The IR band at 715–735 cm^{-1} corresponds to the strongest CuPc absorption. Unfortunately, the oil-based media exhibit strong IR absorptions in the 715–735 cm^{-1} region (Derrick et al. 1999). This overlapping makes the FTIR results not completely reliable for the CuPc polymorph discrimination in the oil-based paints.

It is interesting to note that the discriminating position of the IR band arising in the 765–785 cm^{-1} region of the oil-based paint spectra always supports the crystal form predicted by LDA.

The IR band at 865–885 cm^{-1} can be observed only in the FTIR spectra recorded for the Phthalo Blue GS2 from Daniel Smith, Phthalo Blue 7 from Royal Talens and Phthalo Blue RS3 from W&N. For these three paints, the IR band position recorded in the 865 and 885 cm^{-1} region also matches with the LDA classification.

Nevertheless, for the Royal Talens and W&N paints, the IR band in the 875–876 cm^{-1} region could be also induced by the $CaCO_3$ identified by XRD.

Acrylic Paints

Among the five investigated acrylic paints, the LDA classification leaded to the identification of four β- and one α-CuPc based paints. Indeed, only the Raman spectra recorded for the Phthalo Blue RS1 from Golden, have been classified in the α group. It can be seen in Table 3 that the XRD results have confirmed the LDA classification in every case.

In comparison, the number of CuPc diffraction peaks recorded for the acrylic paints is much higher than for the oil-based paints. Ten CuPc diffraction peaks arise in the diffractogram recorded for each β-CuPc-based paint. Also, the diffraction pattern of the β-form is immediately recognizable in the four diffractograms. Concerning, the Phthalo Blue Red Shade from Golden, the diffraction pattern that includes six CuPc diffraction clearly indicates the α crystal structure. The FTIR results obtained for this paint are much less evident about the α identification. Only the IR band at 721 cm^{-1} indicates the α-form. The two other CuPc IR bands (765–785 and 865–885 cm^{-1}) for which the position is useful for identifying the crystal structure do not appear in the spectrum.

For the four phthalo blue based paints classified as β by LDA, the discriminating IR band positions recorded at 730–732 and 778–782 cm^{-1} support the β-form.

The CuPc IR band at 865–885 cm^{-1} is missing from the FTIR spectrum obtained for the Phthalo Blue GS1 from Golden but, for the three other paints, the IR band recorded at 875–876 cm^{-1} also suggests the β structure. However in the cases of the Liquitex Phthalo Blue2 and the paint from Creative, the IR absorption at 875–876 cm^{-1} is more probably induced by the chalk identified in the mixture by XRD.

Conclusions

In our previous work, we have shown how LDA using intensity ratios as variables allowed us to find a rule for allocating new Raman spectra recorded for an unknown PB15 sample to the correct crystal form. The present study tends to confirm that the developed procedure is a valuable non-destructive method for identifying the CuPc crystal structure in unknown paint samples irrespectively to the nature of the binding medium.

The LDA classification obtained for the SF18 sample underlines that Sam Francis, as a great colorist, has exploited the brightness and the transparency of the α CuPc blue to prepare his famous intense and transparent blues. The XRD measurements and the new Raman analysis performed on Sam Francis's paints have outlined the presence of PG7 in SF60 and, have revealed the combination of phthalo green, cobalt blue and titanium dioxide in SF42. The detection of these pigments underlines the interest of XRD, as well as the importance to carry out Raman measurements on different pigment grains.

On the basis of the LDA results, CuPc phthalo blue-based paints from the GCI Reference Collection includes ten β- and three α-CuPc specimen. For the Daniel Smith Iridescent Blue, belonging to the Luminescent Oil Colours range, it has not been possible to resolve the CuPc crystal form through LDA classification or XRD analysis. Actually, the XRD measurements did not allow the CuPc crystal form identification in three paint samples out of nine.

However, for the six other oil-based paints and for all the acrylic paints, the number and the intensity of the CuPc diffraction peaks were sufficient for identifying the crystal structure. For each of these paints, the polymorphic identification by means of XRD is always in accordance with the LDA classification. These corroborating results have shown the efficiency of the procedure combining RS and LDA to predict the crystal form of a phthalo blue contained in acrylic paint samples as well as in oil-based paint samples.

Acknowledgements This work was supported by the Inter-University Attraction Pole program of the Federal Belgian Science Policy (NACHO P6/16), the University of Liège, and the Institut Interuniversitaire des Sciences Nucléaires. The authors wish especially to thank Sylvia Lycke from the Raman Spectroscopy Research Group (Ghent) for the technical support and, Rachel Rivenc and Art Kaplan from the Getty Conservation Institute (Los Angeles) for the sample information.

References

Ashida M (1966) The orientation overgrowth of metal-phthalocyanines on the surface of single crystals. i. Vacuum-condensed films on muscovite. Bull Chem Soc Jpn 39:2625–2631

Bernstein J (2002) Polymorphism in molecular crystals. University Press, Oxford

Bouchard M, Rivenc R, Menke C, Learner T (2009) Micro-FTIR and micro-Raman study of paints used by Sam Francis. e-Preserv Sci 6:27

Brostoff LB, Centeno SA, Ropret P, Bythrow P, Pottier F (2009) Combined X-ray diffraction and Raman identification of synthetic organic pigments in works of art: from powder samples to artists paints. Anal Chem 81(15):6096–6106

Brown C (1968) Crystal structure of β-Copper phthalocyanine. J Chem Soc A 2488–2493

Colour Index Dyes and Pigments (1971) Colour index, 3rd edn. Society of Dyers and Colourists/American Association of Textile Chemists and Colorists, Bradford

Dagron B (1998) Inventaire du matériel de l'atelier du peintre Vassily Kandisky. Techné 8:64–76

Defeyt C (2010) Influence des solvants aromatiques sur les propriétés optiques du bleu de phtalocyanine de cuivre en milieu pictural. CeROArt. Special issue

Defeyt C (2013) Copper-phthalocyanine blue pigments in conservation. Université de Liège. http://hdl.handle.net/2268/149555

Defeyt C, Vandenabeele P, Gilbert B, Van Pevenage J, Cloots R, Strivay D (2012) Contribution to the identification of α-, β- and ε-copper phthalocyanine blue pigments in modern artists' paints by X-ray powder diffraction, attenuated total reflectance micro-Fourier transform infrared spectroscopy and micro-Raman spectroscopy. J Raman Spectrosc 43:1772–1780

Defeyt C, Van Pevenage J, Moens L, Strivay D, Vandenabeele P (2013) Micro-Raman spectroscopy and chemometrical analysis for the distinction of Copper phthalocyanine polymorphs in paint layers. Spectrochim Acta Part A Mol Biomol Spectrosc 115:636–640

De Keijzer M (2002) The history of modern synthetic inorganic and organic artists' pigments discovered in the twentieth century. In: Mosk JA, Tennent NH (eds) Contributions to conservation research from the ICN. James & James, London, pp 42–54

Derrick MR, Stulik D, Landry JM (1999) Infrared spectroscopy in conservation science. The Getty Conservation Institute, Los Angeles.

Duda R, Stork D, Hart P (2000) Pattern classification. Wiley, New York

George D, Mallery P (2001) SPSS for windows. Allyn & Bacon, Needham Heights

Herbst W, Hunger K, Wilker G (2004) Industrial organic pigments: production, properties, applications, 3rd edn. Wiley-VCH, Weinheim

Khandekar N, Mancusi-Ungaro C, Cooper H, Rosenberger C, Eremin K, Smith K, Stenger J, Kirby D (2010) A technical analysis of three paintings attributed to Jackson Pollock. Stud Conserv 55:204–215

Lomax SQ (2010) The application of X-ray powder diffraction for the analysis of synthetic organic pigments. Part 1: dry pigments. J Coat Technol Res 7:331–336

Lutzenberger K (2009) Kuntstlerfarben im wandel. Herbert Utz Verlag, Munchen

Miller J, Miller J (2005) Multivariate analysis. In: Miller JN, Miller JC (eds) Statistics and chemometrics for analytical chemistry. Pearson Education Limited, Harlow, pp 213–240

Moser FH, Arthur L (1963) Phthalocyanine compounds. Reinhold, New York

New York Times (1935). Blue pigment discovery is the first in 100 years, 25 Aug 1935

Pfeiffer F (1962) Pigmentary copper phthalocyanine in the "R" form and its preparation. Patent US 3,051,721, 28 Aug 1962

Robertson JM (1935) An X-ray study of the structure of the phthalocyanines. Part I. The metal-free, nickel, copper, and platinum compounds. J Chem Soc 7:615–621

Robinson MT, Klein GE (1952) Unit cell constants of α-Copper phthalocyanine. J Am Chem Soc 74:6294–6295

Rohrs S, Schwabe A, Simon S (2012) Forensic investigations of imitated expressionist paintings. In: Book of abstracts, 10th biennial international conference of the infrared and Raman users group, Barcelona. University of Barcelona, Barcelona, pp 190–191

Ruhrberg K, Schneckenburger M, Fricke C, Honnef K (2005) L'art au XXe siècle, vol 1. Taschen

Scherrer NC, Zumbuehl S, Delavy F, Fritsch A, Kuehnen R (2009) Synthetic organic pigments of the 20th and 21st century relevant to artist's paints: Raman spectra reference collection. Acta A Mol Biomol Spectrosc 73:505–524

van der Weerd J, van Loon A, Boon J (2005) FTIR studies of the effects of pigments on the aging of oil. Stud Conserv 50:1–33

Wiswall R (1949) Solvent stable metal phthalocyanine pigments and methods of making the same. US 2,486,351, 25 Oct 1949

Charting the Development of Oil-Based Enamel Paints Through the Correlation of Historical Paint Technology Manuals with Scientific Analysis

Maria Kokkori, Francesca Casadio, Ken Sutherland, and Marc Vermeulen

Abstract This paper presents an overview of developments in oil-based enamel paint technology in the first half of the twentieth century and discusses the selection of ingredients that had a significant impact on the properties of the paints: pigments, extenders, driers, oils and resins. A review of period literature published in French and English is complemented by analysis of historical oil-based Ripolin enamel paint samples produced in France in the years 1910–1950 and a small selection of artists' paint tubes of the same era from the Art Institute of Chicago's reference collection. A range of analytical techniques including x-ray fluorescence spectroscopy (XRF), Fourier transform infrared spectroscopy (FTIR), pyrolysis gas chromatography-mass spectrometry with thermally-assisted hydrolysis and methylation (THM-Py-GCMS), scanning electron microscopy coupled with energy dispersive x-ray spectroscopy (SEM-EDX) and thermogravimetric analysis have been used for the characterisation of the paint samples.

Keywords Oil-based enamel paints • Paint technology • Ripolin

Introduction

Alongside traditional artists' oil tube paints, the twentieth century saw the rapid development of oil-based, ready-mixed house paints (Standeven 2011). These non-artists' paints are sometimes referred to as 'architectural', 'oleoresinous', 'industrial', 'protective', 'decorative', 'ready-to-use', 'enamel' or 'ripolin' (Sabin 1927). Ripolin is the trade name for a specific French brand of commercial paints manufactured for household and other uses; however, the term 'ripolin' was often used to refer to enamel paints in general (Labordere and Anstett 1913). From

M. Kokkori (✉) • F. Casadio • K. Sutherland
The Art Institute of Chicago, 111 S. Michigan Ave., Chicago, IL 60603, USA
e-mail: mkokkori@artic.edu

M. Vermeulen
IRPA-KIK Royal Institute for Cultural Heritage, Brussels, Belgium

K.J. van den Berg et al. (eds.), *Issues in Contemporary Oil Paint*,
DOI 10.1007/978-3-319-10100-2_7

1897 until the end of the Second World War, Ripolin production took place on the premises of the famed fine artists' materials manufacturer Lefranc, in Issy-les-Moulineaux, a suburb of Paris. The close links between the production of oil-based tube paints and house paints is underscored by the fact that the administrators and President of the Ripolin factory were chosen from among the Lefranc family.

Artists including Pablo Picasso, Wassily Kandinsky, Francis Picabia, László Moholy-Nagy, René Magritte, Aleksandr Rodchenko, Nathan Altman and Le Corbusier, among others, are documented to have used commercially-available, non-artists' paints, such as Ripolin and other brands of house, architectural, car and boat paints, because of their desirable surface qualities and handling properties. Recent research at the Art Institute of Chicago for the first time fully explores the composition of French oil-based Ripolin enamel paints, providing a useful complement to ongoing research on the evolving formulations of artist's oil paint tubes of the first half of the twentieth century.

Oil-Based Enamel Paint Formulations

The term *oil enamel* was used to describe ready-to-use paints that contained linseed oil or a blend of oils with resins and pigments, and connoted a hard, glossy surface analogous to porcelain enamel. Enamel paints were formulated to have a number of desirable characteristics including gloss and colour retention, elasticity, opacity, consistency, spreading capacity, good levelling, and durability (Jennings 1919). Driers and thinners were used as required in order to achieve satisfactory application and performance characteristics.

Pigments, Extenders, and Driers

In the first half of the twentieth century the white pigments most commonly encountered in house paint recipes were lead white, zinc white, zinc sulfide, lithopone and titanium dioxide white. Different grades and modifications of each of these pigments affected the properties of the resulting paint. Zinc oxide and leaded zinc oxide, in addition to their use as primary pigments, were used as modifiers in conjunction with various other pigments in order to control such factors as dirt collection, chalking, fading and mildew (Nelson 1935a, 1940). Zinc-based pigments were very important in house paint formulations because of their high durability, achieved by the production of acicular particles and the elimination of colloidal fines (Eide and Depew 1936). Although French Ripolin paints were consistently produced exclusively with zinc oxide as the white pigment before 1950 (Gautier et al. 2009), other French manufacturers of enamel paints made use of other whites. This has been demonstrated by the analysis of three cans of white Valentine paint from the Parisian studio of Wassily Kandinsky (roughly dated to the 1940s) that were shown to contain titanium dioxide white (anatase) in combination with minor amounts of zinc white (McMillan et al. 2013), as well as analysis of oil-based enamel paints used by Pablo Picasso in Antibes in 1946, which were shown to contain zinc white in mixtures with barium sulfate, lithopone or titanium dioxide white (anatase) (Casadio and Gautier 2011; Casadio et al. 2013). Zinc pigments gained popularity in the paint industry over lead white because of health hazards associated with the latter (Truelove 1922; Nelson 1935b). For white house paint formulations, zinc white was often recommended as a fungicide and in some can linings as a sulfide scavenger (Trott 1928). SEM examination of Ripolin white enamel paint samples highlighted the predominant presence of particles of zinc oxide of small diameter (150–300 nm) with rare occurrences of acicular crystals. The latter, on the other hand, are more prevalent in samples of Lefranc zinc white artist's tube oil paints of the same period, which are also characterised in general by pigments of larger particle sizes (Figs. 1 and 2).

The size and form of the pigment particles were shown to have a great influence upon the life of enamel paints. Authors referred to a 'French' process for zinc oxide pigments, also called the 'seal' type, which produced a relatively uniform pigment particle size distribution resulting in a 'very fluffy product with exceptionally good suspended qualities, excellent packaging qualities, good leveling, holding-out qualities, high gloss and gloss retention, good colour, resistance to yellowing or change in tint, and washability, used in enamels of the highest grade.' (Nelson 1940) High-resolution, high sensitivity synchrotron radiation XRF nanoprobe analysis of the zinc oxide pigment particles employed for both the Ripolin enamels and Lefranc artists' tube paints have shown them to be of the highest purity, i.e. free of lead and cadmium impurities, and only containing trace amounts of iron, which have been demonstrated to correspond to the 'white seal' grade of the time and the purest white colour, typically reserved for pharmaceutical uses and gloss enamel paints of the highest quality (Casadio and Rose 2013).

Extenders were used to improve durability, eliminate running of paint and to reduce costs. Mica and magnesium silicate were preferred over other extenders such as barites, whiting, china clay, calcium sulfate, and ground silica. Magnesium silicate was widely recommended for use in trim paints to provide resistance to chalking, cracking and fading (Parker 1943). Interestingly, oil-based French Ripolin enamel paints before 1950 have been found to contain very few extenders, with small amounts of barium sulfate detected only in colour charts dating to the 1930s and 1940s (Gautier et al. 2009), while on the other hand artist's tube paints from Lefranc and a few other French manufacturers of the time have been often shown to contain extenders based on barium sulfate and calcium compounds (calcite and gypsum) and other whites such as lead white and titanium dioxide white (Casadio and Rose 2013). Importantly, the aluminium stearates and magnesium silicates that have been identified as problematic with regard to the condition of paintings made with modern oils have not been documented in Ripolin enamel paints (Noble and Boon 2007). The use of few or no extenders ensured very intense colours for the

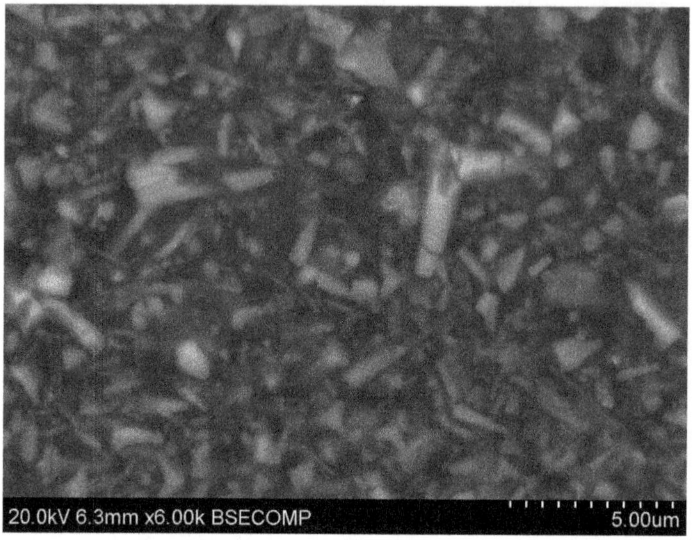

Fig. 1 Backscattered electron image of a Lefranc zinc white tube paint sample

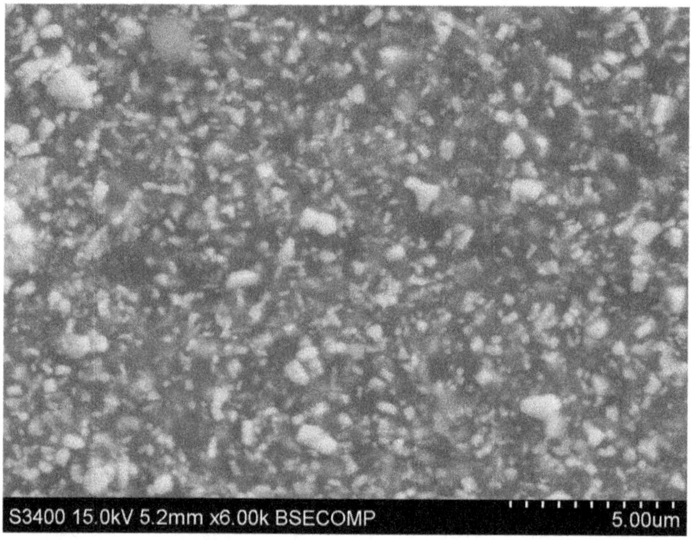

Fig. 2 Backscattered electron image of a Ripolin zinc white house paint sample

Ripolin paints, with superior hiding power (requiring a single coat of paint as opposed to the multiple applications necessary for other brands of enamels). This, together with a wide availability in a much broader range of ready-mixed colours than many of its competitors, likely contributed to the enormous commercial success of this brand of paint.

Driers were mainly used in exterior house paints to set or harden the oil in the paint film so that it would not collect dust or dirt, and could withstand abrasion. Metallic soaps including stearates, palmitates, oleates and naphthenates of aluminium, calcium and zinc, were frequently used to accelerate the oxidation, polymerisation and gelation of the oil and produce desired flatting effects (Licata 1933). For example, lead and manganese linoleate or naphthenates were recommended for use to the extent of 0.6 % lead and 0.02 % manganese as metal, calculated relative to the non-volatile vehicle in the paint. Cobalt was not highly recommended as a drier for exterior paints since its presence was associated with chalking and fading. Linoleate driers were considered as satisfactory driers because they had good mixing and wetting properties. They also permitted a higher thinner content compared to other driers and yielded satisfactory brushing with good leveling and gloss. Analysis with micro-FTIR spectroscopy of Ripolin and artist's tube paints did not allow the detection of features specific to individual driers, most likely because in the percentages used they are below the detection limit of the technique. Ripolin enamel paints containing zinc white display broad FTIR bands centered at 1,580 and 1,440 cm^{-1} that are related to zinc carboxylates. These are most likely due to reaction of the oil-based binding medium with the zinc white pigment particles, and not original components of the paints. On the other hand for tube paints, while the broad zinc carboxylate bands remain, a sharp peak at 1,541 cm^{-1} that is characteristic of zinc stearate is clearly visible (Fig. 3). In addition to zinc driers, preliminary investigations with inductively-coupled plasma mass spectrometry highlighted the presence of driers based on lead, manganese and cobalt in both tube and Ripolin enamel paints (Muir et al. 2011).

In house paints the volume relationship of pigment to binder was considered more pertinent than the weight relationship because of its predominating influence

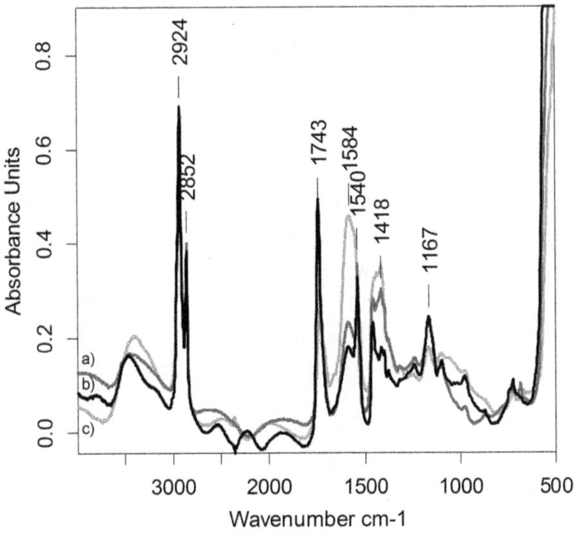

Fig. 3 FTIR spectra of (**a**) reference Lefranc tube paint sample of zinc white (#2007.48); (**b**) reference Lefranc tube paint sample of zinc white (#2007.70); (**c**) reference Ripolin house paint sample of zinc white (#HP002)

on the physical and optical properties of the paints (Pickett 1939). Qualitatively, even a simple visual assessment of the particle distribution and pigment/binder ratios in the SEM images of tube and Ripolin paints illustrated in Figs. 1 and 2 makes it immediately apparent that Ripolin paints have a higher proportion of binder to pigment, when compared to the tubes. This was confirmed with thermogravimetric analysis (Muir et al. 2011) that showed as much as a 30/70 % w/w ratio of pigment to binder in Ripolin ultramarine blue paints, as compared to almost the opposite ratio of 65/35 % pigment to binder in similarly pigmented tube paint.

The essential steps in the preparation of house paints as described in the technical literature included fine grinding of pigments, mixing, thinning and tinting. The primary purpose of the mixing was to distribute the ingredients into a uniform paste and to wet the pigments with the vehicle as thoroughly as possible. Grinding was associated with dispersing and reducing in size the individual pigment particles. The ground paste was mixed with additional quantities of oils, varnishes, driers, and thinners.

Binders and Resins

Traditionally household paints contained linseed oil alone, but as oils from an increasing array of plant and animal sources became available to paint manufacturers throughout the twentieth century, the use of oils in admixture became common to obtain optimal results. Authors recommended the use of tung oil in various combinations with linseed oil (Jolly 1930; Bohannon 1922) as well as other drying oils such as perilla, oiticica, soya bean, fish and dehydrated castor oil (Damitz et al. 1943). With respect to linseed oil, the refined oil was recommended because of its good wetting and grinding characteristics, however, it was characterised by low viscosity, easy brushing, poor levelling and relatively low gloss. A combination of refined with boiled linseed oil was also available where the heat-treated (boiled) oil was added to improve durability, levelling and gloss. Heat-treated oils have higher viscosities and were used in oil enamel paints to improve application and performance characteristics. Enamels based on heat-treated oils had good gloss and colour retention, and produced flexible, durable films. They were also able to hold more pigment than the resin-based enamels, which meant that fewer coats were required (Nylen 1965).

Historically, the highest quality enamels were known to originate from Holland, where oxidised linseed oil was employed as the vehicle (Chatfield 1947). In the Netherlands, as well as in England and Germany, manufacturers also prepared a linseed oil of heavy consistency known as blown oil by boiling at moderate temperatures, without the addition of drying mediums, and blowing air through the oil during the boiling process (Andés 1901).

Mass-spectrometric analysis (electrospray and direct-temperature-resolved mass spectrometry) of Ripolin paints suggested a high degree of cross-linking and chemical modification of the oil medium resulting from pre-polymerisation, and

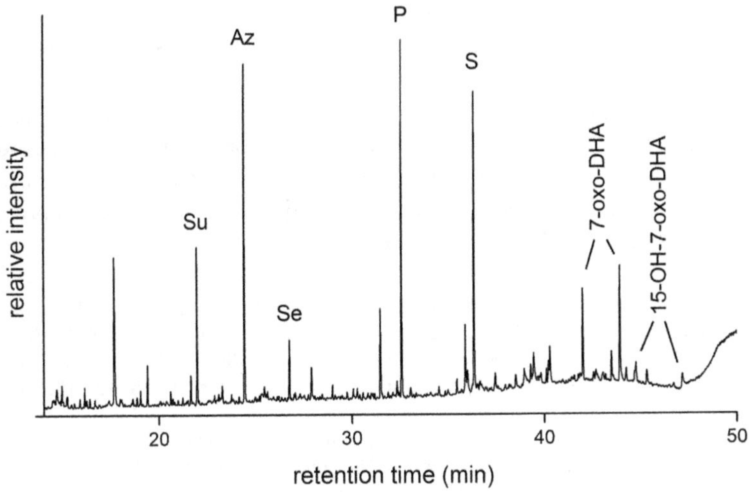

Fig. 4 THM-Py-GCMS total ion chromatogram of a reference Ripolin house paint sample of black paint (#HP064). *Su* suberic acid, *Az* azelaic acid, *Se* sebacic acid, *P* palmitic acid, *S* stearic acid, *7-oxo-DHA & 15-OH-7-oxo-DHA* oxidised forms of dehydroabietic acid; all compounds detected as methyl derivatives

THM-Py-GCMS analysis confirmed the presence of drying oil, in most cases combined with diterpene (Pinaceae) resin (Fig. 4) (Kokkori et al. 2013) a practice also described in the literature (Horton Sabin 1904; Uebele 1913; Cruickshank 1915). Castor oil was identified only in two samples of red Ripolin paint pigmented with alizarin; the identification of other types of drying oil such as soybean, perilla or fish oil recommended in the technical literature was not possible because of the lack of specific chemical markers for these materials. Aside from some variations in the proportions of oil and resin in the binders relating to the requirements of specific colours, the composition of Ripolin paints appears to have remained fairly consistent in the first half of the twentieth century. On the other hand, preliminary analyses of binding media in artists' tube paints carried out for this study suggested a higher degree of variability in formulations with respect to oil types and additives. The organic components of the Ripolin and tube paints will be discussed in more detail in separate papers (Kokkori et al. 2013 (*in preparation*), 2014).

Conclusions

The work presented here considers the composition of Ripolin paints, a very important brand of early twentieth century French oil-based enamel, as compared to artists' tube paints of the same period. Scientific data are interpreted in the larger context of industrial technical literature of the time. While much useful information has been uncovered in the process, a more in-depth study of the composition of modern oil paints needs to take place in order to fully understand works of art that contain them, and to inform their preservation. Open questions remain, in particular, the investigation of zinc carboxylates, their structure in relation to the particle size and shape of zinc oxide pigment, their mobility and their tendency to form aggregates. Additionally, a focus on the development of new strategies for the identification of different oils and resins in modern paints, especially those that may have undergone chemical modification as a consequence of the industrial processes of paint preparation, will be of great value.

Acknowledgements The authors thank Frank Zuccari, Jaap Boon, Inge Fiedler, Kim Muir, and the staff of the Ryerson & Burnham Libraries at the Art Institute of Chicago. SEM-EDX work was performed at the Electron Probe Instrumentation Center (EPIC) facility of the Northwestern University Atomic and Nanoscale Characterization Experimental Center (NUANCE). NUANCE Center is supported by the NSF funded Nanoscale Science and Engineering Center (NSF-NSEC), and Materials Research Science and Engineering Center (NSF-MRSEC), Keck Foundation, the State of Illinois, and Northwestern University. The financial support of the Stockman Family Foundation, the Andrew W. Mellon Foundation and Grainger Foundation is also gratefully acknowledged.

References

Andés LE (1901) Drying oils, boiled oil, and solid and liquid driers. A practical work for manufacturers of oils, varnishes, printing inks, oil-cloth and linoleum, oil-cakes, paints, etc. Scott, Greenwood, London, pp 18–20

Bohannon (1922) Present-day methods of varnish manufacture. The Decorator (January 22), p 237

Casadio F, Gautier G (2011) Picasso at work: making the case for a scientific re-evaluation of the materials of the Antibes cycle. In: Raeburn M, Radeuil N (eds) Picasso express. Musee Picasso, Antibes, pp 37–63 (French), pp 135–150 (English)

Casadio F, Rose V (2013) High-resolution fluorescence mapping of impurities in historical zinc oxide pigments: hard X-ray nanoprobe applications to the paints of Pablo Picasso. Appl Phys A 111(1):1–8

Casadio F, Miliani C, Rosi F, Romani A, Anselmi C, Brunetti B, Sgamellotti A, Andral J, Gautier G (2013) Non-invasive in situ investigation of the Picasso paintings in Antibes: new insights into technique, condition and chronological sequence. J Am Inst Conserv 52(3):184–204

Chatfield HW (1947) Varnish constituents. L. Hill, London, p 11

Cruickshank Smith J (1915) The manufacture of paint: a practical handbook for paint manufacturers, merchants, and painters. Scott, Greenwood & Son, London, pp 206–231

Damitz FM, Murphy JA, Mattiello JJ (1943) Varnishes. In: Mattiello JJ (ed) Protective and decorative coatings, vol 3. Wiley, New York, pp 199–206

Eide AC, Depew HA (1936) Evaluation of zinc oxide for paint. Am Paint J (April 13) 20(27):7–9; 20(28):51–56

Gautier G, Bezur A, Muir K, Casadio F, Fiedler I (2009) Chemical fingerprinting of ready-mixed house paints of relevance to artistic production in the first half of the twentieth century. Appl Spectrosc 63:597–603

Horton Sabin A (1904) The industrial and artistic technology of paint and varnish. Wiley, New York, pp 140–145

Jennings AS (1919) Paints and varnishes. Pitman & Sons, London, pp 5–11

Jolly VG (1930) Oil varnishes: their manufacture, properties, and defects. The Decorator (January), pp 522–526

Kokkori M, Sutherland K, Boon JJ, Vermeulen M, Casadio F (2013) Synergistic use of THM-Py-GCMS, DTMS, ESI-MS and vibrational spectroscopy for the characterisation of the organic fraction of modern house paints. Technart 2013 Conference. Analytical Spectroscopy in Art and Archaeology, Rijksmuseum, Amsterdam (in preparation)

Kokkori M, Casadio F, Boon JJ (2014) A complete study of early 20th century oil-based enamel paints: Integrating industrial technical literature and analytical data. In ICOM-CC 17th Triennial Conference Preprints, Melbourne, 15 -19 September 2014, ed. J. Bridgland, art. 0101, 8 pp. Paris: International Council of Museums. (ISBN 978-92-9012-410-8)

Labordere P, Anstett R (1913) Notes on the testing of anti-corrosion paints. Chem Eng XVII(1):5

Licata FJ (1933) Properties and uses of aluminum in the paint and varnish industry. Official Dig Federation Paint Varnish Prod Clubs 5:160

Mcmillan G, Casadio F, Fiedler I, Sorano-Stedman V (2013) An investigation into Kandinsky's use of Ripolin in his paintings after 1930. J Am Inst Conserv 52(4)

Muir K, Gautier G, Casadio F, Villa A (2011) Interdisciplinary investigation of early house paints: Picasso, Picabia and their "Ripolin" paintings. In: Bridgland J (ed) ICOM Committee for Conservation preprints (CD-ROM), 16th Triennial Meeting Lisbon. Lisbon: Critério-Artes Graficas, Lda. 10

Nelson HA (1935a) Study of paint durability based on some physical properties of the pigments used. Official Dig J Paint Tech Eng Federation Paint Varnish Prod Clubs (January) 142:7–14

Nelson HA (1935b) Zinc sulfide pigments for interior paints. Official Dig J Paint Tech Eng Federation Paint Varnish Prod Clubs 7:177

Nelson HA (1940) The versatile paint-making properties of zinc oxide. New Jersey Zinc Company, New York, pp 13–33

Noble P, Boon JJ (2007) Metal soap degradation of oil paintings: aggregates, increased transparency and efflorescence. In: Helen Mar Parkin (ed) AIC paintings specialty group postprints: papers pres. at the 34th annual meeting of the AIC of Historic & Artistic Works providence, Rhode Island, 16–19 June 2006. AIC, Washington, DC, pp 1–15

Nylen P (1965) Modern surface coatings: a text-book of the chemistry and technology of paints, varnishes, and lacquers. Interscience, New York, pp 85–91

Osmond G, Boon JJ, Puscar L, Drennar J (2012) Metal stearates distributions in modern artists' oil paints: surface and cross sectional investigation of reference paint films using conventional and synchrotron infrared microspectroscopy. Appl Spectrosc 66(10):1136–1144

Parker DH (1943) Exterior trim paints. In: Mattiello J (ed) Protective and decorative coatings, vol 3. Wiley, New York, pp 315–323

Pickett CF (1939) Factors influencing initial gloss and gloss retention of paint films. Official Dig J Paint Tech Eng Federation Paint Varnish Prod Clubs (June) 187:310–319

Sabin AH (1927) The industrial and artistic technology of paint and varnish. Chapman & Hall, London, pp 194–205

Standeven H (2011) House paints 1900–1960: history and use. Getty Conservation Institute, Los Angeles

Trott LH (1928) Zinc oxide and its application to paint. New Jersey Zinc Company, New York, pp 3–19

Truelove RH (1922) Oils, pigments, paints, varnishes, etc. Pitman & Sons, London, pp 57–61

Uebele CL (1913) Paint making and colour grinding. Trade Papers, London, http://chestofbooks.com/home-improvement/repairs/painting/Paint-Making-Colour-Grinding/index.html

Hans Hofmann's Last Lesson: A Study of the Artist's Materials in the Last Decade of His Career

Dawn Rogala

Abstract This paper summarises the analysis of more than 500 paint and fibre samples from the late-career work of Abstract Expressionist painter and teacher Hans Hofmann (1880–1966) and finds a correlation between condition issues in Hofmann's work and the combination of new paint materials with traditional art practice. The results of this study help to inform the conservation of Abstract Expressionist and other works that incorporate both traditional and modern paint media.

Keywords Hofmann • Abstract Expressionism • Modern art conservation • Modern materials • Painting technique

Introduction

> You cannot help it. You belong to a certain time. You are yourself the result of this time. You are also the creator of this time. (Hans Hofmann 1966)[1]

The late-career work of renowned Abstract Expressionist painter and teacher Hans Hofmann (1880–1966) shows us why and how we need to change our thinking about the conservation of modern art. The use of new art-making materials and processes is one of the great legacies of Abstract Expressionism, but the notion that twentieth-century modernist innovators wholly abandoned traditional painting practice is a myth that sometimes compromises the approach to conservation of these works. A mixture of old and new methods and media is more representative of period art-making. Many of the problems encountered by conservators in the treatment of modern paintings are directly related to the mixture of familiar

[1] Jaffe (1971), 37.

D. Rogala, Ph.D. (✉)
Paintings Conservator, Museum Conservation Institute, Smithsonian Institution, 4210 Silver Hill Road, Suitland, MD 20746, USA
e-mail: RogalaD@si.edu

K.J. van den Berg et al. (eds.), *Issues in Contemporary Oil Paint*,
DOI 10.1007/978-3-319-10100-2_8

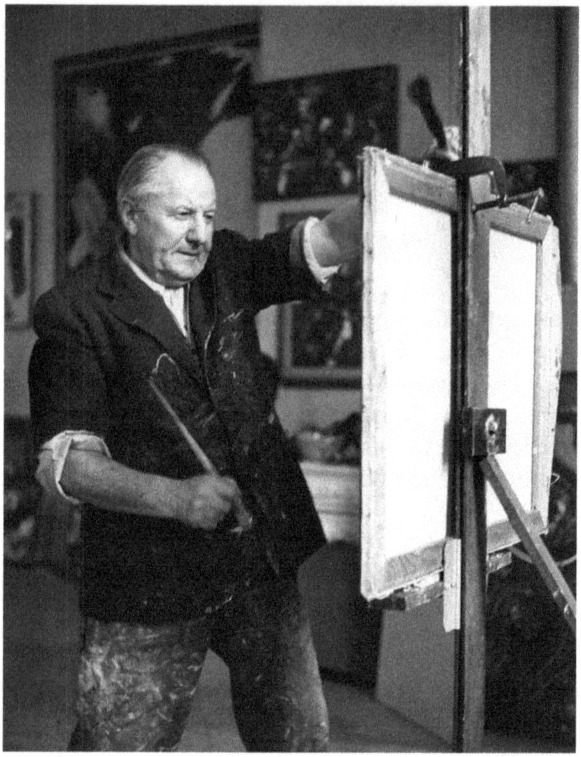

Fig. 1 Hans Hofmann at work in his studio, 1952/Kay Bell Reynal, photographer (Photographs of artists taken by Kay Bell Reynal), Archives of American Art, Smithsonian Institution

techniques and new paint formulations used by Hofmann and other Abstract Expressionists. The study of Hofmann's palette is an appropriate and informative addition to the research canon. Understanding the transitions in painting practice during this period has ramifications for the future conservation treatment of Abstract Expressionist and later works that incorporate both traditional and modern paint media (Fig. 1).

This paper is based on doctoral research undertaken as part of the Preservation Studies Program at the University of Delaware (USA), in which a representative catalogue of Hofmann's late-career materials was built from the analysis of over 500 paint and fibre samples. The research aimed to assess relationships between Hofmann's materials, technique, and the impact of his choices on the long-term stability of his work. The research builds on the author's earlier analysis of mid-twentieth century commercial house paints and assessment of the condition issues related to commercial paint and ground layers observed in works from the Smithsonian Institution's Hirshhorn Museum and Sculpture Garden. A range of complementary analytical techniques were used to examine 26 paintings and eight palettes chosen as exemplars of Hofmann's work from the years 1953–1965, years just prior to and after the 1958 closing of Hofmann's art schools in New York City and Provincetown, Massachusetts. The primary goal for analysis was the broad characterisation of materials to confirm Hofmann's use of industrial materials in his ground layers and to identify Hofmann's use of newly developed pigments or paint media in his paintings.

Study Group Selection

> Hofmann is certainly one of the most uncompromising representatives of what some people call the spatter-and-daub school and I, more politely, have christened Abstract Expressionism. (Robert Coates, *The New Yorker* 1946)[2]

Artist selection. Hans Hofmann was an influential artist and teacher positioned at the center of the Abstract Expressionist community at a pivotal period of change in artistic production. His authority with practitioners and critics and his esteemed teaching role in the United States and made Hofmann simultaneously the influencer and the benefactor of the leading edge of modern art practice. Hofmann studied in Munich and Paris during formative modernist periods and his influence both in Europe and the USA was significant. Hofmann's paintings are a synthesis of Abstract Expressionist style, material constructions that symbolise the changing relationship between artists and art-making. "Hofmann's powerful handling . . . called attention to the physical nature of the paint matter," recalled *New York Times* critic and former Museum of Modern Art curator Sam Hunter, "and thus propelled it forward." (Yohe 2002, 30) The study of Hofmann's late-career paintings provides insight into not only the style but also the substance of the era's artistic and material innovation. "Hofmann's art," Hunter noted, "is surely one of the remarkable examples in the century of modern artistic style of consciousness reporting objectively on itself." (Yohe 2002, 31)

Painting selection. The bulk of the paintings studied were selected from the Hofmann collection at the University of California Berkeley Art Museum and Pacific Film Archive (Berkeley, California), the world's largest and most comprehensive public collection of the artist's work, with additions from the Albright-Knox Art Gallery (Buffalo, New York), the Memorial Art Gallery (Rochester, New York) and the Museum of Modern Art (New York, New York). Palettes were selected from the holdings of the Renate, Hans and Maria Hofmann Trust (New York, New York). Twenty-six paintings and eight palettes were chosen for study and analysis as exemplars of Hofmann's late-career work. The study group includes 23 works on canvas from the years 1953–1965. Two works on canvas from the late 1940s were included for comparison, as were eight palettes on various supports found in Hofmann's studio at the time of his death. Paintings on panel were not included. The paintings selected for this study represent the range of Hofmann's late-career materials and paint application methods. Table 1 presents a chronological listing of the study group paintings.

---

[2]Coates (1946), 83.

Table 1 Chronological listing of Hofmann paintings selected for study group

Year	Title	Collection	Acc. no.	Size
1947	*Delight*	MoMA	2.1956	50.0 × 40.0″ (126.9 × 101.6 cm)
	Ecstasy	BAM	1963.2	68.0 × 60.0″ (172.7 × 152.4 cm)
	The Third Hand	BAM	1966.48	60.1 × 40.0″ (152.7 × 101.6 cm)
1953	*Le Gilotin*	BAM	1965.15	58.0 × 48.0″ (147.3 × 121.9 cm)
1954	*Scintillating Space*	BAM	1966.47	84.1 × 48.4″ (213.6 × 122.9 cm)
1955	*Exuberance*	AKAG	1955:8	50.0 × 40.0″ (127.0 × 101.6 cm)
1957	*Sommernachtstraum*	AKAG	1958:4	52.0 × 60.0″ (132.1 × 152.4 cm)
1958	*Equinox*	BAM	1965.12	72.1 × 60.3″ (127.0 × 101.6 cm)
	Morning Mist	BAM	1966.45	55.1 × 40.4″ (140.0 × 102.6 cm)
1959	*Above Deep Waters*	BAM	1965.13	84.2 × 52.0″ (213.9 × 132.1 cm)
	Indian Summer	BAM	1965.11	60.1 × 72.2″ (152.7 × 183.4 cm)
	Ruby Gold	MAG	60.37	55.4 × 40.5″ (140.7 × 102.9 cm)
	The Vanquished	BAM	1966.49	36.1 × 48.1″ (91.7 × 122.2 cm)
1960	*Bald Eagle*	BAM	1964.3	60.3 × 52.3″ (153.2 × 132.8 cm)
	In the Wake of the Hurricane	BAM	1965.6	74.3 × 60.0″ (153.2 × 132.8 cm)
1961	*Combinable Wall I and II*	BAM	1963.10	84.5 × 112.5″ (214.6 × 285.8 cm)
	Tormented Bull	BAM	1963.6	60.1 × 84.3″ (152.7 × 214.1 cm)
1962	*Heraldic Call*	BAM	1965.17	60.3 × 48.4″ (153.2 × 122.9 cm)
	Magnum Opus	BAM	1963.7	84.1 × 78.1″ (213.6 × 198.4 cm)
	Memoria in Aeternum	MoMA	399.1963	84.0 × 72.1″ (213.4 × 183.2 cm)
1963	*Polyhymnia*	BAM	1964.1	72.1 × 60.3″ (183.1 × 153.2 cm)
1964	*The Clash*	BAM	1965.8	52.1 × 60.3″ (132.3 × 153.2 cm)
	Imperium in Imperio	BAM	1966.43	84.1 × 52.0″ (213.6 × 132.1 cm)
	And Out of the Caves the Night Threw a Handful of Pale Tumbling Pigeons in the Light	BAM	1965.4	84.1 × 60.3″ (213.6 × 153.2 cm)
	Silent Night	BAM	1965.5	84.0 × 78.3″ (213.4 × 198.9 cm)
1965	*Struwel Peter*	BAM	1966.5	72.1 × 60.3″ (183.1 × 153.2 cm)
1966	Palette on plywood	Estate	M536-12	7.5 × 7.5″ (19.1 × 19.1 cm)
	Palette on board	Estate	M593-12	4.8 × 9.0″ (12.1 × 22.9 cm)
	Palette on board	Estate	M537-10	41.0 × 8.0″ (104.1 × 20.3 cm)
	Palette on board	Estate	M536-53	6.5 × 8.0″ (16.5 × 20.3 cm)
	Palette on glass	Estate	no #	24.0 × 24.0″ (61.0 × 61.0 cm)
	Palette on paint can lid	Estate	M536-49	6.0″ (15.2 cm) diameter
	Palette on board	Estate	M536-45	6.0 × 11.8″ (15.2 × 29.8 cm)
	Palette on board	Estate	M536-03	4.0 × 11.0″ (10.2 × 27.9 cm)

AKAG Albright-Knox Art Gallery (Buffalo, New York), *BAM* University of California Berkeley Art Museum and Pacific Film Archive (Berkeley, California), *Estate* Renate, Hans and Maria Hofmann Trust (New York, New York), *MAG* Memorial Art Gallery, University of Rochester (Rochester, New York)

Primary sources of information related to the paintings included the conservation records of the Berkeley collection held at the Berkeley Art Museum and the San Francisco Museum of Modern Art. The conservation staff at the San Francisco Museum of Modern Art have been responsible for the preservation of the university's Hofmann paintings since 1974 and maintain an archive recording nearly 40 years of conservation assessment and treatment for the collection. The Berkeley Art Museum also maintains a curatorial archive related to the collection's conservation. Two unpublished surveys of the Berkeley Art Museum's Hofmann paintings produced by conservators Thornton Rockwell (1982) and Carolyn Tallent (1988) were also consulted.

Analysis of Materials

A total of 284 paint material and 28 fibre samples were taken from the study group paintings and palettes. Some paint samples presented multilayered stratigraphy and were mounted as cross-sections to allow each paint layer to be analysed separately. A total of 519 discrete paints were analysed for this research. Information regarding ground layer stratigraphy and identification of inorganic pigments was obtained using optical microscopy and scanning electron microscopy with energy dispersive X-ray spectroscopy (SEM-EDX). Fourier transform infrared spectroscopy (FTIR) and gas chromatography-mass spectrometry (GCMS) were used to identify binders and organic pigments.[3] The sample preparation methods and instruments used are listed in the appendix. The results of each analytical method are discussed below in order of application.

Optical Microscopy. Optical microscopy was primarily utilised for initial examination and photo-documentation of the study group samples. All 284 paint material samples and 28 fibre samples were examined and photographed using both normal and ultraviolet illumination sources. 83 of the 284 paint samples presented multilayered stratigraphy and were selected for mounting and analysis as cross-sections.

Examination of warp and weft fibres from the study group paintings indicates a distinct late-career shift in Hofmann's choice of fabric support. Cotton canvases were identified on two 1947 paintings from which fibre samples were available, while linen fibres were identified in the canvases from all 14 of the late-career paintings from which fibre samples were available.

Pyrolysis-Gas Chromatography-Mass Spectrometry. GCMS analysis was utilised in order to differentiate between modern and traditional binding media in Hofmann's materials. GCMS binding media analysis was performed on 106 samples. A pyrolysis (Py) attachment was used for all GCMS analysis, to meet the

[3]Analysis conducted by the author at the scientific laboratories of the National Gallery of Art, Washington and the Museum Conservation Institute, Smithsonian Institution.

thermal dissociation requirements of the high molecular weight and heavily cross-linked polymers found in modern paints such as alkyds and modern industrial blends that contain alkyd or acrylic modifiers. Py-GCMS analysis was used where materials were expected to contain modern polymer formulations, such as ground layer paints suspected to contain house paints (Rogala et al. 2010) and those paints exhibiting the flow characteristics or surface sheen associated with industrial paints (Tallent 1988; Arslanoglu et al. 2013). All white and black paints were analyzed using Py-GCMS in order to observe any patterns in use between ground and compositional painting materials. Selected coloured paints were also included in Py-GCMS analysis.

Py-GCMS analysis of Hofmann's paint materials confirmed Hofmann's preferential use of oil-based paint materials and his late-career shift to oil-based alkyd paints in his ground and priming layers. Py-GCMS chromatograms of all 106 samples contained peaks consistent with the palmitic, stearic, oleic, and azelaic fatty acids in oil paint. All three paintings produced in 1947 gave positive results for oil grounds using Py-GCMS analysis. All 17 of the late-career paintings from which ground layer samples were available also included peaks consistent with the phthalic anhydride component of alkyd paints.[4] Py-GCMS analysis was not possible on the limited ground layer sample from the 1960 painting *Bald Eagle*, but SEM-EDX results showing a bulked titanium white layer are consistent with the SEM-EDX results for the alkyd white grounds found in the other paintings, suggesting that a similar ground was used.[5] Alkyd grounds were also found on two primed palettes from Hofmann's studio. Paint under the lip of an impromptu palette made from a paint can lid was also found to be alkyd paint, and the embossed markings on the can lid were traced to a batch of alkyd interior house paint produced on March 17, 1957 by Benjamin Moore & Company. Brochures for this paint were found among Hofmann's studio papers (Rogala 2005). Table 2 presents a summary of all binding media identified during analysis of Hofmann's late-career ground and priming layer materials.

Black paints in alkyd binding media were also found in *Bald Eagle, Tormented Bull*, and *Heraldic Call*. No coloured alkyd paints were identified. No acrylic or water-based paints were identified in any of the samples analyzed using Py-GCMS.

Scanning Electron Microscopy-Energy Dispersive Spectroscopy. SEM-EDX analysis was utilised to identify elements associated with inorganic pigments in Hofmann's paints. EDX was utilised as a qualitative technique to group the samples into inorganic pigment families and highlight potential organic pigments requiring additional analysis.[6] Backscatter electron images were taken of all loose and mounted paint samples, and EDX inorganic materials analysis was performed on all discernable paint layers within those samples.

[4]Tallent (1988) suggests that the oil ground layer found on *Tormented Bull* is a commercially applied ground. Hofmann's applied layer of alkyd should therefore be considered a ground layer.

[5]Hofmann used zinc white compositional paints. Zinc white makes a poor alkyd, so a zinc-lacking paint is also a good indicator of an alkyd in Hofmann's paintings.

[6]EDX analysis was also performed on visible inclusions within individual paint layers.

Table 2 Summary of ground and priming layer analysis

Year	Title	Acc. no.	Sample no.	Result	Method(s)
1947	*Delight*	MoMA 2.1956	N21	Oil	Py-GCMS
	Ecstasy	BAM 1963.2	Esc01	Oil	Py-GCMS
	The Third Hand	BAM 1966.48	C071	Oil	Py-GCMS
1954	*Scintillating Space*	BAM 1966.47	C072	Alkyd	Py-GCMS
1955	*Exuberance*	AKAG 1955:8	B03	Alkyd	Py-GCMS, ATR-FTIR
			B04	Alkyd	Py-GCMS
1957	*Sommernachtstraum*	AKAG1958:4	B17	Alkyd	Py-GCMS, ATR-FTIR
1958	*Equinox*	BAM 1965.12	C057	Alkyd	Py-GCMS
1959	*Above Deep Waters*	BAM 1965.13	C121	Alkyd	Py-GCMS, ATR-FTIR
	Ruby Gold	MAG 60.37	R07	Alkyd	Py-GCMS
			R11	Alkyd	Py-GCMS
1960	*Bald Eagle*	BAM 1964.3	C026b	Alkyd	SEM-EDX
	In the Wake . . .	BAM 1965.6	C180	Alkyd	Py-GCMS
1961	*Combinable Wall I/II*	BAM 1963.10	C085	Alkyd	Py-GCMS, ATR-FTIR
	Tormented Bull	BAM 1963.6	C059	Alkyd[a]	Py-GCMS
1962	*Heraldic Call*	BAM 1965.17	Hera01	Alkyd	Py-GCMS
	Magnum Opus	BAM 1963.7	C151	Alkyd	Py-GCMS, ATR-FTIR
			C153	Alkyd	Py-GCMS, μFTIR
	Memoria in Aeternum	MoMA 399.1963	N04	Alkyd	Py-GCMS, ATR-FTIR
1963	*Polyhymnia*	BAM 1964.1	C137	Alkyd	Py-GCMS, ATR-FTIR
1964	*The Clash*	BAM 1965.8	C003	Alkyd	Py-GCMS, ATR-FTIR
	And Out of the Caves . . .	BAM 1965.4	C169	Alkyd	Py-GCMS, ATR-FTIR
	Silent Night	BAM 1965.5	C101	Alkyd	Py-GCMS, ATR-FTIR
1965	*Struwel Peter*	BAM 1966.5	C145	Alkyd	Py-GCMS, μFTIR
1966	Palette on board	Estate M536-45	S26	Alkyd	Py-GCMS, ATR-FTIR
	Palette on board	Estate M536-03	S29	Alkyd	Py-GCMS, ATR-FTIR

AKAG Albright-Knox Art Gallery (Buffalo, New York), *BAM* University of California Berkeley Art Museum and Pacific Film Archive (Berkeley, California), *Estate* Renate, Hans and Maria Hofmann Trust (New York, New York), *MAG* Memorial Art Gallery, University of Rochester (Rochester, New York), *MoMA* Museum of Modern Art (New York, New York), *Py-GCMS* Pyrolysis-Gas Chromatography-Mass Spectrometry, *SEM-EDX*: Scanning Electron Microscopy-Energy Dispersive Spectroscopy, *ATR-FTIR, μFTIR*: Attenuated Total Reflectance- and Micro-Fourier Transform Infrared Spectroscopy

[a]This alkyd priming was applied by Hofmann over a commercially applied ground layer

White. The pigments identified in Hofmann's ground and compositional materials are consistent with reports of the artist's shift from oil- to alkyd-based ground layers. Mixed zinc and titanium white pigments (distinguishing elements: Zn, Ti) consistent with zinc oxide oil paint formulations were identified in the ground layers of all three of the study group paintings produced in 1947. No zinc pigments were identified in the ground layer materials of the 17 late-career paintings from which ground layer samples were available, but all of available late-career ground layer samples were found to contain titanium white paints consistent with alkyd white paints. All but one of the study group paintings contained compositional applications of zinc/titanium mix white paints consistent with Hofmann's reported use of Permalba (De Kooning 1950). All but one of the study palettes included similar zinc and titanium containing pigment mixtures. *Heraldic Call* contained no compositional white paints.

Yellow. Cadmium-containing yellows (distinguishing elements: Cd, S) were the primary yellow pigments found in Hofmann's late-career work. Cadmium-containing yellows appear in nearly every study group painting and palette, alone or mixed with zinc/titanium white or ultramarine blue.[7] Yellow ochre (distinguishing element: Fe) appears infrequently throughout the study group. Cadmium yellow and yellow ochre both appear in the 1947 painting *Delight*, in *Sommernachtstraum* (1957), *Morning Mist*, (1958) *Combinable Wall I and II* (1961) *Silent Night* (1964), and the unnumbered Estate palette on glass. In *Combinable Wall I and II* a secondary shade of cadmium yellow included a calcium-based extender. In *Imperium in Imperio* a secondary shade of cadmium yellow included a barium-based extender.[8] Zinc chromate yellow (distinguished by the presence of Zn and Cr) appears twice in the study group. Cadmium yellow and zinc yellow both appear in *Ecstasy* (1947) and *Equinox* (1958).

Orange. Cadmium-containing orange (distinguishing elements: Cd, S, Se) is the only orange pigment identified in the study group samples. Cadmium orange appears consistently in the paintings and palettes from 1954, occasionally mixed with zinc/titanium white. *Ruby Gold* (1959) included a secondary shade of cadmium orange mixed with a calcium-based extender and *Combinable Wall I and II* (1961) included a secondary shade of cadmium orange mixed with a barium-based extender.

Red. Cadmium-based red (distinguishing elements: Cd, S, Se) appears consistently throughout the study group paintings and palettes, sometimes mixed with zinc/titanium white, and regularly in compositions containing a range of other cadmium-based colours. A barium-based extender was mixed with secondary shades of cadmium red in *Above Deep Waters* and *Indian Summer* (both 1959), and in one of the palettes (Estate M536-53). Supplemental reds in *Combinable Wall I*

[7]Sampling from palettes was limited to priming layers colours used for mixing. Other colours in consistent late-career use may also appear on the palettes.

[8]Precise identification of extenders would require further analysis.

and II (1961) were mixed with zinc/titanium white or with the titanium white ground material. Cadmium red and an oxide red-brown both appear in *Sommernachtstraum* (1957).

Green. A variety of green pigments were identified in the paintings, including artist's mixtures of cadmium yellow and ultramarine blue. A more homogenous green containing components of cadmium green (elements: Cd, Zn, S) appears consistently in the paintings and palettes beginning in 1953. The cadmium-containing green often appears on the same composition as phthalocyanine green (distinguishing elements: Cl, Cu), and may be the phthalocyanine-containing "[permanent] green light" that appears on a pigment list found among the artist's papers.[9] The cadmium-containing green appears alone on the 1955 painting *Exuberance*, *Sommernachtstraum* (1957) *Morning Mist* (1958) *Memoria in Aeternum* (1962) and *Polyhymnia* (1963). Phthalocyanine green appears alone on the 1947 paintings *Ecstasy* and *The Third Hand*, and on *The Clash* (1964). A second cadmium-containing green in *Above Deep Waters* (1959) is mixed with a barium extender. Viridian, also known as chrome green (distinguishing element: Cr) appears infrequently; a chrome-based green was found alone in samples from *Delight* (1947) and together with cadmium and phthalocyanine greens in *Silent Night* (1964). Similarly, a greenish iron oxide pigment (distinguishing element: Fe) appears infrequently; together with cadmium green in *Ruby Gold* (1959) and with cadmium and phthalocyanine greens in *And Out of the Caves the Night Threw a Handful of Pale Tumbling Pigeons in the Light* (1964). A possible cobalt green (distinguishing elements: Co, Zn) appears with cadmium green in *Memoria in Aeternum* (1962) and with cadmium, phthalocyanine, and chrome greens in *Silent Night*.

Blue. Ultramarine blue (distinguishing elements: Na, Al, Si) is the primary blue pigment appearing in Hofmann's late-career work. Ultramarine blue pigment appears in every painting except the black-and-white composition *Tormented Bull* (1961). Ultramarine blue appears alone and mixed with a variety of colours. Phthalocyanine blue (distinguishing element: Cu) appears with ultramarine blue in the 1947 painting *Ecstasy*, and on *Le Gilotin* (1953) and *Morning Mist* (1958). Phthalocyanine blue also appears on the unnumbered glass palette. Cobalt blue (distinguishing elements: Co, Al) was used in four paintings from the 1960s: *Bald Eagle* (1960), *Magnum Opus* (1962), and *Imperium in Imperio* and *Silent Night* (both 1964). Cerulean blue (distinguishing elements: Co, Sn) appears in only one painting, *Equinox* (1958). An iron-containing blue (possibly Prussian Blue) appears in one painting, *Sommernachtstraum* (1957), and one of the palettes (Estate M536-45).

Violet/Magenta/Purple. Both inorganic and organic purple pigments appear in Hofmann's paintings. Cobalt violet (distinguishing elements: Co, P) was found in many of his later paintings, from *Le Gilotin* (1953) to *Struwel Peter* (1965).

[9]Phthalocyanine pigments contain both organic and inorganic components and can therefore be identified using multiple analytical techniques—in the case of this study, by both FTIR and SEM-EDX.

A similar violet pigment with inconclusive EDX spectra but exhibiting similar topographical characteristics in SEM and closely similar FTIR spectra was found in *Ruby Gold* and *Indian Summer* (both 1959), *In the Wake of the Hurricane* (1960), and *And Out of the Caves the Night Threw a Handful of Pale Tumbling Pigeons in the Light* (1964) Ultramarine violet (distinguishing elements: Na, Al, S) was found in one painting *Morning Mist* (1958). EDX analysis of magenta paint samples from 12 of the 23 study group paintings identified levels of aluminum or barium consistent with substrate used in natural and synthetic dye-based pigments. Further information regarding these pigments was obtained through organic analysis (see FTIR analysis below).

Black. Hofmann's late-career paintings exhibit limited use of black pigment. Ivory black, also known as bone black (distinguishing elements: Ca, P) was found in samples from eight paintings, the earliest occurrence in *Ecstasy* (1947), and the last in *The Clash* (1964). Ivory black frequently appears mixed with phthalocyanine green in Hofmann's paintings. *Heraldic Call* (1962) and the previous year's *Tormented Bull* contain both oil and alkyd blacks. Neither black in *Tormented Bull* has a discernable EDX signature, thus indicating the presence of carbon black pigment. Blacks without discernable EDX signatures also appear in *Delight* (1947), *Ruby Gold* and *The Vanquished* (both 1959), and *Bald Eagle* (1960), although carbon black cannot be reliably characterised by EDX analysis and may appear in other paintings. *Bald Eagle* was painted with two black pigments and two paint mediums; ivory black in oil and carbon black in alkyd paint media.

Fourier Transform Infrared Spectroscopy. FTIR analysis was used to identify organic pigments, classes of organic materials, and functional groups indicative of some inorganic materials. FTIR analysis was performed on 172 loose paints and 76 cross-sections. The method used did not allow for layer-specific analysis of cross-section samples. Organic pigment identification was based on a comparison of study group spectra with the spectral databases of the Infrared & Raman Users Group (IRUG). Results listed below are organised according to international Colour Index (CI) classifications (The Society of Dyers and Colourists 1956, 3267). Table 4 presents a summary of the organic pigments identified in the paintings investigated.

Organic pigments, blue and green. Two phthalocyanine-based pigments were identified in the paintings.[10] CI Pigment Blue 15 was found in two paintings—*Le Gilotin* (1953) and *Morning Mist* (1958)—and two of the Estate palettes. Samples of pre-1953 phthalocyanine containing pigments were present in too low a concentration to be confirmed using FTIR although samples of phthalocyanine blue from *Ecstasy* (1947) showed an elemental profile consistent with other paints identified as phthalocyanine blue by FTIR. The designations PB15.1-15.6 noted in Table 4 reflect the related IRUG spectra for structural variations in the pigment, although structural identification of pigments was beyond the scope of the present study.

[10]Phthalocyanine pigments contain both organic and inorganic components and can therefore be identified using multiple analytical techniques—in the case of this study, by both FTIR and SEM-EDX.

Table 3 Pigments identified using Energy Dispersive Spectroscopy

Title	Acc. no.	Zinc white	Titanium white	Cadmium yellow	Yellow ochre	Zinc yellow	Cadmium orange	Cadmium red	Red ochre	Cadmium green	Phthalocyanine green	Viridian/ Chrome green	Iron oxide green	Cobalt green	Ultramarine blue	Phthalocyanine blue	Cobalt blue	Cerulean blue	Prussian blue	Ultramarine violet	Cobalt violet	Ivory black	Carbon black[b]
Delight	MoMA 2.1956	×		×	×			×					×		×								×
Ecstasy	BAM 1963.2	×		×		×		×			×				×	×						×	
The Third Hand	BAM 1966.48	×		×				×			×				×								
Le Gilotin	BAM 1965.15	×		×				×		×	×				×	×					×	×	
Scintillating Space	BAM 1966.47	×	×	×			×	×		×	×				×								
Exuberance	AKAG 1955:8	×	×	×				×		×					×								
Sommernachtstraum	AKAG 1958:4	×	×	×	×		×	×	×	×					×				×				
Equinox	BAM 1965.12	×		×		×	×	×		×	×				×			×			×		
Morning Mist	BAM 1965.45	×		×	×		×	×		×					×	×				×			
Above Deep Waters	BAM 1965.13	×	×	×			×	×		×	×				×							×	
Indian Summer	BAM 1965.11	×		×			×	×		×					×								
Ruby Gold	MAG 60.37	×	×	×			×	×		×			×		×						×		×
The Vanquished	BAM 1966.49	×		×			×	×		×	×				×						×		×
Bald Eagle	BAM 1964.3	×	×	×			×	×		×	×				×		×				×	×	×
In the Wake of the Hurricane	BAM 1965.6	×	×	×			×	×		×					×						×		
Combinable Wall I and II	BAM 1963.10	×	×	×	×		×	×		×	×				×							×	

(continued)

Table 3 (continued)

Title	Acc. no.	Zinc white	Titanium white	Cadmium yellow	Yellow ochre	Zinc yellow	Cadmium orange	Cadmium red	Red ochre	Cadmium green	Phthalocyanine green	Viridian/ Chrome green	Iron oxide green	Cobalt green	Ultramarine blue	Phthalocyanine blue	Cobalt blue	Cerulean blue	Prussian blue	Ultramarine violet	Cobalt violet	Ivory black	Carbon black[b]
Tormented Bull	BAM 1963.6	×	×																				×
Heraldic Call[a]	BAM 1965.17		×					×														×	
Magnum Opus	BAM 1963.7	×	×	×				×							×		×						
Memoria in Aeternum	MoMA 399.1963	×	×	×			×	×		×				×	×								
Polyhymnia[a]	BAM 1964.1	×	×					×		×					×							×	
The Clash	BAM 1965.8	×	×	×			×	×			×				×						×		
Imperium in Imperio[a]	BAM 1966.43	×		×			×	×		×	×				×		×						
And Out of the Caves the Night Threw a Handful of Pale Tumbling Pigeons in the Light	BAM 1965.4	×	×		×			×		×	×		×								×		
Silent Night	BAM 1965.5	×	×	×	×		×			×	×	×		×	×		×						
Struwel Peter	BAM 1966.5	×	×	×			×	×		×					×						×		

Palette on plywood[a]	Estate M536-12	×																					
Palette on board[a]	Estate M593-12	×									×												
Palette on board[a]	Estate M537-10	×		×			×	×		×	×				×								
Palette on board[a]	Estate M536-53	×					×	×															
Palette on glass[a]	Estate, no #	×		×	×					×					×	×							
Palette on paint can lid[a]	Estate M536-49	×	×												×								
Palette on board[a]	Estate M536-45	×	×												×				×				
Palette on board[a]	Estate M536-03		×								×												

AKAG Albright-Knox Art Gallery (Buffalo, New York), *BAM* University of California Berkeley Art Museum and Pacific Film Archive (Berkeley, California), *Estate* Renate, Hans and Maria Hofmann Trust (New York, New York), *MAG* Memorial Art Gallery, University of Rochester (Rochester, New York), *MoMA* Museum of Modern Art (New York, New York)

[a]Not all paint materials were sampled from this study group item

[b]Carbon black is not detectable using EDX and may be present along with other black pigments

Table 4 Pigments identified using Fourier-Transform Infrared Spectroscopy

Year	Title	Acc. no.	Phthalocyanine Blue	PB 15	PB 15:1	PB 15:2	PB 15:4	PB 15:6	Phthalocyanine Green	PG 7	Other[b]	Alizarin	PR 83	Other[c]	Rhodamine	PR 81
1947	*Delight*	MoMA 2.1956												×		
	Ecstasy	BAM 1963.2												×		
	The Third Hand	BAM 1966.48											×			
1953	*Le Gilotin*	BAM 1965.15		×	×					×	×		×			
1954	*Scintillating Space*	BAM 1966.47								×						
1955	*Exuberance*	AKAG 1955:8														
1957	*Sommernachtstraum*	AKAG 1958:4									×					
1958	*Equinox*	BAM 1965.12								×						
	Morning Mist	BAM 1966.45		×	×						×		×			
1959	*Above Deep Waters*	BAM 1965.13														
	Indian Summer	BAM 1965.11								×						
	Ruby Gold	MAG 60.37														
	The Vanquished	BAM 1966.49														
1960	*Bald Eagle*	BAM 1964.3														
	In the Wake of the Hurricane	BAM 1965.6														
1961	*Combinable Wall I and II*	BAM 1963.10									×					
	Tormented Bull	BAM 1963.6														

1962	*Heraldic Call*[a]	BAM 1965.17														
	Magnum Opus	BAM 1963.7												×		
	Memoria in Aeternum	MoMA 399.1963														
1963	*Polyhymnia*[a]	BAM 1964.1												×		
1964	*The Clash*	BAM 1965.8											×			
	Imperium in Imperio[a]	BAM 1966.43									×					
	And Out of the Caves the Night Threw a Handful of Pale Tumbling Pigeons in the Light	BAM 1965.4														
	Silent Night	BAM 1965.5								×						
1965	*Struwel Peter*	BAM 1966.5														
1966	Palette on plywood[a]	Estate M536-12														
	Palette on board[a]	Estate M593-12		×	×	×	×	×								
	Palette on board[a]	Estate M537-10								×						
	Palette on board[a]	Estate M536-53														×
	Palette on glass[a]	Estate, no #		×	×			×								
	Palette on paint can lid[a]	Estate M536-49														×
	Palette on board[a]	Estate M536-45														
	Palette on board[a]	Estate M536-03								×	×					×

AKAG Albright-Knox Art Gallery (Buffalo, New York); BAM: University of California Berkeley Art Museum and Pacific Film Archive (Berkeley, California), *Estate* Renate, Hans and Maria Hofmann Trust (New York, New York), *MAG* Memorial Art Gallery, University of Rochester (Rochester, New York), *MoMA*: Museum of Modern Art (New York, New York)

[a]Not all paint materials were sampled from this study group item

[b]Samples not directly identified as the pigment but matching FTIR spectra for other samples containing the pigment

[c]Matched to other samples identified as PR83 or presenting SEM-EDX results consistent with an alumina substrate

Organic materials consistent with IRUG standards for phthalocyanine-based CI Pigment Green 7 were identified in samples from ten paintings from 1953 to 1964 and two of the Estate palettes. Samples of phthalocyanine green from seven additional paintings and one additional palette were not available for FTIR analysis but had consistent SEM-EDX results to support the identification of phthalocyanine green paints on a total of 14 paintings and three palettes.

Lake pigments, alizarin. Alizarin-containing pigment is one of two magenta colours that appear in Hofmann's late-career work. FTIR analysis identified materials consistent with IRUG standards for CI Pigment Red 83 (an anthraquinone-based alizarin substitute for natural madder pigment) in samples from *Delight* and *The Third Hand* (both 1947) and three later works—*Le Gilotin* (1953), *Morning Mist* (1958), and *The Clash* (1964). Samples of magenta paints from there other paintings were too small to definitively identify the synthetic organic pigment used but provided FTIR matches to other samples identified as PR 83 or exhibited SEM-EDX results consistent with an alumina substrate. No examples of PR83 were found in the palettes.

Toners. A second magenta colour appears exclusively on Hofmann's late-career palettes. Xanthene-based rhodamine pigment CI Pigment Red 81 does not appear in the paintings but was identified in all three magenta paint samples available for testing from the Estate palettes using FTIR.

Discussion of Analyses

The analysis carried out in this study suggests that Hofmann's palette stays remarkably consistent from the late 1940s to the mid 1960s. The study group paintings employ the same colours recommended by Hofmann in his various student supply lists, including his life-long preference for zinc white (and zinc-titanium blends) over lead white (Hofmann Papers). The most significant shift in materials appears in Hofmann's late-career switch to exclusively alkyd grounds.[11] In 1960, splashed alkyd blacks appear, replacing the dripped inks and black oils of Hofmann's earlier work. Hofmann's palette is largely based on the modern colours that were favoured by the avant-garde throughout Hofmann's career. A handful of preferred colours appear in almost every painting in the study group. The cadmium reds and yellows embraced by Matisse and Fauves appear in every painting in the study group.[12] Hofmann's use of cadmium orange and synthetic phthalocyanine green is also consistent throughout this period, while his use of phthalocyanine blue appears to cease in 1958. Cobalt violet and alizarin crimson—Hofmann's purple and magenta staple colours—are also contemporary colour formulations. Cobalt violet

[11] *Heraldic Call* is the one exception to this, and it may be that the white background expanse was built over a commercially primed painting, as seen in *Tormented Bull*.

[12] Cadmium yellow is visible on *Heraldic Call*, but not sampled.

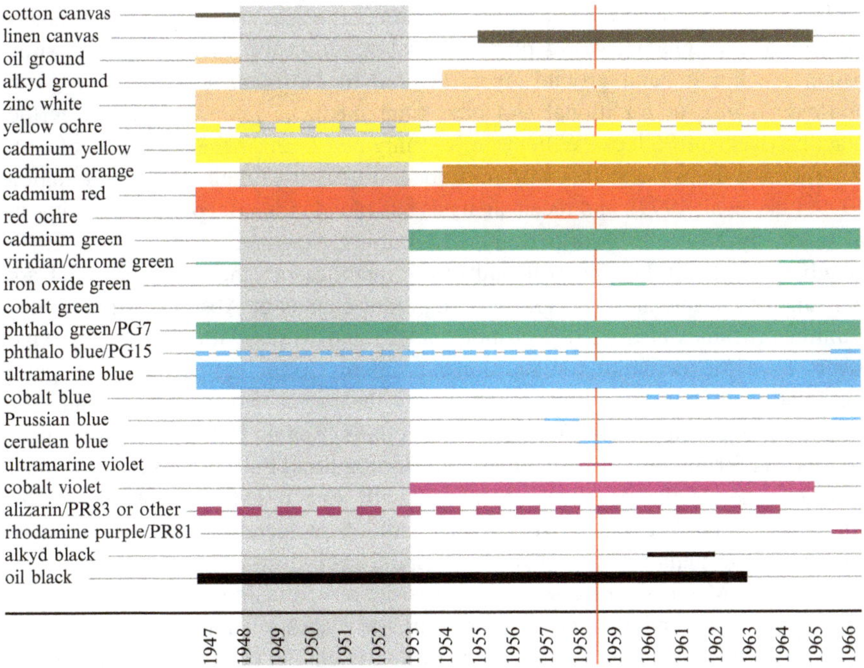

Fig. 2 Timeline of Hofmann's late-career materials. If use of a particular pigment is consistent, a solid line is drawn from beginning to end dates; sporadic use is a dashed line between beginning and end dates; the thickness of the line denotes the relative number of paintings on which the paint is found during this period. Not all colours were sampled from the study group paintings; colours not found in the study group works may exist in other paintings from the same year. *Red line* indicates the year Hofmann closed his schools and began to paint full time. No study paintings appear within grey timeframe

does not appear in the few study group paintings from the late 1940s but archival palette lists and anecdotal recollections note Hofmann's consistent use of the this colour throughout his years in the U.S. and has likely been used for other paintings made in this early period (Kiesler Papers). PR 83 is the only identifiable synthetic alizarin-based pigment used in Hofmann's paintings in the period investigated. PR 81 appears frequently but exclusively on the palettes found in Hofmann's studio at the time of his death. PR 83 was not found in any of the paintings after 1964, nor on the palettes. It is not clear whether this substitution was intentionally made by Hofmann or is the result of a formulation change by the manufacturer (Fig. 2).

Relationship of pigments and paint media to condition issues. Caretakers of Hofmann's work are faced with complex problems of lifting paint, blind cleavage, and slow-drying or efflorescing paint layers that appear to be related to Hofmann's use of new materials. Below are two examples of recurring condition issues in Hofmann's work that are not related to the paints themselves, but to Hofmann's application of new paint materials with traditional painting technique.

The alternating heavy paint layers, underbound washes, and splashed enamels of Hofmann's signature works place unusual stress on the underlying zinc white paints and house paint ground layers favored by Hofmann and his colleagues. Hofmann's mix of traditional and new paint media is representative of period practice. Jackson Pollock, Willem de Kooning and Franz Kline, for example, all incorporated industrial paints with traditional oil paints in their work (Coddington 1999; Lake et al. 1999), and recent analysis has found industrial paint ground layers in works by Kline, Ellsworth Kelly, and Barnett Newman (Rogala et al. 2010; Wijnberg et al. 2011). The stiff and brittle qualities of zinc oxide oil paint are not suited to the flexible and load-bearing requirements of Abstract Expressionist paintings (Maines et al. 2011; Osmond 2012) and works with industrial ground layers have the potential for additional problems from ineffective house paint formulations of zinc oxide marketed during the period when Hofmann and his colleagues experimented with house paint grounds (Rogala 2011). The white alkyd house paint formulations also popular with Hofmann and his Abstract Expressionist colleagues avoid the problems associated with zinc oxide pigment but exhibit similar problematic characteristics related to high pigment volume concentration and the brittle nature of an alkyd paint binder (Feller and Matous 1964; Hagan et al. 2007). Mechanical stresses on composite structures containing these kinds of inappropriate underlayers can result in the widespread ground failure and lifting paint layers common to the work of mid-twentieth century artists including Hofmann. The problem is neither the material nor the method of ground layer application, but the application of *this particular material* as a ground layer.

Exudates ranging from efflorescing fatty acids to expressed liquid oils are another example of problems that arise in Hofmann's work in relation to new materials. For example, unusual drying problems have been observed in relation to synthetic formulations of alizarin crimson (PR83) in every study group painting in which that pigment appears. Hofmann painted the majority of his late-career work while at his studio in Provincetown, Massachusetts and according to Herbst and Hunger (2004, 511), the alkaline-based formulation of PR83 breaks down on exposure to common organic solvents, such as benzine that Hofmann mixed with his paints to speed their drying in the coastal environment (De Kooning 1950, Hofmann Papers). These deterioration issues in Hofmann's paintings may therefore not be related to the new pigment formulation, but the result of incompatibility between traditional painting practices and newly formulated painting materials. Further investigation is needed to fully characterise the causes of this type of deterioration and its occurrence in paintings by other artists.

Conclusion

The analysis of painting materials used for Hofmann's late-career paintings undertaken in this study highlighted the relationship between aspects of the condition of his work and his combined use of new paint materials with traditional art practices

A representative catalogue of Hofmann's materials was characterised by paint and fibre analysis and reflects Hofmann's selective adoption of modern materials throughout his career. Examples of condition issues related to Hofmann's mix of new paint materials with traditional painting technique were investigated, including the use of stiff and brittle zinc oxide-containing house paints as ground layers and the degradation of synthetic organic pigments by common nonpolar solvents. In both cases, the deterioration was caused by an unanticipated incompatibility material and method.

Hofmann's oeuvre is an example of period artistic method that has wider implications for the evaluation of conservation methodology in the treatment of Abstract Expressionist and later works that incorporate both traditional and modern paint media. The legacy of Hofmann and his Abstract Expressionist colleagues would be well served by the continued study of that era's materials and practices.

Acknowledgments The author wishes to express her gratitude for the support of many people and institutions throughout this project. Albright-Knox Art Gallery: Laura Fleischmann. Museum Conservation Institute, Smithsonian Institution: Dr. Robert Koestler, Dr. Paula DePriest, Jennifer Giaccai, Nicole Little. Museum of Modern Art: Jim Coddington, Michael Duffy. National Gallery of Art: Dr. Christopher Maines, Jay Krueger. Renate, Hans and Maria Hofmann Trust: Robert Warshaw, James Yohe. San Francisco Museum of Modern Art: Jill Sterrett, Paula De Cristofaro, Alina Remba. University of California Berkeley Art Museum and Pacific Film Archive: Dr. Lucinda Barnes, Lisa Calden. University of Delaware: Dr. Joyce Hill Stoner, Debra Hess Norris, Dr. Murray Johnston, Dr. Roberta Tarbell. University of Rochester Memorial Art Gallery: Marjorie Searl and Jessica Marten. Funding, UD: Paul Coremans Endowment Art Conservation Research Fellowship, College of Arts and Sciences Dean's Doctoral Scholarship, and Professional Development, Collections-Based Research, and Global Research Awards. Funding, other: Daughters of the American Revolution Centennial Scholarship and National Endowment for the Humanities Material Culture Studies Graduate Research Fellowship; a 2011 Save America's Treasure grant brought the bulk of Berkeley's Hofmann paintings out of long-term storage and provided a brief window of unprecedented access to the full collection.

Appendix: Experimental Conditions

Cross-sections. Samples were positioned on pre-cast epoxy half-tablets and adhered using a small droplet of cyanoacrylate adhesive. The sample and half-tablet are then transferred into a silicon rubber mold for embedding. The epoxy resin, Trabond 2113 two-part system, was mixed and poured over the adhered sample and half-tablet. The liquid resin was held at room temperature for 1 h, then cured at approximately 45 °C for 3 h. Once de-molded, the tablets were trimmed with a mill to expose the sample. Aluminum oxide abrasive was used to polish the face edge the surfaces using an aliphatic hydrocarbon as a lubricant. (Some samples were dry polished later for FTIR analysis.)

Pyrolysis-Gas Chromatography-Mass Spectrometry. Samples were analyzed using a Varian Saturn 2000 GC/MS equipped with a CDS Pyroprobe 2000. Each sample was derivatised using two microliters of tetramethylammonium hydroxide

(TMAH) put onto the sample in a quartz boat. The boat was placed into the coiled platinum probe of a CDS Pyroprobe 2000 filament pyrolysis unit, and the probe was then placed into a helium-purged CDS 1500 Valved Interface attached to the Varian GC. The interface was held at a constant 310 °C and purged with helium for 10 s before opening the valve to the GC column. The sample was then heated with the pyroprobe to a temperature of approx. 600 °C for 10 s. The pyrolysis products were transferred directly to a capillary column (ZB-5 ms; 30 m $\times$ 0.25 mm i.d.; 0.25 micron film thickness; He flow of 1.2 ml/min; splitless.) in a Varian 3800 gas chromatograph (GC) equipped with electronic flow control. The GC oven was programmed with an initial temperature of 30 °C, which was held for 5 min. The temperature was increased at a rate of 10 °C per minute to 300 °C and held for 10 min. The Varian 3800 GC was interfaced to a Varian Saturn 2000 ion trap, the transfer line being held at 270 °C. Operating conditions for the trap were: trap 150 °C, manifold at 80 °C; electron multiplier 1,500 V; scan range 45–650 amu; scan time 1 s; data analysis Saturn GC/MS Workstation 6.5 software and the NIST 2005 spectral libraries.

Scanning Electron Microscopy-Energy Dispersive Spectrometry. Samples were imaged and analyzed using a Hitachi S3700-N scanning electron microscope and a Bruker XFlash energy dispersive spectrometer with Quantax 400 software. The samples were received after they had been prepared for and imaged by polarised light microscopy. They were carbon coated before analysis. The SEM was operated at 20–25 kV, at full vacuum. Elemental maps were generated over 300 s real time (with 0–18 % dead time). Spot and area analyses were conducted for 200 s live time. Analyses were conducted at a working distance between 9.8 and 10.2 mm. EDX analysis was performed on at least three disparate points within each paint layer; inclusions were also analysed.

Fourier Transform Infrared Spectroscopy. ATR-FTIR (Attenuated Total Reflectance) analysis of loose samples was performed using a Thermo Nicolet 6700 Fourier transform infrared (FTIR) spectrometer with a Golden Gate ATR with diamond crystal, single bounce (45°) sampling accessory and DTGS detector. Spectra were obtained from 64 scans taken at a resolution of 4 cm^{-1}. Samples for ATR-FTIR were placed directly on the diamond crystal of the ATR accessory. For small samples a piece of aluminum foil was used to back the sapphire anvil to eliminate any sapphire absorption in the IR spectrum. μFTIR (Infrared Microscope) analysis of embedded cross-sections was performed using a Thermo Nicolet 6700 Fourier transform infrared spectrometer with a Continuum microscope, MCT/A detector and a single bounce diamond crystal μATR objective. Spectra were obtained from 128 scans taken at a resolution of 4 cm^{-1} between 4,000 and 625 cm^{-1}. Samples for μFTIR were examined directly using a variable aperture to choose the area of interest. All spectra were ATR corrected and examined using OMNIC v 7.2 and compared to IRUG 2000 and local spectral libraries.

References

Arslanoglu J, Centeno SA, Digney-Peer S, Duvernois I (2013) "Picasso in the Metropolitan Museum of Art": an investigation of materials and techniques. J Am Inst Conserv 52(3): 140–155

Ball P (2001) Bright earth: art and the invention of color. Farrar, Straus and Giroux, New York

Coates R (1946) The art galleries, at home and abroad. New Yorker 22(7):83

Coddington J (1999) No chaos, damn it. In: Varnadoe K, Karmel P (eds) Jackson Pollock: new approaches. Museum of Modern Art, New York, pp 101–116

De Keijzer M (2002) The history of modern synthetic inorganic and organic artist's pigments. In: Mosk JA, Tennant NH (eds) Contributions to conservation: research in conservation at the Netherlands Institute for cultural heritage. ICN, Amsterdam, pp 42–53

De Kooning E (1950) Hans Hofmann paints a picture. Art News 48(10):38–41, 58–59

Feller R, Matous J (1964) Critical pigment volume concentration and chalking in paints. Bullet Am Group – IIC 5(4):25–26

Goodnough R (1951) Pollock paints a picture. Art News 50(3):38–41, 60–61

Goodnough R (1952) Kline paints a picture. Art News 51(8):36–39, 63–64

Hagan E, Charalambides M, Learner T, Murray A, Young C (2007) Factors affecting the mechanical properties of modern paints. In: Learner T (ed) Modern paints uncovered. Getty Conservation Institute, Los Angeles, pp 227–235

Hans Hofmann papers [ca. 1904]–1978, bulk 1945–1965. Archives of American Art, Smithsonian Institution, Washington, DC

Hess T (1953) De Kooning paints a picture. Art News 52(1):30–33, 64–67

Herbst W, Hunger K (2004) Industrial organic pigments: production, properties, applications, 3rd edn. Wiley-VCH Verlag GmbH & Co., Weinheim

Jaffe I (1971) A conversation with Hans Hofmann. ARTFORUM 9(5):34–39

Lake S, Lomax SQ, Schilling M (1999) A technical investigation of Willem de Kooning's paintings from the 1960s and 1970s. In: Preprints from the 12th ICOM-Committee on conservation triennial meeting in Lyon, France, 29 Aug–3 Sept 1999, pp 381–385

Lake S, Ordonez E, Schilling S (2004) A technical investigation of paints used by Jackson Pollock in his drip or poured paintings. In: Roy A, Smith P (eds) Modern art, new museums: Contributions to the 20th international institute of conservation congress in Bilbao, 13–18 Sept 2004, pp 137–141

Learner T (2004) Analysis of modern paints. Getty Conservation Institute, Los Angeles

Lillian and Frederick Kiesler papers [circa 1910]–2003, bulk 1958–2000. Archives of American Art, Smithsonian Institution, Washington, DC

Lomax SQ, Learner T (2006) A review of the classes, structures, and methods of analysis of synthetic organic pigments. J Am Inst Conserv 45(2):107–125

Maines C, Rogala D, Lake S, Mecklenburg MF (2011) Deterioration in abstract expressionist paintings: analysis of zinc oxide paint layers in works from the collection of the Hirshhorn Museum and Sculpture Garden, Smithsonian Institution. In: Vandiver P, Reedy CL, Li W, Ruvalcaba Sil JL (eds) Materials issues in art and archaeology IX: symposium held 29 Nov–3 Dec 2010, Boston, MA

Osmond G (2012) Zinc white: a review of zinc oxide pigment properties and implications for stability in oil-based paintings. Aust Inst Conserv Cult Mater Bullet 33:20–29

Pomerantz L (1962) Is your contemporary painting more temporary than you think?: Vital technical information for the present-day artist. International Book Company, Chicago

Rockwell T (1982) Conservation problems present in paintings by Hans Hofmann. Unpublished report, Elise S. Haas Conservation Department, San Francisco Museum of Modern Art

Rogala D (2005) Hans Hofmann from the ground up: looking at the artist's preparatory methods as a window to condition. Unpublished report, Art Conservation Department, Buffalo State College, State University of New York

Rogala D (2011) Industrial literature as a resource in modern materials conservation: zinc oxide house paint as a case study. In: Postprints from the 39th American Institute for Conservation of Historic and Artistic Works Annual Meeting in Philadelphia, Pennsylvania (in press)

Rogala D, Lake S, Maines C, Mecklenburg MF (2010) Condition problems related to zinc oxide underlayers: examination of selected abstract expressionist paintings from the collection of the Hirshhorn Museum and Sculpture Garden, Smithsonian Institution. J Am Inst Conserv 49(2):96–113

San Francisco Museum of Modern Art (1966-present) Unpublished records related to the conservation of Hofmann works in the collection of the Berkeley Art Museum. Records of the Elise S. Haas Conservation Department

Society of Dyers and Colourists (1956) Colour index, vol 3, 2nd edn. Society of Dyers and Colourists, Bradford

Standeven H (2011) House paints 1900–1960: history and use. Getty Conservation Institute, Los Angeles

Stella F (1999) Am Herit 50(7):14–17

Tallent C (1988) Investigation of the painting materials and techniques of Hans Hofmann—preliminary report. Unpublished report, Intermuseum Conservation Association

Toch M (1925) The chemistry and technology of paints, 3rd edn. D. Van Nostrand Company, New York

University of California, Berkeley, Berkeley Art Museum and Pacific Film Archive (1966—present) Unpublished records related to the conservation of Hofmann works in the collection

Wijnberg L, Bracht E, van den Berg KJ, de Keijzer M (2011) A study of the grounds used by three post-war American artists (1954–1975): Barnett Newman, Ellsworth Kelly and Brice Marden. In: Preprints from the 16th ICOM-Committee on Conservation Triennial Meeting in Lisbon, Portugal, 19–23 Sept 2011

Yohe J (ed) (2002) Hans Hofmann: revised and expanded. Rizzoli International Publications, Inc., New York

Théorème de Gödel by Georges Mathieu, 1957. Study and Restoration: Consolidation Through Cohesive Regeneration Using a Solvent

Hélène de Ségogne

Abstract This paper presents the study and restoration of Georges Mathieu's 1957 painting *Théorème de Gödel* (oil on canvas, 146 cm × 89 cm), in the Palais des Beaux Arts of Lille, France. The painting shows several signs of degradation which result from a combination of factors such as: the artist's action and painting technique: *tubisme* producing thick black impastos; the composition and interaction between the binding media, linseed oil and the ivory black pigment; the environmental conservation conditions.

Both mechanical and optical damages were identified, such as loss of cohesion and adhesion in the impastos resulting in cracks, severe cleavages and lacunas, linseed oil drips and surface soiling.

Since the painting is still not dry in the thick impastos, the deterioration is ongoing. A method was developed using little or no adhesive with the possibility of reversibility for future interventions. Pure ethanol was used to reactivate the cohesion and the adhesion of the cracked impastos.

Keywords Linseed oil • Drips • Cracks • Cleavage • Impasto • Loss of adhesion • Ivory black pigment • *Tubisme* • Ethanol

Artwork Description: Action Painting and Impastos

The artwork by Georges Mathieu, *Théorème de Gödel*, an oil on canvas painted in 1957, was presented by the Palais des Beaux Arts of the city of Lille, for its study and conservation. The examination of the artwork allowed to identify its constituent materials along with the pictorial movement and process of Georges Mathieu which enabled to understand the mechanisms that triggered the current alterations. The principal conservation issue raised by the work was the stabilization of the ongoing degradation process which were causing loss of cohesion and adhesion of the paint.

H. de Ségogne (✉)
ALMAFLO ART, Paris, France
e-mail: helenedesegogne@hotmail.com

K.J. van den Berg et al. (eds.), *Issues in Contemporary Oil Paint*,
DOI 10.1007/978-3-319-10100-2_9

Georges Mathieu was one of the most celebrated and controversial French artists in the period following the Second World War. His renown was established in the 1950s and was based on the novel aspect of his artistic expression as well as his fierce implication in the promotion of the artistic style which he himself named *Abstraction Lyrique* in 1947.[1]

Within the *Abstraction Lyrique* group were artists such as Hartung, Wols and Soulages, who evolved towards an abstract language comprising of a gestural writing using limitless new processes. This involved small projections of paint onto the canvas, called *tachisme,*[2] or large and ample brush strokes, or again the application of paint directly from the tube: *tubisme*.

These terms are used to describe: an art that is informal, abstract and liberated from all restrictions, unconsciously freeing the spontaneity of movement and emotion.

Georges Mathieu, a self-taught artist, very quickly found his style which he stayed faithful to all throughout his career. The innovative character of Georges Mathieu's paintings relies upon: his gesture, expressed from a deep inner feeling combined with what he refers to as "the movement of his cosmic energy"; his *tubisme* process; and lastly, his quickness in execution. As an example of this rapidness, the large 3×6 m painting *Les Capétiens partout*, kept at the Musée National d'Art Moderne, Centre Pompidou, was completed in less than 40 min.

The title of the artwork, *Théorème de Gödel,* refers to the name of a mathematical theorem on the subject of incompleteness. This title reveals George Mathieu's particular interest for physics, mathematics and metaphysics (Mathieu 1973).

Condition Report and Study of the Alterations: Still Drying and Tacky 55 Years After Creation

The artwork's constituent materials were derived from the artist's own writings (Mathieu 2006, p. 90). He would purchase his stretchers and primed canvas at Artes, a branch of the Berge firm which is today grouped under the name Sauer Raphael.[3] This canvas has a layer of priming composed of lead white in oil.[4] The paints were of

[1]Mathieu 2006.

[2]*tachisme* from french "tache" meaning stain; see e.g. Un art autre? Artistes autour de Michel Tapié. Catalogue d'exposition Christie's, Paris 31 janvier–29 février 2012; Mathieu Rétrospective, Catalogue d'exposition, Galerie Nationale du Jeu de Paume, Paris 17 juin–6 Octobre 2002.

[3]Located in St Brieuc (France).

[4]'lead oxide base' according to Monsieur Sennelier, former owner of Artes, which should be taken as lead white.

the Lefranc brand. The consultation of the Lefranc archives, the material catalogue and colour chart led us to think that the artist used student quality paints, which may contain more oil than the artist quality counterparts.[5] The colour was visually assimilated with ivory black. The binder of ivory black of Lefranc was linseed oil.[6] Mathieu did not use any siccative. Furthermore, *Théorème* is not varnished, as are most of his other paintings.

Based on the video films made of his performances we can assume that *Théorème de Gödel* was accomplished in only 4 min, following these steps:

- A base layer was created by soaking the canvas with an earth colour paint diluted in petroleum spirits giving this brownish tint on top of the white priming.
- Black paint was then applied using a flat brush in large strokes. This paint was mixed with a highly diluted small quantity of white paint, generating the grey drips and the transparent grey tinted areas.
- On top of this, the artist frenetically pressed the tube of black paint directly onto the canvas with wide movements forming these arabesques, sinuous lines and circles. The tube-like impastos were more or less flattened depending on the pressure exerted by Georges Mathieu. Several superposed lines created from the tube form impastos up to 1 cm thick.

This process created drastic differences in thickness of the paint layers and thus varying physico-chemical states of drying, depending on the thickness. The thin pressed areas are hard because the polymerisation of the oil paint film is completed while the parts which are up to 1 cm thick remain supple and sometimes tacky because core drying has still not been achieved (Boon and Hoogland 2014).

On a black and white photograph dating from the 1970s, more than 13 years after the completion of the work, there are no oil drips visible. They are, however, visible on the coloured photograph from the RMN in 1998 (Fig. 1).

The identification of the alterations from 2012 shows that the drips have been physically stable since 1998 because they are identical in shape and length although they are much darker today. Furthermore, more paint loss has been observed.

The condition report for *Théorème de Gödel* reveals a degraded and weakened painting in an unstable state of conservation. The alterations are both optical and mechanical: soiling, drips of oxidized discoloured oil; cracks, fissures, lifting and cupping of paint (up to 0.5 cm) causing paint loss and canvas deformations.

[5]the diameter of rolled impastos is similar to student quality tubes (archives catalogue Lefranc; located at ColArt International SA 5, rue René Panhard, ZI Nord 72021 Le Mans Cedex- France.

[6]Catalogue information des archives Lefranc Bourgeois.

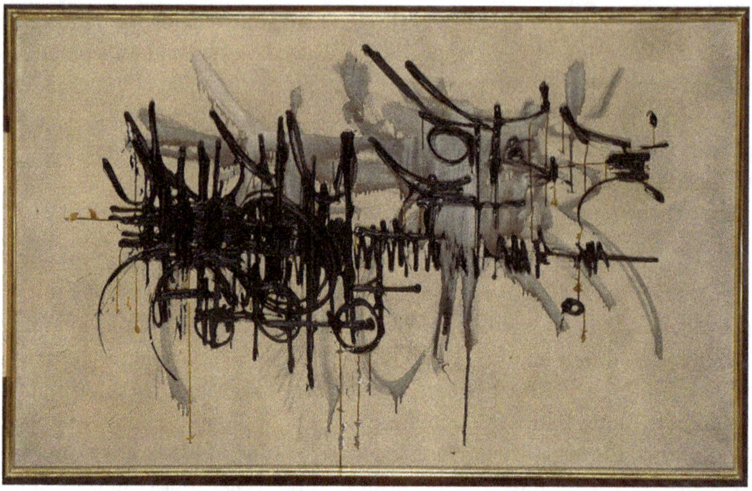

Fig. 1 *Théorème de Gödel*, Georges Mathieu, 146 cm × 89 cm, kept at the Palais des Beaux Arts of Lille, France (RMN: Réunion des Musées Nationaux 1998)

Diagnosis: An Ongoing Deterioration Revealing a Very Fragile Artwork

The alterations on the painting are the result of multiple factors. The interactions of the main paint constituents linseed oil and ivory black pigment have played a role in the physical and chemical properties.

Furthermore, the direct application of paint from the tube with superpositions of more or less crushed paints forming very thick and irregular impastos has also affected the properties of the constituents involved. In addition, the non-ideal environmental conservation conditions of the painting may have contributed to its present state.

Origins of Deteriorations

The mechanical alterations originate from the drying process of the linseed oil paint and ivory black which is a slow drying pigment with high oil absorption (O'Malley and Moffat 2001).

The cross-linking of the oil paint occurs through oxidation of the superficial layer which hardens, retracts and slows down the oxidation and cross-linking inside the impastos (Masschelein Kleiner 1993). This generates differences in stress between the superficial and the internal part which remains in a viscoelastic state, causing the apparition of cracks and degrading the film's cohesion until fissuring and loss of adhesion (Figs. 2 and 3).

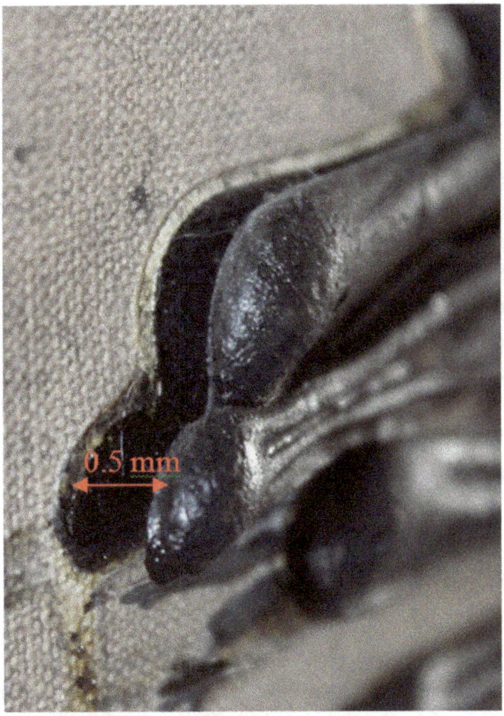

Fig. 2 Detail: 5 mm lifting of an impasto at the priming interface (Hélène de Ségogne)

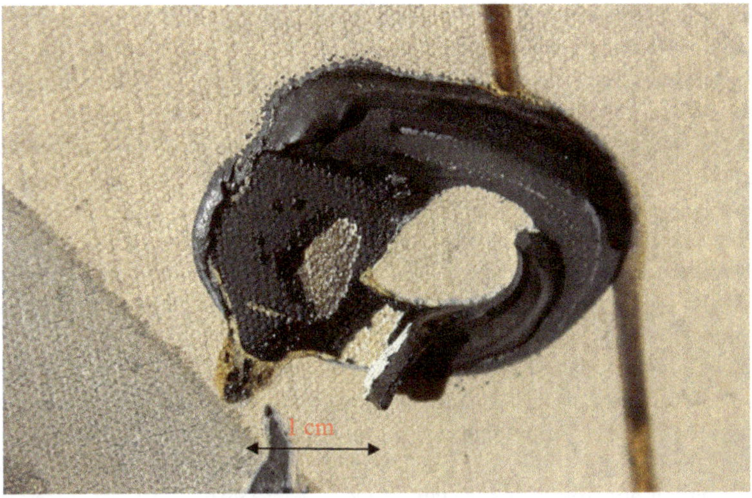

Fig. 3 Detail: fracturing and lifting of pictorial matter tearing off priming and overpaint (Hélène de Ségogne)

During polymerization the superficial film can become porous and permeable. While retracting, the film causes the canvas to distort and permits loss of binding agent in the form of drips of oil, also known as the syneresis phenomenon (De Pracontal 2008).

This loss of binder modifies the pigment volume concentration (PVC) and causes a reduction of elasticity and plasticity of the film in the exuded zones while amplifying the loss of internal cohesion (Roche 2003, p. 83). This leads to a decrease in adhesion inevitably followed by paint loss.

Parallel to this, the viscous internal layers have "sedimented" creating a separation of oil and pigment. The use of a student quality paint containing proportionally more oil than an artist quality one may have favoured this sedimentation.

Finally, it is possible that the storage of the work near a heat source[7] has increased the thermal expansion coefficient of the oil, encouraging its expulsion through the superficial polymerized yet porous layer thus forming the visible drips (Roche 2003, p. 87).

Restoration Aims Following an Axiological Reflection

This diagnosis and the knowledge of the artists' materials (Mathieu 2006) allowed us to establish the aims of the conservation treatment which were defined following an 'axiological' reflection (Verbeeck-Boutin 2009).[8] This was done while considering the intentional values of Georges Mathieu during the creation of *Théorème de Gödel* and the intentional, aesthetical, historical and artistic values that we bring today as conservators, curators or spectators.

Théorème de Gödel has to be 'read' as the testimony of the creating act of Georges Mathieu, that is his performance.

It also has to be understood as a message of the *Abstraction Lyrique* movement which praises spontaneity of action where chance plays a leading role.

The process of applying paint on the canvas through *tubisme,* also has to be considered as an intentional declaration of Georges Mathieu who seeks closure with all past references and classical pictorial processes. His performance and his novel process with the tube are therefore the two principal factors which give its meaning to the artwork.

The evolution of material over time, such as the oil drips, the traces of passing time and the alterations, the lifting of paint, the cracks and paint losses, while bothering the aesthetic apprehension of the work, do not alter its comprehension nor the intention of Georges Mathieu. This is the reason it was decided to leave the oil drips in their apparently stable state, since at least 1998.

[7] According to Mrs Nicole Jouvenet, Member of Société des Amis des Beaux Arts de Lille.

[8] Axiology could be defined as a theory of values in philosophy, both in the fields of aesthetics and ethics.

However without intervening on the significant lifting of paint, the short term risk of generating new paint loss was to high which would compromise any possible manipulation of the work. In the aim of transmitting this cultural heritage piece to future generations while keeping the perception of its meaning fully comprehensible to the public, it is necessary to intervene on the cracks and lifting of paint.

By undertaking punctual curative operations on the lifted and cracked zones we offer a compromise in order to stabilise the artwork while conserving its expressive potential as the intention of Georges Mathieu. The artwork, which stands for multiple values, can then be passed on to future generations.

Developing an Appropriate Consolidating Method

Before establishing a conservation protocol, our concern was to develop a consolidation method to restore the mechanical stability of the artwork. This needed to be done without altering the materiality of the work. For example, through the introduction of new materials which while degrading, would become less reversible over time, rendering future interventions difficult.

The issue of loss of cohesion and adhesion is indeed recurrent in the oil painting work of Georges Mathieu. The conservation reports consulted at the Centre de Recherche des Musées de France (C2RMF)[9] and at the Musée d'Art Contemporain of Liège in Belgium[10] reveals that most paintings have already been consolidated once or twice with varied methods and adhesives and that these have not been successful over time.

After testing the solubility of the paint, ethanol was selected to consolidate the paint layers, as it has proven successful to consolidate paintings from the second half of the twentieth century with similar problems of cracking and lifting impastos.[11]

Regenerating the cohesion and adhesion by introducing ethanol in the lifted zones presents the advantage of not creating a new film at the interface of the films requiring adhesion. Ethanol also presents the quality of having a low surface tension and good wetting abilities.

Out of precautionary principle we have also locally tested adhesives which are commonly used in conservation, in low concentrations with the possibility of reversibility for future interventions.

We tested a set of natural adhesives such as sturgeon glue, rabbit skin glue in sheets as well as acrylic adhesives among those Primal™ E330S (interesting because soluble in ethanol) and Plextol™ B500.

[9] *Pentasilée*, 1958; *Bataille d'Hastings* 1956; *Les Capétiens partout*, 1954; *Lothaire se démet de la Haute Lorraine*, 1954; *La mort d'Ascelin de Rochester*.

[10] *Pompes funèbres du cœur de Condé*, 1960.

[11] Recently, Aurélia Chevalier, paintings conservator based in Paris, has been successful in these treatment in the conservation of paintings by Joan Mitchell.

These tests were carried out on paint reconstructions which were created by replicating the paint layers of *Théorème de Gödel*. For our model paintings we searched for materials which were as close as possible to the original ones used by Mathieu. This involved a canvas with a lead white containing priming and a modern LeFranc ivory black in linseed oil paint . After having subjected these model paintings to artificial ageing for over 200 h by alternating seven cycles of heat and cold and varying hygrometry[12] we observed physical modifications of the canvas, the priming and the pictorial layers and obtained similar alterations to the ones noted on *Théorème de Gödel.*

Comparative testing of elasticity, colour matching, gloss level and removal of the adhesives were also carried out. Considering coloration and gloss, the test results revealed that films of adhesive Plextol™ B500 and Primal™ E330S presented less optical change than an animal glue film.

Following the test results, our choice was oriented towards the regeneration of the film's cohesion using ethanol. Using this technique, the materiality of the artwork is barely modified by the introduction of a new material. The adhesion restored by the ethanol also allows future interventions to be made.

Whenever ethanol seemed insufficient in consolidating the lifting of thick paint, Primal™ E330S diluted to 5 and 10 % in ethanol was chosen, out of the different adhesives tested of similar adhesive strength.

The flexibility that the film presents is an interesting quality in our case and the limited quantity of adhesive applied corresponds to the ethics that we set ourselves.

Treatment: Putting the Preliminary Findings into Practice

Surface dusting and cleaning of the painting were carried out before any other operation in order to prevent any dirt from setting itself underneath the lifted areas of paint.

The consolidation was undertaken with a mini low-pressure table, one area at a time, by injecting ethanol with a hypodermic syringe and applying localized pressure. The lifted zones that were still flexible, were repositioned by a finger exerted pressure. Small fabric bags filled with microspheres of lead, perfectly adopting the required shape, were then placed on the impastos with a intermediate sheet of Melinex™.

Reintegration Procedure

When paint losses were located at an end point and/ or when white priming was teared off the support, this created a striking optical interruption in the line of the

[12]Each cycle consisted of three phases: a hot and dry phase; a cold and wet phase; a hot and humid phase. Each phase lasted 24 h.

impasto. In this case, an infill was applied using a mixture of calcium carbonate and rabbit skin glue. This filling is easily reversible in the case of a future intervention and reduces the optical dissonance by re-establishing an adequate colour and volume stability.

Colour retouching was carried out using ivory black Winsor & Newton watercolours.

Reduction of localised deformations and consolidation were carried out satisfactorily with the ethanol treatment. However in a few areas it was necessary to add Primal™ E330S at a 5 % concentration because the action of ethanol appeared insufficient.

The intervention was able to remove the deformations of numerous lifting, recreating a correct adhesion and stabilising the painting for the years to come, provided that the work be conserved in a stable environment around 18 °C and 50–55 % relative humidity.

Conclusions

The conservation intervention described in this article prevents further paint loss which would compromise the longevity and the comprehension of the artwork.

References

Boon J, Hoogland F (2014) Investigating fluidizing dripping pink commercial paint on Van Hemert's Seven-series works from 1990–1995, these proceedings, Chapter 16

de Pracomtal (2008) Patrice, Lumière, matière et pigment. Gourcuff, Paris, p 440

Maschelein Kleiner Liliane (1993) Liants, vernis et adhésifs anciens. IRPA, Bruxelles, p 50

Mathieu G (1973) de la révolte à la Renaissance. Ed. Gallimard, Paris

Mathieu G (2006) Le privilège d'être. Ed. Complicités, Paris

O'Malley M, Moffat E (2001) Paul E. Borduas, the blacks and whites of the later paintings (1955–1966), in Deterioration of artists' paints: effects and analysis. Extended abstracts of the presentations. ICOM-CC Working Groups Paintings 1 & 2 and The Paintings Section, UKIC. British Museum, London. 10–11 Sept 2001. Alan Phenix (ed), 2001, ICOM and UKIC, pp 84–87

Roche A (2003) Comportement mécanique des peintures de toiles. CNRS Editions, Paris

Verbeeck-Boutin M (2009) De l'axiologie", CeROArt [online], 4 |, last consulted 27 Sept 2013. URL: http://ceroart.revues.org/1298

Non-traditionally Painted Oil Painting: How to Treat It Properly? Josip Vaništa's *Cakes* (1955)

Mirta Pavić

Abstract Modern and contemporary oil paintings that have been executed in seemingly standard technology still hold potential pitfalls for a conservator. Paintings may consist of the expected paint layers, preparation and color, but their content and application do not fit into an expected framework. One example is the painting *Cakes* by Josip Vaništa from 1955, currently in the collection of the Zagreb Museum of Contemporary Art (MSU). Pitfalls can be found in its several layers of color, where there was a previous painting underneath the final painted layers. Such a procedure is not unusual for this versatile artist who created art objects in various media and approached painting more as an act or behavior than to produce an esthetic product. While in the Gorgona group (1959–1966), an assembly of significant Croatian neo-avant-garde artists, he used to express his views on life and art through action and dialogue within the group itself as well as with the audience. This oil painting technically falls within the standards of a traditional approach to oil painting, but from the point of view of technology and execution, conservation practice had to be adjusted to these specific challenges. The most obvious example of this emerged in the process of cleaning the painted layers. It was necessary to find a proper solution for varied surfaces on the same painting. Although the entire unvarnished surface had the same pH level, distinct chromatic entities displayed different solubility in aqueous solutions. So accordingly, those needed to be cleaned with agents of different density and composition. In addition, the non-invasive intervention of stretching the painting onto a working frame accomplished several goals: safe manipulation and protection during treatment, strip lining and flattening.

Keywords Painting • Oil on canvas • Contemporary art • Ethics • Esthetics • Technology

M. Pavić (✉)
Museum of Contemporary Art, Zagreb, Croatia
e-mail: mirta.pavic@msu.hr

K.J. van den Berg et al. (eds.), *Issues in Contemporary Oil Paint*,
DOI 10.1007/978-3-319-10100-2_10

Introduction

Oil on canvas is undoubtedly one of the most widely used painting techniques that even the most casual audience is familiar with. In the realm of modern and contemporary art, this technique is diminishing; artists are replacing it with acrylic, industrial paints, pigments or utterly unexpected materials, both organic and inorganic. Technically and philosophically, oil on canvas may be considered by some to be an "overly classical" medium in comparison to other materials that emerge in contemporary art. Conservators of modern and contemporary artworks are aware of the significance in the artists decision in choosing a material that bears the meaning of an art object. The medium of oil on canvas may appear traditional, and may be viewed as a somewhat "non-charismatic" choice of material that could and should express everything that artists wish to say about the contemporary world. Conservation of contemporary art has been exposed to increasingly probing research, exploration and cooperation in the past 20 years, and not only are new materials being accepted and analyzed, but old techniques and technologies are also both applied and examined in a different way. Nearly 30 years ago, the renowned restorer Heinz Althöfer (1985) developed a categorization of contemporary artworks in order to allow for easier understanding, exploration and simplification of the conservation protocols. One of his categories is *objects in traditional or similar technique.* It is this category that the 1955 painting from the holdings of the Museum of Contemporary Art in Zagreb, entitled *Cakes* by Josip Vaništa (Fig. 1) belongs to.

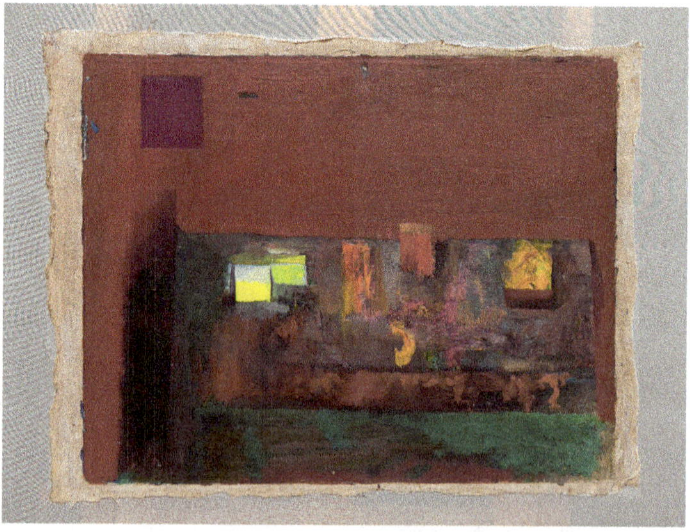

Fig. 1 Josip Vanista, *Cakes*, 1955 oil/canvas 50 × 64 cm Inv. No. MSU **159**

Here is how Dr. Althöfer (1985) explains this category:

> The traditional materials and techniques of painting, as well as the corresponding and traditional restoration methods, will be tested in respect to ageing, reversibility and other basic demands, in order to find an answer to the question of how far traditional art production and the corresponding restoration methods are applicable to modern art. At the same time, it should be examined to see how traditional restoration techniques could be applied to contemporary objects in order to avoid new unreliable methods and materials.

Therefore, since traditional conservation methods may not be appropriate for new artistic materials, the new materials that we apply in restoration must also be carefully examined. Their reliability and advantages must be verified.

Since the conservation-restoration field of expertise started to evolve, there have been opposed schools of thought: one of which was open to restoring with new materials, whereas the other was more traditional. In her article "Hell vs. Ruheman, the Metaphysical and the Physical; Controversies about the Cleaning of Paintings", Prof. Joyce Hill Stoner (2001) writes about two prominent restorers who were active before, during and after World War II and each of whom expounded their own convincing view on the topic. Ruheman and Hell dealt chiefly with issues of cleaning and removing oxidized varnish, which is frequently not a problem with contemporary paintings since they were often painted without a final varnish layer, as was the case with Vaništa's painting *Cakes*. However, this is precisely why dirt that accumulates on the surface may penetrate paint layers more easily. It is evident that cleaning ethics are as pertinent today as it was in the times of Hell and Ruheman. Ruheman was particularly thorough in cleaning and very skillful in retouching, while Hell was more inclined to preserving harmony and unity and thus opposed to radical cleaning. Ruheman's notion that one should gather as much information as possible on the artist to study their work before a conservation treatment fits perfectly into the philosophy of contemporary art conservation. The context, understanding the material and spiritual aspect of an art object, is of prime importance when making appropriate decisions on its conservation procedure. *Cakes* is an excellent example of a seemingly simple oil on canvas that acquires an entirely new dimension of consideration if we are familiar with Vaništa's artistic work and views. Since it dates back to 1955, this painting could have conceivably been conserved and restored by either Hell or Ruheman. It would have been interesting to see their individual manners of approach. Vaništa's work in the anti-group Gorgona, as defined by its members, speaks volumes about their artistic attitude that are anything but traditional. In an effort to explain Gorgona's work, Davor Matičević (2011), a curator and former director of the Zagreb Museum of Contemporary Art writes: *"Aspects of their activity soon turned into conceptual works or they could be conceptualized. In writing, thought, forms or objects, even in a painting of blackness, they strived to express something that is elusive, that belongs to the sphere of intellectual relationships... The concept of Gorgonic humor was created and I believe that it was also possible to recognize Gorgonic clothing style by its neutrality"*. Although *Cakes* formally predates the foundation of Gorgona, the painting's appearance and title provide an insight into Vaništa's unconventional spirit. If we go back to Althöfer's aforementioned categorization of contemporary artworks, this painting partly falls into another Althöfer category: *monochrome paintings*. It is here that Althöfer emphasizes the problem of finding a suitable agent for removing dirt (fingerprints) and treating the non-varnished surface of painting. This problem became apparent when we undertook the treatment of *Cakes*. The aim of this study was not only to find and apply an appropriate treatment to this particular painting but also to understand where the differences are in comparison to a traditional oil painting, created with a more "disciplined" technology, and how to preserve the spiritual component of the art work in the conservation process.

Treatment

The cleaning of *Cakes* has already been emphasized as the most delicate phase of the conservation treatment. But before cleaning, there was a previous step that needed to occur in order to avoid any risk of mishap during the treatment. The painting was stretched on a working frame by applying a strip lining using the *nap-bond* technique of Vishwa Raj Mehra. A thin non-woven polyester cloth is used for strip lining because it is inert, resistant to distortion and the influence of moisture. This method is useful and often applied in our department because it diminishes the risk of damage while handling the painting and keeps it constantly taut. The stretching ensures safe handling in every phase of treatment, including consolidation, lining, filling in the damaged parts and retouching. After the treatment it is also easy to attach the painting onto its original stretcher with the extended edges of polyester fabric.

This was a variation of the method introduced in 1974 by Vishwa Raj Mehra (1975, 1995) at the Greenwich conference (Von Imhoff 2009) when he inaugurated a low-pressure perforated vacuum table and acrylic adhesives that do not require heat activation, which is less stressful on a painting that is being flattened or lined.

If we take another look at more recent conservation history, we can see that the Greenwich conference greatly influenced developments the conservation field. On this occasion, two groups spontaneously emerged in the course of a debate on using new materials in lining paintings, and they called themselves Trads and Mods – the very names reflecting their views. This was also when Gustav Berger introduced Beva 371. All these events have contributed to the consideration and questioning of decision-making models, as well as ethical questions with the introduction of new materials. Just as Hell's and Ruheman's lines of reasoning are still pertinent, the same can be said of the Greenwich conference agenda, and especially of the materials that have been applied in conservation ever since. Among the more recent materials that are used in lining and consolidation, like synthetic water-soluble polymers and nonionic cellulose ethers that do not present a health risk, the acrylic dispersions that were introduced by Viswa Raj Mehra are still applied in painting conservation.

In the case of the treatment of Vaništa's painting, the next step was to locally fix small sections where color was peeling off the surface with Plextol B 500 (Fig. 2), a material that Mehra was the first to use for this purpose.

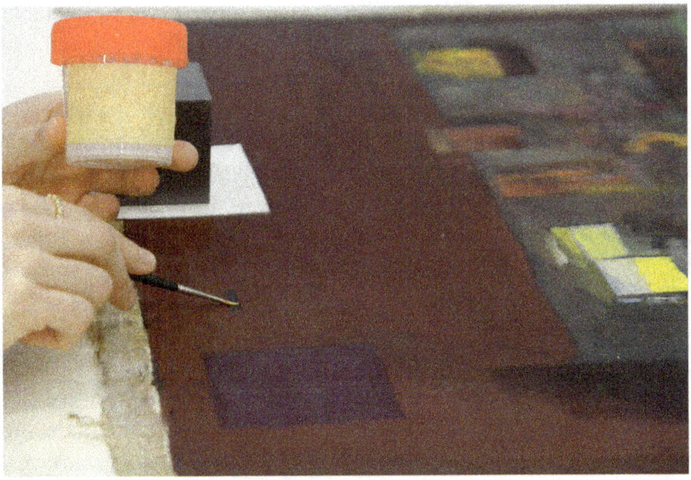

Fig. 2 Consolidating the paint layer locally with Plextol B

Fig. 3 Painting's surface in UV light, showing indication of an earlier painting

An image of the painting underneath UV light (Fig. 3) confirmed that the canvas was used secondarily, i.e. that a previous painting was painted over. It was obvious to the naked eye that in places color was applied in several layers, which was a process that was typical of Vaništa, and they did not adhere to each other as homogeneously and firmly as might have been the case with another paint. Over time, the oil paint became more fragile as components dried out, affecting elasticity and cohesive properties. This has caused paint to peel off the ground in several places that became visible as minute light specs. However, all the components were still stable enough so that it was not necessary to consolidate the entire painting by introducing a new material into its structure.

Small consolidated areas had to be leveled with the original painted layer. Since mechanical damage was minor, a 7 % Mowiol filler proved to be most suitable. Its application is simple because it is easy to control density and achieve the desired viscosity, it can be applied with a brush, it is easy to treat and it can also be easily removed from the surface. But before filling the tiny damaged gaps that were already fixed and stable, cleaning was carried out.

After testing to ascertain a suitable cleaning agent according to the appropriate pH of the surface color, it was decided to apply two different cleaning agents for different sections of the painting. Tests for an appropriate cleaning agent included solutions, emulsions, organic solvent based gels, synthetic solvent based gels and Pemulen gel. Even though the pH of all the painted layers was the same, the various paint compositions reacted differently to the agent used to clean up surface deposits. The monochrome red brown surface was cleaned by Pemulen gel, pH 7, whereas the multicolored painting section with a sensitive pigment, which is easily soluble in any water-based agent, was cleaned by using a nonionic gel based on synthetic solvent. This exemplifies the viewpoint that *Cakes* should not be viewed only as a whole – it is also necessary to individually view the distinctly separate chromatic entities with their individual physical compositions and chemical makeup. Owing to research and achievements by conservator chemists such as Richard Wolbers (2000), even the most delicate issues in cleaning modern and contemporary paintings are rendered solvable without risking damage to an artwork. In the history of conservation, cleaning has been subject to so much consideration because of its nature as an irreversible procedure.

As Salvador Munos Vinas (2005) puts it, since reversibility and minimal intervention are often impractical, a conservator has to be aware of the consequences of irreversibility. This is why it is crucial to be certain that after cleaning a painting's surface, no changes or unseen chemical reactions have taken place, in case the cleaning agent has penetrated deeper than the surface layer. These reactions can have a very slow and indiscernible effect, but may cause permanent damage in the long run.

In the last phase retouching was done with Gouache paints as a base and the final layer was executed with pigment bound by a Klucel solution in distilled water because the original oil paint is matte.

Conclusion

When it comes to oil on canvas, all of us have an approximate idea of what it should look like, but in modern and contemporary art, it is precisely the motive, or rather its absence, that can make a painting technique less relevant or less recognizable. Often even experts cannot easily identify a material without chemical analysis. In several cases we have found the media to be incorrectly documented in the MSU data base after performing sample analysis for conservation purposes. The days when oil on canvas led to known and tested conservation treatments are long gone, despite of the fact that the technique has been known for centuries. Previously conservators who were working with traditional oil paintings had established approaches for conserving paintings from earlier periods. The treatment of *Cakes* necessitated a different approach, as a consequence of Vaništa's desire to make this technique functional, without wasting time on a thorough technological preparation and a proper painting procedure. The goal is to use a medium in a way that is right for the artist.

In conclusion, the conservation of contemporary oil paintings on canvas differs in certain aspects from the conservation of traditional oil paintings. It is not prudent to always apply the same methodology and technology to both old and new oil paintings but it is certainly a prerequisite to know this methodology and technology in order to deal with contemporary conservation issues successfully. Ethically, they can be more nuanced and complex, and from a technological point of view, it is necessary to understand the variety of conservation materials as well as their characteristics and applicability in every individual case.

As the case in point, Vaništa's painting had to be handled more carefully because it is as not resilient as an old canvas painting. Stretching it on a working frame helped protect unevenly thick paint layers from developing folds or cracks in critical areas. Careful testing and finding a mild but efficient cleaning system resulted in an appropriate appearance of the non-varnished painted layer and protected it from potential discoloration during cleaning, however minimal it may have been. The chosen methodology rendered the painting presentable again, without jeopardizing its original appearance that mediates the author's intention (Fig. 4).

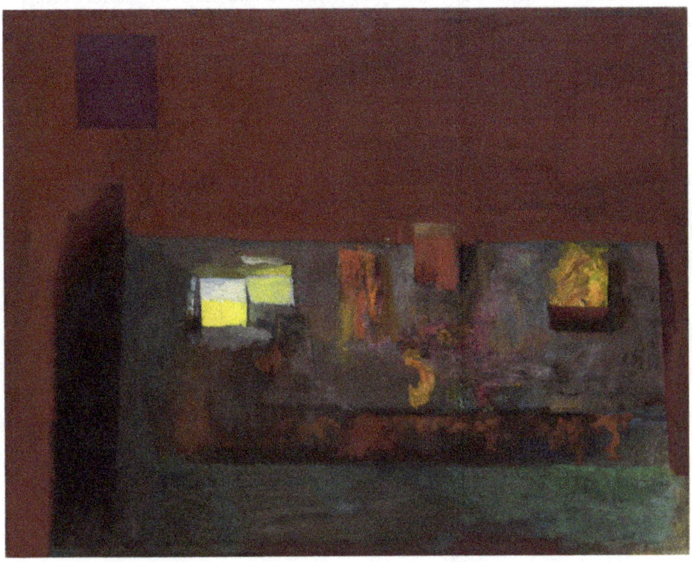

Fig. 4 *Cakes* after the treatment

Appendix: List of Used Materials

Nonwoven polyester fabric, acrylic adhesive Lascaux 498 HV, Plextol B 500, blotting-paper, pH paper indicator, distilled water, non-ionic solvent gel (100 ml Shellsol T, 2 g Carbopol EZ 2, 20 ml Ethomeen C-12), Pemulen gel (pH 7), 7-percent Mowiol solution with Bologna chalk, gouache, pigment bounded with Klucel.

References

Althöfer H (1985) Restaurierung moderner Malerei: Tendenzen, Material, Technik In Zusammenarbeit mit Hiltrud Schinzel (Modern art and the new restoration; tendencies, material, technique), Munich. http://d-nb.info/850376351/04

Caple C (2000) Conservation skills; Judgment, method and decision making. Routledge, Oxford

Chiantore O, Rava A (2005) Conservare l'arte contemporanea. Electa, Milano, pp 18–198

Hill Stoner J (2001) Hell vs. Ruheman, the metaphysical and the physical; controversies about the cleaning of paintings. In: Past practice, future prospects Journal. British Museum, London, pp 109–114

Matičević D (2011) Suvremena umjetnička praksa Ogledi 1971–1993 (Contemporary art practice assays 1971–1993). Diurieux/Muzej suvremene umjetnosti/Hrvatska sekcija AICA, Zagreb, pp 96–110

Mehra VR (1975) Comparative study of conventional relining methods and materials and research towards their improvement. Interim Report to the ICOM Committee for Conservation, 4th Triennial Meeting, Venice, Paper no 75/11/5

Mehra VVR (1995) Foderatura a freddo. I testi fondamentali per la metodologia e la pratica. Nardini editore, Firenze

Munos Vinas S (2005) Contemporary theory of conservation. Elsevier Butterworth-Heinemann, Oxford

Schadler-Saub U, Meyer A (eds) (2010) Munos Vinas S, The artwork became a symbol of itself: reflections on the conservation of modern art (Theory and practice in the conservation of modern and contemporary art). Archetype Publications, London, pp 9–19

Vaništa J (2013) Ukidanje retrospektive/Abolition of retrospective, Catalogue for Vaništa's exhibition in the MSU 2103 edited by Nada Beroš Muzej suvremene umjetnosti, Zagreb

von Imhoff H-C (2009) Aspect and development of conservator-restorer profession. Article in Issue 8, February. www.e-conservationline.com

Wolbers R (2000) Cleaning painted surfaces, aqueous methods. Archetype Publications, London

Conservation Issues in Several Twentieth-Century Canadian Oil Paintings: The Role of Zinc Carboxylate Reaction Products

Kate Helwig, Jennifer Poulin, Marie-Claude Corbeil, Elizabeth Moffatt, and Dominique Duguay

Abstract In an on-going study of the materials and techniques of twentieth-century Canadian painters, similar conservation issues in oil paintings by various artists have been noted. These include delamination and lifting paint, zinc soap protrusions and surface efflorescence or accretions. Examples of these phenomena are presented. Delamination in an oil painting from 1956 was found to be related to an underlayer with a high concentration of zinc fatty acid salts (zinc soaps). In two paintings that date from 1936 and 1937, zinc soaps have aggregated and formed protrusions that have broken through the paint surface. The protrusions were analysed using a combination of SEM-EDX, GCMS and FTIR. The FTIR spectra were compared to those of synthetic zinc palmitate, stearate, azelate and oleate. The combined GCMS and FTIR results indicate that the protrusions contain primarily zinc palmitate and stearate. Peak splitting in the FTIR spectrum, which is not observed in synthetic zinc palmitate, stearate or binary palmitate-stearate salts, is likely due to structural distortion. The final example describes a disfiguring surface accretion on a 1952–1954 painting caused by the reaction of zinc with a low molecular weight carboxylic acid (2-hydroxypropanoic or lactic acid).

Keywords Zinc soaps • Zinc fatty acid salts • Zinc palmitate • Zinc stearate • Zinc oleate • Zinc azelate • Zinc lactate • FTIR • GCMS

Introduction

Zinc white based oil paints almost always contain some proportion of zinc fatty acid salts (zinc soaps) that are formed as the paint dries and ages. Zinc soaps may also be present as additives in oil paints. While the presence of a certain amount of metallic soaps can help to create a well-dried paint film, high concentrations of

K. Helwig (✉) • J. Poulin • M.-C. Corbeil • E. Moffatt • D. Duguay
Canadian Conservation Institute, 1030 Innes Road, Ottawa, ON K1B 4S7, Canada
e-mail: kate.helwig@pch.gc.ca; jennifer.poulin@pch.gc.ca; marie-claude.corbeil@pch.gc.ca; elizabeth.moffatt@pch.gc.ca; dominique.duguay@pch.gc.ca

K.J. van den Berg et al. (eds.), *Issues in Contemporary Oil Paint*,
DOI 10.1007/978-3-319-10100-2_11

zinc soaps have been associated with various types of deterioration in modern oil paintings. Zinc soaps sometimes produce efflorescence or disfiguring protrusions at the surface of paintings (van der Weerd et al. 2003; Keune 2005; Osmond et al. 2005; O'Donoghue et al. 2006; Faubel et al. 2011). High concentrations of zinc soaps within the oil paint matrix have also been linked to a lack of cohesion and brittleness in the paint (Rogala et al. 2010; Maines et al. 2011).

Understanding the composition of zinc soaps can help to determine their characteristics, as well as how and why they form. To this end, studies of synthetic zinc soaps have been undertaken. Robinet and Corbeil (2003) published FTIR (Fourier transform infrared spectroscopy), Raman and XRD (X-ray diffraction) data for four synthetic zinc soaps (stearate, palmitate, oleate and linoleate). FTIR spectra for zinc fatty acid salts of hydrocarbon chain lengths from C4 to C18 have also been reported by Taylor and Ellis (2007).

The crystal structures of zinc aliphatic carboxylate salts of various chain lengths have been determined, either from single crystal studies or powder diffraction data (Lacouture et al. 2000; Peultier et al. 1999; Mesbah et al. 2006, 2007). In all cases, the zinc is bonded to the carboxylate groups through a bidentate bridging structure. The zinc is in a tetrahedral environment with the four ligands coming from four different carboxylate groups. The polymeric sheets are stacked perpendicular to the longer axis and linked by Van der Waals interactions. The distance between the zinc atom planes (also called the interlayer distance) has been found to increase linearly with chain length (Mesbah et al. 2011). The crystal symmetry and stacking of the lamellar sheets varies, depending on the length of the hydrocarbon chain (Mesbah et al. 2007).

The composition and structure of zinc soaps present in oil paints are less well understood than those of their synthetic analogues. Published analyses of zinc soaps in oil paintings have used primarily FTIR or Raman spectroscopy, often combined with imaging techniques (van der Weerd et al. 2003; Keune 2005; Osmond et al. 2005, 2012; O'Donoghue et al. 2006; Faubel et al. 2011). Certain of these studies also reported selected results using secondary ion mass spectrometry (SIMS), which has the advantage of spatially resolving components in the zinc soap aggregates. However, fewer analyses of zinc soaps have been undertaken using SIMS than of the corresponding lead soaps. A recent study (Maines et al. 2011) used Py-GCMS (pyrolysis-gas chromatography-mass spectrometry) to determine the fatty acid profile of zinc white paints containing zinc soaps. In this case, though, the fatty acid profile comprises the glyceryl esters in the cross-linked oil paint as well as free acids and zinc salts.

In an on-going study of the materials and techniques of twentieth-century Canadian painters, conservation issues in oil paintings related to high concentrations of zinc fatty acid salts have been noted. Examples of these phenomena are presented in this paper. Delamination in an oil painting from 1956 was found to be related to a zinc white underlayer with a high soap concentration. In two paintings that date from 1936 and 1937, zinc soaps have aggregated and formed protrusions that have broken through the paint surface. The results of analysis of these materials add new information to the literature on zinc soap formation in modern oil paints. The final example describes a disfiguring surface accretion produced by the reaction of zinc not with fatty acids, but rather, with a low molecular weight carboxylic acid. Zinc lactate was identified in a surface layer on an oil painting from 1952 to 1954. Once it reacted with zinc in the paint layer, this volatile carboxylic acid was trapped at the surface of the painting.

Methods of Analysis

Samples were analysed using a combination of FTIR, GCMS, scanning electron microscopy combined with energy dispersive X-ray spectrometry (SEM-EDX), and in a few cases XRD. The material in the protrusions was manually isolated from the surrounding paint prior to analysis. Components in the samples were identified by comparison to published reference data and to data obtained from pure materials, either purchased or synthesised in our laboratory. Commercial zinc lactate trihydrate was obtained from City Chemical LLC (CAS 312619-27-5, Lot # 31G118, 98 % purity). The zinc oleate (abbreviated $Zn(C18{:}1)_2$) was synthesised as part of the work by Robinet and Corbeil (2003). Binary zinc palmitate-stearate salts ($Zn(C16)_{2x}(C18)_{2-2x}$) and zinc azelate (Zn(di-C9)), as well as new batches of zinc palmitate ($Zn(C16)_2$) and zinc stearate ($Zn(C18)_2$), were synthesised for this study. The purity of the zinc salts was determined by GCMS, FTIR, and in the case of the binary salt Zn(C16)(C18), by XRD. Experimental details are given in the Appendix.

Results and Discussion

Delamination and Lifting Paint

Delamination between paint applications has been observed in a number of modern oil paintings by Canadian artists. One example is *Rouge sur blanc*, a 1956 oil painting by Jean McEwen (accession # A92 447 P1, Musée d'art contemporain de Montréal). This painting was created by superimposing multiple paint layers, some with significant impasto. The paint is cracking and lifting in a numerous areas.

Analysis of paint samples and cross-sections showed that the delamination is occurring at the level of a thick, zinc white paint layer. Based on FTIR, the zinc white paint was found to contain a high concentration of zinc soaps. A partial cross-section from the painting is shown in Fig. 1. The sample separated into two parts within the zinc white layer, due to its lack of cohesion. The cross-section illustrates the lower part of the sample. A calcium carbonate in drying oil ground layer (layer 1) is followed by a thin lead white paint (layer 2) and an orange-red paint containing Hansa Yellow (PY 3), cadmium red, barium sulfate and a trace of lead white (layer 3). The upper layer (layer 4) is the zinc white paint, composed of

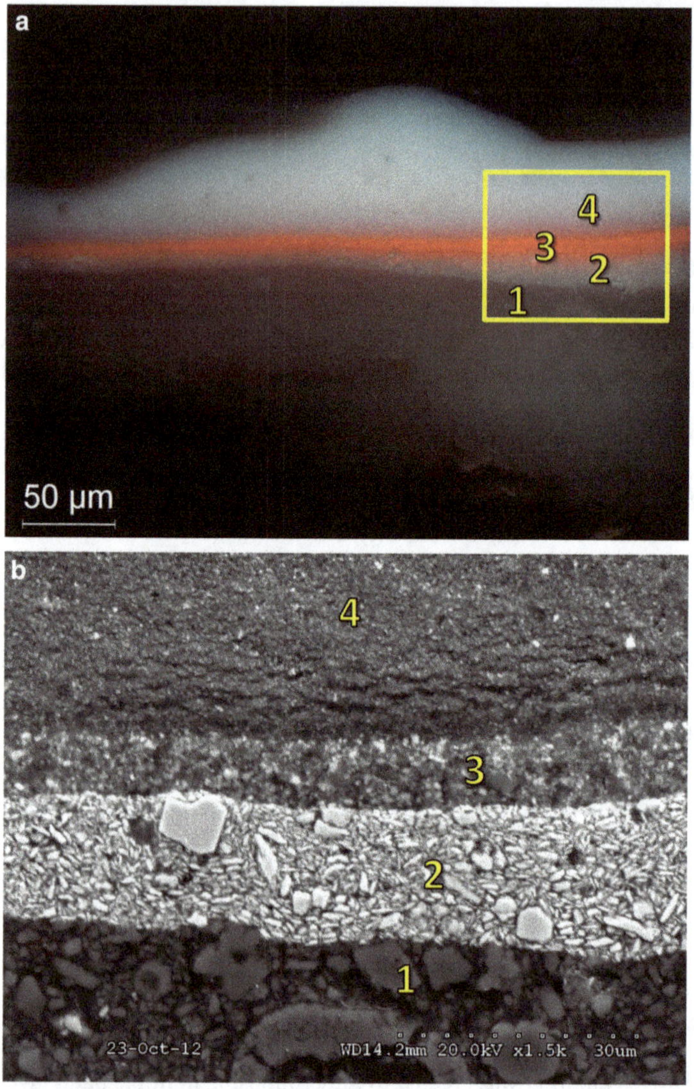

Fig. 1 Incident light (**a**) and backscattered electron (**b**) images of a cross-section from *Rouge sur blanc*. The ground layer (1) is followed by a thin lead white paint (2) and an orange-red paint (3). The upper layer (4) is composed of zinc white and a trace of barium sulfate with a high concentration of zinc soaps. The backscattered electron micrograph shows horizontal micro-fissuring of the zinc white paint near the interface with the orange-red paint

zinc white and a trace of barium sulfate with a high concentration of zinc soaps. The backscatter electron micrograph illustrated in Fig. 1 shows horizontal micro-fissuring of the white paint near the interface with the orange-red paint, perhaps related to the lamellar structure of the zinc soaps. The lack of cohesion in the zinc white underlayer that has led to lifting and cracking in *Rouge sur blanc* is consistent with a recent study of deteriorated mid-twentieth-century Abstract Expressionist paintings with similar underlayers (Rogala et al. 2010).

Zinc Soap Protrusions

Description of the Paintings and Protrusions

Similarly to *Rouge sur blanc*, both the 1936 painting *Equations in Space* by Lawren Harris (National Gallery of Canada, accession number 15673) and the 1937 painting *Storm over the Fields* by Carl Schaefer (Art Gallery of Ontario, accession number 53/53) show areas of cracking and lifting paint. In addition, these two paintings exhibit disfiguring, waxy zinc soap protrusions. In both cases, the zinc soap agglomerates are large enough to have broken through the surface, causing the upper paint layers to crack and lift. Figure 2 shows incident light and backscattered electron images of a cross-section from *Equations in Space*. This area of the protrusion appears homogeneous, without the re-mineralised core that has been observed in some cases (Keune 2005). Re-mineralisation is thought to be related to the degree of maturity of the soap protrusion and is often observed in lead soap agglomerates and protrusions (Noble and Boon 2007).

In the case of *Equations in Space*, the paint beneath the protrusions was not available for sampling. A cross-section from an edge of the painting showed that the preparation layers are composed of talc, titanium white and anhydrite, and are not a source of zinc. Above the ground layers, the cross-section showed several underlayers with high concentrations of zinc white and zinc soaps, which are undoubtedly the source of zinc in the protrusions. The upper blue paint seen in Fig. 2 is composed of lead white, Prussian blue, cobalt blue and vermilion and does not contain zinc white or zinc soaps. Based on the analysis of cross-sections from adjacent areas, the source of the zinc in the protrusions from the second painting, *Storm over the Fields,* is the lowest two ground layers of the painting. These layers contain barium sulfate, talc, zinc soaps and possibly a small amount of zinc sulfide or zinc oxide.

SEM-EDX

Elemental analysis of several of the white, waxy protrusions on *Storm over the Fields* showed that they contain primarily carbon, oxygen and zinc, with traces of lead and possibly sulfur. The waxy protrusions on *Equations in Space* contain primarily carbon, oxygen and zinc, with traces of other elements (aluminum, lead, chlorine and possibly sulfur).

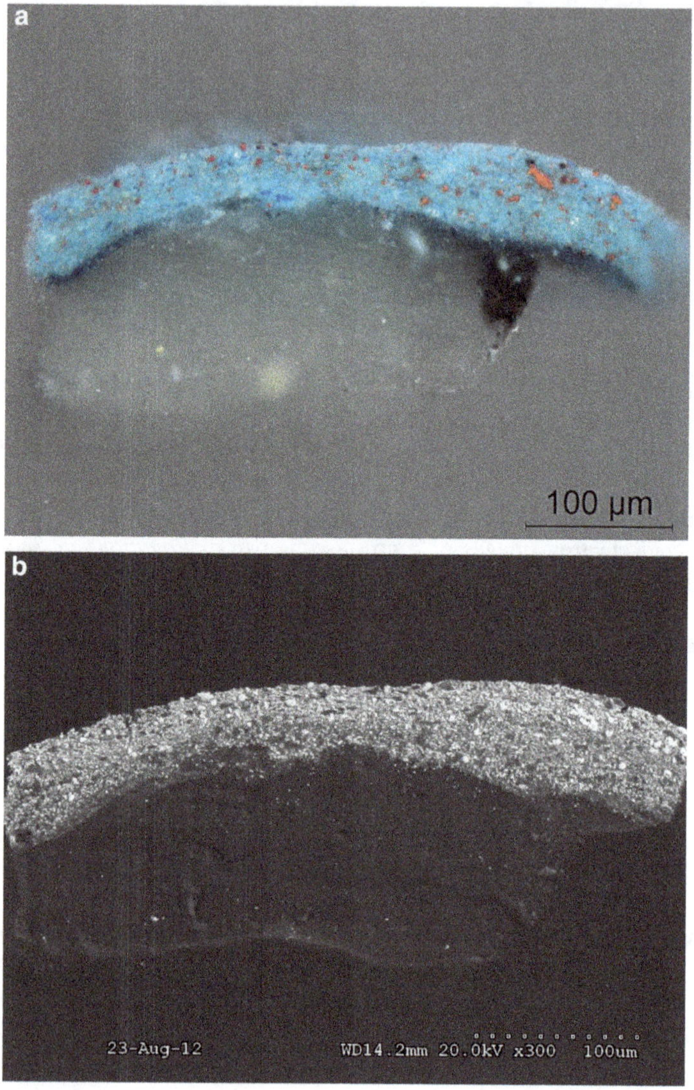

Fig. 2 Incident light (**a**) and backscattered electron (**b**) images of a cross-section from *Equations in Space*. The sample is from the side of a protrusion adjacent to the area where it has broken through the surface of the blue paint

GCMS

The fatty acid profiles for the zinc soap protrusions on the two paintings are shown in Table 1. Fragments of the samples were extracted and derivatised by two different methods, methylation and trimethylsilylation, and both sets of results are presented in the Table. The methylation allows all fatty acids in the sample to be determined,

Table 1 Fatty acid profiles of the protrusions measured by GC-MS[a]

	Relative % fatty acid; *Equations in Space* (MethPrep)	Relative % fatty acid; *Equations in Space* (BSFTA)	Relative % fatty acid; *Storm over the Fields* (MethPrep)	Relative % fatty acid; *Storm over the Fields* (BSFTA)
Suberate (di-C8)	5.3	2.4	6.6	5.9
Azelate (di-C9)	15.4	8.7	25.8	20.2
Myristate (C14)	4.6	7.5	3.1	4.1
Palmitate (C16)	48.0	56.5	35.0	39.4
Stearate (C18)	20.1	23.4	22.1	21.6
Oleate (C18:1)	6.6	1.4	7.5	8.8
	P/S = 2.4	P/S = 2.4	P/S = 1.6	P/S = 1.8
	A/P = 0.3	A/P = 0.2	A/P = 0.7	A/P = 0.5
	O/S = 0.3	O/S = 0.03	O/S = 0.3	O/S = 0.4

[a]The relative percentages of fatty acids were calculated from the integrated peak areas of the derivatised fatty acids. Only fatty acids present with relative concentrations above 1 % are listed in the table

including free fatty acids, salts of fatty acids and glyceryl esters from the drying oil medium. The silylation, on the other hand, only derivatises the free fatty acids or fatty acid salts. The results obtained using the two methods are similar, indicating that most of the fatty acids are present as free acids or salts and that there is little oil in the protrusions. Small differences could be related to inhomogeneity of the fatty acids throughout the protrusion (since different fragments of the sample were used for the two methods). Analysis of multiple samples, or use of a method that allows spatial resolution of the fatty acids within the protrusion, would be required to determine the homogeneity of the fatty acid profile.

The protrusion from *Equations in Space* shows a slightly higher ratio of palmitate to stearate (P/S) than the protrusion in *Storm over the Fields*, which may indicate that a different type of oil was the source of fatty acids in the two paintings. In general, however, the overall pattern of fatty acids in the two protrusions is comparable. In both cases, the major fatty acids are palmitic (C16) acid, stearic (C18) acid and azeleic (C9 dicarboxylic) acid. The other fatty acids identified—myristic (C14) acid, oleic (singly unsaturated C18:1) acid and suberic (C8 dicarboxylic) acid—are present at relative concentrations below 10 %.

These results are consistent with what has been observed in lead soap protrusions on much older paintings, where the three major fatty acid methyl esters were also palmitate, stearate and azelate (Higgitt et al. 2003). The relative amount of oleate in the zinc soap protrusions is of interest in light of recent research on samples from mid-twentieth-century Abstract Expressionist paintings and from cast paint films in the Smithsonian Institution's Materials Study Collection. This study showed that zinc white paints contained higher amounts of oleate (as measured by Py-GCMS) than paints without zinc white and the authors suggest that it is in the form of zinc oleate (Rogala et al. 2010; Maines et al. 2011). However, since the fatty acid profile of the paint—measured after tetramethylammonium hydroxide (TMAH) derivatisation—comprises free fatty acids, fatty acids that exist as glyceryl esters in the cross-linked oil paint and fatty acid zinc salts, the presence of zinc oleate soaps is not certain. As seen in Table 1, the abundance of oleate in the zinc soap protrusions studied here represents less than 10 % of the total fatty acid content, indicating that in these cases, zinc oleate is not likely a dominant component.

FTIR

The FTIR spectra of the protrusions from the two paintings are shown in Fig. 3. A scale expansion from 1380 to 1050 cm^{-1} for the protrusion from *Equations in Space,* which includes the series of weak CH_2 progression bands in this region, is inset. The spectra from the two protrusions are very similar to one another and are dominated by bands due to zinc fatty acid salts. Small amounts of free fatty acid and ester are also present (evidenced by the weak carbonyl bands at 1717 and 1737 cm^{-1} and the C–O stretch of the ester near 1175 cm^{-1}). A notable feature in both spectra is the fact that the asymmetric and symmetric carboxylate bands of the soaps (ν_aCOO and ν_sCOO) and the strong CH_2 deformation band (δCH_2) show peak splitting. The ν_aCOO absorption is split into three bands (a weak band at 1593 cm^{-1} and a strong doublet at 1550/1530 cm^{-1}) while the δCH_2 and ν_sCOO peaks are doublets (1469/1459 and 1409/1400 cm^{-1} respectively).

Figure 4 shows FTIR spectra of a series of synthetic zinc fatty acid salts: zinc palmitate ($Zn(C16)_2$), zinc stearate ($Zn(C18)_2$), zinc oleate ($Zn(C18{:}1)_2$) and zinc azelate (Zn(di-C9)). Beneath each spectrum, a scale expansion from 1380 to 1050 cm^{-1} is given. Although the FTIR band positions for the zinc stearate, palmitate and oleate have been previously published (Robinet and Corbeil 2003; Taylor and Ellis 2007), the spectra are presented here to facilitate comparison to the spectra from the protrusions. Examination of the spectra in Fig. 4 shows that neither zinc oleate nor zinc azelate are major components of the protrusions. The FTIR spectrum of zinc oleate exhibits peak splitting of the ν_aCOO, ν_sCOO and δCH_2 bands similar to those of the protrusions. However, the presence of a high concentration of zinc oleate can be ruled out since the spectra of the protrusions do not show the characteristic C═C—H stretch of zinc oleate at 3003 cm^{-1} and they show a different pattern of CH_2 progression bands. The spectrum of zinc azelate is very different from those of the long chain monocarboxylate zinc soaps and does not match the spectra of the protrusions. Key differences between the protrusions and the zinc azelate spectrum are in the CH_3 and CH_2 stretching region between 3000 and 2800 cm^{-1}; the pattern and number of the CH_2 progression bands; and the ρCH_2 bands from 775 to 720 cm^{-1}. As well, the zinc azelate shows a different pattern of peak splitting of the νCOO bands than that seen in the protrusions. A comparison of the spectra of the synthetic zinc soaps to those of the protrusions in the region of the low intensity CH_2 progression bands, which are sensitive to the hydrocarbon chain length, shows that they match well to zinc palmitate with a possible small contribution from zinc stearate.

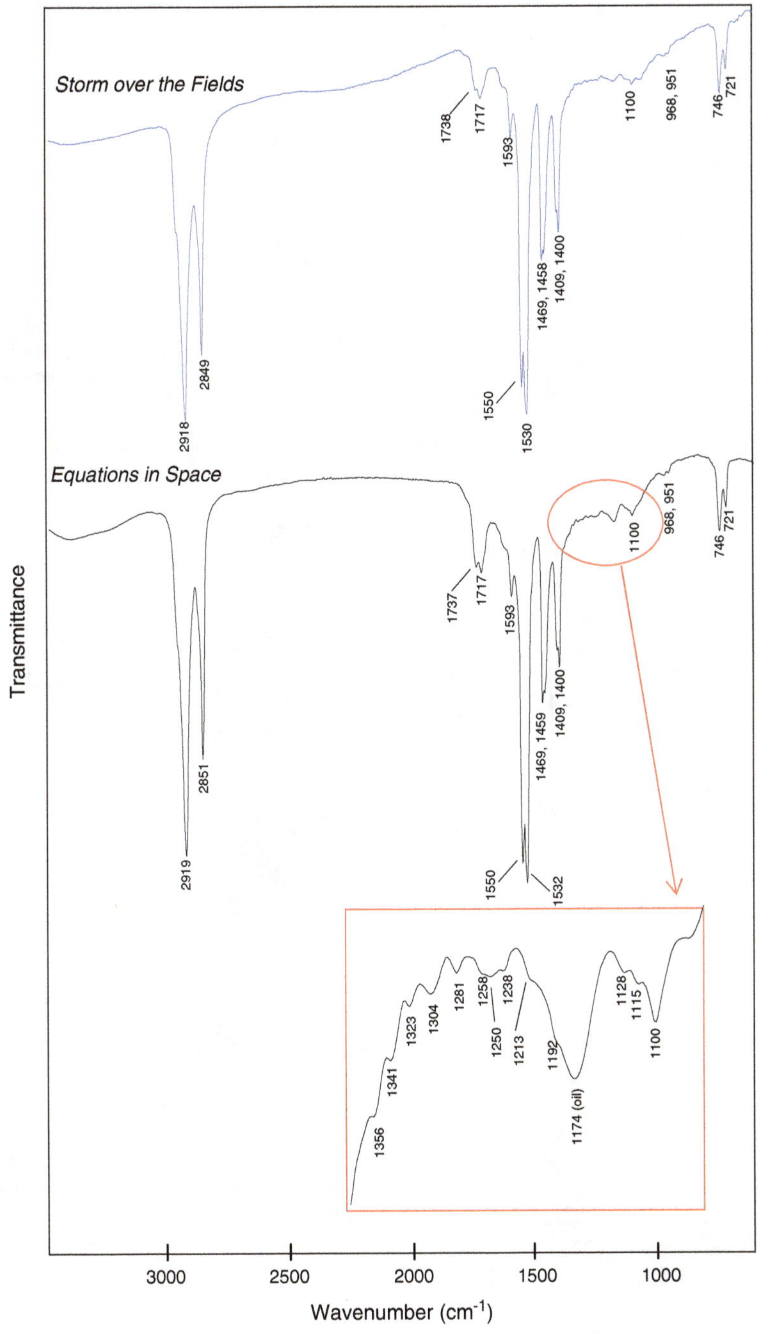

Fig. 3 FTIR spectra of the protrusions from *Storm over the Fields* (upper trace) and *Equations in Space* (lower trace). A scale expansion from 1380 to 1050 cm^{-1} for the protrusion from *Equations in Space* is given below the full spectrum

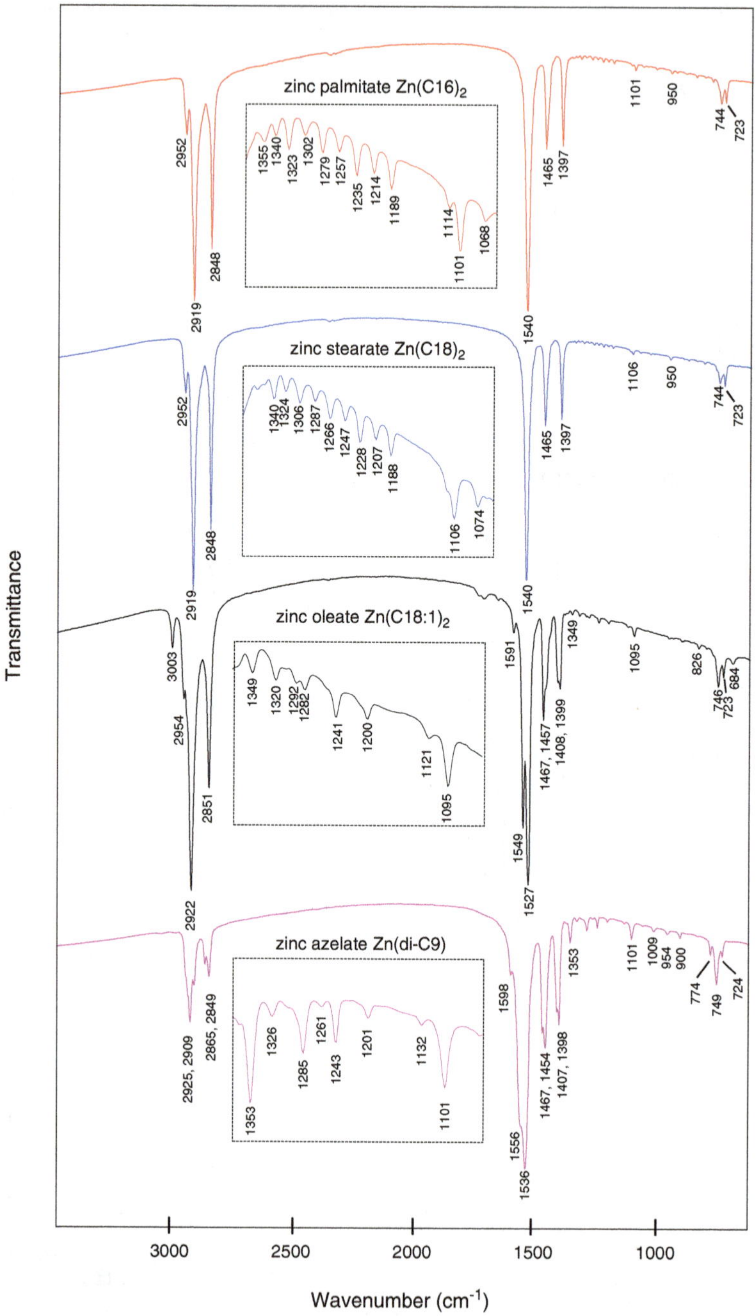

Fig. 4 FTIR spectra of a series of synthetic zinc fatty acid salts: zinc palmitate ($Zn(C16)_2$), zinc stearate ($Zn(C18)_2$), zinc oleate ($Zn(C18{:}1)_2$) and zinc azelate (Zn(di-C9)). A scale expansion from 1380 to 1050 cm^{-1} is given below each spectrum

The presence of primarily zinc palmitate and stearate in the protrusions, and the absence of a significant amount of zinc azelate, is consistent with results of analyses of lead soap protrusions. Studies of lead soaps show that the aggregates contain primarily lead palmitate and stearate and that the azelate is likely in the form of free acid and not lead azelate (Higgitt et al. 2003; Keune 2005; Boon et al. 2005). The complicating factor in the zinc soap protrusions studied here, however, is the peak splitting observed in the FTIR spectra that is not seen in the spectra of synthetic zinc palmitate and stearate.

Peak splitting in specific zinc carboxylate salts has been reported and is related to the presence of non-equivalent zinc-oxygen bonds that lower the symmetry around the zinc ion. Non-equivalent bonds can occur either due to different types of coordination of the zinc (Zeleňák et al. 2007) or to a distortion of the tetrahedral structure for a single type of coordination (see for example Ishioka et al. 2000; Taylor and Ellis 2007). In the case of the zinc soap protrusions, distortion of the tetrahedron is a likely cause of the peak splitting, since, as described in the introduction, the bridging bidentate coordination around the zinc is the same for all the saturated zinc fatty acid salts that have been studied to date.

To investigate this further, several binary stearate–palmitate salts were synthesised and analyzed by FTIR. It has been shown that when a 1:1 mixture of palmitic acid and stearic acid react to form a zinc salt, the binary salt Zn(C16)(C18) is thermodynamically favoured and will form (Mesbah et al. 2011). This binary salt has a bridging bidentate structure in which each zinc atom is coordinated to two C16 carboxylate ions and two C18 carboxylate ions (Mesbah et al. 2011). When the palmitate–stearate ratio is greater than one, a mixture of zinc palmitate $Zn(C16)_2$ and a binary salt of the general formula $Zn(C16)_{2x}(C18)_{2-2x}$ is formed (where x represents the relative amounts of stearic and palmitic acid available for reaction; $x = 0$ for pure stearic acid and $x = 1$ for pure palmitic acid). Given that the protrusions contain both palmitic and stearic acid, binary soaps are expected to be present. It was hypothesized that the presence of binary soaps could cause sufficient structural distortion to lead to the observed peak splitting in the FTIR spectrum.

The upper trace in Fig. 5 shows the FTIR spectrum of the binary salt Zn(C16)(C18). The salt was synthesized with equal proportions of palmitic and stearic acid and its composition was confirmed by XRD (see Appendix). The lower trace shows a 1:1 physical mixture of zinc stearate and zinc palmitate. The two spectra are virtually identical. The progression bands in the binary salt indicate the presence of both zinc palmitate and stearate and no peak splitting of the νCOO bands was observed. This was also found to be true of a binary zinc salt obtained from a 2:1 mixture of palmitic to stearic acid (a ratio close to that found in the protrusions). So, although it seems that structural distortion of zinc palmitate and stearate may be the root of the observed peak splitting in the protrusions, the presence of binary soaps is not the cause.

The formation of zinc soap protrusions in a painting has a complex mechanism and occurs over a long timescale that would be difficult to replicate in the laboratory. In the case of the protrusions studied here, we can conclude that the structure is not analogous to zinc soaps produced by standard synthesis. Analysis by FTIR shows

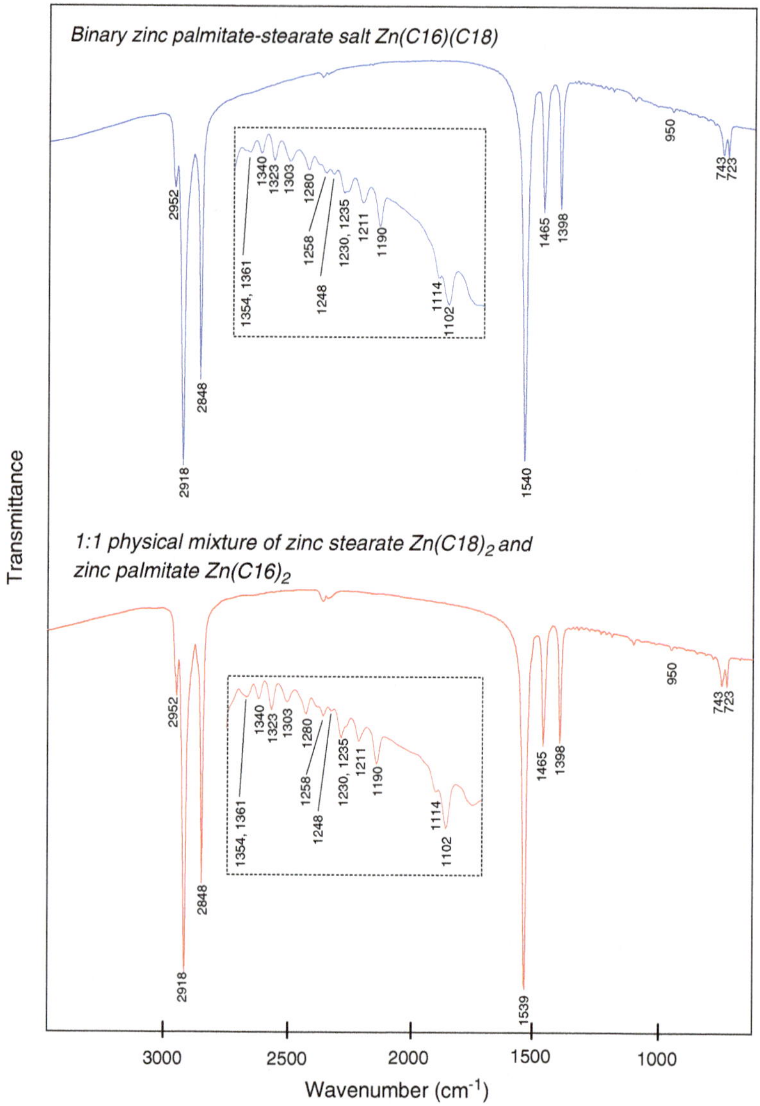

Fig. 5 The upper trace shows the FTIR spectrum of the binary salt Zn(C16)(C18). The lower trace shows a 1:1 physical mixture of zinc stearate and zinc palmitate. A scale expansion from 1380 to 1050 cm^{-1} is given below each spectrum

that small amounts of oil and free fatty acids are also present in the protrusions, and these minor components may affect the structure of the zinc soaps.

When zinc soaps are intimately mixed within an oil matrix, the FTIR spectrum of the paint often shows only the ν_aCOO, and sometimes the δCH_2 and ν_sCOO bands of the soaps, due to peak overlaps with pigments and the oil medium. In these cases, it is important to keep in mind that these bands alone are not enough to allow identification of the type of zinc soap. Although peak splitting of these bands is seen in synthetic zinc oleate, the analysis of the protrusions described in this work has shown that zinc soaps containing primarily saturated fatty acids can also produce peak splitting when they form naturally in paint layers. Other FTIR spectra of zinc soaps with band splitting similar to the protrusions presented here have been reported, although without corresponding fatty acid profiles (van der Weerd et al. 2003; Corbeil et al. 2011; Osmond 2014). Further work may determine the prevalence of this type of zinc soap protrusion and shed more light on its formation.

Zinc Lactate in a Surface Accretion

Description of the Painting and Surface Accretion

Surface efflorescence or accretions are often due to fatty acids or fatty acid metal salts (Ordonez and Twilley 1997; Noble and Boon 2007). A more unusual type of surface phenomenon was observed on *The Encounter* (accession # 2001/83, Art Gallery of Ontario), an oil on Masonite painting from 1952 to 1954 by Andor Weininger, a Hungarian-born artist who worked in Canada during the 1950s. A water-soluble, disfiguring surface layer is present on the painting. Analysis was undertaken to determine the composition and possibly the source of the material.

SEM-EDX

Elemental analysis of a sample of the surface accretion showed that the major elements were carbon and oxygen and that the only metal in the layer was zinc. Small amounts of chlorine and sulphur and traces of silicon may be indications of pollutants and dirt at the surface.

GCMS

Because very little sample was available, pyrolysis-gas chromatography-mass spectrometry was undertaken using direct sample introduction with a thermal separation probe (Py(TSP)-GCMS), a technique that allows samples as small as a few micrograms to be analysed. The analysis showed that both fatty acids and lactic acid (2-hydroxypropanoic acid) are present in the surface accretion. The major fatty acids identified were palmitic (C16), myristic (C14) and stearic (C18). Analysis of multiple samples showed some variation in the relative proportions of lactic acid to the fatty acids. This is likely due to inhomogeneity in the composition of the accretion, which is more apparent when using very small samples.

FTIR

The FTIR spectrum of the surface accretion is dominated by bands due to zinc lactate, which formed from the interaction of lactic acid with the reactive zinc white paint below. The zinc salt of lactic acid gives a characteristic infrared spectrum that is very different from the more commonly identified zinc fatty acid salts. The FTIR spectrum of the surface accretion is shown in the upper trace of Fig. 6; the lower

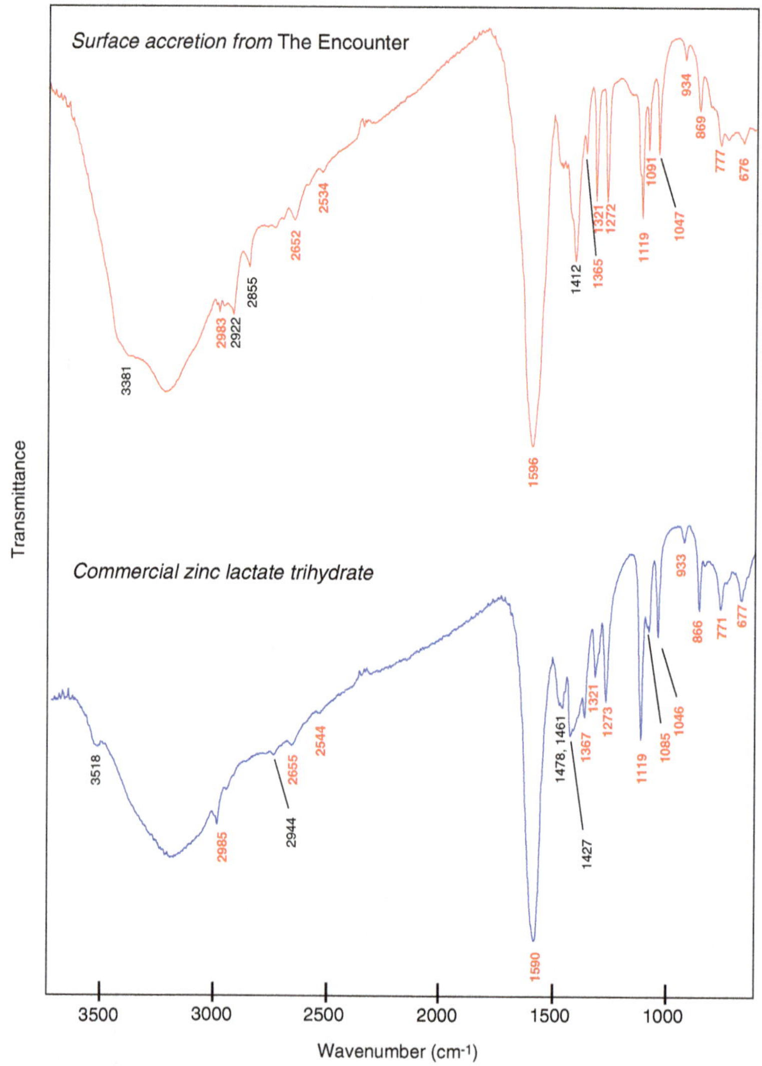

Fig. 6 The upper trace shows the FTIR spectrum of the surface accretion from *The Encounter*. The lower spectrum is that of commercial zinc lactate trihydrate. The series of bands that allowed identification of the lactate salt are labelled in red on the spectra

trace shows the spectrum of commercial zinc lactate trihydrate. The majority of the bands in the two spectra match well; differences in the region around 1430–1400 and in the OH stretching region around 3500 cm^{-1} could be due to the presence of other components in the accretion or to differences in the hydration state between the commercial material and the zinc lactate found on the painting. The peaks at 2922 and 2855 cm^{-1} in the spectrum of the accretion, and not present in the commercial zinc lactate, are due the C–H stretches of the fatty acid component.

As seen in Fig. 6, zinc lactate shows the asymmetric stretch of the carboxylate ion (ν_aCOO) at around 1590 cm^{-1}. Because this broad asymmetric stretch is similar to that seen in other zinc salts—for example zinc formate (Baraldi 1979) or the zinc fatty acid salts that occur within the drying oil medium of some aged paints (van der Weerd et al. 2005; Corbeil et al. 2011)—it is important to consider other characteristic bands for the identification of the lactate salt. The FTIR spectra of metal lactate salts have a number of sharp, intense bands related to the alpha-hydroxy carboxylate structure that allow identification; these are labelled in red on the spectra in Fig. 6. The lactate salts of other metals, for example ferrous lactate (Bio-Rad 2012), give similar spectra, so confirmation of the metal bound to the carboxylate is important.

Zinc lactate has not been previously identified on an oil painting and its presence may be related to reaction of the paint to pollutants. Lactic acid and other low molecular weight acids from both industrial and natural emissions constitute a major class of atmospheric pollution (Dabek-Zlotorzynska and McGrath 2000; Gallagher et al. 2008). Once the lactic acid is in contact with the paint, it reacts with the zinc and is trapped at the surface. A similar phenomenon has been observed for formic acid; zinc formate has been reported at the surface of a work by Salvador Dali (Keune and Boevé 2014) and also within a paint layer in a 1964 painting by Jack Chambers (Helwig et al. 2013).

Conclusions

Reaction products of zinc with carboxylic acids have led to various types of deterioration in twentieth-century Canadian oil paintings. In a painting from 1956, delamination and lack of cohesion within a thick, white layer were found to be related to a high concentration of zinc soaps. In two paintings that date from 1936 and 1937, zinc soaps have aggregated and formed protrusions. Analysis indicates that the protrusions contain primarily zinc palmitate and stearate. Peak splitting in the FTIR spectrum, which is not observed in synthetic zinc palmitate, zinc stearate or binary palmitate-stearate salts, is likely due to structural distortion. Finally, a disfiguring surface accretion on a painting dating from 1952 to 1954 was found to contain a short chain carboxylate salt, zinc lactate (zinc 2-hydroxypropanoate). Knowledge of the compounds present in deterioration phenomena such as those presented here is an important step in the design of appropriate conservation treatments for modern oil paintings.

Acknowledgements Élisabeth Forest (Centre de conservation du Québec) provided samples from the painting *Rouge sur blanc* as well as information about its condition. Maria Sullivan (Art Gallery of Ontario) noted the formation of protrusions on *Storm over the Fields* and provided sample material. Stephen Gritt (National Gallery of Canada) brought the protrusions on *Equations in Space* to our attention. Marie-Eve Thibeault (formerly of the Art Gallery of Ontario) examined the surface accretion on *The Encounter* and provided sample material. The authors would also like to thank Gillian Osmond (Queensland Art Gallery, Australia) and Scott Williams (Canadian Conservation Institute) for helpful discussions.

Appendix

Cross-section samples were mounted in polyester resin, ground and polished using standard petrographic techniques, and observed using incident light and fluorescence microscopy with a Leica DMRX microscope. Small fragments of unmounted samples were analysed using a combination of the techniques described below.

For FTIR, samples were mounted on a low pressure diamond anvil microsample cell and analysed using a Bruker Hyperion 2000 microscope interfaced to a Tensor 27 spectrometer. The samples were analysed in transmission mode by co-adding 200 scans and using a 4 cm^{-1} resolution. Spectra were collected from 4000 to 430 cm^{-1} using a wide band MCT detector or from 4000 to 550 cm^{-1} using a midband detector.

GCMS analysis was performed on small fragments of the protrusion samples (approximately 20–30 μg) using an Agilent 6890 gas chromatograph interfaced to an Agilent 5973 quadrupole mass spectrometer. Samples were extracted and derivatised using two separate methods: methylation and trimethylsilylation. For silylation, the samples were treated with 30 μL of *N,O*-bis(trimethylsilyl)trifluoroacetamide (BSTFA; Pierce) for 30 min at 60 °C to form trimethylsilyl (TMS) derivatives. For methylation, the samples were treated with a 1:1 solution of 15 μL of Meth-Prep™ II (m-(trifluoromethyl)phenyl trimethylammonium hydroxide, TMTFTH, 0.2 N in methanol; Grace Davison Discovery Science) and 15 μL toluene for 60 min at 60 °C to form methyl derivatives. Because only micro-samples of the zinc lactate surface accretion (approximately 3 μg) were available, Py(TSP)-GCMS was employed. For this method, an Agilent TSP was used for direct sample introduction to an Agilent 7890 gas chromatograph interfaced to an Agilent 5975 triple-axis quadropole mass spectrometer. Samples were derivatised using 1.5 μL hexamethyldisilazane (HMDS; Supelco) to form trimethylsilyl derivatives. The TSP was ramped from 50 to 450 °C at a rate of 900 °C/min and held at that temperature for 3 min before cooling to 250 °C for the remainder of the run.

SEM-EDX was undertaken using an Hitachi S-3500 N scanning electron microscope integrated with a lithium-drifted silicon, light element X-ray detector and an Oxford Inca X-ray microanalysis system. The SEM was operated at an accelerating voltage of 20 kV and a working distance of 15 mm. Using this technique, elemental analysis of small volumes, down to a few cubic micrometers, were obtained for elements from boron (B) to uranium (U) with a sensitivity of about 1 %.

XRD patterns of selected paint samples were obtained with a Bruker D8 Discover with GADDS (General Area Detector Diffraction Solution) equipped with a rotating anode and cobalt target. The patterns were measured at 40 kV and 85 mA, using a 0.5 mm collimator.

The synthesis of the zinc fatty acid salts was based on a published method (Robinet and Corbeil 2003). A solution of the fatty acid in anhydrous ethanol was gently heated until dissolved and subsequently decanted into an aqueous sodium hydroxide solution held at 60 °C. The solution was heated under reflux at 98 °C overnight. The resulting sodium salt was filtered, dried and subsequently dissolved in ultrapure water and added to an aqueous zinc chloride solution, where zinc chloride was present in excess. In the case of the binary salts, the intermediate sodium salt was not isolated, and the zinc chloride was added directly to the aqueous sodium fatty acid salt solution. The purity of the synthetic fatty acid salts was confirmed by comparison of the FTIR spectra to published data (Robinet and Corbeil 2003; Price and Pretzel 2007; Taylor and Ellis 2007) and by a determination of the fatty acids present using GCMS. In the case of the 1:1 binary salt, where a published infrared spectrum was not available, the composition was confirmed by X-ray diffractometry using a Rigaku Ultima IV equipped with a scintillation counter. The diffractogram was measured in step mode from 1° to 40°, 2θ (step size: 0.02°, count time: 1 s). A single set of long spacing lines, mid-way between those of zinc palmitate and zinc stearate were obtained, consistent with the published data for this compound (Mesbah et al. 2011).

References

Baraldi P (1979) Thermal behaviour of metal carboxylates: metal formates. Spectrochim Acta 35A:2003–2007

Bio-Rad Laboratories Inc (2012) Merck-BioRad-Sadler Spectral Library (Accessed from Know-ItAll Informatics System, IR/NIR Edition)

Boon JJ, Keune K, Zucker J (2005) Imaging analytical studies of lead soaps aggregating in preprimed canvas used by the Hudson River School painter FE Church. Microsc Microanal 11(Suppl 2):444–445

Corbeil M-C, Helwig K, Poulin J (2011) Jean Paul Riopelle: the artist's materials. Getty Conservation Institute, Los Angeles

Dabek-Zlotorzynska E, McGrath M (2000) Determination of low-molecular-weight carboxylic acids in the ambient air and vehicle emissions: a review. Fresenius J Anal Chem 367(6):507–518

Faubel W, Simon R, Heissler S, Friedrich F, Weidler PG, Becker H, Schmidt W (2011) Protrusions in a painting by Max Backmann examined with confocal u-XRF. J Anal Atomic Spectrom 26(5):942–948

Gallagher M, Wysocki CJ, Leyden JJ, Spielman AI, Sun X, Preti G (2008) Analyses of volatile organic compounds from human skin. Br J Dermatol 159(4):780–791

Helwig K, Thibeault M-E, Poulin J (2013) Jack Chambers' mixed media paintings from the 1960s and 1970s: painting technique and condition. Stud Conserv 58(3):226–244

Higgitt C, Spring M, Saunders D (2003) Pigment-medium interactions in oil paint films containing red lead or lead-tin yellow. Nat Gall Tech Bull 24:75–95

Ishioka T, Maeda K, Watanabe I, Kawauchi S, Harada M (2000) Infrared and XAFS study on structure and transition behavior of zinc stearate. Spectrochim Acta 56A:1731–1737

Keune K (2005) Binding medium, pigments and metal soaps characterized and localised in paint cross-sections. FOM Institute for Atomic and Molecular Physics, Amsterdam, pp 113–169

Keune K, Boevé G (2014) It's surreal: zinc-oxide degradation and misperceptions in Salvador Dalí's *Couple with Clouds in their Heads*, 1936. This proceedings, Chapter 19

Lacouture F, Peultier J, François M, Steinmetz J (2000) Anhydrous polymeric zinc (II) octanoate. Acta Cryst C56:556–557

Maines CA, Rogala D, Lake S, Mecklenburg M (2011) Deterioration in abstract expressionistic paintings: analysis of zinc oxide paint layers in works from the collection of the Hirshhorn Museum and Sculpture Garden, Smithsonian Institution. Mater Res Soc Symp Proc 1319:275–284

Mesbah A, Juers C, Lacouture F, Mathieu S, François M, Rocca E, Steinmetz J (2006) Structures cristallines de carboxylates métalliques assurant la protection du zinc et du magnésium contre la corrosion. Matériaux 2006, 13–17 Novembre 2006 – Dijon, France, np

Mesbah A, Juers C, Francois M, Rocca E, Steinmetz J (2007) Structures of magnesium and zinc long aliphatic chains carboxylates. Z Kristallogr Supp 26:593–598

Mesbah A, Jacques S, Rocca E, Francois M, Steinmetz J (2011) Compact metal-organic frameworks for anti-corrosion applications: new binary linear saturated carboxylates of zinc. Eur J Inorg Chem 2011:1315–1321

Noble P, Boon JJ (2007) Metal soap degradation of oil paintings: aggregates, increased transparency and efflorescence. AIC PSG Postprints 19:1–15

O'Donoghue E, Johnson AM, Mazurek J, Preusser F, Schilling M, Walton MS (2006) Dictated by media: conservation and technical analysis of a 1938 Joan Miró canvas painting. Stud Conserv 51(2):66–68

Ordonez E, Twilley J (1997) Clarifying the haze: efflorescence on works of art. Anal Chem 69(13):416A–422A

Osmond G (2014) Zinc white and the influence of paint composition for stability in oil based media. This proceedings, Chapter 18

Osmond G, Keune K, Boon J (2005) A study of zinc soap aggregates in a late 19th century painting by R.G. Rivers at the Queensland Art Gallery. AICCM Bull 29:37–46

Osmond G, Boon JJ, Puskar L, Drennan J (2012) Metal stearate distribution in modern artists' oil paints: surface and cross-sectional investigation of reference paint films using conventional and synchrotron infrared microspectroscopy. Appl Spectrosc 66(10):1136–1144

Peultier J, Francois M, Steinmetz J (1999) Anhydrous polymeric zinc(II) heptanoate. Acta Cryst C55:2065–2065

Price B, Pretzel B (2007) Infrared and Raman Users Group spectral database. Edition 2007, vol 2. Infrared and Raman Users Group, Philadelphia

Robinet L, Corbeil M-C (2003) The characterization of metal soaps. Stud Conserv 48(1):23–40

Rogala D, Lake S, Maines C, Mecklenburg M (2010) Condition problems related to zinc oxide underlayers: examination of selected abstract expressionist paintings from the collection of the Hirshhorn Museum and Sculpture Garden, Smithsonian Institution. J Am Inst Conserv 49:96–113

Taylor RA, Ellis HA (2007) Room temperature molecular and lattice structures of a homologous series of anhydrous zinc(II) *n*-alkanoate. Spectrochim Acta Part A 68:99–107

van der Weerd J, Geldof M, van der Loeff LS, Meeren RMA, Boon JJ (2003) Zinc soap aggregate formation in 'Falling Leaves' (Les alyschamps) by Vincent van Gogh. Kunsttechnologie Konservierung 17(2):407–416

van der Weerd J, van Loon A, Boon JJ (2005) FTIR studies of the effects of pigments on the aging of oil. Stud Conserv 50:3–22

Zeleňák V, Vargová Z, Györyová K (2007) Correlation of infrared spectra of zinc(II) carboxylates with their structures. Spectrochim Acta Part A 66:262–272

Zwei Jünglinge und Zwei Mädchen, A Tempera Painting by Otto Mueller Circa 1917. *Paint Analysis and Reconstruction*

Hannie Diependaal, Klaas Jan van den Berg, Meta Chavannes, Cosima Damiana Calvano, Inez Dorothé van der Werf, and Louise Wijnberg

Abstract The present study focuses on material aspects of the painting Zwei Jünglinge und zwei Mädchen (c.1917) by Otto Mueller (1874–1930) from the collection of the Stedelijk Museum in Amsterdam.[1] Research was performed in the context of the recent conservation treatment of this painting. Friable, porous paint layers gave rise to investigation of the paint composition. The study coincided with a strong international interest for Mueller's contemporaries and German Expressionism, which was reflected in the reconstruction of the 'Sonderbund Ausstellung 1912' in Cologne and the interdisciplinary research project Keiner hat diese Farben wie Ich – Studien zur Maltechnik Ernst Ludwig Kirchners.[2] These projects

[1]This research was performed by Hannie Diependaal during an internship at the Stedelijk Museum Amsterdam in the context of a thesis to finish the Advanced Master's programme Conservation of Cultural Heritage, specialty Conservation of Paintings at the University of Amsterdam. The entire study covers two German Expressionist paintings, respectively by Otto Mueller and E.L. Kirchner (1880–1938).

[2]Sonderbund Ausstellung '1912-Mission Moderne', Wallraf-Richartz-Museum, Köln, 31-8-2012/30-12-2012; Keiner hat diese Farben wie Ich – Studien zur Maltechnik Ernst Ludwig Kirchners, project started in 2009, realised by Staatliche Akademie der Bildenden Künste Stuttgart, Kirchner Museum Davos, Doerner Institut, Bayerische Staatsgemäldesammlungen, München, Schweizerisches Institut für Kunstwissenschaft, Zürich.

H. Diependaal (✉)
Independent Paintings Conservator, Amsterdam, The Netherlands
e-mail: hannie@hdiependaal.nl

K.J. van den Berg
Cultural Heritage Agency of the Netherlands, Amsterdam, The Netherlands

M. Chavannes • L. Wijnberg
Painting Conservation Stedelijk Museum Amsterdam, Van Baerlestraat 31, 1071 AN Amsterdam, The Netherlands
e-mail: M.Chavannes@stedelijk.nl; L.Wijnberg@stedelijk.nl

C.D. Calvano • I.D. van der Werf
Department of Chemistry, University of Bari, Bari, Italy

K.J. van den Berg et al. (eds.), *Issues in Contemporary Oil Paint*,
DOI 10.1007/978-3-319-10100-2_12

provided a framework for the present research. In particular the binding medium of Mueller's matte paint was studied through Thermally-assisted Hydrolysis Methylation (THM) – Gas Chromatography Mass Spectrometry (GCMS) and Matrix Assisted Laser Desorption Ionisation (MALDI) – Time of Flight (TOF) – Mass Spectrometry (MS). Initial results were further explored by paint reconstructions. The findings led to the insight that some materials identified through analyses are not likely to have been part of the original paint composition, but are probably present as a result of conservation treatment.

Keywords Tempera • Matte paint • Otto Mueller • German Expressionism • Reconstruction • Binding medium • MALDI TOF MS analysis

Mueller's Matte Paint

Zwei Jünglinge und zwei Mädchen, *c.*1917 (Fig. 1) painted on jute is illustrative of the matte aesthetic that Otto Mueller tried to achieve. The surface of the painting is soft and pastel-like and unvarnished with occasional glossy patches. Today the paint is locally cracked and sensitive to water in combination with mechanical action. Despite several conservation treatments in the past, including a gelatine

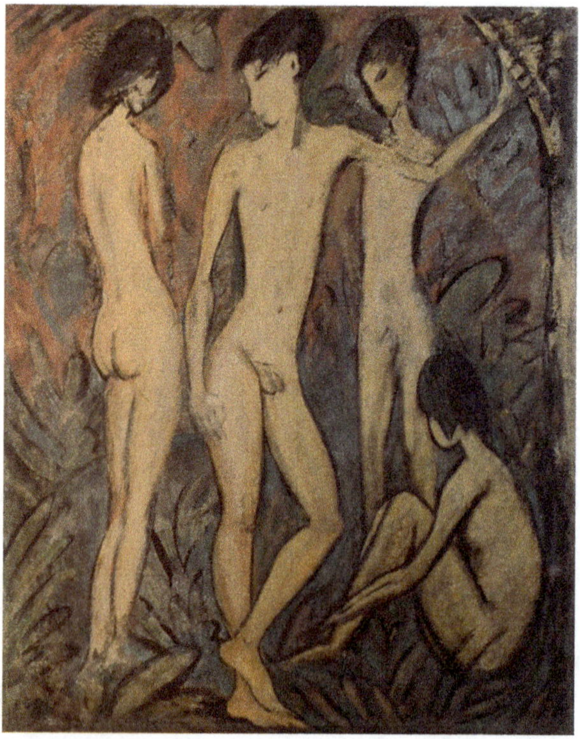

Fig. 1 Otto Mueller, *Zwei Jünglinge und zwei Mädchen*, c.1917, tempera on jute, 108.5 × 89.5 cm, before treatment, collection Stedelijk Museum Amsterdam. inv.nr. A21703 (Photo H. Diependaal)

spray-coating of the painting's surface and a lining with a polyvinylacetate glue, the paint layers were delaminating at the start of the present project. The object needed treatment prior to an upcoming exhibition.

Written sources indirectly provided better understanding of the artist's intention and the nature of his paint.[3] Otto Mueller himself rarely gave away details on his painting technique. In contrast contemporaries Ernst Ludwig Kirchner (1880–1938) and Erich Heckel (1883–1970) were much more outspoken. Both painters were founding members of *Die Brücke,* the group of German Expressionist painters joined by Mueller in 1910.

Like his fellow-members of *Die Brücke* Otto Mueller experimented with paint preparation and application.[4] Around 1907 he took a radical decision when he had destroyed his early work which was executed in oil paint.[5] His subsequent choice for a matte painting technique was significant and valued by Kirchner and Heckel on first sight. They invited Otto Mueller to become a member of *Die Brücke* after attending an exhibition in Berlin where several paintings by Mueller had caught their attention. In these paintings they recognised a vitality, a sensuality and an aesthetic akin to their own.[6]

Three years later Kirchner wrote: '*Er* [Mueller] *brachte uns den Reiz der Leimfarbe',* referring to a glue bound paint.[7] In 1919 Otto Mueller himself stated that he had been inspired by ancient Egyptian art.[8] Alternatively the paint used by Otto Mueller has been characterised as *tempera*, which in the first decades of the twentieth century had become a broad term for a wide variety of emulsion paints. These tempera paints were described by Wilhelm Ostwald in comprehensible terms in *'Malerbriefe',* first published as newspaper articles in 1903 and 1904.[9] His articles were popular and could have reached the *Brücke* artists although Heckel in retrospect remarked that all had to be learned by trial and error.[10]

The above sources do not offer an unambiguous clue to the binding medium used by Otto Mueller, however they indicate the importance of the new aesthetics for *Die Brücke* members. These served the artists to distance themselves from academic

[3]Citations from letters, diaries, writings and quotes from the *Brücke* artists in relation to the material aspects of their paintings have amongst others been published by Buchheim (1963), Schick and Skowranek (2011), Von Saint George (2003).

[4]Buchheim, p. 86, Von Saint-George, p. 311, Pirsig-Marshall, p. 152.

[5]Pirsig-Marshall (2011, p. 144).

[6]Hohenzollern and Lüttichau (2003, pp. 47–48). Mueller's paintings were shown in 1910 in the '*Kunstausstellung Zurückgewiesener der Berliner Secession'.*

[7]E.L. Kirchner, *Chronik der KG Brücke*, 1913 in: Scotti (2005, pp. 64–65).

[8]Buchheim, p. 86 reports a rare statement from the artist recorded in 1919 in a Berlin catalogue of an exhibition at Paul Cassirer's gallery: 'Mir vorbildlich, auch für das rein Handwerkliche, war und ist noch jetzt die Kunst der alten Ägypter.'

[9]Pohlmann, p. 19. *Malerbriefe* were published in the *Münchener Allgemeinen Zeitung.*

[10]Von Saint-George, p. 306.

values.[11] Following in the footsteps of Impressionist painters they experimented with binders to achieve glowing colours and a matte surface, which had come to represent modernity.[12]

Binding Medium Analysis

Returning to the object, questions concerning the nature of the paint in relation to the choice of materials for conservation still had to be answered. Was this paint the *Leimfarbe* that Kirchner mentioned in 1913 or an alternative? What was the original appearance? Analyses of three tiny paint samples scraped off the object's edges were carried out to disclose the actual paint composition (Table 1).

Table 1 Analytical results of samples from *Zwei Jünglinge und zwei Mädchen*

	THM-pyrolysis-GCMS					
Samples	P/S[a]	A/P[b]	A/Sub[c]		MALDI-TOF -MS	Micro-Raman spectroscopy
(1) White ground layer	1.5	1.9	3.1	Oxidized rapeseed oil, drying oil (probably linseed oil)	Mammal animal glue	Calcite matrix ($CaCO_3$), finely dispersed pigments (synthetic ultramarine)
(2) Black-brown paint[d]	–	–	2.7	Beeswax and venetian turpentine, some drying oil, small amount of rapeseed oil		
(3) Red and black paint with white ground	1.5	2.6	3.0	Drying oil and rapeseed oil, beeswax, colophony	Mammal animal glue, egg yolk, egg white	

[a]Palmitic acid/Stearic acid
[b]Azelaic acid/Palmitic acid
[c]Azelaic acid/Suberic acid
[d]Sample contained unpigmented waxy material, probably not original. See text

[11]E.L. Kirchner, '*Die Brücke, Manifest 1906*', '*Chronik der KG Brücke*', in Scotti (2005, pp. 64–65).

[12]Callen, p. 746. A reluctance to use oil paint had already developed in the late nineteenth century. It grew out of ill-understood degradation of oil paintings, the cause of which was blamed on the binding medium in the paint. Pohlmann p. 17–19 gives many references for the contemporary discussion.

Results

The off-white ground was found to contain chalk with some finely dispersed pigment, including ultramarine. MALDI-TOF-MS analysis of the protein fraction revealed the presence of mammal collagen peptides, which indicates that mammal animal glue has been used. The results from the THM-GCMS pyrogram suggested that significant amounts of oxidised rapeseed oil are present, in combination with a drying oil, probably linseed oil.

A variety of materials was identified in the upper paint layers: typical oxidation products for drying oil were identified, as well as the main degradation products of erucic acid which is particularly abundant in rapeseed oil (47 % w/w). The MALDI-TOF-MS analysis of the protein fraction showed peaks of collagen (mammal animal glue), but also egg yolk and egg white peptides. Significant amounts of beeswax were identified as well as colophony. (Venetian) turpentine was indicated by peaks of diterpenoids epimanool, and larixyl acetate in the THM-GCMS pyrogram.

Discussion

The analytical results of the samples taken from *Zwei Jünglinge und zwei Mädchen* were somewhat puzzling. The binding medium of the paint layers did not comply with the expected glue binder. Perhaps the artist used a more complex paint mixture. On the other hand some of the materials could have been introduced after completion of the painting during past conservation treatments, which were documented to some extent only since 1953.

For reference the results were compared to recent studies concerning paint composition of fourteen paintings by Kirchner and four paintings by Heckel.[13] Comparative chemical analysis of Mueller's paint was first published while the present project was underway and concerns a later painting by Otto Mueller, dating 1927.[14] These studies supplied valuable background for the interpretation of the present analytical outcome, in particular for the ground paint, as discussed below.

---

[13]Dietemann et al. (2013), Von Saint-George (2003) and unpublished results of analyses of *Skizzierender Künstler mit zwei Frauen* (1913) by Ernst Ludwig Kirchner, collection Stedelijk Museum Amsterdam inv.nr. A 4767, which was sampled in the context of the present project.

[14]Deimel and Kolodziej (2012) report briefly that similar findings of two further paintings by Otto Mueller from the Saarland Museum justify the conclusion that Mueller hardly changed his painting technique from 1919 to 1927.

Grounds

Heckel postulated that the *Brücke* artists prepared their grounds themselves out of chalk and animal glue with the occasional admixture of a drying oil.[15] The analytical results of sample 1 in the present study indicate a lean ground composed primarily of chalk in a medium of animal glue with a small amount of oil. This finding thus sustains Heckel's remark.

In fact the analysis showed the presence of two types of oil in Mueller's ground paint. Apart from the drying linseed oil a non-drying rapeseed oil was identified. This type of oil was not recognised in the binding medium of the reference paintings, neither in the grounds, nor in the upper layers. Yet rapeseed oil seems not to be an uncommon ingredient for this period. Rapeseed oil was described as a semi-drying adulteration of linseed oil in 1909 by Alexander Eibner.[16] The artist may have added the rapeseed oil unintentionally, when using a falsified linseed oil. With a P/S ratio of 1.5 the oil content in Mueller's ground is quite different from technical oils with a P/S ratio <1.0 which were identified in significant quantities in seven grounds applied by Kirchner.[17] These technical oils, or chemically enhanced semi-drying oils, were not denominated as such by Kirchner, nor Heckel. Both painters simply referred to the use of (linseed) oil or stand oil in their grounds.[18]

Then again, in 1934, Max Doerner in 'The Materials of the Artist' refers to rapeseed oil as a non-drying oil used in half-chalk grounds to enhance flexibility. Thus Mueller could have deliberately added a rapeseed oil to his ground paint.

As a third possibility it is proposed that he did not add any oil to the ground at all. This possibility is paralleled by the analytical results of the ground of the 1927 painting by Mueller consisting of chalk in a protein binder.[19]

The relatively small amount of oil in the ground layer of *Zwei Jünglinge und zwei Mädchen* could have been absorbed during the application of upper paint layers, rather than mixed into the ground paint from the start. In the red and black coloured upper layer (sample 3) the quantity of rapeseed oil is somewhat lower with respect to the ground layer. The relative amounts of rapeseed oil as compared to linseed oil found in samples 1 and 3, apparently illogical, might be explained by a phenomenon of preferential penetration of the slower drying rapeseed oil into the porous chalk ground.

Perhaps Otto Mueller's paint was not entirely composed of raw materials. A semi-drying oil is not an unusual addition to the binder of tube paints. Analysis of certain tube paints samples described elsewhere has shown the presence of rapeseed

[15] Von Saint-George, p. 311 Remark made by Heckel in a letter to Max Saerlandt, 1924.

[16] Eibner, p. 311.

[17] Dietemann, p. 44, Table 1.

[18] Schick, p. 34, Von Saint-George, p. 311.

[19] Deimel/Kolodziej, p. 24. Analysis of paint samples from 'Zigeuner mit Sonnenblume', 1927, collection Saarland Museum.

oil in various colours produced by various manufacturers in the early twentieth century.[20] To date it is assumed that the non-drying oils were added to the tube paints to improve the shelf-life or the consistency of the paint.

Upper Paint Layers

Is it likely that Otto Mueller used tube paints? Considering the range of materials found in the upper paint layers of the *Stedelijk Museum's* painting in Amsterdam it is not precluded that prefabricated paints were used, either straight out of the tube or as tube paints mixed on the palette with further ingredients into a tempera paint. (Fig. 2)

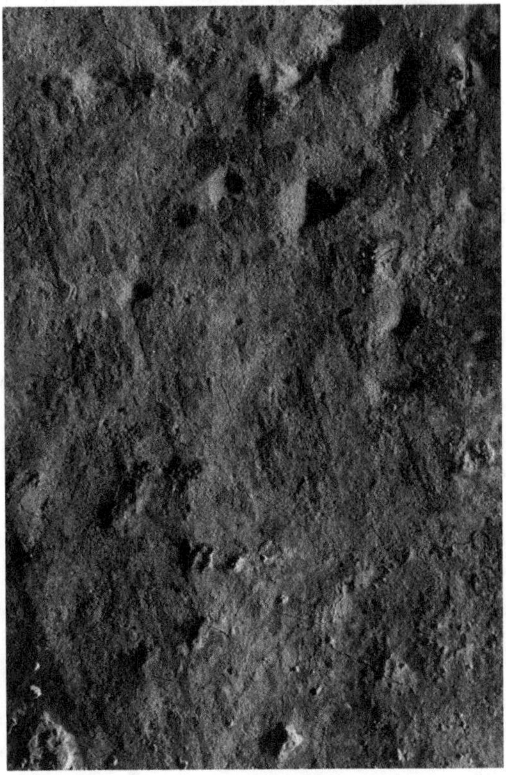

Fig. 2 Macro-photo of the upper paint layers of *Zwei Jünglinge und zwei Mädchen* after treatment (Photo H. Diependaal)

[20] Diependaal et al. (2013, pp. 182–185 Table 3, T1, T19, T 28, T 38, T51, T56, T58); Hermens et al., pp. 115–116 Table 2 T1, T2, T8.

With respect to his individual painting technique Mueller referred to ancient Egypt in a rare utterance.[21] The technique of the ancient Egyptians was much debated at the time, following the discovery of Fayum mummy portraits in the late nineteenth century. According to Eibner these portraits were either painted in encaustic, with a warm wax binder, or in tempera. The latter he specifies as '*Leimmalerei*' in the particular case of the Fayum portraits.[22] Elsewhere Eibner explains that the modern (1909) concept of tempera should be understood as an emulsion paint, based on oil, water and an emulsifier to hold the two immiscible parts together. This could be egg, casein, gum or salts from either fatty or wax acids.[23]

Wilhelm Ostwald agrees with Eibner that tempera is a paint which is water soluble when fresh, but insoluble once dried. As a binder in this tempera Ostwald suggests casein, or alternatively a glue which is made insoluble by adding iron sulphate or a chromium acid salt. Another option would be to paint with a glue bound pigment and cover the painting with formalin dust after drying. Ostwald continues by stating that yet another important form of tempera is emulsion paint, for which he recommends a wide variety of materials: on the one hand slimy watery parts like gum, glue or egg white; on the other fatty parts like oils, resins and balms.[24]

In 1940 Kurt Wehlte in his book '*Temperamalerei*' also presents tempera as an emulsion paint, either water soluble or insoluble; either lean or fat. Wehlte offers a wide range of recipes and methods, one of which is an oil tempera mixed on the palette.[25] This procedure entails tube paints, to be mixed by the artist with egg tempera and painted out with a fluid dammar-turpentine mixture, eventually enriched with a wax salve (*Wachssalbe*).[26] According to Wehlte this method was followed by the painter Max Kaus (1891–1977), who met Otto Mueller in 1920.

A variation on the oil tempera is described by Wehlte as well. For this tempera the binder is prepared in a flask, from egg white and egg yolk, animal glue, linseed oil varnish and resin or linseed oil and Venetian turpentine with the optional addition of beeswax, all vigorously shaken into an emulsion.[27] (Fig. 3)

The components identified in the paint layers of *Zwei Jünglinge und zwei Mädchen* (Table 1) are all mentioned in one or more of the above recipes. Whether or not Mueller applied these materials himself is a question still to be answered.

[21]Buchheim, p. 86 Statement made by Mueller first published in 1919.

[22]Eibner, p. 333.

[23]Eibner, p. 269.

[24]Wilhelm Ostwald, *Malerbriefe, Brief XV*, in Pohlmann, pp. 126–128.

[25]Wehlte, pp. 71–72.

[26]Wehlte, p. 72: *Wachssalbe* is described as beeswax mixed with turpentine oil *au bain Marie*.

[27]Wehlte, pp. 66–69.

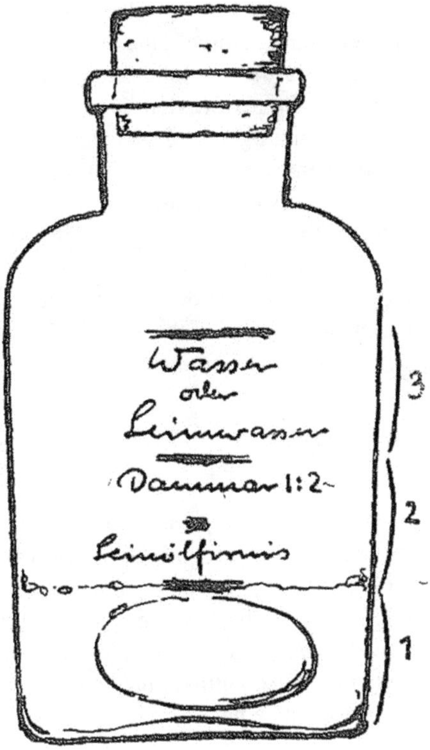

Fig. 3 Drawing of a flask with indication of proportions of ingredients for oil tempera. (Illustration from Kurt Wehlte, *Temperamalerei,* 1940)

Reconstruction

The binding medium analysis of the upper paint layers disclosed materials that could have been introduced during past conservation treatments. This speculation was supported by the analytical results of the 1927 painting which showed pigments bound in no more than a protein glue with a small amount of soap or saponified oil.[28] In order to discriminate between original and later material the matte paint in *Zwei Jünglinge und zwei Mädchen* was empirically studied through reconstruction.

Wehlte's recipes were prepared and painted out. Comparison of the reconstruction with the painting's surface gloss and texture suggested that Otto Mueller did not use the full method, nor the complete recipe described by Wehlte for oil tempera. The wax and resin identified in the red and black paint layers (sample 3) were more likely introduced as a consolidation adhesive in a treatment that was not documented. This deduction was sustained by ultraviolet fluorescence investigation and solubility tests on the sampled spot which indicated that the wax and resin detected were present very superficially. The same research method indicated the

[28]Deimel/Kolodziej, p. 24.

patchy presence of what was likely the gelatine introduced during former treatment. Being a mammal animal glue, analogous to the binding medium of a '*Leimfarbe*', this gelatine further complicated the interpretation of the analytical results.

A wax-resin mixture dominates the brown-black paint sample 2 of the *Stedelijk Museum's* painting to such an extent that a correct calculation of molar ratios of the oil component was not possible. What appeared to be an original paint blob is now held for an early conservation material.

On the other hand animal glue, egg, rapeseed oil and linseed oil can be considered to be the constituents of Mueller's tempera, which is not a *Leimfarbe* in the strict sense of the word. The paint composition may be regarded as proof of the experimental attitude of the artist in this period.

Conclusion

Otto Mueller's matte paint has been described as '*Leimfarbe*' or tempera. Binding medium analysis has identified a variety of materials in this case study. Reconstruction of a number of tempera paint recipes and complementary research clarified the interpretation of chemical analysis. The three samples taken from *Zwei Jünglinge und zwei Mädchen* suggest that Mueller's tempera around 1917 consists of pigments bound in mammal animal glue, egg and mixed oil. It cannot be precluded that Otto Mueller used tube paints, although it seems unlikely. Future research of additional paint samples from the object might establish a more profound conclusion.

The findings led to valuable insight into the studio practice of the artist, the original appearance of the painting's surface, as well as the painting's conservation history, which is indispensible for an efficient decision-making process during conservation treatment.

References

Buchheim LG (1963) Otto Mueller. Leben und Werk. Buchheim, Feldafing

Callen A (1994) The unvarnished truth: mattness, 'primitivism' and modernity in French painting, c.1870–1907. Burlington Mag 136(1100):738–746

Deimel K, Kolodziej D (2012) Untersuchung, Konservierung und Restaurierung des Bildes 'Zigeuner mit Sonnenblume'von Otto Mueller im Saarland. Museum. Museum Aktuell (November):22–25

Diependaal H et al (2013) Breitner's paint box and the colours of Standage. Zeitschrift für Kunsttechnologie und Konservierung 27(1):171–191

Dietemann P et al (2013) Ernst Ludwig's binding media – material, surface and colour perception. Zeitschrift für Kunsttechnologie und Konservierung 27(1):43–55

Eibner A (1909) Malmaterialenkunde als Grundlage der Maltechnik. Springer, Berlin

Hohenzollern JG, Lüttichau MA (eds) (2003) Otto Mueller. Prestel, München

Pirsig-Marshall T (2011) Otto Mueller and the Brücke: a creative dialogue. In: Weikop C (ed) New perspectives on Brücke expressionism, bridging history. Ashgate, Farnham, pp 143–163

Pohlmann A (2005) Wilhelm Ostwald und die Situation der Maltechnik in Deutschland um 1900. In: Ostwald W (ed) Maltechnische Schriften 1904–1914. Mitteilungen der Wilhelm Ostwald Gesellschaft zu Grossbothen e.V, 10(22):15–44

Schick K, Skowranek H (eds) (2011) Keiner hat diese Farben wie ich. Kirchner malt. Hatje Cantz, Ostfildern

Scotti R (ed) (2005) E.L. Kirchner und die Brücke: die Jahresmappen der Brücke 1906–1912 und die Chronik der Brücke KG 1913 aus der Sammlung Eberhard W. Kornfeld, Bern-Davos. Kirchner Verein, Davos

Von Saint-George C (2003) Studie zur Maltechnik Erich Heckels. Technologische Untersuchung an ausgewählten Gemälden aus dem Museum Ludwig in Köln. Zeitschrift für Kunsttechnologie und Konservierung 17(2):305–324

Wehlte K (1940) Temperamalerei. Einführung in Werkstoffe und Malweisen. Otto Maier, Ravensburg

Metal Soaps and Visual Changes in a Painting by René Magritte – *The Menaced Assassin*, 1927

Michael Duffy, Ana Martins, and Jaap J. Boon

Abstract The *Menaced Assassin* by René Magritte is a major surrealist painting from the collection of the Museum of Modern Art (MoMA). A recent conservation treatment coupled with scientific analysis has characterised some puzzling and visually disturbing surface phenomena once attributed to mold growth.

A detailed examination of the painting under the microscope has evidenced that the formation of radiating dark micro cracks is responsible for the overall speckled appearance of the painting. These web-like cracks often surround a small white crystalline particle and are generally accompanied by aggregates of very small white particles sitting on the top of the surface. The examination also revealed the presence of larger flat pustules of a translucent and softer material with no apparent visual impact. An overall whitish haze is visible as well over the areas painted in red and brown earth tones.

Samples of both white and translucent material were analysed by Fourier transform infrared spectroscopy (FTIR) and confirmed the presence of metal soaps. Past exhibition conditions of the painting and composition of the ground layer may have played a role in the formation of the soaps, in particular the coarse calcium carbonate rich priming layer and presence of kaolin as identified by FTIR and scanning electron microscopy combined with electron dispersive spectroscopy (SEM-EDX).

Keywords Lead soaps • Zinc soaps • Magritte • FTIR • SEM-EDX • Pustules • Micro-cracks

M. Duffy (✉) • A. Martins
The Museum of Modern Art, 11 West 53rd Street, New York, NY 10019, USA
e-mail: michael_duffy@moma.org; ana_martins@moma.org

J.J. Boon
JAAP Enterprise for Art Scientific Studies (Formerly MOPAS Group AMOLF), Nieuwendammerdijk 79, 1025 LD Amsterdam, The Netherlands
e-mail: boon@amolf.nl

K.J. van den Berg et al. (eds.), *Issues in Contemporary Oil Paint*,
DOI 10.1007/978-3-319-10100-2_13

Introduction

The Menaced Assassin (1927) by René Magritte is a major surrealist painting from the collection of The Museum of Modern Art (Fig. 1). It was painted for Magritte's first one-man show at Le Centaur in Brussels. It is sometimes considered to be a pendant to a painting of the same size and date titled *The Secret Player* (1927, Royal Museum of Fine Arts Belgium).

MoMA acquired *The Menaced Assassin* in 1966. Prior to this, both paintings were on long term loan from the previous owner to the Casino Communal in Knokke Belgium where the environmental conditions are presumed to have included exposure to high humidity for approximately 10 years. Upon acquisition of *The Menaced Assassin*, MoMA Conservation department records noted an overall spotty surface appearance, which was attributed to mold growth. To address this condition, in 1967, the painting was cleaned, lined with wax resin adhesive, and surface coated with a synthetic resin varnish. In subsequent treatments the "mold" spots were concealed with inpainting as they were judged to be visually disturbing.

In preparation for the exhibition *Magritte: The Mystery of the Ordinary 1926–1938* opening in September 2013, the painting underwent treatment to remove the

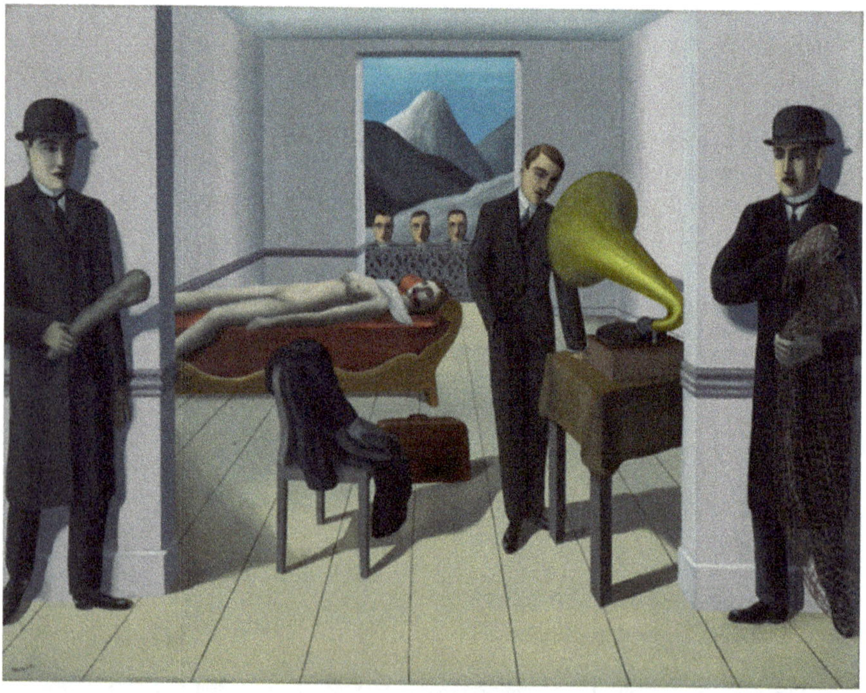

Fig. 1 René Magritte. *The Menaced Assassin*. Brussels, 1927. Oil on canvas, 59 1/4″ × 6′ 4 7/8″ (150.4 × 195.2 cm) (The Museum of Modern Art. Kay Sage Tanguy Fund. © 2013 Charly Herscovici, Brussels/Artists Rights Society (ARS), New York)

degraded varnish, inpainting, and excess yellow wax resin from the surface. This treatment also allowed for a closer examination of the surface phenomena once thought to be mold but now determined to be a condition related to the Magritte's choice of materials.

Examination

Magritte used preprimed, commercially prepared canvas cut from a roll for *The Menaced Assassin*. Microscopic examination of cross-sections indicated the priming consists of two white preparatory layers. Reflected IR does not reveal any obvious under-drawing. Radiography and transmitted IR imaging showed that Magritte applied his paint in broad strokes for the background while leaving space for the figures and furniture in reserve. Adjustments to the length of the table cloth and the diameter of the phonograph speaker were made but otherwise no major changes to the composition are evident. Magritte used a fluid paint imparting a smooth surface with low impasto. Glossy areas in the male figures hats and suits indicate that he used a natural resin or medium rich paint to highlight these areas. He also applied or mixed in a resin to highlight the faces of the three heads at center. These areas have a greenish fluorescence under ultraviolet illumination suggesting an aged natural resin component. No overall resin coating was applied by the artist. The speckled appearance resembling mold growth is shown in Fig. 2. It is most prominent and visible in the darker colours but is also seen in the light gray background suggesting an overall condition issue. A whitish haze is also observed over the brown and red earth coloured areas. Aside from the spotty appearance, the overall condition of the paint film is very good. There are a few localised mechanical cracks but no extensive drying cracks or other paint film defects.

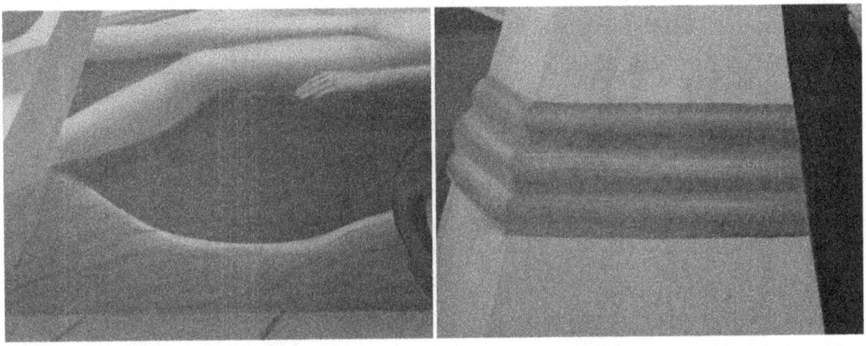

Fig. 2 Details of *The Menaced Assassin* showing the dense population of spots on the overall painting. The dimensions of the visible spots are mostly between 1 and 3 mm diameter

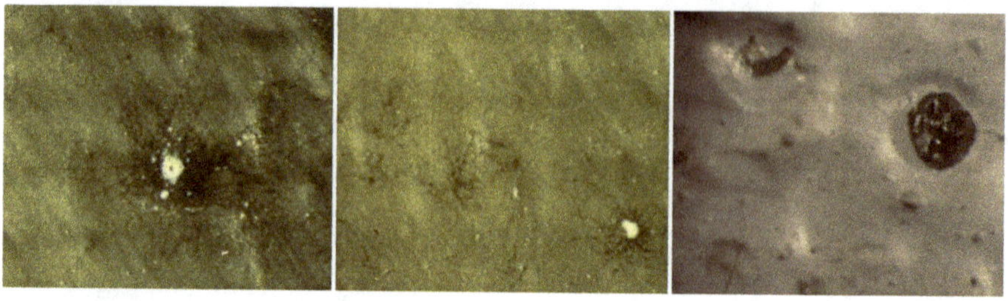

Fig. 3 Magnified view of some representative spots (from *left* to *right*): web like network of dark micro cracks with or without large white opaque formation in the center (~20 μm) and smaller opaque micron size white droplets and small round and transparent to semi opaque pustules (~200 μm). Images obtained with a Dinolite ×200 microscope

Further examination of the surface was thus carried out under magnification and revealed three distinct phenomena (see Fig. 3):

- Web like network of radiating dark micro cracks (originally confused with mold hyphae)
- Opaque white droplets associated with the dark cracks (often one large round particle in the middle of the network surrounded by smaller particles)
- Small round pustules – transparent to semi opaque

Analysis

Cross sections of the two preparatory layers and scrapings of both types of surface deposits and pustules were examined and/or analysed by microscopy, SEM-EDX and FTIR. In situ analysis with a handheld XRF was also done to help elucidate the composition of the paints. The paint layer is smooth and thin and shows no areas of loss to take samples for cross sections.

Examination of a cross section from the two preparatory layers (taken from the edge of the painting) under normal light and UV illumination showed the presence of a first layer containing small particles that fluoresce green under UV as well as large particles, some opaque, and some more translucent and a few blue particles too small to identify by FTIR. This first layer is covered with a thin white more opaque coating that also contains small translucent particles.

The composition of the two layers was further examined by SEM and EDX mapping. The first layer is in average 100–150 μm thick and contains mostly calcium based large and coarse particles (possibly carbonate) mixed with Zn containing particles (possibly zinc oxide), as well as Al and Si (possibly kaolin). This layer is coated with a thin intermediate layer (less than 20 μm) of lead white mixed with kaolin (Fig. 4).

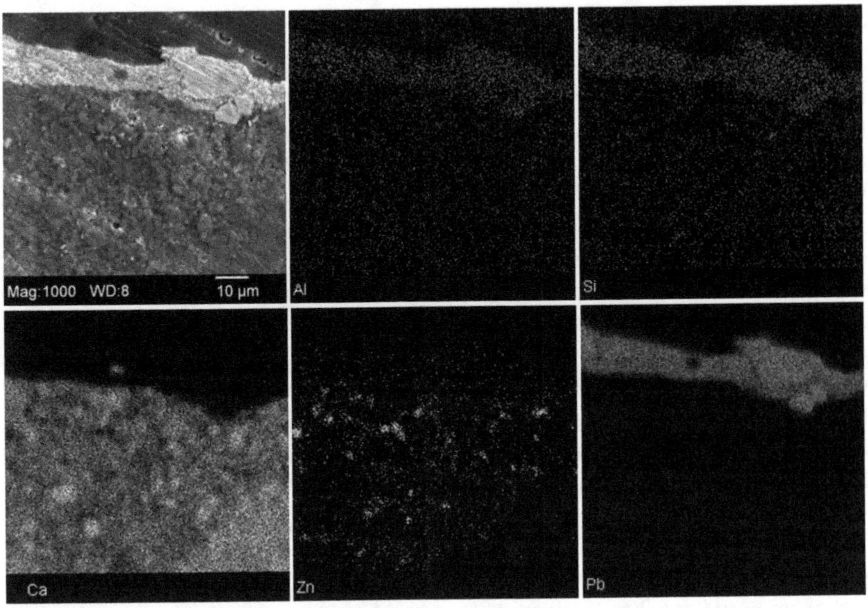

Fig. 4 SEM image of the cross section showing the ground and an intermediate layer that form the preparation (*left top corner*) and maps of the elements identified by EDX (*top*: Al, Si, bottom: Ca, Zn and Pb). SEM-EDX was performed with an Evex Mini SX-3000 instrument in high vacuum mode. The cross section was coated with gold and both imaging and elemental mapping were conducted at an accelerated voltage of 20 kV. The presence of these elements was confirmed by handheld XRF performed with a Bruker Tracer III-SD operated at 40 kV and 3 μA under He purge for the detection of Al and Si

A sample from the ground was also analysed by FTIR and confirmed the presence of calcium carbonate ($CaCO_3$ 2,509, 1,794, 875 and 713 cm^{-1}) and oil (fatty acid chains 2,923, 2,851 cm^{-1}). A second sample was particularly rich in beeswax residues from the wax lining. The presence of kaolin was not confirmed by FTIR but Al and Si were clearly detected by XRF analysis.

Characterisation of the Deposits and Pustules Material

Samples of the small white deposits and large translucent pustules were analysed by FTIR. The spectra obtained show the typical bands of metal monocarboxylates (Noble et al. 2002; Robinet and Corbeil 2003; Noble and Boon 2007; Centeno and Mahon 2009; Ferreira et al. 2011) namely the COO^- asymmetric and symmetric stretch around 1540 cm^{-1} (sometimes seen as a doublet) and 1400 cm^{-1} respectively, as well as bands characteristic for the fatty acid chains and free fatty acids (shoulder at 1,710 cm^{-1} corresponding to the C═O vibration). Identification of the metal counterpart (Zn, Pb or Ca) requires further analysis with other techniques.

Summary

Microscopic examination of *The Menaced Assassin* by René Magritte has revealed that the peculiar freckled appearance of the painting previously confused with mold is in fact due to the presence of radiating micro cracks possibly saturated with varnish. The occurrence of these cracks is apparently related to the formation of metal soaps that accumulate at the surface of the painting in the form of aggregates of very small white particles, larger translucent pustules and a whitish haze. The simultaneous occurrence of these three types of manifestations is somehow unusual and needs to be further investigated.

The formation of soaps is a well known conservation issue in both old and modern paintings and is due to the reaction of metals from pigments in the ground layer or in the paints, with free fatty acids from the oil binding medium. Calcium carbonate, lead white and zinc white can lead to the formation of metal soaps and have all been identified in the ground layer, but further analysis is needed to precise which of the metals are involved in the saponification.

Other components and factors can contribute to the formation of soaps. The ground layer for this painting for example is unusually rich in coarse calcium carbonate particles when compared to other Magritte paintings also examined in preparation for this exhibition. The ensuing porosity of the ground may have facilitated the absorption of free fatty acids. The presence of kaolin is also known to lower the local pH and promote some of the reactions involved during the saponification (Ferreira et al. 2011). Environmental conditions can play a role as well (Cotte et al. 2006). The high humidity of the previous collector's uncontrolled environment might also have contributed to the speckled appearance of the surface of *The Secret Player*, 1927 (Royal Museums of Fine Arts of Belgium) although no white deposits on the surface are evident on this painting. Comparison with another Magritte painting from 1927, *Meaning of Night* (The Menil Collection, Houston) that exhibits a similar patchy appearance in dark passages, but again no white droplets on the surface, suggests that the formation of the soaps is dependent on a combination of external forces and material composition.

Once the removal of the discoloured varnish and wax resin residue was complete the nuances of Magritte's original palette could be appreciated. Even after treatment, the presence of the soaps and associated patchy appearance was still visually distracting in areas. With this in mind, a decision was made in consultation with the exhibition curator to do a minimum of inpainting in the most affected areas. This solution allowed for the appreciation of the composition without completely disguising the paint film defects. With the work on display, monitoring of the surface will continue. It is hoped that in the museum environment, the factors contributing to the soaps formation have been arrested and no further conservation intervention will be required to stabilise *The Menaced Assassin* for the near future.

Acknowledgements Cindy Albertson, Anny Aviram, Jim Coddington, Chris McGlinchey, and Gillian Osmond.

References

Centeno SA, Mahon D (2009) The Chemistry of aging in oil paintings: metal soaps and visual changes. In: The Metropolitan Museum of Art Bulletin, New Series, vol 67, no 1, Scientific Research in the Metropolitan Museum of Art, pp 12–19

Cotte M, Checroun E, Susini J, Dumas P, Tchoreloff P, Besnard M, Walter P (2006) Talanta 70:1136–1142

Ferreira ESB, Boon JJ, Stampanoni M, Marone F (2011) Study of the mechanism of formation of calcium soaps in an early 20th century easel painting with correlative 2D and 3D microscopy. Preprints of 16th triennial ICOM-CC conference, Lisbon, pp 1604–1608

Noble P, Boon JJ (2007) Metal soap degradation of oil paintings: aggregates, increased transparency and efflorescence. In: AIC annual meeting: paintings specialty group postprints, pp 1–15

Noble P, Boon JJ, Wadum J (2002) Dissolution, aggregation and protrusion. Lead soap formation in 17th century grounds and paint layers, in ARTMATTERS – Netherlands Technical Studies in Art I, pp 46–61

Robinet L, Corbeil MC (2003) The characterization of metal soaps. Stud Conserv 48:23–40

Fleurs Grises – A Delicate Surface. Difficult to Clean – Essential to Protect

Susanne Stangier

Abstract Surface cleaning of a painting by Nicolas de Staël proved to be much more challenging than expected. Though the painting was executed in oil paint and the paint layer was dried properly, a traditional cleaning with fluids and cotton swabs was impossible. The very vulnerable paint surface with an extreme relief and widely undamaged impasto asked for a new combination of cleaning methods to make a safe and responsible surface cleaning possible. The selected method will be discussed, as will the re-framing and preventative measures that might protect the surface in the future.

Keywords Palette knife • Surface cleaning • Impasto • Paint relief • Patina • Aqueous cleaning systems • Preventive conservation • École de Paris

Introduction

The "simple" surface cleaning of a painting by Nicolas de Staël (1914–1955) turned out to be an extraordinary challenge. Different aspects such as the undamaged, original paint relief and the presence of dirt particles that some interpreted as 'patina', rather than dirt, had to be taken into account during the preliminary investigation that preceded the treatment. The role of preventive conservation to protect the extreme vulnerable surface against dirt in the future had to be addressed.

S. Stangier (✉)
Restauratie Atelier Stangier, Guttecoven, The Netherlands
e-mail: susanne@stangier.nl

K.J. van den Berg et al. (eds.), *Issues in Contemporary Oil Paint*,
DOI 10.1007/978-3-319-10100-2_14

Artist and Painting

The painting *Fleurs Grises*[1] (Fig. 1) was painted by Nicolas de Staël, 2 years prior to his death, in 1953[2] in Paris. He painted *Fleurs Grises* with oil on canvas, probably on a commercially prepared support. De Staël, born in 1914 in Russia, attended the Académie des Beaux-Arts in Brussels, and later settled in Paris where he became a highly regarded artist of the École de Paris.[3] The style of work coming out of this school was characterised by the division of space in different areas and harmony of colours in figurative as well as non-figurative motifs. Comparable with cubism, the

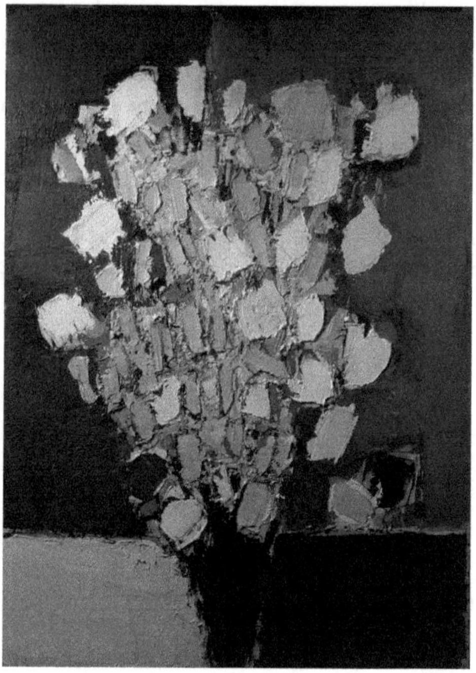

Fig. 1 Nicolas de Staël, *Fleurs grises*, 1953, oil on canvas, 99.7 × 72.8 cm (Cultural Heritage Agency, Amsterdam, the Netherlands, inv. AB 17683)

[1]Nicolas de Staël, *Fleurs grises*, 1953, oil on canvas, 99.7 × 72.8 cm.

Cultural Heritage Agency of the Netherlands, Amsterdam, the Netherlands, inv. AB17683. The painting stays on long term loan at the Stedelijk Museum Amsterdam and is currently not on display.

[2]His family had to flee during the Russian Revolution and de Staël got adopted after the early death of both his parents by a Russian family in Brussels. Dobbels (2000), pp. 242.

[3]After World War II Paris became the European center of abstract art. As part of the non geometrical abstract art, the artists of the École de Paris developed an European equivalent to the American abstract expressionism. Paris Central (2009, p. 7).

motifs are translated into planes, but here less tight and stiff and in a different colour scheme, developing from figurative to abstract depiction.[4]

Many of de Staël's works are to be found in this exciting grey area between figurative and abstract painting. For this period his typical paint application with thick layers and a distinct relief makes his works highly recognisable, just as his characteristic colour scheme. De Staël's paintings stand out with a strong and clear tonality, the colours remain pure and intense and often display a restricted colour palette. *Fleurs Grises* is one of the many flower still life's in de Staël's oeuvre in the years 1952–1953. The depiction of a bunch of flowers in a vase can be found repeatedly in different colour variations.[5] The size and motif vary but the abstraction of the flowers, the application of the paint with the palette knife and the reduced colour scheme are similar. The flowers are reduced to thick patches of paint and the background is a more or less monochrome area.

The paint surface of *Fleurs Grises* fits perfectly into this characterisation. The flower-piece is built up using paint of different thickness, often applied with the palette knife. The heavy impasto has most of its very delicate details preserved. At the edges of the single brush strokes, that symbolise the flowers, underlying paint layers, sometimes intense blue or orange coloured, are clearly visible. This layering of the paint intensifies the depth in the composition in a exciting way and strengthened the effect on three dimensionality. The dominating colours in the painting are kept in the range of white, light and dark grey, black and a blue grey.

After almost 50 years of private ownership the painting entered the Dutch state collection in 2001. Treatment was needed prior to display but a closer investigation of the painting in 2005 already concluded that "*Fleurs Grises* could not be cleaned with the commonly known cleaning methods because of the high risk of paint loss in the original paint layer".[6] A decision in favor of a treatment was first taken when the painting was to be presented at the exhibition "Paris Central" in 2009.[7]

Treatment

The painting entered the state collection close to its original state, still retaining the original stretcher and only minor retouching along the edges of the painting. The thick and heavy impasto of the delicate surface showed surprisingly few

[4]Paris Central (2009, p. 10).

[5]Some of the similar flower paintings by de Staël from that period are Fleurs, 1953; Fleurs noires, 1953; Fleurs rouges, 1952, Fleurs dans un vase bleu, 1953 and Fleurs blanche et rouges dans un vase gris, 1953. Dobbels 2000.

[6]ICN project "moderne schilderijen schoonmaken", research project on the cleaning of modern paintings. *Fleur Grises* was chosen as one of three case studies for this project. However, the preliminary investigation was not followed up by further research or treatment. www.moderneschilderijenschoonmaken.nl, 2005.

[7]Exhibition "Paris Central, Vrije stad, vrije kunst in de jaren'50", Cobra Museum voor Moderne Kunst Amstelveen, 24.10. 2009–17.01 2010.

damages. Local areas, however, suffered from lifting paint and required treatment. Consolidation was necessary to protect the original against actual paint loss. Lots of surface dirt was present in all parts of the distinct paint relief with a distorting effect on the overall appearance as a result. The painting was most probably never surface cleaned or even dusted before.

A series of cleaning tests was established to find out more about the solubility of the dirt, the reactivity of paint layer, ground and support to water and aqueous cleaning systems, and the possibilities of consolidation. Because the paint surface was built up with fine, vulnerable details in the impasto, a traditional cleaning with cotton swabs all over the surface was immediately ruled out. Also different types of dry cleaning methods that used sponges, cloth or erasers[8] were not suited to clean this intriguing paint surface in a controlled manner. To reduce the physical action on the paint surface during the cleaning process, the dirt had to be softened first. Afterwards dirt could be loosened and residues of the cleaning process, mainly moisture and dirt, could be removed from the surface. Earlier tests had confirmed that the paint layers were not especially sensible to water, so aqueous systems could be used for both surface cleaning and consolidation without harming the original.

In the areas with lifted paint surface, cleaning was carefully combined with consolidation.[9] Fortunately the lifted paint was situated in an area with relatively few impasto and good accessibility. Where lifted paint occurred in a very thick and stiff paint layer it was decided to flatten the raised areas only to a responsible degree and fill the open cracks with glue, rather than force the paint chip to lay down to the exact level of the surrounding paint layer. The risk of damaging the original paint layer was judged to be too high. The applied treatment assured that all original material was well attached to the ground and support of the painting. The accepted compromise did not esthetically disadvantage the appearance of the painting.

Having investigated a cleaning method together with a range of materials to be used, surface cleaning still was the most delicate part of the treatment. Surface dirt had to be removed to correct changes in colour, tonality, surface gloss and texture. At the same time it was necessary to try and differentiate between surface dirt and 'patina' on the aged surface.[10] Probably soon after completion, small dirt particles, possible soot, became attached to the relatively fresh paint and even sunk into the top layer. They unintentionally became part of the original and removal turned out to be impossible for the particles that were imbedded into the paint layer. Some of the tiny black dots, especially visible in the light colour areas, were therefore

[8]RCE (formerly ICN) project dry cleaning, the research project investigated different materials for dry cleaning. www.cultureelerfgoed.nl searching for dry cleaning (Daudin-Schotte et al. 2013, p. 209).

[9]Consolidation took place with Lascaux 4176 and a heat spatula set on proximately 40 C. Lascaux 4176 is a finely dispersed, aqueous dispersion of an acrylic copolymer, MSDS: www. talasonline. The glue combines the qualities of relatively high strength with high viscosity, a fairly mat dried film and fast drying properties.

[10]Brachert 1985, p. 11 "… selbst Schmutz kann mitunter noch als Patina angesprochen werden".

left on the surface, accepted as part of the original rather than interpreting them as dirt. Their influence on the overall appearance of the painting was very limited. Besides acceptance of the imbedded dirt particles on the cleaned surface, special attention had to be given to the fact that the visual effect of the cleaning should not be different between the easily accessible areas of paint and those that are much more difficult to reach and extremely vulnerable. Only dirt embedded into the surface was considered to be 'patina', dirt that wasn't sunk into the original surface but was located in areas that were difficult to reach in the paint relief still had to be removed. The cleaning method that was decided upon included a combination of moisture, an aqueous cleaning system, thin tissues, small pulp pads,[11] tiny cotton swabs and brushwork with a stiff brush to achieve the most satisfying surface after cleaning.

To soften the dirt layer without touching the surface, especially in the areas with the most extreme impasto, an area of 10 × 10 cm was thinly misted with de-mineralised water (Fig. 2). After 2 min, the surface was then brushed with a stiff brush and a 2 % tri- ammonium citrate solution. Thin cotton tissues were used in some areas as an intermediate layer to reduce the direct interaction on the paint surface and to take off dirt together with introduced cleaning material as much as possible. With small circular movements the surface was brushed in different directions to reach every detail of the impasto. Even the deepest parts of the most extreme paint build up could be reached and cleaned safely, varying the pressure that were put onto the impasto depending upon its vulnerability.

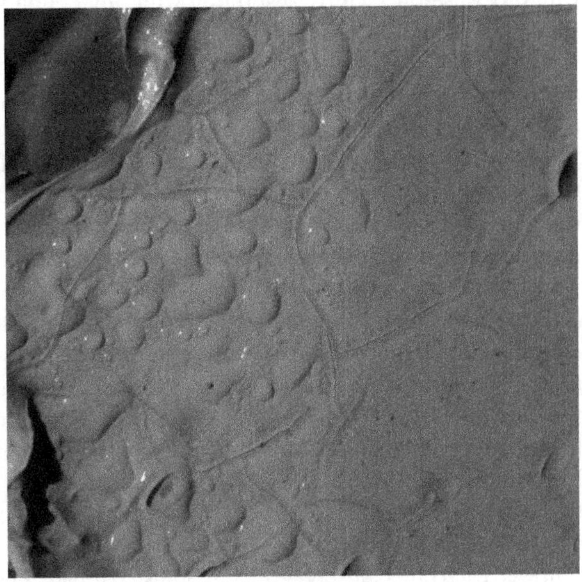

Fig. 2 Detail during cleaning. Clearing of a cleaned area with misted de-mineralised water

[11]Pulp pads = "Zellstoff auf Rolle" and "Zellstofftupfer" from Deffner & Johann, www.deffner-johann.de. Pulp is a lignocellulosic fibrous material prepared by chemically or mechanically separating cellulose fibres from wood, fibre crops or waste paper.

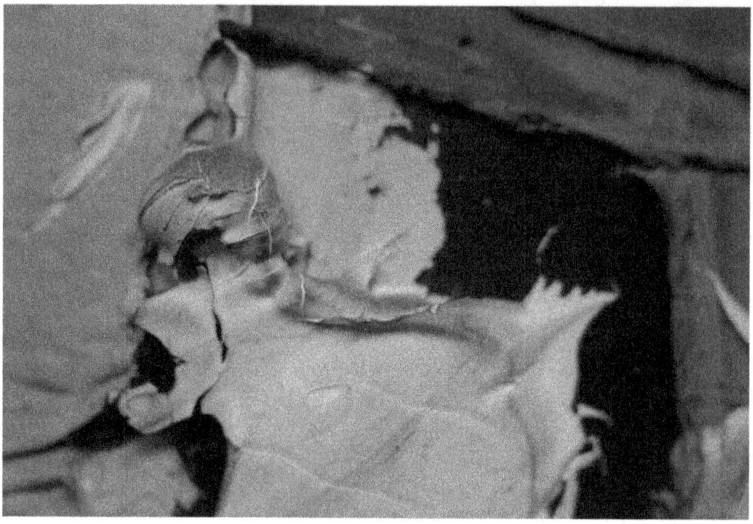

Fig. 3 Detail of the impasto, before surface cleaning

Moisture and loosened dirt were removed from the surface with small pulp tissue pads and tiny cotton swabs. The pads were put onto the surface and softly pushed with the brush into the paint relief. From the deepest impasto moisture together with dirt was collected by absorption with tiny cotton swabs. During cleaning the painting was laid on a table so the introduced moisture stayed in the area where it was applied. The horizontal position of the painting also gave more accessibility to the difficult impasto in all kind of directions without moving the painting itself. This procedure had to be repeated two to three times, depending on the surface texture.

The result of the surface cleaning was spectacular (Figs. 3 and 4). To clear the treated area of cleaning agent and residues, it was misted twice with de-mineralised water and dried again (Fig. 2). Building up a routine in the cleaning procedure made it possible to work different areas at the same time, making use of the waiting time while the dirt in a new area could be softened.

Surface cleaning, combined with the necessary consolidation, and the standard conservation of support and stretcher that included dusting the back, securing the keys and adding a backing board, completed the treatment. When the painting was purchased in 2001 it was framed in a wooden, gilded ornament frame. After treatment it was decided to reframe the painting in that same frame. Because the heavy impasto extended right to the edges the rebate of the frame had to be adapted to the extraordinary paint surface. A counter mould of the paint relief along each edge was build up from balsa wood and evazote[12] rubber so rebate and paint surface followed each other closely. With this adjusted rebate it was assured that the pressure from the actual framing was not only put on singular spots in the highest impasto but instead of it divided over the whole length of all four edges.

[12]EVAZOTE® Foam is a closed cell cross-linked ethylene copolymer foam.

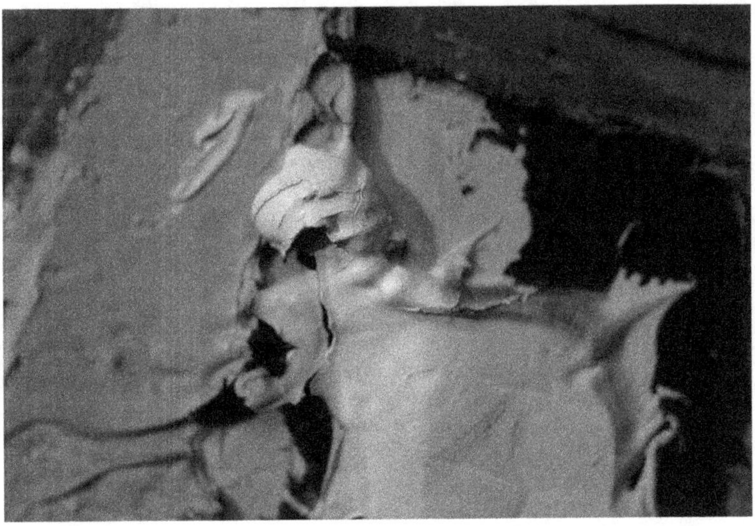

Fig. 4 Detail of the impasto, during surface cleaning

Preventive Conservation

Because of the extraordinary surface and its high vulnerability different preventive measures were suggested before the painting was given on loan again. Especially the necessity of repeated cleaning in the future was of great concern. Unfortunately for this painting, it was decided to present it without any additional surface protection. An extra exhibition frame with higher sidebars that would extend over the level of the paint surface to protect the surface against some dirt and movements close to the surface, as well as protective glazing in front of the painting were both judged to disturb the overall appearance of the painting too much.

The desire for some kind of protection is universal for many contemporary art objects with vulnerable surfaces, sometimes even with non-drying and sticky materials right at the surface. Protective glazing seems in many cases from the conservators point of view inevitable or at least an acceptable compromise.

Conclusion

Only this time-consuming and strenuous cleaning process made it possible to clean the painting safely with its delicate surface. The result was exciting and brought important changes to the appearance of the painting, justifying the high risk of this treatment. Preventive measures, however, should play a role in the decision making process to protect the painting against dirt and damages, but also future treatments.

References

Brachert T (1985) Patina: von Nutzen und Nachteil der Restaurierung. Callwey, München

Daudin-Schotte M, Bisschoff M, Joosten I, van Keulen H, van den Berg KJ (2013) Dry cleaning approaches for unvarnished paint surfaces. In: Mecklenburgh M (ed) Proceedings from of the cleaning 2010 conference, Valencia, 26–28 May 2010. Smithsonian Institute, pp 209–219

Dobbels D (2000) Nicolas de Staël. Hazan, Paris

Paris Central (2009) Vrije stad, vrije kunst in de jaren'50, Cobra Museum voor Moderne Kunst Amstelveen, 2009, exhibition catalogue

Derivatisation Technique for Infrared Spectroscopy – Characterisation of Oxidative Ageing Products in Modern Oil Paint

Stefan Zumbühl, Nadim C. Scherrer, and Wolfgang Müller

Abstract In conservation science the Fourier transformed based infrared spectroscopy (FTIR) is a standard technique to the study of historic paint layers. There are, however, clear limitations when analysing modern oil paint. The spectral information of aged oils is affected from overlapping signals of the binder, the oxidative ageing products, soaps formed during ageing, as well as several pigments and fillers. The distinction of the ageing products such as ketone and carboxylic acid functional groups pose the next problem, as these interfere with the triglyceride esters. Gaseous sulfur tetrafluoride SF_4 was used to discriminate overlapping signals in aged oil paint. This derivatisation technique makes it possible to convert carboxylic acids and carboxylate salts into acyl fluoride, while ketones and ester groups remain intact. IR spectra after SF_4 treatment show a characteristic splitting of the carbonyl band, allowing the characterisation of the different functional groups formed during oxidative ageing. Furthermore, it is possible to increase the spectral selectivity thanks to the successful elimination of spectral interferences caused by organometallic and inorganic compounds. This sample pre-treatment can be applied to both micro-samples and polished cross-sections, enabling 2D localisation of these compounds in aged oil paint samples using infrared imaging FTIR-FPA (focal plane array). Both artificially aged and real paint samples of twentieth century paintings were investigated to characterise degradation processes in oil paint, including the dripping phenomenon of modern oil paints.

Keywords Infrared spectroscopy • Infrared imaging • Overlapping signals • Derivatisation technique • Sulfur tetrafluoride • Acyl fluoride • Liquefied paint • Sunflower oil • Safflower oil

---

S. Zumbühl (✉) • N.C. Scherrer
Bern University of Applied Sciences, Fellerstrasse 11 CH-3027, Bern, Switzerland
e-mail: stefan.zumbuehl@bfh.ch; nadim.scherrer@bfh.ch

W. Müller
H. Schmincke & Co. GmbH & Co. KG, Otto-Hahn-Str. 2, D-40699 Erkrath, Germany
e-mail: mueller@schmincke.de

K.J. van den Berg et al. (eds.), *Issues in Contemporary Oil Paint*,
DOI 10.1007/978-3-319-10100-2_15

Introduction

Fourier Transform infrared spectroscopy (FTIR) is a long established standard technique in conservation science (Derrick et al. 1999). With the arrival of 2D FTIR-FPA imaging at a spatial resolution down to 1 micron, this analytical technique is reaching even more analytical power. For example, it makes it possible to identify and localise organic compounds *in-situ* within micro-samples. But there are clear limitations to FTIR when analysing binding media in paint samples. For example, the interpretation of standard infrared spectra on modern oil paint samples suffers from overlapping signals of the binder, the oxidative ageing products, soaps formed during ageing, as well as several pigments and fillers. The chemical alteration of oil paint is a complex process due to interdependent chemical reactions running either simultaneously or sequentially (Zumbühl et al. 2011). The chemical pathways are well investigated and described in several overviews (Kamal-Eldin 2003; Frankel 1998; Chan 1987; Richardson and Korycka-Dahl 1984). Of particular interest in this context are oxidative cross-linking reactions, oxidative fragmentations, hydrolysis, and soap formation (Van den Berg 2002; Van den Berg et al. 1999). All these different reactions lead to various functional groups producing characteristic IR-absorptions. With respect to the interpretation of ageing products, standard FTIR spectroscopy faces a fundamental problem due to interfering signals from all the oxidation products such as ketones and carboxylic acids with the oil triglyceride esters. Additionally, artist's paints are complex material systems containing colorants, mineral fillers and different kinds of additives, all contributing to the spectral signal. In order to improve the selectivity of the FTIR technique with respect to the study of the oxidative ageing products in modern artist's oil paints, an additional sample preparation step was introduced to selectively derivatise specific compounds or functional groups, applicable both to micro-samples (μ-FTIR) and polished cross-sections (2D FTIR-FPA imaging) (Zumbühl et al. 2014).

Derivatisation with Sulfur Tetrafluoride SF_4

The selective alteration of functional groups using reactive gases is an established technique and was introduced for the FTIR analysis of oxidation products in polymeric materials (Wilhelm and Gardette 1994; Mailhot and Gardette 1992; Carlsson et al. 1988). The same technique was used by Mallégol et al. for the characterisation of the reaction products formed after peroxide homolysis in non-pigmented drying oils (Mallégol et al. 1999, 2000a, b, c). The reagent SF_4 reacts selectively with carboxylic acids R(O)OH to form an fairly stable acyl fluoride R(O)F, leading to a characteristic splitting of the carbonyl band. The conversion of the carboxylic acids into acyl fluoride leads to a shifted carbonyl vibrations $\nu(C{=}O)$ at 1,841 cm^{-1}, while ketone $\nu(C{=}O)$ at $\approx$1,720 cm^{-1} and ester $\nu(C{=}O)$ at $\approx$1,740 cm^{-1} groups remain intact (Wilhelm and Gardette 1994).

Moreover, the elimination of the carboxylic O—H signal makes it possible to detect alcoholic hydroxyl functionalities ν(O—H) [br] at ≈3,500 cm^{-1} selectively (Zumbühl et al. 2014). This makes it possible to investigate the formation of carboxylic acids during oxidation and hydrolysis. Besides the interferences by organic compounds mentioned above, there are further overlapping bands generated by inorganic pigments and fillers. Furthermore, ageing will generate organometallic reaction products formed between the binder molecules and the cations of the pigments. These are known to stabilise the aged binder network (Van den Berg 2002; Van den Berg et al. 1999). The dominant changes during ageing are due to the soap formation, causing broad signals of the carboxylates at around ≈1,600 cm^{-1} ν_a ($C(O)O^-$) [s], ≈1,450 cm^{-1} ν_s ($C(O)O^-$) [m–s] and ≈850 cm^{-1} sci ($C(O)O^-$) [m] (Socrates 2004) covering the signals from the fatty acid backbone at 1,464 cm^{-1} δ(CH_2) [s], 1,378 cm^{-1} wag(CH_2) [m] and 1,418 cm^{-1} wagR-(C—H_2)—$C(O)O^-$ [m] (Lazzari and Chiantore 1999; Meilunas et al. 1990). Applying this derivatisation to aged oil paint samples makes it possible to eliminate these signals, since soaps such as zinc or lead carboxylates R(O)OM are transformed into acyl fluoride R(O)F (Zumbühl et al. 2014). Furthermore, pigments such as the often used lead white exhibits a dominant and broad signal of the $CO_3{}^{-2}$ ion ν_3 at ≈1,400 cm^{-1} [vs] (Meilunas et al. 1990; White 1974), which strongly limits the observation of the organic binder compounds in paint samples. Using the SF_4 derivatisation technique, the basic lead carbonate $2PbCO_3{\cdot}Pb(OH)_2$ can be transformed into the non-IR active lead fluoride β-PbF_2 (Zumbühl et al. 2014). Following the treatment, the signals of the pigment are no longer present and the IR spectrum of the oil binder is free of disturbing signal overlaps (Fig. 1).

Application of the Derivatisation Technique

With the spectral elimination of alteration products and pigments in paint micro-samples, the identification of complex binder systems is possible. One focus was set on the formation of different oxidation products and the characterisation of the material polarity during natural and accelerated ageing of oil paint (Zumbühl et al. 2011). This is of interest because the chemical pathways are strongly influenced by the paint formulation and the variable ageing conditions. Being able to detect minor binder components within the 2D structure of drying oil based paint is a very powerful research tool. The SF_4 derivatisation technique was successfully applied to the detection of small protein concentrations in tempera paint (Zumbühl et al. 2014). The investigation of such emulsified systems made of oil and protein is not possible with standard FTIR, as the broad bands of the carboxylate anion ν_a ($C(O)O^-$) at ≈1,600 cm^{-1} (ageing products) interfere with the protein bands amide I (ν(C═O)) at ≈1,650 cm^{-1} and amide II (δ(N—H) and ν(N—C)) at ≈1,540 cm^{-1}. Furthermore, the amide II band can be superimposed by the very broad and strong band of the carbonate $CO_3{}^{-2}$ anions ν_3 ≈ 1,400 cm^{-1} generated by pigments or fillers. Besides the derivatisation of micro-samples, this pre-treatment can also be

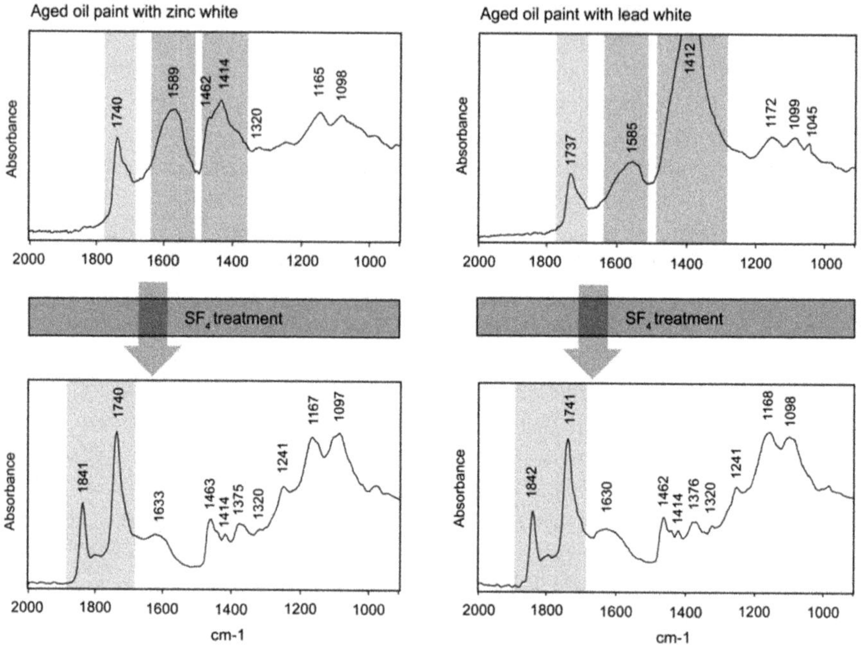

Reaction of sulfur tetrafluoride

$SF_4 + 2\,H_2O \longrightarrow SO_2 + \mathbf{4\,HF}$

Derivatisation of carboxylic acids

$C(O)OH + HF \longrightarrow \mathbf{C(O)F} + H_2O$

Derivatisation of carboxylates

$C(O)OM + HF \longrightarrow \mathbf{C(O)F} + MOH$
$MOH + HF \longrightarrow \mathbf{MF} + H_2O$

Derivatisation of basic lead carbonate

$Pb(OH)_2 + 2\,HF \longrightarrow \mathbf{PbF_2} + 2\,H_2O$
$PbCO_3 + 2\,HF \longrightarrow \mathbf{PbF_2} + H_2O + CO_2$

Fig. 1 Derivatisation of 60-year old oil paint films before (*upper*) and after SF4 treatment (*lower*). Interference of the soaps and the pigments is eliminated (*dark grey*). The different carbonyl groups are now visible as separated bands (*grey*)

applied to polished cross-sections or surfaces (Zumbühl et al. 2014). The technique allows research on the interrelationships of such complex material systems within the original stratigraphic structure. 2D studies on the intact morphology disclose a great deal more information than detached sample analysis with destroyed context. By applying this derivatisation technique in combination with FTIR-FPA (focal plane array) imaging to paint cross-sections, it is possible to map the binder components within a paint layer at a spatial resolution of 1–1.5 μm. An example is presented in Fig. 2, where it was possible to image the oxidation gradient in a modern oil paint film. The reaction of SF_4 is superficial and the penetration

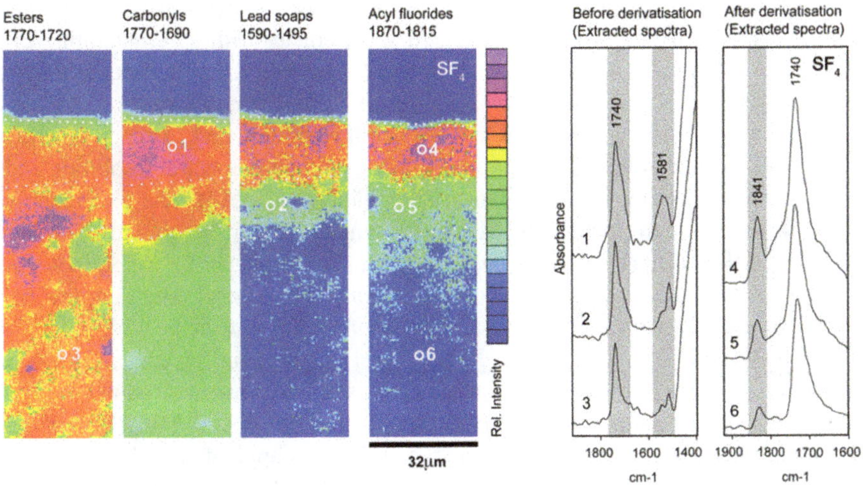

Fig. 2 Derivatisation of oil paint containing lead white. The FTIR-FPA imaging before and after treatment with SF4 delivers information on the spatial distribution of the different oxidation and degradation products formed within a paint layer

depth at maximum a few microns deep so that samples can easily be repolished to remove the derivatised surface layer. The penetration depth using ATR technology in organic layers ($\approx n_D 1,5$) using a germanium crystal (n_D 4.0, beam angle 45°) is in the range of 0.2–0.7 μm, meaning that a reaction depth of 1 μm is sufficient for the method to work successfully (Zumbühl et al. 2014). The standard reaction time of 24 h was in all cases sufficient to achieve an ubiquitous and homogeneous derivatisation across the surface. The present work has applied this derivatisation technique to characterise different functional groups within a paint layer in order to gain information on the drying process of modern oil, including dripping oil paint.

Experimental

Sulfur tetrafluoride SF_4 (ABCR® Dr. Braunagel GmbH & Co. KG, Art No. AB10417) treatment in gas phase was used derivatise different kinds of oil paint micro-samples as well cross-sections. The micro-samples were rolled on to a CVD diamond window prior to placement into the gas reaction chamber. Stratigraphic samples were embedded in epoxy resin Araldite 2020 (2 component XW396/397) and sections were prepared using dry polishing cloths of the type Micro-mesh®. The derivatisation treatment using the active gas SF_4 was performed in an all-teflon® chamber (volume 10 cm^3) (Carlsson et al. 1988; Mallégol et al. 2000c). All samples were exposed to the SF_4 gas at room temperature for 24 h. Before and after the reaction, the chamber was flushed with nitrogen N_2 and the eliminated gas was adsorbed to active charcoal. FTIR measurements were performed on a Bruker® Hyperion 3000/Tensor 27 infrared spectrometer with a germanium ATR crystal (∅ 250 mm). The pixel resolution is 0.5 μm with the FPA detector made of 64 × 64 detector elements, covering an area of 32 × 32 μm. Micro-samples were analysed in transmission mode at 4 cm^{-1} spectral resolution and 32 scans. The 2D FTIR-FPA imaging was performed in ATR mode at 8 cm^{-1} spectral resolution and 64 scans. More details can be found in (Zumbühl et al. 2014). The electrospray ionisation mass spectrometric ESI-MS measurement was performed with a Bruker maXis 4G UHS-TOF (Ultra High Resolution Time of Flight) mass spectrometer at the University Duisburg-Essen (Institute for Organic Chemistry). The samples was extracted in a chloroform / methanol mixture and measured in the positive mode.

To investigate the oxidation gradients in very thick paint layers, accelerated ageing tests were performed. The paint products used were linseed oil paint (Kremer GmbH & Co. KG) containing basic lead white (Kremer 460007) and zinc white (Kremer 463207) without dryer content. The paint was applied at a film thickness of 2 mm to a silicon-coated Hostaphan®-Foil RNT 36 (Kremer GmbH & Co. KG) and pre-dried in darkness under room condition (23 °C/50%RH) for 1 week. The samples were aged under window glass filtered fluorescence light (True Lite® 5500 K and Philips® UV-20 W/08 F20 T12 BLB generating ≈5,800 lm/m^2, 557 mW lm^{-1}/≈3,200 mW/m^2) at ≈40 °C, 55%RH and atmospheric oxygen content over a time period of 6 weeks. To investigate the dripping process of oil paint (applied in thick layers), different paint samples were produced in 2003 by H. Schmincke & Co. GmbH & Co. KG, Erkrath (Germany) according to paint formulation of 1994. The paint was applied in variable thickness up to 5 mm on a paint board. The binder of the blue paint (Prussian blue) contains ≈66 % linseed oil, 23 % sunflower oil and 11 % safflower oil. The binder of the light pink paint (titanium dioxide and cadmium pigments) contains ≈13 % linseed oil, ≈70 % sunflower oil and ≈17 % safflower oil. The light pink paint contains Co- and Zr-dryer, whereas the blue paint is free of additives. Pure oil samples used in paint formulation of 1994 were applied at a thickness of 60 μm in 2005 (vertical store of the paint bord results in slots of the oils). All samples were aged under room conditions. A sample of liquefied paint was taken from a 10 years old painting by the German artist Jonathan Meese.

Results and Discussion

Ageing of modern oil paint is fundamentally different to ageing of the classical paint systems. This is strongly related to the material composition of modern artist's paints. On the one hand, the ageing process is strongly influenced by the binder composition. Modern oil paints often consist of a variable mixture of different oils in order to control the drying behaviour of the different colours. In addition to this, general regulations with respect to the quality and stability modern pigments have changed. As a result, modern oil paints exhibit a different degradation behaviour when compared to classical oil paints. There is an increasing number of incidents where such modern paints liquefied within a few years after application (in thick layers) and drying (see Boon and Franken et al. in this contribution). This phenomenon was observed both in oil paint and in binder samples applied without pigments, fillers or additives. It can thus be assumed that the process is related to the binder composition. All incidents of paint liquefaction upon ageing so far were related to a high content of sunflower and/or safflower oil. To achieve a characterisation of these cases, SF_4 sample pre-treatment in combination with FTIR spectroscopy came into application. Since drying oils are highly oxidative, the ageing of oil leads to characteristic changes in the IR spectra. The kind of stable functional groups formed is of direct relevance to predict a specific ageing behaviour (Mallégol et al. 1999). A comparison of two sample paints out of the same product series having variable oil composition, permits to draw preliminary conclusions on specific degradation processes. While the first paint (blue) is bound with ≈66 % linseed oil, the second one has been made with ≈87 % sunflower and safflower oil (light pink). These two paints exhibit a completely different ageing behaviour: while the linseed oil based paint is still solid after 7 years of natural ageing, the sunflower/safflower oil bound paint has liquefied.

In both samples, the increase in oxidation products is evident in the FTIR spectra from the formation of carbonyl groups ν(C═O) [vs] within 1,720–1,740 cm^{-1} in varying extent. But the softened paint shows a distinctive splitting of the ν(C═O) peak. FTIR results suggest that distinct reaction pathways are evident and that different functional groups are being formed during oxidation. An increase of the ν(C═O) carbonyl intensity refers to new carbonyl containing functional groups. Furthermore, a broadening of the triglyceride ester signal ν(C═O) [vs] (≈1,720–1,740 cm^{-1}), forming a shoulder at higher frequencies (≈1,770 cm^{-1}), indicates the formation of lactones, whereas a broadening to lower frequencies (≈1,700–1,720 cm^{-1}) indicates the formation of ketones and carboxylic acids (Mallégol et al. 2000a; Meilunas et al. 1990). After selective fluorination of the carboxylic acids groups during derivatisation with SF_4, the intensity of the ν(C═O)F absorption at ≈1,840 cm^{-1} is an indicator for the acidity of the aged paint. It is evident, that the sunflower and safflower rich oil paint is much more acidic than the linseed oil paint (Fig. 3). The corresponding analysis of the dripping phase of aged oil paints and the liquefied pure oils all showing a very high carboxylic acid content (high signal intensity of the acyl fluorides at ≈1,840 cm^{-1} relative to the ester band at ≈1,740 cm^{-1}). Moreover, the elimination of the carboxylic O—H signal during fluorination of the acids makes it possible to detect the alcoholic hydroxyls ν(O—H) [br] at ≈3,500 cm^{-1} selectively. Since there are no significant hydroxyl absorptions present in the liquid oil phase (Fig. 4), it can be concluded that the carboxylic acids are mainly formed during oxidation rather than during hydrolysis.

The formation of carboxylic acid groups is caused by the fate of thermo semi-stable peroxides R—O—O—R' formed during oxidation (Fig. 5) (Mallégol et al. 2000a). The homolytic cleavage of these primary reaction products leads to the formation of alkoxy radicals R—O·, which initiate different kinds of secondary

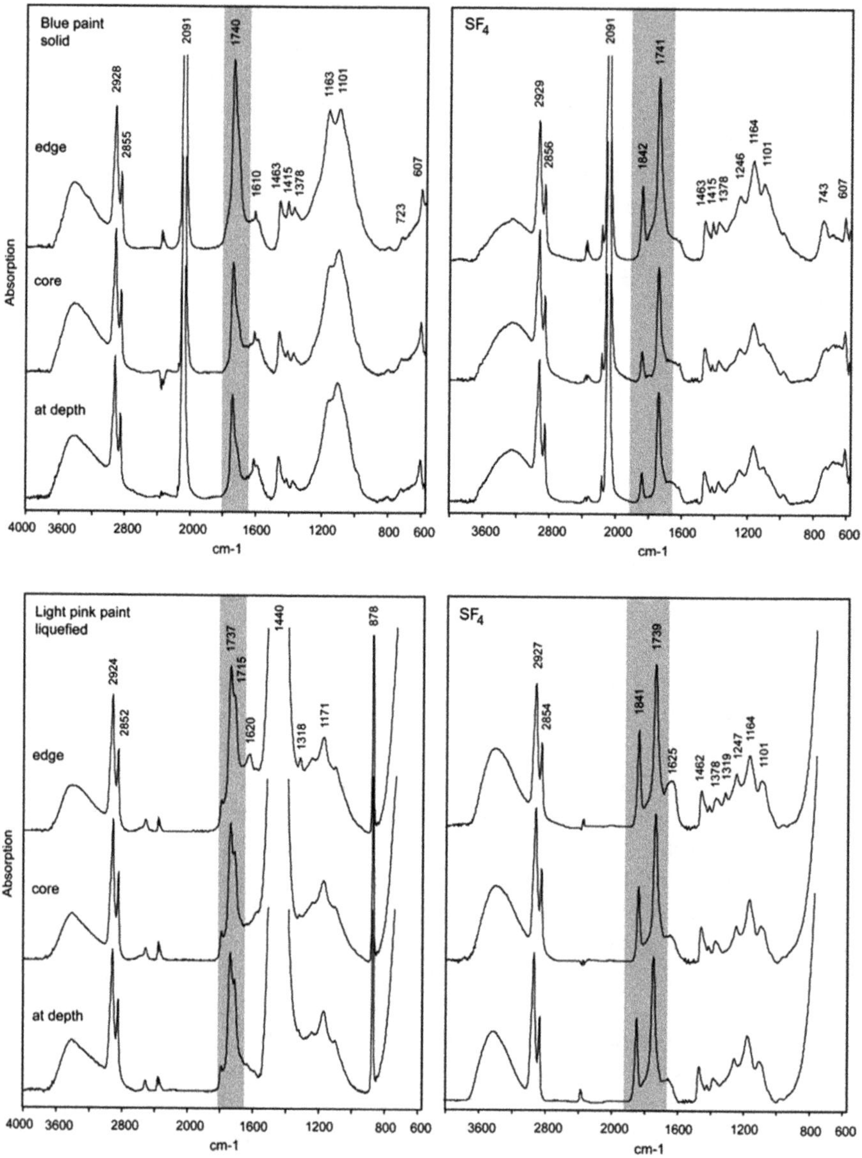

Fig. 3 Infrared spectra before (*left*) and after derivatisation with sulfur tetrafluoride (*right*). The upper spectra represent a solid paint rich in linseed oil, whereas the lower series represents paint rich in sunflower/safflower oil that has liquefied. Applying SF4 derivatisation removes signal interferences and reveals the carbonyls as split bands (*grey*). The signal response of the acylfluoride at 1,840 cm^{-1} suggests that the liquefied paint contains a lot of carbonic acids

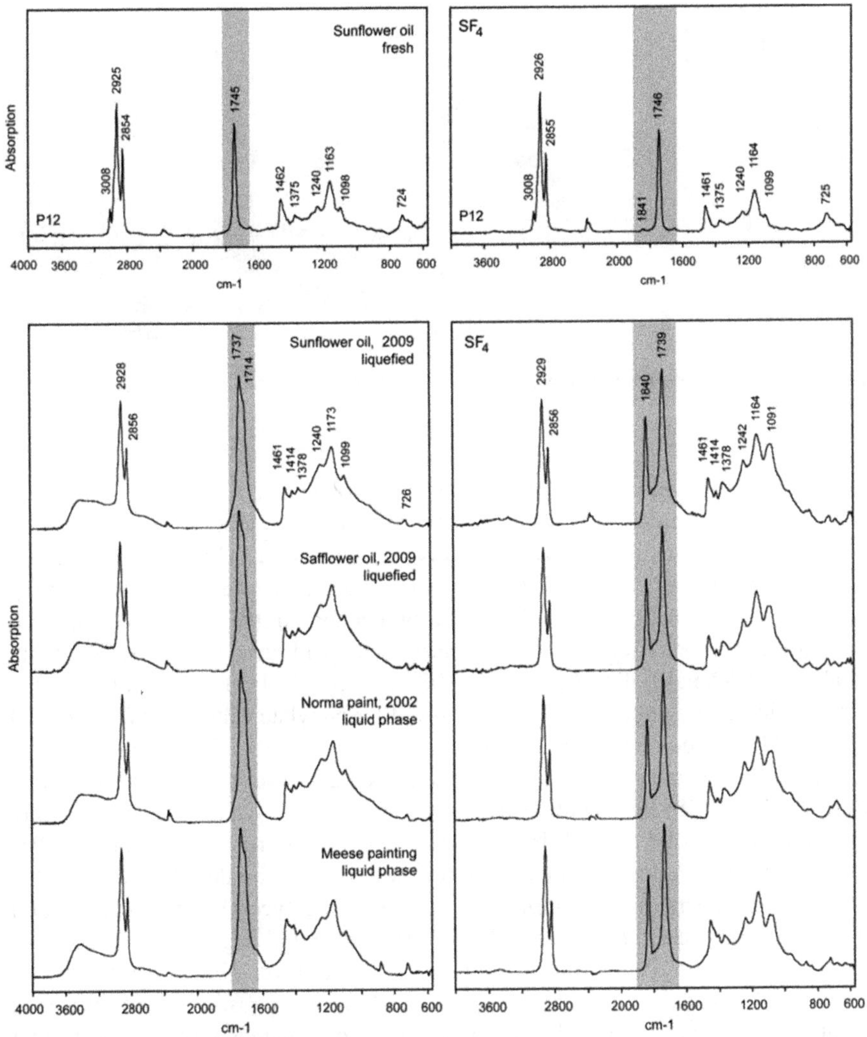

Fig. 4 Infrared spectra of liquefied oils before (*left*) and after derivatisation with SF4 (*right*). The derivatised sample reveals the carbonyls as separate groups, with the acylfluoride at 1,840 cm^{-1} pointing at a very high content of carbonic acids

reactions (Mallégol et al. 2000a; Gladovic et al. 1997; Chan 1987; Gardner 1987; Grosch 1987; Schieberle et al. 1984). The peroxide homolysis can be induced by temperature and may be assisted by metal catalysts (Van den Berg 2002; Mallégol et al. 2000b; Black 1978; Bawn 1957; Morley-Smith 1957; Banks et al. 1954). The degradation of unsaturated oils is fundamentally caused by the β-elimination (fragmentation reaction), which is the dominant reaction pathway at room temperature (Mallégol et al. 2000b, c; Lazzari and Chiantore 1999; Chan 1987;

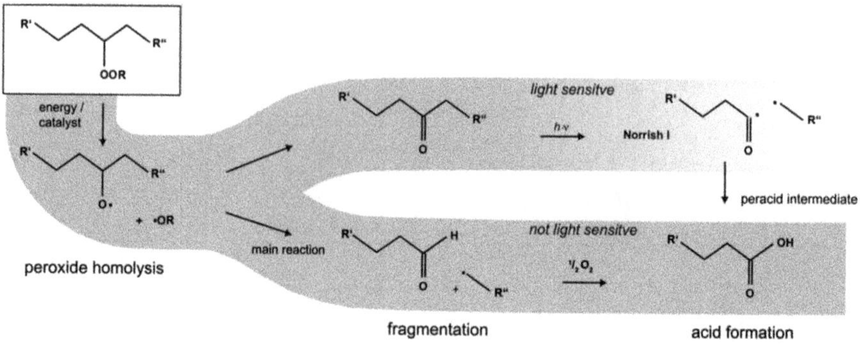

Fig. 5 Survey of the reaction pathways of lipid oxidation and fragmentation as described in the literature

Grosch 1987; Muizebelt 1978). The implication of this reaction can also be derived from the high content of oxalates, as oxalic acid is generated similarly by the decomposition of hydroperoxides of oxidised double bonds. Our data indicate that this kind of degradation process is a particularly distinctive feature of semi-drying oils. The process finally ends in the total fragmentation of the fatty acid chains and the parallel formation of end-standing carboxylic acid groups (Fig. 5) (Zumbühl et al. 2011; Kamal-Eldin 2003; Chan 1987). In classical oil paint, however, the increase in acidity enhances the formation of carboxylate salts (soaps) (Zumbühl et al. 2011; Van den Berg 2002; Van den Berg et al. 1999). These organometallic reaction products are important to stabilise the aged triglyceride network (Van den Berg 2002; Van den Berg et al. 1999). In the presence of bivalent cations, this kind of cross-links can be formed. In absence of a cation source, which often applies to modern oil paint, the fragmentation reaction induces the destabilisation of the network and a decrease of the molecular mass of the triglyceride molecules. In semi drying oils, this reaction leads to a large variety of liquid triglycerides fragments. This molecular decomposition in sunflower and safflower oil was confirmed by electrospray ionisation mass spectroscopy ESI-MS. In contrast to the non-liquefied paint the liquid phases contain high amounts of compounds with low molecular masses below the fatty acid triglyceride molecules (200–900 Da) (Fig. 6). The chemical nature of these liquid triglycerides fragments was not investigated in more details. Furthermore, thermogravimetric TGA measurements showed, that up to 30 % of the binder content can be evaporated within 24 h at no more than 50 °C, indicating a high amount of comparable low molecular weight oil compounds. In contrast the amount of volatilise compounds in the linseed oil rich paint is comparable low (see Franken et al. in this contribution). Even though the fragmentation of the fatty acids is common to all oils, the implications on the complete system are strongly dependent on the type of oil. The network properties are influenced on the balance between stabilising and destabilising reactions. The possibility of recombination of radicals in poorly unsaturated oils is structurally limited and the density of cross-linking is reduced. The decomposition of peroxides

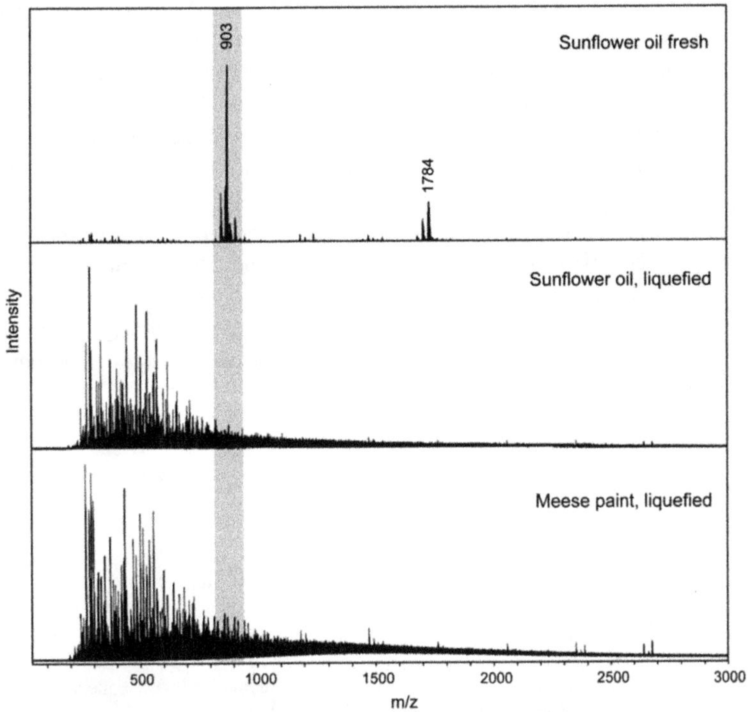

Fig. 6 The electrospray ionisation mass spectra (ESI-MS) demonstrate, that aged sunflower oil is fragmented into lots of triglyceride fragments with masses below the triglyceride monomers (*grey*). A comparable MS pattern was found in a liquefied painting by Jonathan Meese

thus will have a stronger influence on the material properties. Aggravating to this situation will be the potentially increased demand on siccatives with semi drying oils to achieve efficient drying. Those will have a negative influence on the ageing behaviour, as these catalysts decompose peroxides very effectively and as such also act as catalysts in the degradation process. As a result of this, colours with semi drying oils as binder, in thick layers applied, may liquefy after a few years. From this we can infer, that the degradation behaviour of modern oil paint is strongly controlled by the binder composition.

Conclusion

A major drawback of infrared spectroscopy FTIR is the poor selectivity in complex material systems due to signal interferences of different functional groups within similar spectral ranges. Furthermore, signal interferences with the oil binder are caused by oxidative ageing products and carboxylate salts formed during ageing, as well as several pigments and fillers. This is of great limitation, in particular to the study of modern oil paint. Sample pre-treatment with sulfur tetrafluoride SF_4 to derivatise certain compounds substantially increases the selectivity of the IR technique by eliminating signal overlaps. The method allows the spectral separation of carbonyls of different functional groups and the elimination of signal interferences caused by ageing products, pigments and fillers. This derivatisation technique was applied in the present case in particular to document the acidity of the different aged oils. The data was used to gain information on potential degradation mechanisms in modern oil paint. As demonstrated, the SF_4 technique considerably enhances the power of standard FTIR and is of great relevance to art conservation science. Its applicability to polished cross-sections in combination with high resolution 2D FTIR-FPA imaging can be regarded as a fundamental advancement.

References

Banks GL, Chalk AJ, Dawson JE, Smith JF (1954) Catalysis of olefin autooxidation by heavy-metal ions in non-polar media. Nature 174:274–275

Bawn CEH (1957) The role of peroxidic catalysts in coating compositions. J Oil Colour Chem Assoc 40:1027–1034

Black JF (1978) Metal-catalysed autooxidation – the unrecognized consequences of metal-hydroperoxide complex formation. J Am Chem Soc 100(1):527–535

Carlsson DJ, Brousseau R, Zhang C, Wiles DM (1988) Chapter 27: Identification of products from polyolefin oxidation by derivatization reactions. In: Benham J, Kinstle JF (eds) Chemical reactions on polymers. ACS symposium series 364. American Chemical Society, Washington, DC, pp 376–389

Chan HW-S (1987) Autooxidation of unsaturated lipids. Academic, London

Derrick MR, Stulik D, Landry JM (1999) Infrared spectroscopy in conservation science. scientific tools for conservation. The Getty Conservation Institute, Los Angeles

Frankel EN (1998) Lipid oxidation. The Oil Press LTD, Dundee, Scotland

Gardner HW (1987) Reactions of hydroperoxides – products of high molecular weight. In: Chan HW-S (ed) Autooxidation of unsaturated lipids. Academic, London, pp 51–93

Gladovic N, Zupancic-Kralj L, Plavec J (1997) Determination of primary oxidation products of linoleic acids and triacylglycerols. J Chromatogr A767:63–68

Grosch W (1987) Reactions of hydroperoxides – products of low molecular weight. In: Chan HW-S (ed) Autooxidation of unsaturated lipids. Academic, London, pp 95–149

Kamal-Eldin A (2003) Lipid oxidation pathway. AOCS Press, Champaign

Lazzari M, Chiantore O (1999) Drying and oxidative degradation of linseed oil. Polym Degrad Stab 65:303–313

Mailhot B, Gardette J-L (1992) Polystyrene photooxidation – I. Identification of the IR-absorbing photoproducts formed at short and long wavelengths. Macromolecules 25:4119–4126

Mallégol J, Gardette J-L, Lemaire J (1999) Long-term behavior of oil-based varnishes and paints – I. Spectroscopic analysis of curing drying oils. J Am Oil Chem Soc 76(8):967–976

Mallégol J, Gardette J-L, Lemaire J (2000a) Long-term behavior of oil-based varnishes and paints. Fate of hydroperoxides in drying oils. J Am Oil Chem Soc 77(3):249–255

Mallégol J, Gardette J-L, Lemaire J (2000b) Drier influence on the curing of linseed oil. Prog Org Coat 39:107–113

Mallégol J, Gardette J-L, Lemaire J (2000c) Long-term behavior of oil-based varnishes and paints. Photo- and thermooxidation of cured linseed oil. J Am Oil Chem Soc 77(3):257–263

Meilunas RJ, Bentsen JG, Steinberg A (1990) Analysis of aged paint binders by FTIR spectroscopy. Stud Conserv 35(1):33–51

Morley-Smith CT (1957) Metal-drier catalysis. J Oil Colour Chem Assoc 40:1035–1050

Muizebelt WJ (1978) Oxidative crosslinking of alkyd resins studied with mass spectrometry and NMR using model compounds. J Coat Technol 70:83–93

Richardson T, Korycka-Dahl D (1984) Lipid oxidation. In: Fox FP (ed) Developments in dairy chemistry, vol 2, Developments series. Applied Science Publishing, London, pp 241–263

Schieberle P, Grosch W, Firl J (1984) Photolysis of unsaturated fatty acid hydroperoxides. In: Bors W, Saran M, Tait D, Gruyter D (eds) Proceedings of the third international conference on oxygen radicals in chemistry and biology, Berlin, 1984

Socrates G (2004) Infrared characteristic group frequencies – tables and charts, 3rd edn. Wiley, West Sussex

Van den Berg JDJ (2002) Analytical chemical studies on traditional linseed oil paints. Academic Proefschrift – PhD thesis, Universiteit Amsterdam

Van den Berg JDJ, Van den Berg KJ, Boon JJ (1999) Chemical changes in curing and ageing oil paints. In: Bridgeland J, Brown J (eds) ICOM-CC, 12th triennial meeting. James & James, London, pp 248–253

White WB (1974) The carbonate minerals. In: Farmer VC (ed) The infrared spectra of minerals, vol 4, Mineralogical society monograph. Bartholomew Press Dorking, Surrey, pp 227–284

Wilhelm C, Gardette J-L (1994) Infrared identification of carboxylic acids formed in polymer photooxidation. J Appl Polym Sci 51:1411–1420

Zumbühl S, Scherrer N, Ferreira E, Hons S, Müller M, Kühnen R, Navi P (2011) Accelerated ageing of drying oil paint – an FTIR study on the chemical alteration. Zeitschrift für Kunsttechnologie und Konservierung 25(2):139–151

Zumbühl S, Scherrer NC, Eggenberger U (2014) Derivatisation technique to increase the spectral selectivity of 2D FTIR-FPA imaging – analysis of binder composition in aged oil and tempera paint. Applied spectroscopy 68(4):458–465

Investigating Fluidizing Dripping Pink Commercial Paint on Van Hemert's *Seven*-Series Works from 1990–1995

Jaap J. Boon and Frank G. Hoogland

Abstract A painting from the *Seven* series by Frank van Hemert made between 1990 and 1995 was starting to show fluorescent exudates and drips after about 7 years. Paint samples were investigated with analytical mass spectrometry, solid state NMR, imaging FTIR and SEM-EDX. Fluorescent exudate of fluidised paint compared to rubbery paint underneath showed differences in FTIR and MS characteristics related to the distribution of polar acylglycerides with ester bonded and free azelaic acid groups. The oil in the Schmincke Norma Fleischfarbe nr 213 tube paint used by van Hemert consists mainly of semidrying oil giving a Drying Index of 72. Imaging FTIR identified aluminium soaps in fluorescent vesicles in the solid rubbery paint. Solid state NMR indicated that these aluminium soaps are degraded and consist mainly of aluminium hydroxy-monostearates. DTMS and ESIMS showed apolarity inside the rubbery paint compared to the exuding material. The drying of the paint produced high relative amounts of smaller oxidised fractions that could not be retained in the paint mass as aging progressed. A physical separation resulted in more polar fractions. The fluorescence of these exuding fractions can be used a tracer for failing or impending failing of oil paint.

Keywords Dripping paint • Failing aluminium stearates • Fluorescent exudates • Phase separation • Mass spectrometry (ESIMS • DTMS • GCMS) • Ester and non-ester bound organic fractions • FTIR-microscopy

J.J. Boon (✉)
JAAP Enterprise for Art Scientific Studies (Formerly MOPAS Group AMOLF), Nieuwendammerdijk 79, 1025LD Amsterdam, The Netherlands
e-mail: boon@amolf.nl

F.G. Hoogland
Formerly MOPAS Group AMOLF, Limmerweg 2, 1935 AM Egmond-Binnen, The Netherlands

K.J. van den Berg et al. (eds.), *Issues in Contemporary Oil Paint*,
DOI 10.1007/978-3-319-10100-2_16

Introduction

Seven years after completion of the paintings from the *Seven* series (made between 1990 and 1995 by Frank van Hemert (Gemeentemuseum Den Haag 1996), the pink paints became tacky, the impasto developed a molten appearance, and began to form drips. Van Hemert used three paints for most of these works: a bright red toluidine oil paint, a dark alizarin oil paint and a pink oil paint, applied directly from the tube to the canvas, and further worked using a palette knife. The paints initially dried but the pink paint fluidised after about 7 years. In 2004 samples were taken for an initial study using mass spectrometry. (Boon et al. 2007; Hoogland and Boon 2007). Tube paint applied to a solid support in 2002 by paint chemist Pieter Keune was made available for analysis in 2004. The thicker impasto paint mass began to flow in 2009 and developed a molten appearance closely similar to that observed on the paintings. Drawdowns and the painting selected for analytical studies remained exposed to natural light until analysis in the present study.

The phenomenon of weeping paint and drip formation is of wider interest to artists, gallery owners and conservators as paintings by other artists show similar deterioration phenomena. In 2001, Duffy and McGlinchey reported dripping paint in a work by Jack Youngerman (Black, Red and White from 1962 @ MOMA). In 2003, Götz mentions in her Master thesis works by Matthieu, Soulages, Immendorf, Richter, Abello, Uhlig, Schultze, Dix and Hoehme with serious drying problems and fluidising paint. Schultz investigated in detail a so-called *fire*-painting *Harvest* (1993) by Otto Piene that started to show drips developing from a thickly painted impasto about 7 year after completion (Schultz 2011). More recently, we were contacted by Paul Walls, an artist in Northern Ireland whose white impasto paints started to drip after about 7 years. Paintings by the German action painter Meese have intentional drips resulting from his technique and unintentional drips that result from unstable paint, observed about 6–7 years after painting (Andreas Franz, 2013, personal communication).

Softening of post-1950 paints has been reported for Riopelle (Corbeil et al. 2004, 2011; Bronken 2010; Bronken and Boon 2014) who worked in France. Boon discussed and presented at the ICOP symposium a black oil painting by Soulages made in 1960 (Art Institute of Chicago) that is extensively weeping (Boon and Lister 2014). A poster at ICOP on a painting by G. Matthieu made with black tube paint in 1957 presented strong evidence of dripping paint (De Ségogne 2014). Many of the dripping paintings appear to be incidents related to poorly formulated paints, however the phenomena of softening and fluidizing oil paints may be part of a more general phenomenon that occurs in late twentieth and early twenty-first century paintings. Our rationale therefore is to contribute to the understanding of the mechanism of the undesirable softening and fluidisation of oil paints as the paintings age.

This paper reports many aspects of the dripping paints from the *Seven* series of works. An earlier version of a summary report (in Dutch) was provided to the lawyers of Van Hemert in 2010 for submission to the Court of Justice in Amsterdam where the artist is suing for damages. The studies performed at AMOLF and reported here are forensic art studies since no information was provided by the Schmincke Company about their pink paint for a long time. This paper characterises the visual qualities of a *Seven* series painting from 1992 and summarises highlights from a larger body of analytical information. A more detailed report on the MS and solid-state NMR studies will be published separately (Boon et al. in prep.).

Painting with Pink Paint Drips from the *Seven* Series

One painting made as part of the *Seven* series in 1992 (Fig. 1a–d) is the focus of the present study. The painting in Fig. 1a has a green wash with open spaces as background on the primed canvas. These open spaces were filled partially with a red wash and the numbers 1, 2, 3, 4, 5, 6. Drips from these washes are intentional. Thicker swatches of red, very dark brown and pink paint were applied as islands in/on the red washed areas directly from the tube. In other paintings from the *Seven* series, the same combination of the paints was used but the swatches are applied directly on the primed canvas. These paintings have the same problem of tacky

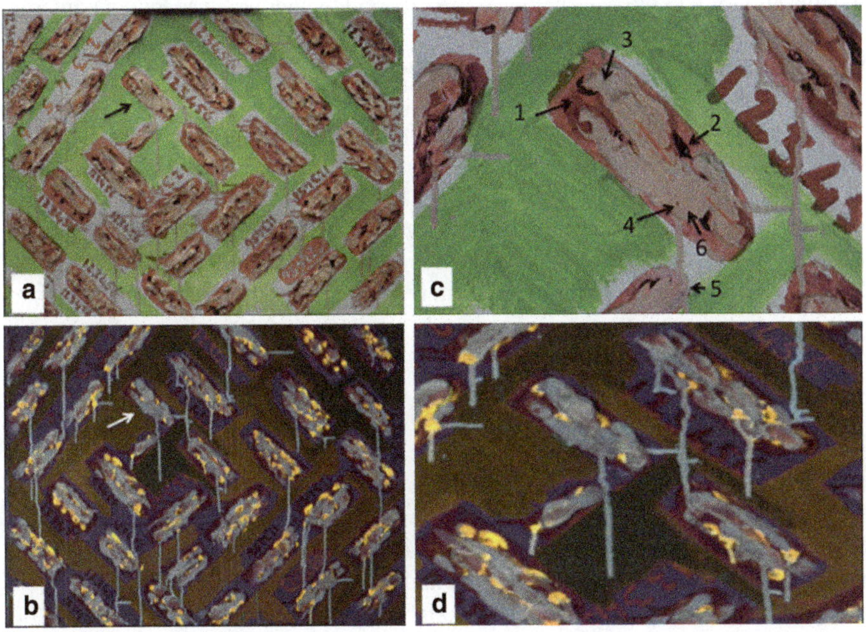

Fig. 1 Painting from van Hemert's *Seven* series made in 1992 that began to drip after about 7 years and photographed in 2005 (**a**). Exudate and drips are fluorescent under UV light (**b**). The *arrow* indicates the swatch that was chosen for sampling. The numbers in (**c**) (VIS) are the code used in the text for the paint and drip sample. (**d**) is the UV fluorescent picture corresponding to (**c**)

pink paints and drip formation. The stable toluidine red and dark brown alizarin paint were tube paints from Lefranc-Bourgois. The pink paint is Schmincke Norma Fleischfarbe nr 213 that the artist used in very large quantities from 1990 till 1995 when Schmincke took it out of production. The pink impasto paints show a molten appearance and drips were forming after a period of about 7 years. In some areas the pink paint drips dissolved some of the red paint. The reverse of these paintings shows a "wet" appearance where exuding medium has penetrated the canvas. This is often observed where the paint was applied more thickly to create the impasto effect. The surfaces of pink painted areas are still tacky and have attracted dust and fibers. Upon examination of the paint in cross section, it became apparent that the exuding fluids were highly fluorescent. Figure 1b, d show the fluorescence of the paints under UV light. The orange fluorescence is typical for the alizarin coloured dark brown paint (2 in Fig. 1c), which is still in the same place. The toluidine red paint (1 in Fig. 1c) is not fluorescent, and the bluish grey fluorescence is associated with the pink paint. Fluorescent exudate develops on the pink paint surface and drips develop when sufficient exudate has accumulated. All drips are connected to areas where pink paint was applied. The thickness of the paint ranges from less than 1 mm to about 5 mm. The relative amounts of dripping paint increased yearly thus increasing the length of some drips by 1–2 cm. Although not systematically investigated, the exudate production and flow increase seems associated with the relative humidity in the environment. The rate of flow increase was low in winter in the dry laboratory conditions of the AMOLF Institute, and higher in summer with a higher relative humidity. Samples of paint and dripping fluid were taken from one of the swatches in the center of the painting indicated with an arrow in Fig. 1a and b. Sample positions are shown in Fig. 1c with numbers and arrows. Pink paint was sampled near the top of the swatch (3) and lower areas (4 and 6) where the drip (5) had developed. Samples from the toluidine red (1) and dark brownish alizarin (2) paint were taken as reference of "normal" dried paint. Since the painter considered the painting a total loss, sufficient sample could be taken for multiple analyses. Remarkably, sample areas from pink paint "repaired" themselves after some time by plastic flow.

Figure 2a, b and c shows images from a remaining original Norma Fleischfarbe nr 213 paint tube used for the painting. A thin (<1 mm) and thicker layer (5–7 mm) of this paint was applied to a solid support in 2002 by the chemist P. Keune who made the paint tube and drawdowns available for analysis in 2004. Samples were taken from the middle of the paint tube, which was opened for that purpose with a scalpel knife and resealed. Photos of the drawdowns were taken in 2005 (Fig. 2b) and 2010 (Fig. 2b). At the end of 2009 about 7 years after making the drawdowns, drips had formed from the impasto paint, which itself also began to show a molten appearance. Samples of the drips were taken in 2010 to compare their MS features with those from 2005. The drip was found to be fluorescent under blue LED-light. The thinly applied layer was not tacky but showed a thin 1 mm wide rim of fluorescent material all around its circumference.

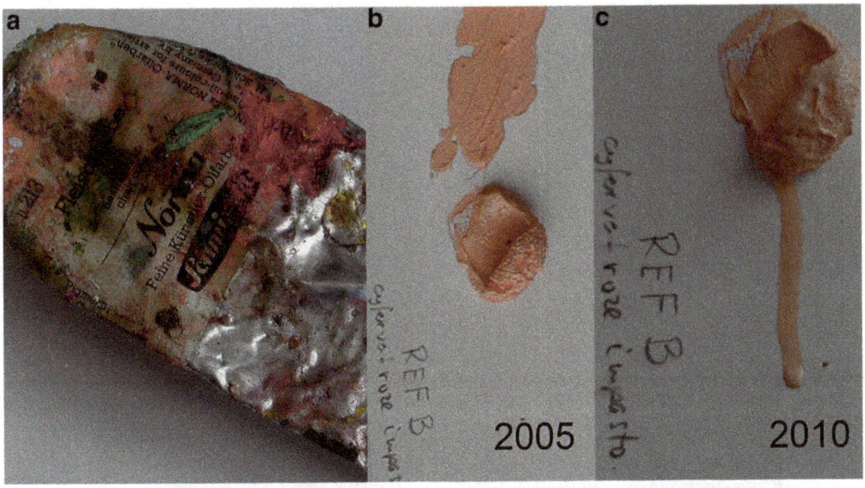

Fig. 2 A tube of Norma Fleischfarbe pink paint (**a**) used by van Hemert for his *Seven* series paintings. The paint was used in 2002 for thin and thick paint application (**b**) on a nonpermeable support (Photos were taken in 2005 and 2010). In late 2009 the thicker impasto swatch (**c**) showed a molten appearance followed by drip formation

Cross Sectional Analysis of the Pink Paint

The pink paint samples were too soft for regular embedding and polishing. Paint from position 4 could be embedded in Technovit LC2000. The resin block was sectioned with a glass knife microtome at a temperature of 5 °C to obtain an about 10 μm thick thin-section of the paint. This section was placed in a diamond cell for imaging FTIR and light microscopy. The remaining embedded section surface was further polished using Argon ion polishing at low kV and examined with light microscopy, SIMS and SEM-EDX to investigate the nature and distribution of pigments, extenders and organic constituents. Figure 3 is a backscatter electron (BSE) image at 16,000 magnification showing tiny rounded particles of Titanium white (EDX Ti, O), high intensity BSE aggregates of Cadmium red (EDX Cd, Se) and Cadmium yellow (EDX Cd, S) and rather fluffy darker grey irregular masses of calcium containing material (EDX Ca, C, O) with a low BSE intensity. The irregular aggregates are relatively large compared to the Titanium white particles. The dark low BSE intensity areas between these particles correspond to binding medium, which is present in relatively thick layers around these particles. This composition correlates well with data from a sample of the pink Norma Fleischfarbe tube paint and one from the 1992 painting analysed with XRD at the University of Amsterdam (Sonneveld 2006). The relative amounts estimated without calibration from the XRD data were 30–50 % Titanium white (rutile form), 40–60 % calcium carbonate, 5–10 % cadmium pigments and <2 % unexplained.

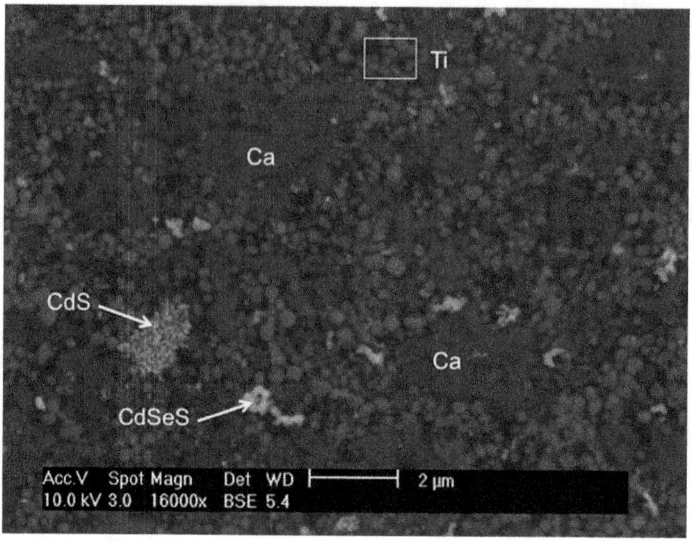

Fig. 3 Scanning electron microscopic image taken under back electron conditions at 16.000 magnification using a FSEM instrument (Fei XL30). The surface was obtained by microtoming and subsequent Argon ion polishing using a JEOL CP system

The thin-section of the pink paint from sample position 4 was investigated with light and FTIR microscopy in reflection and transmission mode. Figure 4a, b illustrate the more yellow-orange visible (A) and more pink-red UV-fluorescence (b) image of the paint. The top layer of the paint shows a bluish fluorescence. Smaller and bigger fluorescent vesicles are seen in the paint mass. Large Cd-yellow pigment particles show a reddish fluorescence. The red and blue squares indicate the areas that were imaged with FTIR microscopy. The corresponding FTIR images in Fig. 4c–e and f show the image maps of the CH asymmetric stretch vibrations of aliphatic chains (of fatty acids) at 2,920 cm^{-1} (c), the asymmetric C═O (carbonyl) stretch vibration at 1,740–1,780 cm^{-1} of ester bonds (acylglycerols) (d), of 1,610–1,630 cm^{-1} interpreted as C—O—C═C (vinyl ether) bonds (Colthup et al. 1964) (e), and 1,470–1,540 cm^{-1} specific for the CO_3 asymmetric stretch vibration in calcium carbonate (f). The images are false coloured from red via yellow and green to blue representing a gradient in relative concentration from high to low. The imaging FTIR methodology and interpretation has been described by Van de Weerd et al. (2005). The surface layer (marked 1) has a lower relative concentration of calcium carbonate, which is an extender in the paint. Higher relative amounts for aliphatic and ester moieties, and CH chains with a vinylether-bonded system are present in the top layer 1. The enrichment points to a compositional difference between top layer and the main paint mass (bulk). The FTIR image maps for the red lined area in Fig. 4g–i were derived from the data cube for 2,920 cm^{-1} (aliphatic CH chains)(g), 1,370–1,540 cm^{-1} from calcium carbonate (h) and 1,581 cm^{-1} (i) from a

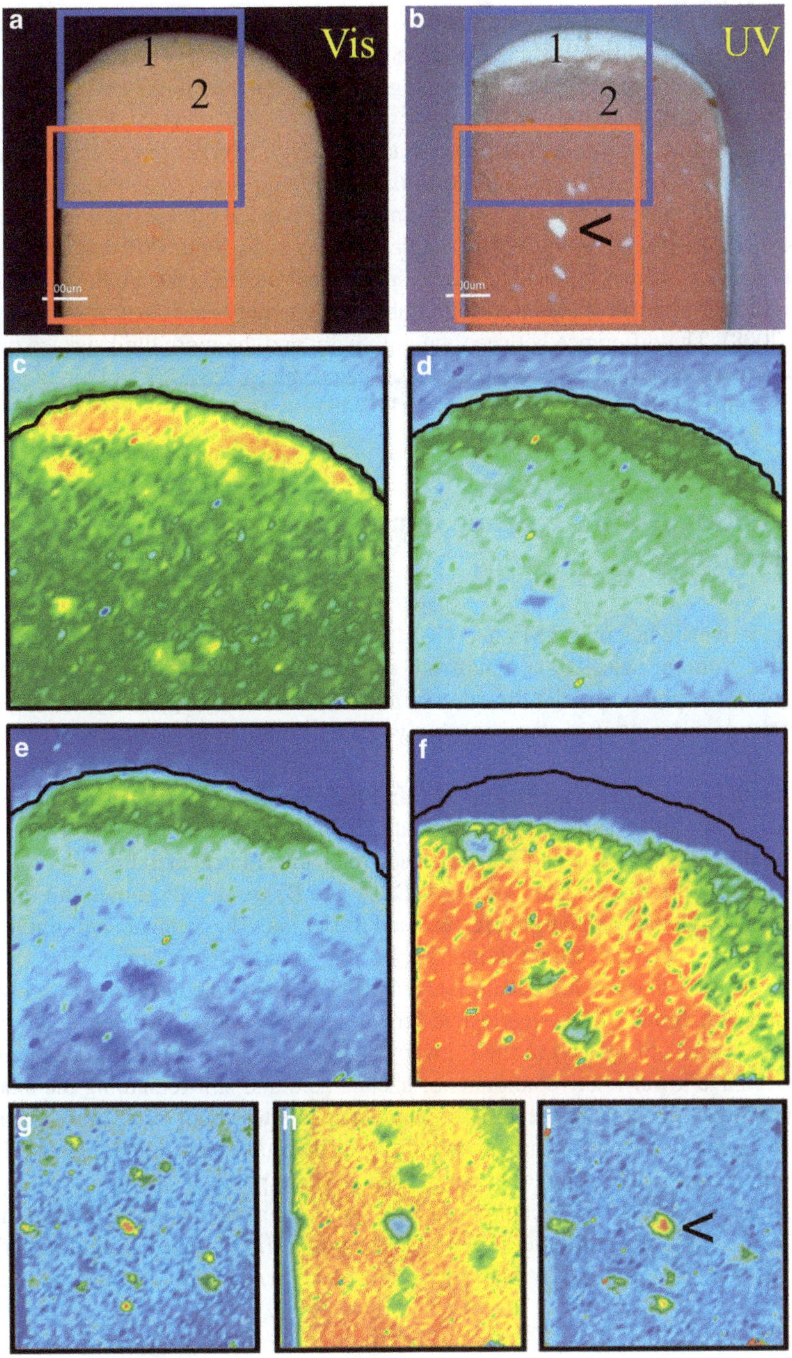

Fig. 4 Light microscopy (**a**: VIS en **b**: UV-fluoresence) and imaging FTIR maps (**c**–**i**) of an Argon ion polished section of van Hemert's paint with exuding material on top. The maps (**c**)–(**i**) correspond to specific wavelength discussed in the text; the red and blue squares to maps (**c**, **d**, **e** & **f**) and (**g**, **h** & **i**) respectively

high intensity peak in aluminium stearate (Al-soap),[1] which is a dispersion and gel-forming agent used in oil paints (Tumosa 2001; Mayer 1991). The FTIR spectrum of the fluorescent spot (arrow) shows a high relative concentration of FTIR features of Al-soaps as well as those from esters.

These results point to a conclusion that the fluidised surface paint contains less solid materials and Al-soaps relative to the main paint mass. Fluorescent vesicles in the bulk of the paint contain Al-soaps and ester bonded oil derived components. Al-soaps were confirmed by Magic Angle spinning solid state NMR but compared to an intact Al-soap reference, the soaps in the paint and the tube are degraded and point to acyl-hydroxy-Al-soap with probably just one fatty acid chain attached (van Eck 2008). The difference between the bulk paint and the more fluid surface paint was further studied with electrospray mass spectrometry (ESIMS) and direct probe temperature resolved mass spectrometry (DTMS).

Identification of the Organic Constituents by Mass Spectrometry

Comparative studies by DTMS and GCMS were performed on the different paints sampled from the selected painting from 1992 (see Fig. 1c), the Fleischfarbe paint tube itself and the tube paint drawdowns after 7 years of drying. Analyses of drips from the drawdowns after seven or more years were limited to studies by DTMS.

The GCMS methodology developed by Van den Berg et al. (2001) was used to quantify biological ester bound and free i.e. non-ester bound fatty acids and diacids as ethylester and silylether derivatives. Figure 5a shows the distribution of the palmitic, stearic, oleic and azelaic acid in the paints indicated in Fig. 1c, the tube paint and the predrip-stage drawdowns (Dd A and B) from 2005 shown in Fig. 2b. Ethyl and silylether derivatives are summed in Fig. 5a. The linolenic and linolenic acid content of the tube paint (about 65 %) is not included in the graph.

All painting samples include azelaic acid (black) as main constituent with lower amounts of palmitic (blue), stearic acid (green) and only a few % of oleic acid (dark pink) that was still present in the drawdowns made in 2005. The reference paints 1 and 2 have lower relative amounts of azelaic acid and still remaining oleic acid (with a much higher % in the alizarin paint 2) even after about 12 years of drying. These results demonstrate that chemical drying had taken place completely. The surprisingly high relative amounts of azelaic acid are possibly due to an unknown added drying catalyst. Only painting sample 3 showed a significantly different ratio of stearic acid to palmitic acid, pointing to the possibility that stearic acid moieties are preferentially retained. In this instance the position of sample 3 at the top of the

[1]Aluminium stearates consist of a technical mixture of soaps with predominantly stearic acid moieties. They will be referred to as Al-soaps.

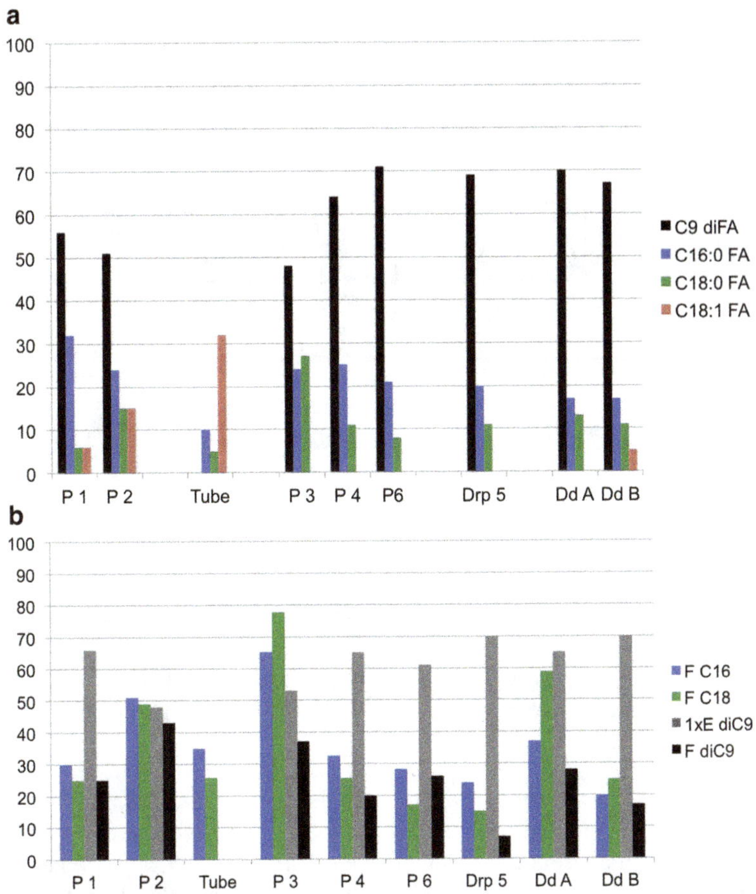

Fig. 5 Graph (**a**) shows distribution of fatty acids (*FA*) and azelaic acid (*diFAC9*) released from paint using the derivatisation methodology of Van den Berg et al. (2001). The numbers correspond to paint sample positions shown in Fig. 1(**c**) indicating Paint (*P*), drip (*Drp*), drawdown thin paint 2005 (Dd **a**), drawdown thin paint 2005 (**b**) and the pink Norma tube paint (Tube). Graph (**b**) shows the distribution of the % non-esterified acid groups as fatty acids and diacids in the samples. The 1xE diC9 is azelaic acid attached to the glycerol moiety while F diC9 is free azelaic acid

swatch probably leads to a preferential loss of fluid exudate due to gravity while Al-stearates remains.

Figure 5b shows the percentage of free (F) i.e. non ester-bound or ionically bound fatty acid moieties as silylethers. The fatty acids in the tube paint are mainly esterified as acylglycerol compounds. The free palmitic and stearic acids are present as metal soap or ionically attached to calcite as suggested by the 13C NMR data (van Eck and Blaakmeer 2011). The C9-diacid with one acid group at the ΔC9 position free and one group still ester bound is a main constituent in all paint samples. "Free" diacid amounts are relatively high in the alizarin paint P2, the pink paint P3 and the

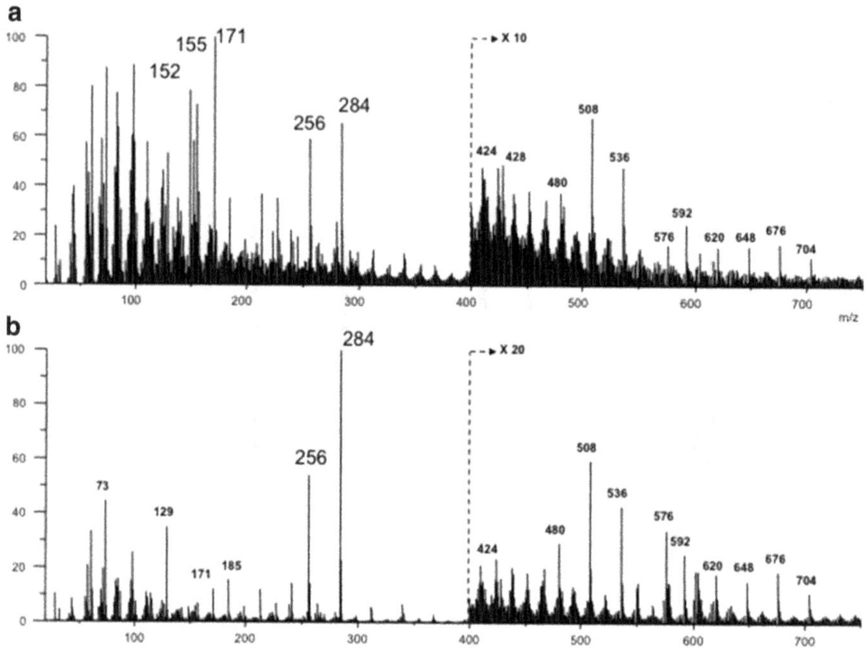

Fig. 6 Direct temperature mass spectrometric summary spectra (*DTMS*) of the fluidised surface paint (**a**) and the underlying rubbery bulk paint (**b**) from the *Seven* series 1992 painting

thin drawdown A. The palmitic and stearic acids are in high concentration in P3 in a free or ionically bound state as metal soap or bound to mineral surfaces. Although the relative amount of azelaic acid is high in the drip material 5, most of that azelaic acid is present in azelaic mono-glycerolester form. Drip 5 had a low % of free azelaic acid (<10 %) whereas paint 3 has a percentage of 35 % free azelaic acid.

Since FTIR microscopy showed a difference between the more fluid paint at the top and the more rubbery bulk paint below, separate sample was obtained from these layers for DTMS and ESIMS analysis. Figure 6a, b shows the DTMS spectra of the surface paint and the underlying bulk paint respectively. These spectra are summation spectra over the mass range before the appearance of cross-linked fractions at higher temperature (beyond scan 60). The lower bulk paint shows high intensity peaks for palmitic (m/z 256) and stearic acid (m/z 284). Stearic acid has a higher intensity than palmitic acid which could correspond to an addition of Al-stearate, a common additive in tube paints and shown to be present by FTIR microscopy. Peaks at m/z 550, 576 and 604–606 correspond to acylglycerolester compounds with combinations of palmitic and stearic acid. In addition we observe peaks at m/z 592, 620, 648, 672 and 704 from beeswax a common suspension agent in tube oil paints (Mayer 1991; Izzo 2011). A main difference between the bulk paint and the more fluid surface paint is the presence of high intensity peaks for C9-diacid (azelaic acid) at m/z 152 and 98, and peaks at m/z 155 and 171 that point to mid-chain-O-functionalised fatty acids like 9,10-epoxy-stearic acid and related keto- or hydroxyl-compounds that lead to a m/z 155 EI fragment ion (Van den Berg 2002). The stearic acid relative intensity is much lower in the surface paint. The intensity of m/z 280 interpreted as a [M-H2O] ion from a keto-stearic acid is relatively high. The combination of higher m/z 280, 155 and 152 is also seen in medium exudate forming from paints by Riopelle (Bronken and Boon 2014), and from black paint in the painting from 1960 by Soulages (Boon, unpublished results). Similar features were observed in the drip that developed from the impasto drawdown material (Fig. 2c) that was sampled in 2010 and later in 2012.

DTMS of the pink Norma Fleischfarbe nr 213 paint showed features of triglycerides with fragment ions of di- and monoacylglycerols and acyl moiety ions at m/z 256, 260, 262, 264 and 284 from palmitic, linolenic, linoleic, oleic and stearic acid (Boon et al. 2010). A small amount of beeswax was also found with peaks at m/z 592, 620, 648, 676, 704 from palmitoyl waxesters with alcohol chain length from C24 to C32. Positive ESIMS of a chloroform-methanol extract (3/1) of the tube paint shows high peaks for triglyceride C55:4 (m/z 873), C55:3 (m/z 875), C57:5 (m/z 897) and C57:6 (m/z 899). The composition of the basic hydrolysates of the extracts analysed by GCMS and ESIMS was 8.2 % palmitic acid, 4.3 % stearic acid, 19.7 % oleic acid, 62.2 % linoleic acid and 5.6 % linolenic acid. So the main triglycerides in the tube paint identified are [3x C18:2 FA] and [C18:1, C18:2 and C18:3 FA] with m/z 897, and [C18:1, C18:1, C18:3 FA] and [C18:1, C18:2, C18:2 FA] with m/z 899. No azelaic acid was found which points to an excellent preservation of the paint in the tube. This distribution is very different from linseed oil with the main triglyceride [C18:3, C18:3, C18:3 FA] at m/z 890 (Van den Berg 2002). The drying index (DI) for the tube paint could be calculated using the relative amounts of fatty acids obtained by GCMS. The value of 72 for the tube paint is rather close to the critical value of 70 when an oil paint will never dry (Mayer 1991). For comparison, a DI for linseed oil would be 140 and for sunflower oil 71 when analysed under the same conditions as the tube paint.

The low DI and the absence of azelaic acid and other diacids suggests that the paint must have been made with a minor amount of a high linolenic moieties containing drying oil. During the ICOP symposium it became known that the paint was made almost completely with semi-drying oil like sunflower oil.[2]

The distribution of fatty acid moieties between surface and lower bulk paint was further investigated in extracts after basic hydrolysis and chemical work-up using negative ion nanospray-ESIMS shown in Fig. 7. Figure 7a shows high intensity peaks for C9-diacid at m/z 187 and relatively low peaks for palmitic (m/z 255) and stearic acid (m/z 283) [all M-H negative ions]. Several oxygen functional stearic acids are observed in the fluid surface layer with one (m/z 297), two (m/z 315), three (m/z 329) or four (m/z 331) oxygen insertions in the fatty acid chain. These peaks were identified on the basis of their exact mass. The hydrolysate ESIMS in Fig. 7b of a sample taken from the bulk of the paint mass illustrates that the ratio of C9-diacid

[2]The use of sunflower oil in this paint has been confirmed by a representative of Schmincke (Dr W. Mueller) during the ICOP 2013 conference. The composition of the paint has been revealed to Schulz and coworkers since then. See Franken et al., Chap. 22, p. 333.

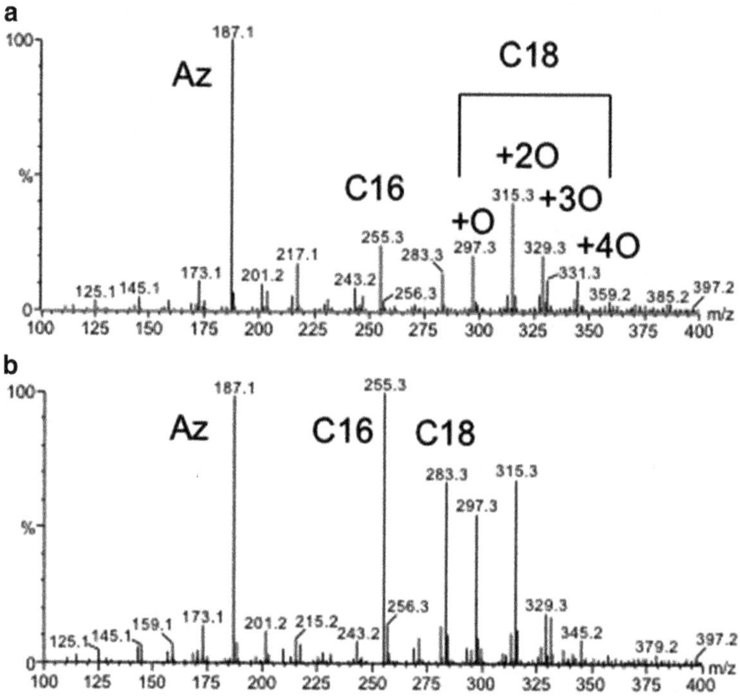

Fig. 7 Negative ion electrospray mass spectrometry (*ESIMS*) of basic hydrolysates of the fluidised surface paint (**a**) and the underlying rubbery paint (**b**)

to palmitic and stearic acid is quite different. Some oxygen-functionalised stearic acids are also relatively higher than the C9 diacid. Examination of the negative ion ESIMS of the ethanol extracts of these paint shows similar patterns but with intact acylglycerides with C8- and C9-diacids, and glycerides with palmitic, stearic and O-stearic acyl moieties (Boon et al. 2007, 2010). The free palmitic and stearic acids are very prominent deeper in the paint where the stearic acid to palmitic acid ratio is >2. In conclusion, the bulk of the paint contains more apolar fatty acid containing compounds while the more fluid surface material primarily contains acyl-glycerol moieties with more polar diacid and O-C18 fatty acid compounds attached.

Discussion

Oil paint, a mixture of drying highly polyunsaturated vegetable oils and inorganic materials, develops into remarkably stable durable material that lasts for many centuries on easel paintings. The chemically drying organic materials play a decisive role in the stabilisation to a dried substance. While there is now a body of evidence for the formation of radicals, reaction with oxygen and formation of cross-links in drying oils (Van den Berg 2002), the molecular level interaction of mineral components in paint including pigment and extenders with the organic materials is less well understood. Drying oils without mineral additions can develop into a viscous mass upon atmospheric exposure so a dried paint could be visualised as a viscous mass that encapsulates the particulate mineral matter. Although this may be true for relatively fresh oil paints, exposure over centuries to atmospheric carbon dioxide and water has an impact on the stability of the biological ester bonds of the original triglycerides. Boon et al. (1997) showed that a large proportion of the acylglycerols in century-old dried oil paint was hydrolysed. The de-esterification over time was further confirmed by Jorrit van den Berg with a specific chemical work-up method combined with GCMS (Van den Berg et al. 1999, 2001). Similar results have been published by Schilling et al. (2007). Hydrolysis of biological ester bonds can completely dissolve the cross-linked oil network, thus presenting a challenge for understanding the preservation of aged oil paintings. Boon et al. proposed in 1997 that the acid groups of the organic matter in aging oil paint play a vital role in stabilisation of the oil paint over time. The relatively large amount of diacids that form by oxidation of the Δ-C9 double bonds can form very stable structures in three dimensions (Boon 2006). Indeed metal soap structures of azelaic acid are remarkably heat stable to nearly 300 °C (Ibidapo 1996). Another factor of importance influencing the stability of oil paints is the interaction of the acid groups in the oil network with surfaces of pigments and extenders. This is particularly important in the early stages of drying when the Δ-C9 double bonds of unsaturated C18 fatty acids turn into diacids with one side attached to the biological ester groups of the triglyceride moieties (Verhoeven et al. 2006). The molecular level interaction of oil derived materials and mineral surfaces has not been investigated for oil paints. Osman and Suter (2002) investigated the interaction of saturated and unsaturated fatty acids with clean calcium carbonate surfaces on a molecular level in the context of the role of fatty acyl derivatised calcium carbonates for ore flotation separation. Important for oil paint studies is their observation that unsaturated fatty acids in interaction with the mineral surface can develop cross-links. Applying this observation to a model for drying oil and calcium carbonate as mineral surface, we postulate that oxidizing polyunsaturated triglycerides will attach via ionic bonds to the mineral surface but will also react with each other via radical cross-linking processes. This ensures a good anchoring of the binding medium components to the available mineral surfaces in the paint. However, the surfaces of calcium carbonate extenders are not always clean and often treated with surface modifiers to increase wettability and flow (Gysau 2006) so interaction to and stabilisation of an extender surface can be inhibited. Oil paint contains many different pigment mineral additions, so a variety of different behaviors are expected and have indeed observed in the organic composition of young ageing oil paints (Keune et al. 2008). Oil paint making is not just mixing drying oil and pigment solids. The Artist's Assistant by Carlyle (2001) gives insight into the paint maker's kitchen and early paint industry in the nineteenth century. In the course of the twentieth century the scale of the paint making process increases significantly and recipes for manufacture remain secret. This factor and that the fact many of the archives of colourmen and small paint industries are lost presents a paucity of historical information about how oil paints were prepared. The archives of Winsor & Newton dating from 1847 onwards have been saved in recent years and made partly available to the public (Carlyle et al. 2011). These records show the increased complexity of paint formulations from the nineteenth century onwards, while simpler mixtures of oil and pigment may have been used in the workshops of the early oil painters. One important surfacing fact is that nondrying oils are mixed with drying oils. Furthermore, tube paints require stabilisation. Mayer (1991, 184–187) discusses several strategies used to stabilise tube oil paints. Addition of waxes and metal soaps leads to buttery paints when used in small amounts. Mayer warns about the percentage by weight and by volume mostly in connection with tinting strength, but light weight solids in the paint can lead to relatively large amounts of Al-stearate for example if measured by volume. Al-soaps are mixed into the dry solids first to increase wettability and to ease oil absorption, which implies that the industry modifies the surface quality of the dry solids in paints. Corkery (1998) mentions that Al-tri-fatty acids are expensive, difficult to make and hard to keep stable. The quality of Al-soaps varies considerably and technical grade may contain a large percentage of free fatty acids (Tumosa 2001). This implies unfortunately that free acids in technical grade Al-soaps will react with surfaces just like the free fatty acids in the experiments of Osman and Suter (2002). So although ideally, high quality artist grade oil tube paint should consist of oil and pigment, in practice the cheaper grades of paint contain extenders and gel-forming agents such as Al-soaps up to high percentages. One may expect therefore that these cheaper formulations can behave in unexpected and different ways.

The analytical data on the Norma Fleischfarbe points to an oil with very small amounts of linolenic acids and large relative amounts of linoleic and oleic acid. The drying index is so close to the critical value that formation of a strong viscous network of linolenic and linoleic fatty acid moieties is nearly impossible. The linolenic and linoleic acyl moieties react relatively rapidly compared to oleic acid moieties that do not contribute to cross-linking but rather act as end-cappers limiting the size of the cross-linking material. Initially good quality Al-soaps should still have stabilised gel formation as the paint was buttery but with advancing degrees of oxidation of the unsaturation and the formation of increasing relative amounts of polar groups, this gel-state is likely to have become compromised. Since almost no unsaturation is present after about 10 years of drying of the softening paints of the painting, it is clear that appearance of fluid exudate is not due to non-dried triglycerides or original oil fractions. Even after a few years of drying, samples from the drawdowns contained hardly any unsaturated fatty acids and the relative amount of azelaic acid moieties is high. The separation of fluid fractions from the bulk of the paint is therefore due to a physical separation between developing polar fractions, the absence of anchor points for acid groups on surfaces and the presence of apolar fractions inside the paint. The relative amount of acid groups generated from the unsaturated fatty acids is very high, which would require a large reservoir of acid neutralizing materials in the paint formulation to compensate for this rapid increase in acidity. Calcium carbonate in the paint could in principle play this role but all calcium containing material in the paint is sequestered in aggregates as SEM at high magnification has demonstrated. Osman and Suter (2002) have shown that fatty acids can attach to clean calcium carbonate surfaces. Such attachment may either have happened in the production phase of the paint or at a later stage when the acidity developed upon oxidation. One factor of importance resulting from the ^{27}Al-NMR studies (Van Eck 2008) is that the Al-soaps utilised were already degraded in the tube paint to hydroxyl-aluminates with one fatty acyl group attached, which implies that there must have a been a large percentage of free stearic acids present in the metal soap added to the pigment and extender solids. Addition of free fatty acids in the degraded Al-soap must have covered a large percentage of the available surface area of the calcium carbonate filler. Apart from the degraded Al-soap, mass spectrometry also discovered beeswax esters that play a role as wettability agents (Mayer 1991) but these also added more apolarity to the paint before any mixing in of the oil. The fluid exudate at the surface and in the drips is more polar than the remaining paint substance underneath the fluidised surface. A lot of that polarity is in the form of acylglycerides with ester bonded azelaic aids and midchain oxidised stearic acids. These compounds are physically separated from the more solid paint mass that shows vesicles of fatty substances clearly excluded from the paint mass. The results of the present study further suggest that these vesicles are remnants of the fluid exudate. Vesicles and surface exudate show the same fluorescence and are therefore a good tracer of physically separating fractions. Similar fluorescence has also been observed on the surface of a painting by Riopelle and in cross sections of its exuding paints (Bronken and Boon 2014). The fluorescence of the exudate from the 7-year dried fluidizing Norma Fleischfarbe is not caused by one compound but is a property of substances that form unusually broad peaks upon HPLC analysis (Van Bommel, 2013, personal communication), which suggest that the fluorophores are attached or part of oligomeric fractions. The distribution of remaining fluorescent vesicles in the paint with FTIR features pointing to Al-soaps suggests that the Al soaps are not well dispersed in the paint initially or have aggregated into these vesicles. The presence of esters in these vesicles furthermore suggests that materials with the original biological ester bonds have accumulated there. This supports the idea that polar fractions develop inside the body of the paint as vesicles that fuse together and find their way to the surface given sufficient time. It is remarkable in this respect that exudates take about the same time of 6–7 years for appearance for paints from Schmincke (van Hemert) and paints from other manufacturers used by Walls and Meese.

The mechanism is complex and probably not related to one factor. The semidrying oils in the paints form many oligomeric polar fractions that should be stabilised by acid neutralizing surfaces in the paint solids. The Al-soaps degraded possibly due to exposure to moisture before use which consequently added free acids and Al-acyl-hydroxides to the solid matter. Beeswax added further apolarity to the surfaces but beeswax is often used without much of a problem. Then oils are mixed into this paint formulation and tubes are filled. Upon use as paint, oxidative cross-linking begins forming a viscous substance that encapsulates the solids first but fail to create molecular bridges sufficiently to stabilise the paint mass. Al-ions from degraded

Al-soaps might drive the oxidation further since they can act as secondary drier (Mecklenburg et al. 2013). The polar Al-OH-soaps appear to function as nuclei for other polar substances in the drying oil paint leading to polar micelles where apolar ester bonded acylglycerides form the core and acid groups from azelaic esters and hydroxyl group from Al-hydroxides are concentrated at the surface. The calcium carbonates in the paints are aggregated and are likely unable to stabilise the ever increasing relative amounts of acidic fractions as oxidation of the semidrying oil progresses. First, thin films of exudate develop at the surface followed by collapse of sharp impasto features where the paint looses consistency. Thicker paints incorporate a larger reservoir of exudate that is physically separated and flows out under the force of gravity. Moisture that could fluidise the polar micelles in the paint facilitates the exudation process. The remaining paint has lost quite some medium and begins to show a rubbery appearance. It is remarkable that this paint is completely soluble in acetone unlike the two "hard" paints from Lefranc-Bourgois.

Van Hemert never stopped using Schmincke paint since he liked their colour and consistency. None of these paint had any drying problems after a long period of time. The behaviour of the pink Norma paint from the period 1990–1995 suggests that there has been a formulation fault due to failing ingredients. Its formulation (ingredients, mixing procedure etc.) was an industrial Schmincke secret. There is some evidence in messages on the Internet about sticky Fleischfarbe Schmincke paint (Petronella 2004) pointing out that van Hemert was not the only one that suffered from a faulty paint. In this message Petronella reports that her pink paint began to flow after about 10 years.

Although there are now many more cases of sticky and dripping paints not all the cases may have the same causal factors, and the unique circumstances of each needs to be considered. In some cases the oil itself may be an issue, in other cases faulty additive(s) may play a role, a lack or a covering of anchor points for acid fractions in the paint is potentially another factor. However, this study suggests that the increase in polarity in relation to failure to anchor these polar fractions inside the paint is the main physical cause of the formation of exudates that create sticky films or even drips depending on the size of the polar fluid-producing reservoir.

Conservators are now wondering how to stop the process. Theoretically an addition of anchor points by addition of metal ions would be a way to bind the acidic fractions. This approach is presently tested on samples from a painting by Riopelle (Bronken, 2013, personal communication). Even exposure to ionizing radiation thus renewing the radical chemistry in the paint has been tried with some success (Schultz 2011). These approaches may work despite considerable experimental difficulties. There is however a caveat: when exudates appear after so many years, the structure of the paint itself has changed too. The distribution of fluorescent polar vesicles inside the pink Norma paint can be seen as evidence for a severe disturbance of the paint consistency. In our study it was noted that the speed of flow was related to moisture suggesting that atmospheric water facilitates the exudation of polar fractions. Drying the dripping painting at 80 °C for several hours indeed stopped dripping in the case of the painting by Piene (Schultz 2011), but it is not clear whether this will remain a permanent change in the paint. One suggestion for treatment could be to keep the painting on a flat surface and develop a treatment to diminish the exudates from the reverse, but this may not always be a practical option. Artists like van Hemert and Walls, and owners of their paintings however consider their works ruined due to loss of intent caused by changes in the surface appearance of the paint.

Appendix

Experimental Details

Microscopy

Light microscopic images (VIS and UV-fluorescent) were obtained with a Leica DMRX microscope (Leica, Wetzlar, Germany). The Bio-Rad Stingray (Bio-Rad, Cambridge, MA), combining the Bio-Rad FTS-6000 spectrometer equipped with a Bio-Rad UMA 500 infrared microscope with a 64×64 mercury-cadmium telluride (MCT) focal plane array camera was used to record the FT-IR images (see van der Weerd et al. 2005). A Philips-FEI XL-30 FSEM coupled to an EDAX energy dispersive analyser was used for SEM-EDX. The ion polished section (Boon et al. 2008) was covered with a few nm of gold in vacuo using a sputter coating system.

Mass Spectrometry

A JEOL SX102-102A mass spectrometer was used for DTMS (Direct Temperature resolved Mass Spectrometry) equipped with a Pt/Rh (9/1) filament (diam.100 μm) probe for in-source MS. Spectra were obtained at 16 eV electron ionisation over a mass range of 1,000 Da at a cycle time of 1 s and acceleration voltage of 8 kV (Boon 1992). Degree of esterification was determined by GCMS after derivatisation of paint samples using the method of Van den Berg et al. (2001). Tube paint was extracted with chloroform/methanol (3/1). Part of the extract was analysed directly with ESIMS. Another part subjected to basic hydrolysis, worked up and analysed by ESIMS or by GCMS after silylation. Paint samples were also weighed (400–600 microg) into 1 mL tapered vials. For ethanol extraction 100 microL ethanol was added with tridecanoic and tetradecanedioic acids as internal standards. The samples were extracted for 18 h under ambient conditions. Extract were analysed directly with ESIMS. Hydrolysis of ethanol extracts: a 0.1 M ethanolic NaOH solution with tridecanoic and tetradecanedioic acids as internal standards was added to the samples. The hydrolysis took place in an oven held at about 80 °C for 1.5 h. Each solution was shaken and centrifuged after extraction or hydrolysis. The extract or hydrolysate was subsequently transferred to a vial containing cation exchange resin (DOWEX HCR-W2 ion-exchange resin, Sigma–Aldrich) and left for 30 min to remove pigment-derived cations. The supernatant solution was removed from the resin and transferred to a new vial. For all ESIMS, Ethanolic ammonium acetate was added until a final concentration of 10 mM ammonium acetate was achieved.

Nano-ESI–QTOF–MS was performed on a Q-ToF2 instrument (Micromass Limited, Wythenshawe, UK). The solutions were sprayed with Econo10 needles or BG-10-58-2-CE-20 needles (New Objective, Woburn, USA) with a flow of approximately 10–80 nl/min. Data were processed using the MassLynx V3.5 software (Micro- mass Limited, Wythenshawe, UK). Mass spectrometer conditions: capillary voltage: 2 kV, cone voltage: 10 V, collision energy: 10 eV, TOF: 9.1 kV and the detector was set at 2,150 V in the positive mode and 2,300 V in the negative mode.

References

Boon JJ (1992) Analytical pyrolysis mass spectrometry: new vistas opened by temperature-resolved in-source PYMS. Int J Mass Spectrom Ion Process 118/119:755–787

Boon JJ (2006) Processes inside paintings that affect the picture: chemical changes at, near and underneath the paint surface. In: Boon JJ, Ferreira ESB (eds) Reporting highlights of the De Mayerne Programme: research programme on molecular studies in conservation and technical studies in art history. NWO, The Hague, pp 21–31

Boon JJ, Lister K (2014) Investigating aged oil-medium phase separating and exuding as drips from a 1960s painting by Pierre Soulages. In: Bridgland J (ed) ICOM-CC 17th Triennial conference preprints, Melbourne, 15–19 September 2014. International Council of Museums, Paris, art. 0901, 7 pp. ISBN 978-92-9012-410-8

Boon JJ, Peulvé S, van den Brink OF, Duursma MC, Rainford D (1997) Molecular aspects of mobile and stationary phases in ageing tempera and oil paint films. In: Maastricht T, Bakkenist T, Hoppenbrouwers R, Dubois H (eds) Early Italian paintings: techniques and analysis. Conservation Institute (SRAL), Limburg

Boon JJ, Hoogland FG, van der Horst J (2007) Mass spectrometry of modern paints. In: Learner TJS, Smithen P, Krueger JW, Schilling MR (eds) Modern paints uncovered. Getty Conservation Institute, Los Angeles, pp 85–95

Boon JJ, van der Horst J et al (2008) Remarkably improved spatial resolution in SEM images of paint cross-sections after argon ion polishing. In: Townsend J (ed) Preparation for painting: the artists choice and its consequences: [Proceedings of Paintings Group Interim Meeting ICOM Committee for Conservation 31 May-1 June 2007 in London]. Archetype, London, pp 42–49

Boon JJ, Hoogland FG, Van Eck E (2010) Natuurwetenschapplijk onderzoek ter verklaring van het plakken en druipen van roze Schmincke verf op schilderijen gemaakt door Frank van Hemert tussen 1990 en 1995. Rapport ter schikking gesteld aan Mr. Maarten Schepel te Den Haag tbv het Gerechtshof te Amsterdam

Boon JJ, Hoogland FG, Van Eck E (in preparation) Solid state ^{27}Al, ^{13}C and ^{1}H NMR of fluidizing paint in Van Hemerts's Seven Series

Bronken Tank IA (2010) Undersøkelse og dokumentasjon av Composition 1952 av Jean-Paul Riopelle. Dissertation, University of Oslo, Norway

Bronken Tank IA, Boon JJ (2014) Hard dry paint, softening tacky paint, and exuding drips on Composition 1952 by Jean-Paul Riopelle. These proceedings, Chapter 17

Carlyle L (2001) The artists assistant. Archetype, London

Carlyle L, Alves PC, Otero V, Melo MJ, Vilarigues M (2011) A question of scale and terminology, extrapolating from past practices in commercial manufacture to current laboratory experience: the Winsor & Newton 19th-century artists' materials archive database. In Preprints of ICOMCC Lisbon, ICOM Committee for Conservation 16th triennial meeting, Lisbon 2011, Bridgland J (ed), Lisbon. (ISBN: 978-989-97522-0-7). Paper 0102

Colthup NB, Daly LH, Wiberley SE (1964) Introduction to infrared and Raman spectroscopy. Academic Press International, Boston

Corbeil MC, Helwig K, Poulin J (2004) Analysis of the painted oeuvre of Jean-Paul Riopelle: from oil to mixed media. In: Roy A, Smith P (eds) Modern art, new museums contributions to the Bilbao congress 13–17 Sept 2004. The International Institute for Conservation of Historic and Artistic Works, London, pp 170–173

Corbeil MC, Helwig K, Poulin J (2011) Jean Paul Riopelle. The artists' materials. Getty Publications, Los Angeles

Corkery R (1998) Artificial biomineralisation and metallic soaps. PhD thesis, Australian National University

De Ségogne H (2014) *Théorème de Gödel* by Georges Mathieu, 1957. Study and restoration: consolidation through cohesive regeneration using a solvent. These proceedings, Chapter 9

Duffy M, McGlinchey C (2001) Weeping Cadmium paint: a case study. In: Deterioration of artists' paints: effects and analysis. Extended abstract of presentation of ICOMCC Paintings Working Group. 10 and 11 Sept 2001, London, pp 81–83

Gemeentemuseum Den Haag (1996) 124 Werken van Frank van Hemert. ISBN 90-6730-109-4. Den Haag, 184 pp

Götz E (2003) Zur Trocknungproblematik pastoser Leinöl Farben. Diplomarbeit Fachhochschule Köln

Gysau D (2006) Fillers for paints. European coatings literature. Vincetz Network, Hannover

Hoogland FG, Boon JJ (2007) Liquefying oil paint in some late-twentieth century oil paintings. In: Learner TJS, Smithen P, Krueger JW, Schilling MR (eds) Modern paints uncovered. Getty Conservation Institute, Los Angeles, pp 282–283

Ibidapo TA (1996) The characterisation of the halatopolymers: a review. Polymer Eng Sci 35:1920–1927

Izzo FC (2011) 20th century artists' oil paints: a chemical-physical survey. PhD thesis, Ca' Foscari University of Venice, 257 pp

Keune K, Hoogland FG, Peggie D, Higgit C, Boon JJ (2008) Comparative study of the effect of traditional pigments on artificially aged oil paint systems using complementary analytical techniques. In: Bridgland J (ed) 15th triennial conference, New Delhi, 22–26 Sept 2008: preprints, vol II. Allied Publishers Pvt. Ltd, New Delhi, pp 833–842

Mayer R (1991) In: Sheehan S (ed) The artist's handbook of materials and techniques, 5th edn. Faber and Faber, London

Mecklenburg MF, Tumosa CS, Vicenzi EP (2013) The influence of pigment and ion migration on the durability of drying oil and alkyd paints. In: Mecklenburg MF, Elena Charola A, Koestler RJ (eds) New insight into the cleaning of paintings. Proceedings from the cleaning 2010 international conference Universidad Politécnica de Valencia and Museum Conservation Institute. Smithsonian Institution Scholarly Press, Washington, DC, pp 59–67

Osman MA, Suter UW (2002) Surface treatment of calcite with fatty acids. Surface properties of the organic monolayers. Chem Mat 14:4408–4415

Petronella (2004) Forum WetCanvas: http://www.wetcanvas.com/forums/archive/index.php/t-206710.html

Schilling MR, Mazurek J, Learner TJS (2007) Studies of modern oil-based artists' paint media by Gass chromatography/Mass spectrometry. In: Learner TJS, Smithen P, Krueger JW, Schilling MR (eds) Modern paints uncovered. Getty Conservation Institute, Los Angeles, pp 129–143

Schultz J (2011) Liquefying oil paint: contemporary paintings in danger. In: Preprints of ICOMCC Lisbon, ICOM committee for conservation 16th triennial meeting, Lissabon 2011, Bridgland J (ed), Lisbon. (ISBN: 978-989-97522-0-7). Paper 1320

Sonneveld EJ (2006) Ongepubliceerd XRD analyse rapport van drie monsters. Project nr 06-001

Tumosa C (2001) A brief history of aluminiumstearate as a component of paint. WAAC Newsl 23:10–11

Van de Weerd J, Heeren RMA, Boon JJ (2005) FTIR studies of the effect of pigments in the ageing of oil. Stud Conserv 50:3–32

Van den Berg JDJ (2002) Analytical chemical studies on traditional linseed oil paints. PhD thesis, University of Amsterdam. Report of MOLART Series: Molart 6 van den Berg JDJ, van den Berg KJ, Boon JJ (1999) Chemical changes in curing and ageing oil paints. In: Bridgeland J (ed) ICOM-CC 12th triennial meeting. James and James, Lyon, pp 248–253

van den Berg JDJ, van den Berg KJ, Boon JJ (2001) Determination of the degree of hydrolysis of oil paint samples using a two-step derivatisation method and on-column GC/MS. Prog Org Coat 41(2001):143–155

Van Eck E (2008) 27Al NMR/1H measurements on paint samples. Unpublished report of the National NMR Center, Nijmegen

Van Eck E, Blaakmeer M (2011). Report on Carbon 13 NMR on van Hemert paint samples. Unpublished report of the National NMR Center, Nijmegen

Verhoeven MA, Carlyle L, Reedijk J, Haasnoot JG (2006) Exploring the application of solid state NMR to the study of deterioration of paintings. In: Boon JJ, Ferreira ESB (eds) Reporting highlights of the De Mayerne Programme: research programme on molecular studies in conservation and technical studies in art history. NWO, The Hague, pp 33–43

Hard Dry Paint, Softening Tacky Paint, and Exuding Drips on *Composition* (1952) by Jean-Paul Riopelle

Ida Antonia Tank Bronken and Jaap J. Boon

Abstract The painting *Composition* (1952) by Jean Paul Riopelle (1923–2002) is part of the Henie Onstad Art Centre's collection in Norway. Examination of the painting revealed the presence of drips of soft medium exuding from the surface. The main aim of this paper is to report chemical characteristics of its hard, soft and dripping paints. Information was collected about Riopelle's use of materials. Data were obtained from microsamples by analysis with Scanning Electron Microscopy with Energy Dispersive X-Ray (SEM-EDX) and Direct Temperature resolved Mass Spectrometry (DTMS). *Composition* (1952) is painted with 26 different colours or shades of paint. All these paints vary in their specific physical condition when tested on the painting. Paints were classified into three categories: hard, soft, softer and dripping. The most problematic of all paints contained cobalt blue pigment.

Dripping exudates show relative high DTMS features for polar compounds such as azelaic acid and midchain-oxygen-functionalised stearic acids. Lower relative amounts are present in softening paints. Hard paints are characterised by higher relative amounts of saturated fatty acids ranging from palmitic (C16) to lignoceric (C24) fatty acid. We infer that a physical separation mechanism is responsible for formation of polar micelles that form from oxidising semidrying oils in the paints used by Riopelle. Results from *Composition* (1952) are compared with another painting by Riopelle from 1956 and one by Pierre Soulages from 1954.

Keywords Jean-Paul Riopelle • Dripping paint • Soft paint • High impasto • Fatty acids • Pigments • Cobalt blue • Bone black • Pierre Soulages

I.A.T. Bronken (✉)
Ida Antonia Tank Bronken (Corresponding A.). The National Museum, Pb. 7014 St. Olavs plass 0130, Oslo, Norway
e-mail: ida.bronken@nasjonalmuseet.no

J.J. Boon
JAAP Enterprise for Art Scientific Studies (Formerly MOPAS Group AMOLF), Nieuwendammerdijk 79, 1025 LD Amsterdam, The Netherlands
e-mail: boon@amolf.nl

K.J. van den Berg et al. (eds.), *Issues in Contemporary Oil Paint*,
DOI 10.1007/978-3-319-10100-2_17

Introduction

Composition (1952) is painted with oil on canvas and measures 131.2 × 251 × 4.5 cm without the frame (see Fig. 1).[1] After stretching, the glue sized and pre-primed canvas was given a second ground. Riopelle then painted the motive using distinctly different colours or hues, mostly wet on wet with a palette knife. The last details he threw or dripped on.[2] In general the painting is multi-layered with high impasto details.

Examination in 2007 revealed the presence of drips of soft medium forming on and streaming over the paint surface of the painting (see Fig. 2).[3] These drips are partly coloured and partly transparent brown. They may have developed slowly over a period of time, but because of the visually complex paint structure and the artist's technique of throwing the paint towards the canvas, it has been difficult in the past to see the difference of intended and solid drips and soft paint with non-intentional drips.

Much is known about Riopelle's use of materials from interviews, written sources and personal information. After Riopelle settled in Paris he and his wife

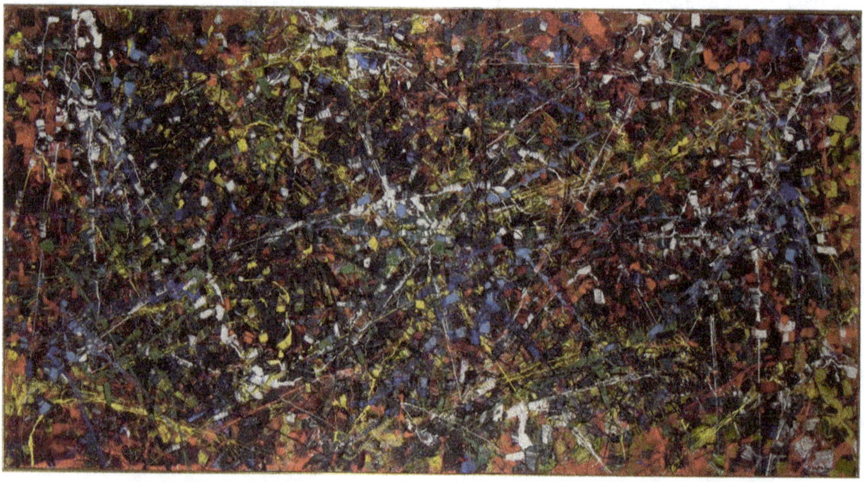

Fig. 1 *Composition* (1952) Jean-Paul Riopelle 131,2 × 251 × 4,5 cm (without frame). *Sans titre* 1952.018H.1952, Le Catalogue Raisonné de Jean-Paul Riopelle (Photo: Øystein Thorvaldsen)

[1] As Composition is a title for many of Riopelle's paintings therefore the date of each painting is necessary to tell them apart.

[2] For more information on painting technique, see

Conservation challenges and physical changes in *Composition* (1952) by Jean-Paul Riopelle. A case study, by the same authors in *Public paintings by Edvard Munch and his contemporaries. Change and conservation challenges* post print Archetype (in press).

[3] The starting point for this project was the first author's MA Thesis (Bronken 2010).

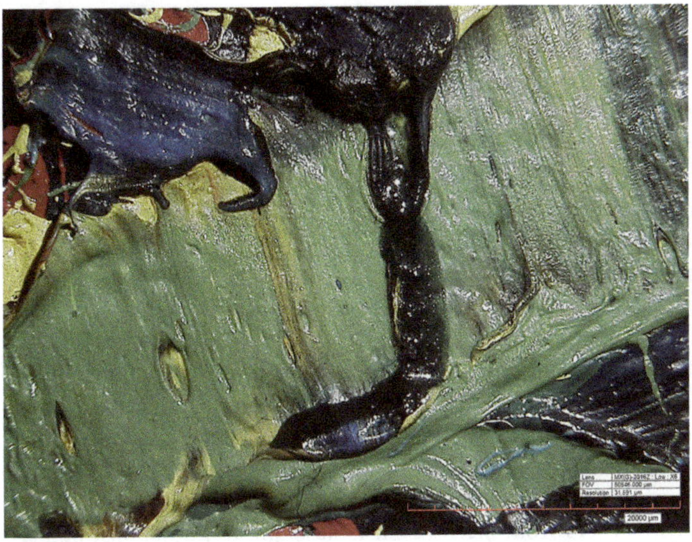

Fig. 2 Detail from *Composition* (1952), drip with partly coloured medium from cobalt blue paint (Blue 2). Photo made with Hirox 8700 (Emilien Leonhardt – Hirox Europe Inc.) mounted on MOPAS stand (Jaap Boon – JAAP Enterprise Inc.)

had to live of an annual allowance from his parents. As he hardly sold any paintings until his breakthrough in 1952, he did at times make his own paint because of his economic situation. Brunet-Weinmann says this was in the late 1940s and Jeffery Spalding refers to René Detroye claiming it was until 1952 (Bronken 2010; Spallding 2002). Viau claims that Riopelle painted directly from the paint tube in 1949, so it is possible he at times could afford to by artist paint (Viau 2002). Riopelle referred to his use of eighteenth century paint recipes in an interview in his studio (Sainte-Marguerite du Lac Masson, October 2, 1985) with Brunet-Weinmann:

> JPR: I was not knowledgeable. I found an old manual (in the stalls) along the Seine riverbank that dated from 1800. Because I did not have any money and the colour-men sold pigment powders, I bought oil, pine resin varnish and started to prepare paints myself.
>
> I made my own paint during the night and painted everything during the day. What I found remarkable is that the products that I made myself never gave me any problems. I never had an accident.
>
> I followed the manual literally as a recipe for cooking: the amount of siccative, the pigment powders, the kind of oil ... I used it so often that I don't know anymore where I put it. I have lost it.
>
> (Translation by Boon from French. Original text cited in Bronken 2010)

Riopelle mentions in this quote that he added siccatives to his self-made paints. However, he did not say what type of siccative or oil he used. In 1952 Riopelle sold all the paintings in his breakthrough exhibition. This changed his economic situation totally. It is still not known if *Composition* (1952) was made before or after this exhibition. We know from publications that Riopelle's choice of brands in the 1950s was the French paint manufacturers Lefebvre-Foinet (Corbeil 2004), and Lefranc Bourgeois (Leclerc 1992). Already in 1952 he bought paint in half a litre and litre cans manufactured at Sennelier (Bronken 2010). According to Mr. Sennelier's daughter her father sold paint to Riopelle in the 1950s. She also informs that in some instances the firm used some grapeseed oil in their paints when linseed oil was difficult to purchase following the years after World War II (Bronken 2010). Grapeseed oil is a semi-drying oil. Ysuelt Riopelle investigated her father's paintings in connection with the publication of his Catalogue Raisonné and she could not find the Lefebvre-Foinet stamp on the reverse of any painted canvas dated before 1955 (Bronken 2010).

Composition (1952) does not have a Lefebvre-Foinet stamp on the reverse of the canvas, but it does not exclude that Lefebvre-Foinet paint could have been used in the making of the painting. Canadian Conservation Institute researchers reported in 2011 that a Lefebvre-Foinet tube from Riopelle's estate contains safflower oil, which is a semi-drying oil (Corbeil et al. 2011). The fact that there were so many different suppliers makes it difficult to determine what he used in 1952. It remains possible that slow drying oils or non-drying oil may have been present in his paints.

In addition to the exuding dripping medium and changing surface gloss the painting also suffers from areas of unstable paint, deformations, numerous paint losses and surface dust. Prior to treatment of the painting the different coloured paints where tested for sensitivity to pressure, heat and solvents. The information gained was then further processed to create three categories for the analytical work. To be able to record the findings, all colours and hues were given a number (see Table 1).

Investigation of the Painting and Sampling

All 26 different main colours and hues used in the painting were investigated and classified into three categories: A: soft and dripping, B: soft and weak, and C: stiff and brittle (see Table 1).

The obviously soft and dripping paints are placed in category A where the softest one is Blue 2. The paints that proved to be in better condition are placed in descending order. Blue 2, Black 2 and Red 2 are the only paints exhibiting dripping or oozing medium. Black 2 and Violet 1 reacted somewhat to water. Both saliva and water picked up colour from Violet 1 when swabbed, and if treated with White Spirit it could only be swabbed once. White Spirit did pick up a colour if swabbed a second time from several paints in category A. Ethanol dissolves paint in category A partly or completely if swabbed. In the cases were ethanol only partly dissolved the layer, the swab was yellowed as it leached medium from the paint layer.

The paints in category B are found to vary strongly in condition. A few paints are so soft that a light prod with a bamboo stick leaves a mark, but other paints did not react. In normal light it is not possible to see oozing medium on the paint surface in this category. One colour – Blue 3 – was thought to be solid, matt and cracking. It was suspected that it might have reached a critical pigment volume concentration (CPVC) and that this alone was the reason for its condition, but it also classified to be soft by DTMS analysis.

Table 1 Condition categories A, B and C of the 26 main colours on *Composition* (1952). Below data on pigments and fillers based on SEM-EDX and XRF on the 26 main colours and two grounds

Categories A-C

A: Soft and dripping	B: Slightly soft or cracking	C: Stiff to brittle paint and grounds		
Blue nr. 2	Blue nr. 1	Blue nr. 4	Green nr. 5	Red nr. 1
Black nr. 2	Yellow nr. 3	Blue nr. 5	Green nr. 6	Grey nr. 1
Violet nr. 1	Yellow nr. 1	Green nr. 1	Yellow nr. 2	White nr. 1
Yellow nr. 5	Black ? nr. 1	Green nr. 2	Yellow nr. 4	White Ground nr. 1
Red nr. 2	Blue nr. 6	Green nr. 3	Orange nr. 1	White Ground nr. 2
	Blue nr. 3	Green nr. 4	Orange nr. 2	

Category A: Soft and dripping	Category B: Slightly soft or cracking	Category C: Stiff to brittle paint and grounds		
Blue nr. 2	**Blue nr. 1**	**Blue nr. 4**	**Green nr. 5**	**Red nr. 1**
Cobalt blue,	Bone black	Ultramarin	Viridian	Zinc oxide
Calcium carbonate,	Ultramarin	Titanium oxide	Potassium zinc chromate	Titanium oxide
Silicon dioxide	Cobalt blue	Zinc oxide	Zinxoxide	Barium sulphate
Black nr. 2	**Yellow nr. 3**	Calcium sulphate	Calcium sulphate	**Grey nr. 1**
Bone black	Potassium zincchromate	**Blue nr. 5**	**Green nr. 6 (XRF only)**	Barium sulphate
Calcium carbonate	Calcium sulphate	Cobalt blue	Emerald green	Calcium carbonate
Mars black	Silicon dioxide	Lead white	Calcium carbonate	Zinc oxide
Violet nr. 1	**Yellow nr. 1**	Zinc oxide	Chrome green	**White nr. 1**
Silicon dioxide	Lead white,	**Green nr. 1**	**Yellow nr. 2**	Zinc oxide
Calcium carbonate	Zinc white	Zinc oxide	Zinc oxide	Titanium oxide
Yellow nr. 5	Calcium sulphate	Calcium sulphate	Calcium sulphate	**White Ground nr. 1**
Calcium sulphate	**Black ? nr. 1**	**Green nr. 2**	**Yellow nr. 4**	Zinc oxide
Silicon dioxide	Calcium sulphate	Zinc oxide	Potassium zincchromate	Calcium carbonate
Zinc oxide	Ultramarin	Calcium sulphate	Calcium carbonate	Potassium zincchromate
Red nr. 2	**Blue nr. 6**	**Green nr. 3**	**Orange nr. 1**	Silicon dioxide
Calcium carbonate	Zinc oxide	Zinc oxide	Calcium sulphate	**White Ground nr. 2**
	Titanium Oxide	Titanium oxide	Zinc oxide	Zinc oxide
	Cobalt blue	**Green nr. 4 (XRF only)**	**Orange nr. 2**	Titanium oxide (more than in White 1)
	Blue nr. 3	Emerald green	Calcium sulphate	
		Calcium carbonate	Zinc oxide	
			Barium sulphate	

The colours within category C proved to be more resilient oil paint judging from their response to pressure and heat. The paint has cracked in some cases, but with a more conchoidal fracture pointing to glass-like brittle behaviour. These paints could withstand heat up to or over 50 °C. White 1 has a film of medium on its surface that is clearly visible with the naked eye, but the paint was solid in all cases observed.

The main observational result from all the tests, and the results from DTMS analysis (see below), is a confirmation of a consistent difference between the three categories whether the paint layers were thin or thick.

The different 26 paints in all areas of the painting appear to be the same colour blends. Some of the greens are very similar in colour, but still can be divided in six distinct different blends when examined closely. So it seems that the paint was prepared either by one manufacturer or by the artist himself. This view is supported by one of Riopelle's own statements:

> ....I never begin a painting without having ready to hand, all the colours I need. (Érouart 1995)

Because the colours show such distinct differences in condition, it became a priority to understand the composition of the materials used to make the different paints. All colours were analysed by handheld X-Ray Fluorescence (XRF) (Table 1). All readings of the paint with XRF show a high intensity peak for zinc possibly from deeper layers or the ground since several paints did not contain any zinc when analysed by SEM-EDX. Hence, we conclude that XRF is not precise enough for our study. Small crumbs were then sampled from several of the paints, two cross sections were made and a canvas sample was taken from the tacking edge with ground. All of these samples were analysed by light microscopy and SEM-EDX (Table 1). The main aim of these analyses was to look for factors that would support the distinction between category A, B and C. To get information about the organic composition at the same time samples of drips, soft paint and solid paint were taken. Samples from drips came from category A. Samples of soft paint came from category A and B. The solid samples considered as healthy paint came from category C.

Analysis by SEM EDX

All stubs and cross sections were analysed by SEM EDX to identify pigments and fillers. The results from the identification of heavy elements are shown in Table 1. No attempt was made to identify the organic dye colours since this requires other methodologies. At this stage of the study, we believe that there is enough information even though organic dyes have not been identified.

Category A paints appear to consist of relatively pure colours. Only Yellow 5 contains some zinc white. Most paints in this category probably contain organic dyes. The worst condition is associated with cobalt blue.

Although cobalt blue is a good initial drier, it is also a poor film former (on Melinex) when used alone in drying oil (Mecklenburg 2012;

Mecklenburg et al. 2013). So far our list of problematic colours in *Composition* (1952) contains pigments that are being described as poor film formers such as cobalt blue, bone black and ultramarine (Mecklenburg et al. 2013). No traces of manganese or lead and just a small amount of zinc white were found in category A samples. So the paint layers on the painting are relatively pure except Yellow 5 that has some traces of white pigment. No signs of added driers but EDX is usually not sufficiently sensitive to observe these. SEM EDX only gave an identification of the fillers in several paints and from this we conclude that organic dyes were probably used.

Tests have shown that especially barium sulphate can result in poor quality paint films (Tumosa and Mecklenburg 2013). We identified four different fillers in *Composition* (1952). Barium sulphate has been used only in the stable category C paints, which contradicts Tumosa and Mecklenburg. Calcium carbonate, silicon dioxide and calcium sulphate were found in category A paints. So the information from literature on fillers does so far not give any clues that could explain the softness in the paint.

One general difference between category A and C is that all the colours in C have some zinc containing compounds like zinc white or potassium zinc chromate, and in some cases lead white. In category A only Yellow 5 contains some zinc, but then comparatively very little. Note that this paint is the only colour in category A that does not drip or bleed. In category B there is not such a distinct difference. So from an inorganic composition point of view we conclude that the paints that display the worst condition have very few or no heavy elements present that are normally associated with good film formation and drying. This implies that polar organic fractions that develop over time may not be stabilised sufficiently by metal organic chemical bonds (Boon 2006; Boon et al. 1997).

Medium Analysis of the Composition 1952 Paints

Visually, the tackiness and dripping of medium from paints suggests issues caused by the present chemical or physical condition of the oil medium inside the paints. We have to assume that the paints in *Composition* (1952) dried and hardened fully after completion. We think that Riopelle would have noticed if the paints did not dry properly. In the course of time some of the paints might have hydrolysed prematurely thus destabilising the viscous cross-linked material that we call dry oil paint. If he was using semi-drying oils, it is indeed possible that some of the triglycerides of these oils could not become part of this cross-linked oil network. These fractions may have segregated after oxidation of their double bonds accumulating into polar micelles and eventually leading to exudation. This process is more prominent if fatty acid moieties can not be stabilised on the surfaces of the pigment or filler particles (Osman and Suter 2002). In order to study the medium in its present state after about 60 years, mass spectrometry was used (Boon et al. 2007). Direct Temperature resolved Mass Spectrometry (DTMS) was used, which is a micro-analytical technique that produces mass spectral fingerprints as a function of temperature. Micro samples homogenised in ethanol is placed by pipetting to a thin platinum-rhodium wire that is resistively heated when the analytical probe is inside the ion source of the mass spectrometer (Boon 1992). During this process loosely bound fractions evaporate first, followed by stronger bound compounds and finally compounds appearing from cross-linked fractions that decompose into smaller molecules. The desorption profile tells more about the condition of the paint sample. The mass spectra indicate what kinds of compounds are evolved (van den Berg 2002).

Five samples from drips, four samples of soft paint and four samples of solid paint were taken from *Composition* (1952) and analysed by DTMS.

Figure 3 shows the total ion current trace and summary spectra excluding Mass spectrometry data from cross-linked fractions of a sample of category C healthy stiff named colour Blue 4, and category A dripping medium from paint named colour Blue 2. The stiff paint shows a different TIC (Fig. 3a) than the material taken from the drip (Fig. 3b). The drip material shows a higher concentration of material which is stronger chemically bound or more polar then the dried healthy stiff paint. This is especially obvious in the 50–60 scan range of the TICs. The corresponding mass spectra in Fig. 3c, d point to a difference in organic composition.

Figure 3c from the stiff paint shows mass peaks of saturated fatty acids like palmitic acid (m/z 256; C16:0), stearic acid (m/z 284; C18:0), arachidic acid (m/z 312; C20:0), behenic acid (m/z 340; C22:0) and lignoceric acid (m/z 368; C24:0 – present but outside the mass range shown). There are also indications for oleic acid (m/z 264; C18:1) and even linoleic acid (m/z 262; C18:2) or their structural isomers. Peaks indicative for azelaic acid (C9 diacid) at m/z 98 and 152 are also observable.

The summary spectrum in Fig. 3d from the dripped exudate is quite different. Here high intensity peaks are observed for m/z 98, 152, 155, 171 while the features of saturated fatty acids are relatively low. The mass speaks m/z 155 and 171 are interpreted as indicative of mid chain functionalised stearic acids with an epoxy, keto or hydroxy function (MS atlas of fatty acids (Van den Berg 2002; Izzo 2011). The mass peak m/z 280 is strongly related to these two peaks and is interpreted as a [M-H2O] of hydroxyl stearic acid compounds. The high m/z 152 and 98 are indicators for azelaic acid. Polar fatty acids and diacids are clearly more prominent, while the most remarkable feature when comparing the two samples is the abundance of saturated fatty acids in the exudate. The ratio of the mass peak indicators m/z 152, 155, 256, 264, 280 and 284 over m/z 129 was calculated to compare DTMS data from a larger number of paint samples.

Figure 4 shows that all drip exudates show relatively high amount of the polar fractions (m/z 152, 155 and 280) while the soft paint have intermediate values compared to the stiff dried paint. Considering that polar fractions appear at the surface and even exude from some of the paints, we postulate that such fractions can not be accommodated sufficiently in the paint mass and are separated according to a physical mechanism possibly after forming polar micelles i.e. acid groups at the surface and apolar aliphatic chain inside.

Composition (1952) is not the only painting by Riopelle with issues of soft, cracking and/or lifting paint. Several publications from Corbeil, Helwig and Poulin

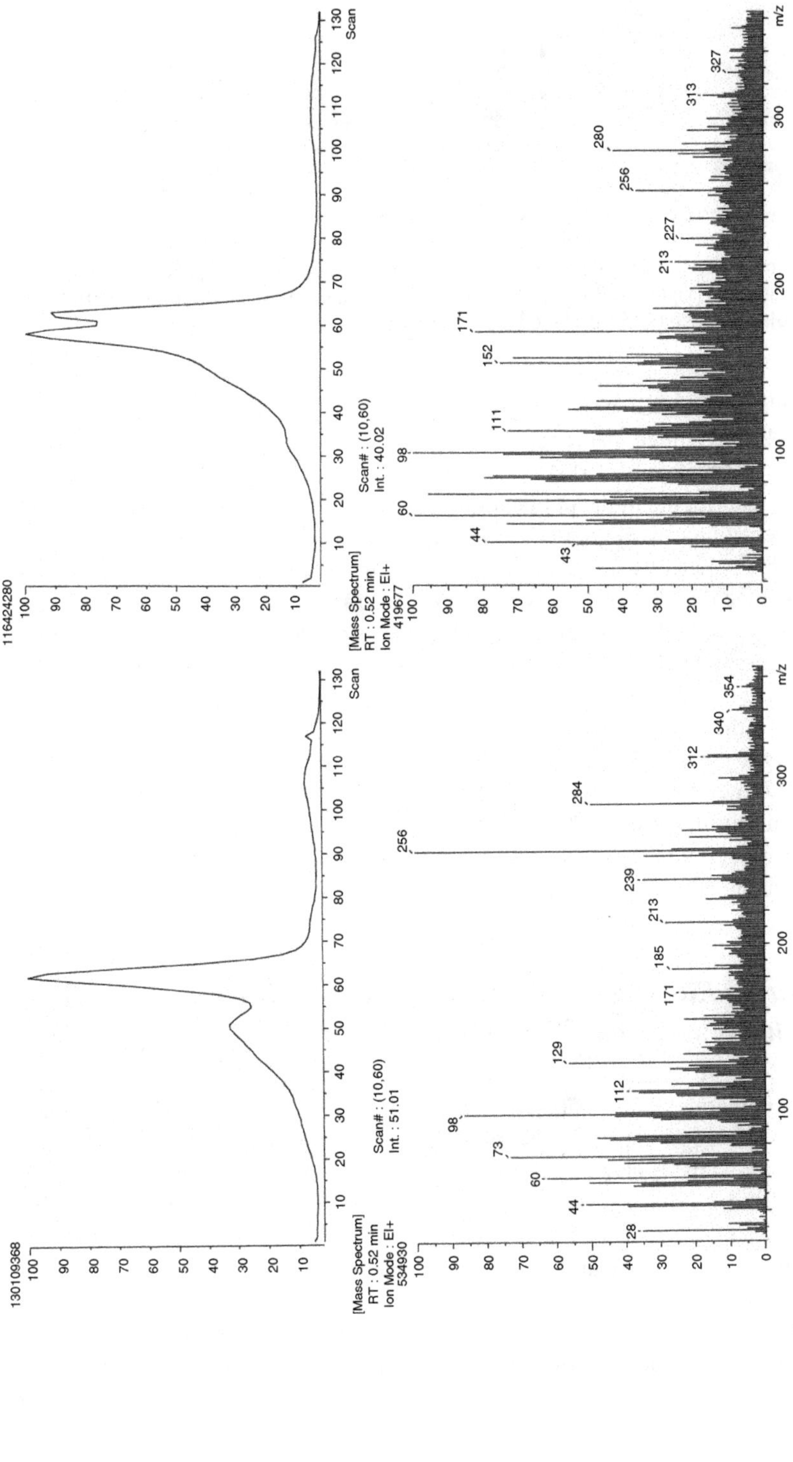

Fig. 3 Total ion current trace and partial low voltage DTMS summary spectra (scan 10–60 summed) from a sample of solid gray 4 (A – TIC and C – MS), and a drip from blue 2 (B – TIC and D – MS)

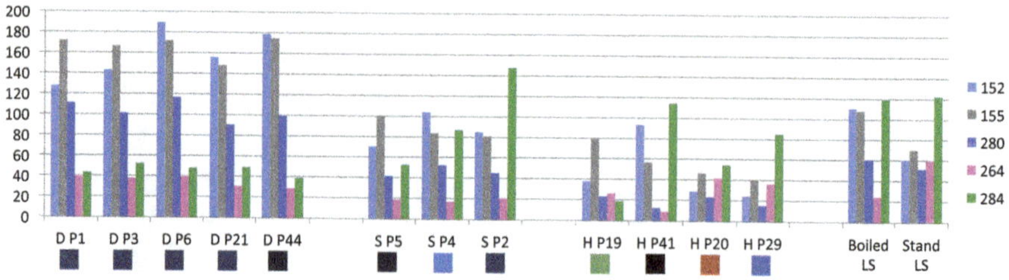

Fig. 4 Discriminating DTMS features of drips (*D*), soft paints (*S*), hard paint (*H*) and 5 years natural light aged linseed boiled and stand oil (Talens). The m/z 152, 155, 280, 264, 284 calculated as a ratio over m/z 129 are plotted (see text). The sample numbers correspond to the colours in Tabel 1: P1, P3, P6 and P21 [Blue 2], P44 and P5 [Black 2], P4 [Blue 3], P2 [Blue 2], P19 [Green 5], P41 [Black 1], P20 [Red 1 background], P29 [Blue 4]

in Canada at the CCI point to problems in the same direction. Between 2002 and 2011 thirty-three paintings in oil from 1944 to 1977 were sampled and analysed (Corbeil et al. 2004; Corbeil 2006; Corbeil et al. 2011). There is no report about dripping in the publications, but several of the condition problems like solvent sensitivity, weak and cracking paint and tacky medium on the surface are comparable to what we are observing in *Composition* (1952). The researchers from CCI conclude that the softness is related to an elevated azelaic/palmitic (A/P) ratio and must be a result of an excessive chain scission during drying of the paint or aging of the paint (Corbeil et al. 2011). They found elevated A/P ratios especially in black and blue colours. It is not clarified why more azelaic acid would be formed in some of the paints. However it correlates well with our observation that polar fractions are more prominent in softer paints. We want to stress also that analysis shows that the paints have initially chemically dried as original features of drying oils are no more present. There is no evidence for extensive hydrolysis of ester bonds. Further hydrolysis of such bonds could however weaken the condition of the paints dramatically and might move paints from category B to A.

Comparing Riopelle's Composition (1956) and Peinture (1954) by Soulages to Composition (1952)

Two other paintings *Composition* (1956) by Riopelle and one sample from *Peinture* (1954) by Pierre Soulages in the Henie Onstad Art Centre collection were also available for study. Both paintings show signs of soft paint.

Riopelle did not make his own paint anymore in 1956 according to written sources. Also the canvas in *Composition* (1956) has a weak outline with the same oval as the manufacture stamp from Lefebvre-Foinet. *Peinture* (1954) by Soulages also had traces of a similar oval stamp on the back of the canvas.

Examination of the paintings shows that the paint has some of the same softness, even if they visually do not look so similar. *Composition* (1956) has stronger impasto

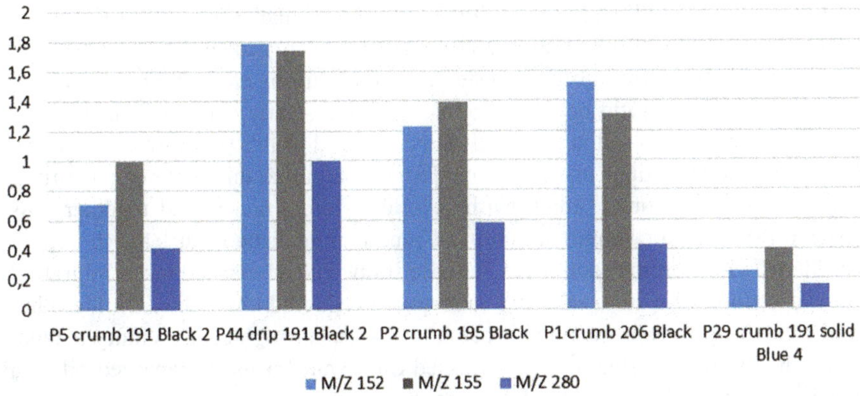

Fig. 5 Discriminating DTMS features of Black paints and a drip from Black paint sampled from *Composition* (1952) and *Composition* (1956) by Riopelle and *Peinture* (1954) by Pierre Soulages. Solid paint Blue 4 from *Composition* (1952) is plotted for comparison

and less layers, and is in general in better condition. *Peinture* (1954) by Soulages is thinly painted and suffers from the 'characteristic' lifting of paint layers (Helou-de La Grandier et al. 2008).

Samples of black paint from these three paintings were analysed with DTMS and compared with sample P29 from healthy paint Blue 4 from *Composition* (1952) (see Fig. 5). The values for azelaic acid (m/z 152) and the oxidised stearic acids (m/z 155) shown in blue in the graphs are higher in the samples of soft paint, and even higher in the drip from a bone black paint layer. These paintings have very challenging condition problems and we believe that these problems are related to the softness of the paint. It is interesting to find that the black paint in *Peinture* (1954) is soft even if it was painted relatively thinly compared with some of the Riopelle paintings. A heavily dripping painting in *Black on Red* from 1960 painted by Soulages and presently in the Art Institute of Chicago, is under study but show the same type of polar fractions exuding from bone black paint (Boon and Lister 2014). Some comparative data are available from 5 year-old dried linseed stand oil and boiled oils (Talens, NL) spread on glass slides and exposed to daylight. Both prepolymerised oils show the m/z 280 feature and higher relative intensities of m/155 and 152 in their DTMS spectra (see Fig. 4). These numbers suggest that such prepolymerised oils may have been added to modify the rheology of paints used by Soulages and Riopelle.

Discussion

One of the first questions of a conservator facing softer paint in an artwork asks is: what type of oil could have been used? A paint layer without good film formation could point to slow- or non-drying oils. Researchers still identify the type of oil used from the palmitic acid: stearic acid (P/S) ratio (Mills and White 1994). However further research into the migration and aging of oil has shown that palmitic acid does evaporate four times faster than stearic acid (Schilling et al. 1999). It has been suggested to use the ratio between azelaic acid and stearic acid (A/S) instead, as the difference in evaporation is at half rate compared to P/S (Ferreira et al. 2005). For unvarnished paintings made with modern oil paint that might contain additives like metal stearates or are made with several different oils mixed together, this means a P/S ratio for identification of oil type is unrealistic or at least should be considered with extreme caution. In our case study with so many signs of migration and evaporation of medium the P/S ratio was considered unreliable. On the other hand when Riopelle was making his paint, he used a late eighteenth century manual so it is likely that he tried to obtain linseed oil. Sennelier used grapeseed oil as a substitute to linseed oil, which tells us that the brand usually used linseed oil. So it is probable that the paint he used in 1952 contained linseed oil.

Indications for very long chain fatty acids found in some of the samples might give a hint to semi-drying oils. Izzo (2011) reports C20-C24 fatty acids in Lefranc Bourgeois paint, which she ascribes to additions of sunflower oil. Such fatty acids are also prominent in rapeseed oil produced in large quantity after the Second World War (Shahidi 1990). It is well known that semi-drying oils never dry and form tacky surfaces. So we can't rule out that Riopelle's paints do contain some semidrying oils or have been enriched with such oils by the artist. Monique Sennelier describes that her father did have difficulty getting enough linseed oil in the years following World War II. Could it be that some suppliers sold low quality "linseed" oil blends?

Whatever the type of oil that has been used it seems very likely that the softness of some colours are a result of a combination of factors, resulting in a variety of symptoms. Returning to matt Blue 3 one of the colours in category B, which was sampled from an area showing weakness, cracking, matt-ness and is relatively thinly applied. At first it was suspected that the binding medium was not the problem in this case. However, after DTMS analysis the levels of organic moieties show the same composition profile as samples of soft paint category B. There is nothing in the application of this paint that would indicate oxygen deprivation during the drying. But both suspected pigments, cobalt blue and ultramarine, are mentioned as poor film formers (Mecklenburg 2012). It is identified mainly cobalt and aluminium content in the sample from preliminary analysis, but this needs to be further investigated.

Another sample believed to be solid and matt was a layer in Black 2 paint. The sample was taken because the area was believed to be an example of paint in good condition, as it did not respond to being prodded by a bamboo stick with light pressure. This sample however shows the same m/z ratios as other soft paints in the DTMS analysis (see Fig. 5). Black 2 contains a lot of bone black pigment. Ferreia et al. (2005) posed the question if their bone black pigmented paint had some kind of antioxidant properties as it was still showing some unsaturation in the film after 350 years. It is uncertain if it is relevant in our case study. Many of the paints however, show some preservation of oleic acid or an isomer (m/z 264 from C18:1).

One factor that comes forward strongly is that paints that drip have less of the kind of pigments that could react to compensate the acidity of the aging oil by creating metal-ionic bridges with the organic moieties in the film to help stabilise the paint when compared to the pigments in the solid paint category C. Lead, zinc, copper and even calcium can help to stabilise the aging film (Keune and Boon 2007; Keune et al. 2008). Lack of this anchoring stabilisation leads to accumulation of polar fractions and results in softening and eventually dripping of the paint. It is remarkable in this respect that soft and dripping paints from different paintings show the same kind of mass spectrometric features.

Maybe the total thickness of the structure could be of importance? According to the CCI-team Riopelle normally did not use either a second ground or a background colour (Corbeil et al. 2011). *Composition* (1952) has two grounds as well as a red background colour. This background colour can be seen on the edges but also in several of the areas of loss further into the motive. Could a combination of the materials in the colours together with the total thickness of the structures have affected both the initial drying process? We are considering that superposition of paints with different PVCs might have adverse effects on drying leaving some of the in between layers that Riopelle applied with paints supersaturated with oil. This is an underestimated aspect of superposition of multiple fresh paint layers.

The findings from our SEM-EDX analysis so far correspond to results published in 2012 and 2013 on how pigments effect the development of film formation in drying oils (Mecklenburg 2012; Mecklenburg et al. 2013). The paints that display the softest paint and dripping in *Composition* (1952) are mainly organic pigments and pigments, like cobalt blue or bone black. They are more or less without the presence of white pigments. It is therefore probable that these pigments are playing a big part in determining the variations between the different colours and their difference in condition.

We can not be certain about what oil has been used as we do not have any analytical technique at present that could provide a certain identification in twentieth century paintings. Analysis with DTMS shows that the soft paints have a higher content of diacids and polar fatty acids compared to hard paint. The drips have an even higher content of polar material. This could be related to the presence of semi drying oils as their triglycerides are too saturated to be accommodated in the rapidly cross-linking material from drying oils. These fractions could in the course of time lead to unbound polar fraction inside the paint mass. Our hypothesis is that the polar oxidation products can not be accommodated in the paint mass and appear at the surface as the paint becomes more apolar with time and possible polar micelles grow. In other paintings (Boon and Hoogland 2014) thickness plays a role but mainly as reservoir while on thinner paints tackiness is more apparent. In some paintings, large amounts of medium also appear at the reverse where it accumulates in the canvas. Such relationships are hard to quantify. However, the exuding materials are highly fluorescent and this is a means for the conservator to monitor accumulation of non-bound medium fraction in paintings. It is also important that this fluorescence can be misinterpret as a local varnish. UV fluorescence of the exuding medium is rapidly apparent when dissolved medium

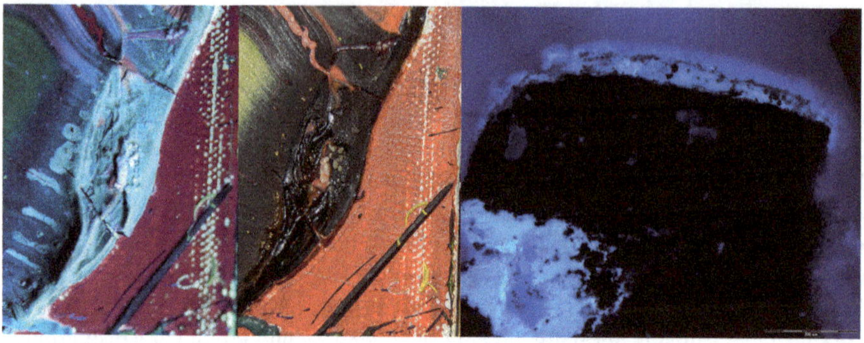

Fig. 6 Partially delaminating black paint with oozing exudate at the surface from Riopelle's *Composition* (1952) seen in UV-light on the *left*, the same area in visual light in the *middle*. To the *right* a detail from a cross section from a crack in the same area in UV-light. The oozing material has a *greenish grey* fluorescence when examined on the painting

is examined under UV light. Figure 6 makes this clear on the painting where an area with black paint is wet with exudate. Under UV light that same area shows a greenish grey fluorescence.

Experimental Details

A Hirox 8700 on a MOPAS 1 stand system was used for digital microscopy. XRF was performed with a Thermo Scientific handheld XRF model Niton XL3t. A Philips-FYE XL-30 FSEM coupled to an EDAX energy dispersive analyser was used for SEM-EDX. Paint crumbs were mounted on carbon sticky tape and covered with a few nm of gold in vacuo using a sputter coating system. A JEOL SX102-102A mass spectrometer was used for DTMS equipped with a Pt/Rh (9/1) filament (diam. 100 μm) for in-source MS. Spectra were obtained at 16 eV electron ionisation over a mass range of 1,000 Da at a cycle time of 1 s and acceleration voltage of 8 kV.

References

Boon JJ (1992) Analytical pyrolysis mass spectrometry: new vistas opened by temperature-resolved in-source PYMS. Int J Mass Spectrom Ion Process 118/119:755–787

Boon JJ (2006) Processes *inside* paintings that affect the picture: chemical changes *at*, *near* and *underneath* the paint surface In: Reporting highlights of the De Mayerne Programme: research programme on molecular studies in conservation and technical studies in art history/ed. Boon JJ, Ester SB Ferreira. – The Hague: NWO, 2006, pp 21–31 (De Mayerne [MolArt])

Boon JJ, Hoogland FG (2014) Investigating the possible mechanism of fluidizing dripping pink paint on Van Hemert's Seven-series works from the period 1990–1995. These proceedings, Chapter 20

Boon JJ, Lister K (2014) Investigating aged oil-medium phase separating and exuding as drips from a 1960s painting by Pierre Soulages. In: Bridgland J (ed) ICOM-CC 17th Triennial conference Preprints, Melbourne, 15–19 September 2014. International Council of Museums, Paris, art. 0901, p 7

Boon JJ, Sophie P, van den Brink Oscar F, Duursma Marc, Rainford David (1997) Molecular aspects of mobile and stationary phases in ageing tempera and oil paint films, In: Early Italian paintings: techniques and analysis, symposium, Maastricht, 9–10 Oct 1996/Bakkenist T, Hoppenbrouwers R, Dubois H (eds). Maastricht: Limburg Conservation Institute, 1997, pp 35–56

Boon JJ, Hoogland FG, van der Horst J (2007) Mass spectrometry of modern paints. In: Learner TJS (ed) Modern paints uncovered. Getty Conservation Institute, Los Angeles, pp 85–95

Bronken IAB (2010) Undersøkelse og dokumentasjon av Composition 1952 av Jean-Paul Riopelle. Dissertation, University of Oslo, Norway

Corbeil M-C (2004) Considerations regarding Riopelle's pictorial technique and conservation. In: Jean Paul Riopelle Catalogue raisonné 1954 – 1959 Tome 2. Montreal, Hibou Éditeurs, pp 19–25

Corbeil M-C (2006) Une étude scientifique de la technique pictural de Jean Paul Riopelle. Techne no 24, pp 47–52

Corbeil M-C, Helwig K, Poulin J (2004) Analysis of the painted oeuvre of Jean-Paul Riopelle: from oil to mixed media. In: Roy A, Smith P (eds) Modern art, new museums contributions to the Bilbao Congress, 1317 Sept 2004. The International Institute for Conservation of Historic and Artistic Works, London, pp 170–173

Corbeil M-C, Helwig K, Poulin J (2011) Jean Paul Riopelle: the artist's materials. Getty Publications, Los Angeles

Érouart G (1995) Riopelle in conversation. House of Anansi Press Limited, Ontario

Ferreira, ESB, van der Horst, J, Boon JJ (2005) Chemical aspects of the binding media of the Oranjezaal ensemble: an insight into 17th century Netherlandish materials and methods. In: Triennial meeting (14th), The Hague, 12–16 Sept 2005: preprints. London, James & James, pp 774–782

Helou-de la Grandière P, Hô L, Anne-Solenn, Mirambet F (2008) Delaminating paint films at the end of 1950s: a case study on Pierre Soulages. In, Preparation for Painting. The artist's choice and its Consequences. Archetype publications, London, pp 156–162

Izzo FC (2011) 20th century artists' oil paints: a chemical-physical survey. PhD dissertation in Chemical Sciences, Ca' Foscari University of Venice, 2011

Katrien K, Hoogland F, Peggie D, Higgitt C, Boon J (2008) Comparative study of the effect of traditional pigments on artificially aged oil paint systems using complementary analytical techniques. In: 15th Triennial Conference, New Delhi, 22–26 Sept 2008: Preprints/ed. Janet Bridgland. Allied Publishers Pvt. Ltd., New Delhi, 2008, vol II, pp 833–842

Keune K, Boon JJ (2007) Analytical imaging studies of cross sections of paintings affected by lead soap aggregate formation. Stud Conserv 52:161–176

Leclerc D (1992) The crisis of abstraction in Canada: the 1950s; with a technical essay by Barclay MH

Mecklenburg MF et al (2012) Technical assessment of cultural objects in the planning of transport. In: Bronken I, Braovac S, Olstad T, Ørnhøi A (eds) Moving collections. Processes and consequences. Archetype Publications, London, pp 1–8

Mecklenburg MF, Tumosa CS, Vicenzi EP (2013) The influence of pigment and ion migration on the durability of drying oil and alkyd paints. In: Marion F, Mecklenburg A, Elena C, Koestler RJ (eds) New insight into the cleaning of paintings. Proceedings from the cleaning 2010 international conference Universidad Politécnica de Valencia and Museum Conservation Institute. Smithsonian Institution Scholarly Press, Washington, DC, pp 59–67

Mills JS, White R (1994) The organic chemistry of museum objects. Butterworth-Heinemann, Oxford

Osman MA, Suter UW (2002) Surface treatment of calcite with fatty acids. Surface properties of the organic monolayers. Chem Mat 14:4408–4415

Schilling MR, Carson DM, Khanjian HP (1999) Gas chromatographic determination of the fatty acid and glycerol content of lipids. IV. Evaporation of fatty acids and the formation of ghost images by framed oil paintings. In, Triennial meeting (12th), Lyon, 29 Aug–3 Sept 1999: preprints, vol 1. James & James, London, pp 242–247

Shahidi F (1990) Canola and rapeseed oil: production, chemistry, nutritional and processing technology. Van Nostrant Rheinhold Publications, New York

Spallding J (2002) Immaculate misconception. In: Riopelle, Connasissance des arts, special issue no 179. Montreal Museum of Fine Arts, Montreal, pp 9–15

Tumosa CS, Mecklenburg MF (2013) Oil paints: the chemistry of drying oils and the potential for solvent disruption. In: Marion F, Mecklenburg A, Elena C, Koestler RJ (eds) New insight into the cleaning of paintings. Proceedings from the cleaning 2010 international conference Universidad Politécnica de Valencia and Museum Conservation Institute. Smithsonian Institution Scholarly Press, Washington, DC, pp 52–58

van den Berg, JDJ (2002) Analytical chemical studies on traditional linseed oil paints. PhD thesis, University of Amsterdam, Amsterdam

Viau, R (2002) Jamais deux fois le même tableaux. In: Vie des arts 187, vol 53, pp 48–49

Zinc White and the Influence of Paint Composition for Stability in Oil Based Media

Gillian Osmond

Abstract Zinc oxide is a prevalent pigment in twentieth century oil based paints. Its reactivity with fatty acids has consequences for paint film properties and stability. Resulting zinc carboxylates are implicated in a variety of deterioration phenomena affecting late nineteenth and twentieth century paintings. Naturally aged artists' oil paints and reference paint film draw downs are investigated using scanning electron microscopy, and conventional attenuated total reflectance and synchrotron transmission Fourier transform infrared spectroscopy. This study reports results for soybean oil paints and Naples yellow hue pigment mixtures with zinc oxide in addition to previously reported findings for linseed and safflower oil based paints involving zinc oxide alone and in combination with lead and titanium whites. A solvent model examining interactions between zinc oxide and stearic acid or aluminium stearate supports observations made in paint films. The significance of aluminium stearate as a stearic acid source in paint formulations is corroborated and is more significant than oil type or pigment combination for formation of high concentrations and more pronounced separation of saturated C16 and C18 chain zinc carboxylates.

Keywords Zinc oxide • Zinc stearate • Aluminium stearate • Zinc soap • Zinc carboxylate • FTIR • Synchrotron • Oil paint

Introduction

Zinc oxide became a frequent component of modern oil based paints following developments in industrial scale production during the nineteenth century. Modern zinc white history and pigment properties have recently been reviewed (Osmond 2012). Zinc oxide was particularly valued in the housepaint industry for its good

---

G. Osmond (✉)
Queensland Art Gallery/Gallery of Modern Art, Brisbane, QLD 4101, Australia

Australian Institute for Bioengineering and Nanotechnology, University of Queensland, Brisbane, QLD, Australia
e-mail: gillian.osmond@qagoma.qld.gov.au

K.J. van den Berg et al. (eds.), *Issues in Contemporary Oil Paint*,
DOI 10.1007/978-3-319-10100-2_18

suspension properties in oil. Although some nineteenth century artists specifically sought zinc white for its resistance to yellowing and blackening, artists generally were less enthusiastic about zinc white oil paint. However, zinc oxide was progressively incorporated into many artists' paints and in both the decorator and artists' paint markets it has typically been used to modify the properties of paints based on other white pigments or colours. Straight zinc white oil paints are less common, although the oil based Ripolin® gloss range much favoured by artists was substantially based on zinc oxide (Dredge et al. 2013; Gautier et al. 2009). Research indicates that even as a minor component zinc oxide strongly influences mechanical responses and water absorbency of paints (Browne 1956; Mecklenburg 2008).

Zinc oxide's effect on paint film properties makes its prevalence in twentieth century paints significant. An emerging body of evidence links zinc oxide to a variety of deterioration phenomena including formation of lumps based on zinc soaps of stearic and palmitic acids, and brittle paint films subject to cracking or cleavage between layers. The reactivity of zinc oxide with fatty acids in oil based paints to form zinc carboxylates appears implicated in each case.

The factors which determine whether zinc carboxylates aggregate to form lumps, migrate to interfaces, or remain dispersed throughout a paint film are the subject of ongoing research. Potential influences on zinc oxide reactivity have recently been reviewed (Osmond 2012). The zinc oxide particle is one variable, with different grades representing a range of particle and crystal properties. Zinc oxide is electrostatically unstable because it has two polar facets at opposing ends of the crystal, one terminating in zinc atoms and one with oxygen (Wöll 2007). Particle size and shape may emphasise the influence of one facet over another, and lattice variations will also influence pigment properties, particularly interactions with light with implications for fluorescence characteristics and photostability (Winter and Whittem 1950). Production method may influence properties; French (Indirect) Process is typically of higher purity, while American (Direct) process pigment is considered less reactive for paint applications because of its larger particle size and stoichiometry (Morley-Smith 1950; International Zinc Association 2009).

Other significant variables include availability and profile of fatty acids in the paint. Mid twentieth century paint producers focussed on reducing acid level and proportion of saturated fatty acids in oils (Elm 1957; Jacobsen and Gardner 1941). Various additives were found to influence solubility and dispersion of zinc soaps in prepared paints, with implications for shelf life (Ware and Christman 1916). More recently in studies of paintings, the significance of co-pigmentation has been raised with many incidences of deterioration being observed in the presence of other metal ions (Osmond 2012).

Isolating the most critical factors for paint stability is difficult given the complexity of typical formulations. Analysis and interpretation of samples from paintings is further complicated by external variables arising from artist technique, painting age, environmental history and conservation interventions, each with potential to influence paint film chemistry. The study of reference paint films and models offers the opportunity to examine specific aspects of paint ageing with fewer variables than unique works of art.

This paper involves a detailed study of paint films which differ only with respect to pigment composition in association with zinc oxide, oil type, the presence of specific additives, commercial brand or casting date. It expands on previously reported results indicating aluminium stearate in paint strongly influences formation and distribution of zinc carboxylates (Osmond et al. 2012). Interactions between pigment and fatty acid source alone are further examined in model systems.

Experimental

Reference Paints

Paints containing zinc oxide (ZnO) were sourced from the Smithsonian's Museum Conservation Institute reference collection. Paints were cast on Melinex substrates between 1978 and 1999 and have aged naturally under stable interior ambient conditions. Cast films include a range of commercially available artists' paints and custom produced controls prepared using specified pigment and oil combinations. Samples discussed include artists' oil paints from Grumbacher, Winsor & Newton and Gamblin, and custom produced control paints containing ZnO alone or in combination with lead white, titanium white or within Naples yellow hue formulations. A variety of linseed oils, safflower and soybean oils, and preparations with and without aluminium stearate are represented. Optical microscopy of paint surfaces was followed by acquisition of spectra from exposed top and protected bottom surfaces using attenuated total reflectance – Fourier transform infrared spectroscopy (ATR-FTIR). Paint cross sections were prepared and examined using optical and scanning electron microscopies with elemental analysis (OM and SEM-EDX) while synchrotron-FTIR (SR-μFTIR) of thin paint cross sections enabled mapping of metal carboxylate distributions at high spatial resolution. Experimental conditions have been detailed previously (Osmond et al. 2012).

Model System

Model experiments were designed to support observations made in paint films by investigating specific component interactions under controlled conditions. A solvent model was chosen and toluene selected on the basis of trials and observed in-vial behaviour in conjunction with published solubility parameters and graphic models charting oil paint film solubility according to swelling response to specific solvents. The solvent was intended to model a 'typical' paint film based on an observed affinity implying comparable intermolecular forces. Toluene has been categorised as high-moderate swelling for a representative paint film (10 year old lead white and yellow ochre in linseed oil) (Phenix 2002). Sensitivity of the representative paint film to toluene was consistent between light- and dark-aged paint samples, indicating its applicability as a model has validity for paints at varying states of oxidation.

Zinc oxide pigment samples were obtained from Kremer Pigmente (Germany), also the source of ZnO used in formulating control paints in the Smithsonian reference collection. Tight particle size distribution was important to minimise size as a variable, so ZnO grades not specifically produced for the pigment market were additionally obtained. Sample compositions were confirmed with X-ray diffraction and particle characteristics investigated using transmission electron microscopy (TEM) and particle sizing techniques. The International Zinc Association (2009) advocates ZnO with specific surface area 5 m^2/g as the preferred 'standard' for paint, corresponding to maximum opacity. Potential model candidates with comparable average particle size were sought, however only nano grades had sufficiently narrow size distributions. 30 nm ZnO, MKN-ZnO-030 (M K Impex Corp., Canada) was selected. Both Kremer and MKN samples comprise ZnO of wurtzite structure. TEM shows Kremer pigment has highly variable size and shape including acicular particles (Fig. 1). MKN particles are nodular and regularly sized with diameters consistently around 30 nm; particle shape is similar to an industrial paint grade pigment, Umicore White seal. Kremer ZnO has specific surface area 5.5–6.5 m^2/g (specified) compared with 16.6 m^2/g for MKN (measured). The larger surface area of 30 nm ZnO will enhance surface activity and reduce time required for reactions of interest to proceed. To ensure comparable interactions occur with paint grade ZnO, selected experiments were conducted simultaneously using Kremer pigment.

Reagent grade stearic acid $\geq$95 % (octadecanoic acid) was obtained from Sigma Aldrich and its composition confirmed with FTIR.

Aluminium stearate was synthesised from sodium stearate and aqueous solution of aluminium sulfate (3:1 stoichiometry) using the modified double decomposition technique described by Corkery (2004). Transmission FTIR confirmed the presence of the intended soap, the absence of reactants and the extent of any free acid. Retention of some stearic acid in the synthesised soap was considered acceptable as it reflects industrial grades of aluminium stearate typically used in paint production.

Zinc oxide and stearic acid or aluminium stearate were combined with toluene in capped glass vials with Teflon liners. Reactants were combined in two ratios representing an excess of stearic acid (HSt) or ZnO assuming

$$ZnO + 2HSt \rightarrow ZnSt_2 + H_2O \quad (1)$$

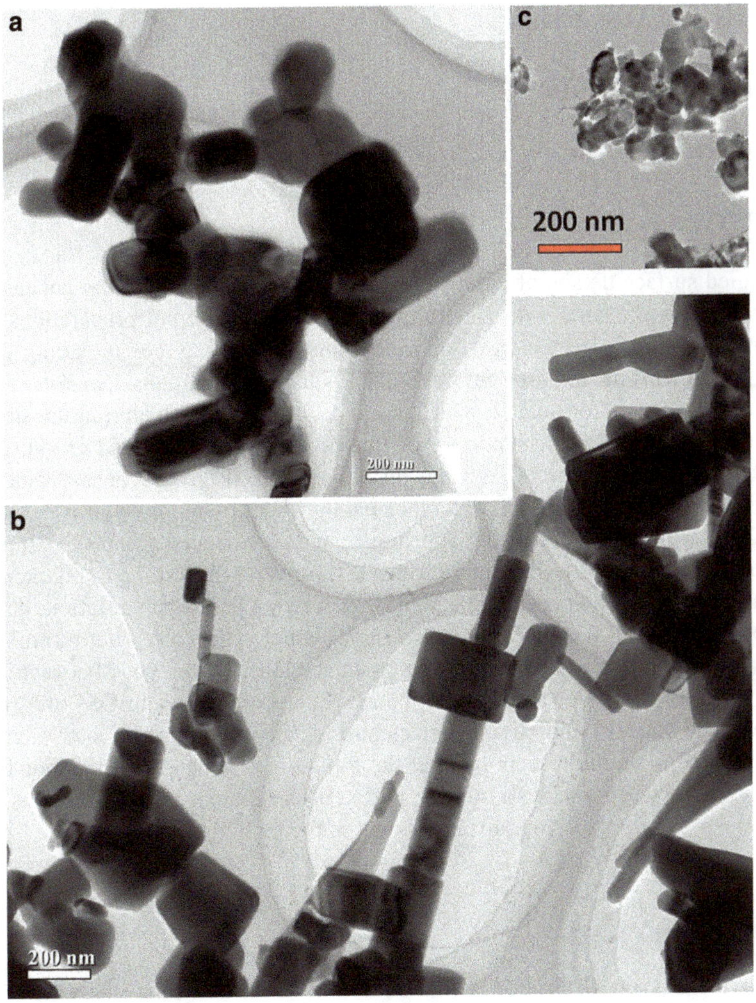

Fig. 1 TEM images of zinc oxide shown at equivalent magnification (**a**) Umicore 'White seal' (**b**) Kremer Pigmente 46300 (**c**) MKN-ZnO-030 (Images: Yanan Guo and John Drennan)

Zinc oxide is essentially insoluble in toluene, while stearic acid solubility was calculated as 42 g/L. In each vial the amount of stearic acid was at or below the solubility limit for the measure of toluene. Vials were manually shaken then left to stand. After 5 min, and subsequent intervals extending from minutes to hours and then days and weeks, each solution was shaken and a drop pipetted from the vial onto a glass slide to dry for in situ OM and FT-Raman, and sampling of solids for transmission FTIR.

Results and Discussion

Reference Paint Films

General Observations

Microscopy

Magnified surface details of paint films reveal differences in physical appearance. Control paints prepared with ZnO and linseed oil of different preparations have features reflecting expected variation in rheology and drying rate, including strong retention of casting features in linseed + litharge formulations, consistent with rapid drying. Paints formulated with ZnO in different oils with aluminium stearate demonstrate the yellowing tendency of linseed oil, while fatty acid efflorescence is notable on the surface of soybean oil based paint (Fig. 2). These distinctions are not apparent in comparable reference paints where titanium white is combined with ZnO, which have consistent appearance and pronounced granular texture. Of the commercially produced paints, surface texture varies according to pigment mix within Grumbacher paints and later Winsor & Newton paints but variations are less pronounced in 1978 Winsor & Newton samples which reflect an earlier formulation. More information is obtained through cross sections. Differences between 1978 and 1980 Winsor & Newton titanium paints are apparent both in OM (ultraviolet fluorescence – UVF) and backscatter electron (BSE) images. The later paint has a more perceptible fluorescent band at the top surface which has correspondingly lower electron density in BSE images. Differences apparent in cross sections from the range of Naples yellow paints are shown in Fig. 3.

FTIR

Zinc oxide has no absorption signal in the mid infrared range between 4,000 and 600 cm^{-1}, but oil based paints containing ZnO are among the most likely to develop strong carboxylate absorption (van der Weerd et al. 2005), reflecting formation of zinc soaps. Carboxylic acids present or formed produce a shoulder at wavenumber ca. 1,705 cm^{-1} to the ν C═O glycerol ester absorption (ca. 1,740 cm^{-1}), accompanied by broad O—H stretch centred 3,400 cm^{-1} and peaks at 1,415 cm^{-1} and 915 cm^{-1}. Soap formation is characterised by disappearance of the O—H stretch and replacement of ν C═O and C—O (1,300 cm^{-1}) bands with COO^- asymmetric and symmetric stretching at ca. 1,550 cm^{-1} and 1,400 cm^{-1}, reflecting the ionised structure of the COO-metal bond (Mehrotra et al. 1995). Broad carboxylate absorption reflects coordination of zinc ions to a range of carboxylate moieties. Individually, zinc stearate and palmitate have the same characteristic sharp peaks at 1,540 cm^{-1} and 1,398 cm^{-1}, and CH_2 bend at 1,465 cm^{-1} (Robinet and Corbeil 2003). Aluminium stearate has characteristic absorptions at 1,588 and 1,469 cm^{-1}.

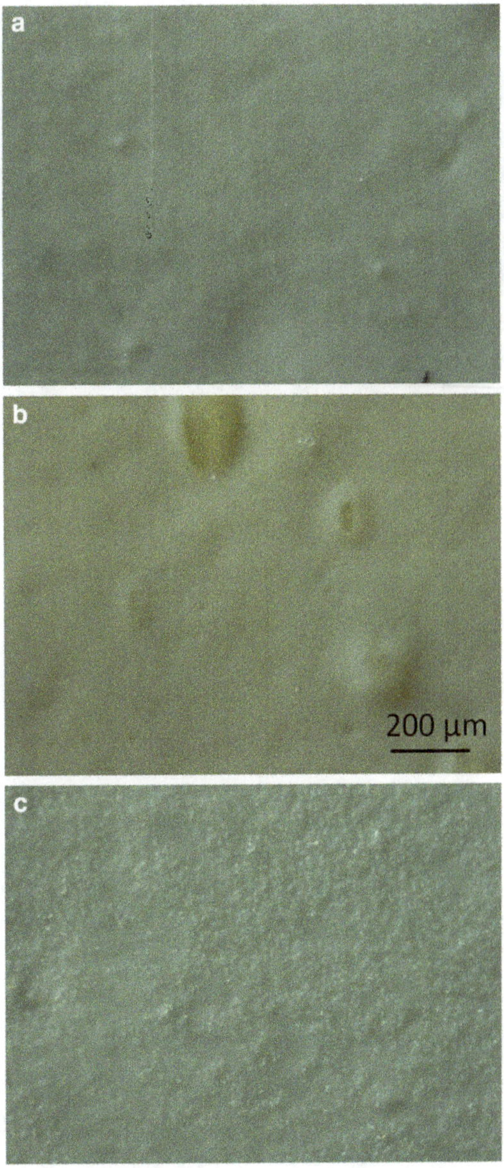

Fig. 2 Macro surface details of zinc white control paints with aluminium stearate cast 1998: (**a**) safflower, (**b**) linseed and (**c**) soybean oils

SR- μFTIR enables chemical maps to be produced by integrating absorbance intensities corresponding to specific vibrational frequencies and characteristic features of spectra represented in the samples. Synchrotron source maps are of higher spatial and spectral resolution than those obtained by conventional FTIR mapping techniques. Zinc oxide pigment is of fine particle size (typically <1 μm diameter, or 1–5 μm aggregated effective particle size), so high spatial resolution

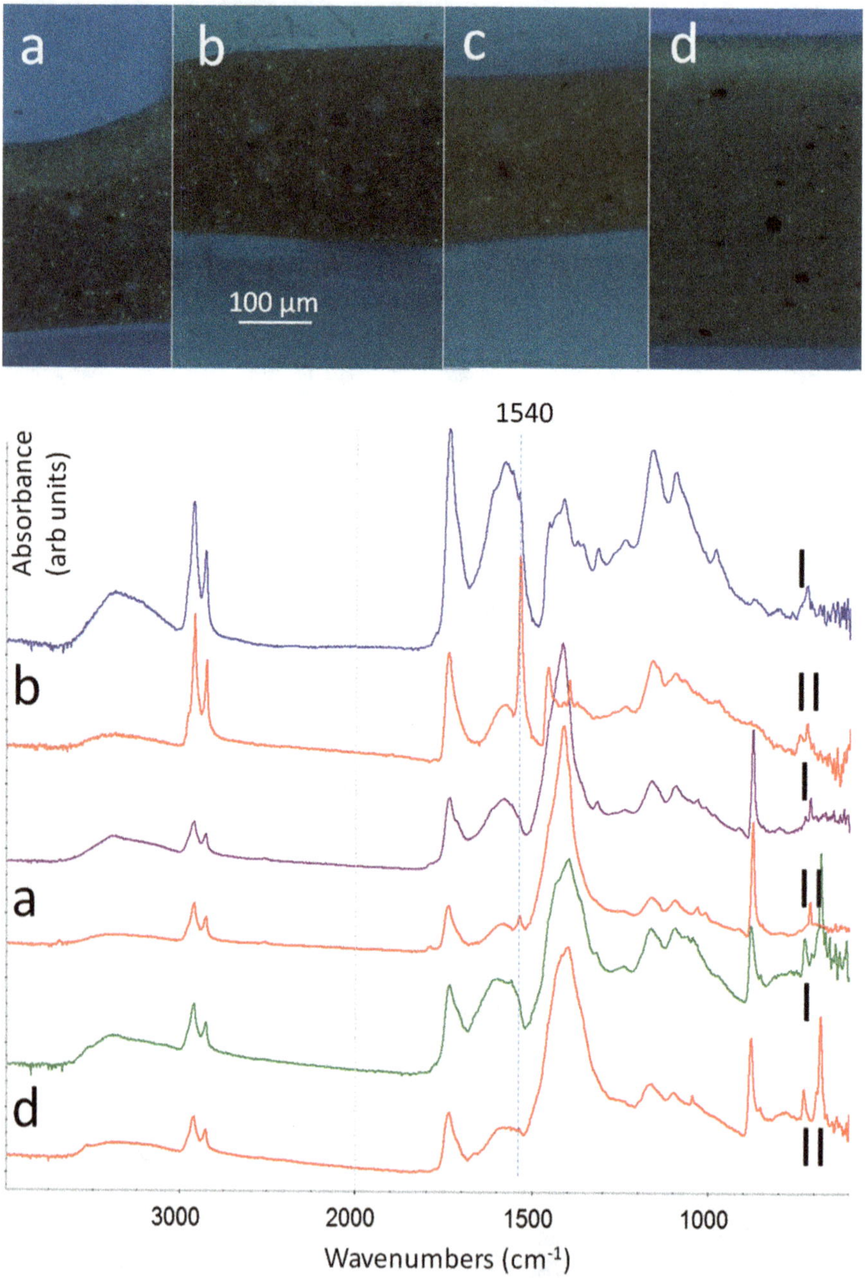

Fig. 3 UVF cross section images and ATR-FTIR spectral pairs from exposed surface (**I**) and protected underside spectra (**II**) of Naples yellow hue paints: (**a**) Gamblin 1999, (**b**) Grumbacher 1999, (**c**) Winsor & Newton 1999 and (**d**) Winsor & Newton 1978. Characteristic cross section fluorescence suggests more zinc oxide is present in Gamblin and Grumbacher paints. Differences in IR absorption between surfaces do not correlate with fluorescent surface bands. ν_a COO^- zinc stearate/palmitate (1,540 cm^{-1}) features strongly in the Grumbacher underside spectrum which contains aluminium stearate

is critical to characterisation of the chemical signature of reactive sites at an early stage and allows correlation with high resolution backscatter electron images and elemental maps.

Across the range of ZnO-containing paints, ATR-FTIR measurements of exposed surfaces show varying intensity of carboxylate absorption but a broad peak 1,560–1,620 cm^{-1} is consistent irrespective of commercial origin, pigment combination, oil type, film age or presence of additives. The presence of specific zinc soaps is not indicated other than a small side peak at 1,540 cm^{-1} in spectra from soybean oil based paint.

Previously reported results for white paints based on linseed and safflower oils showed a significant number of samples recorded unexpected differences in carboxylate profiles between exposed top and protected underside surfaces (Osmond et al. 2012). Upper surfaces invariably absorbed broadly while the underside of paints incorporating aluminium stearate recorded a defined peak at 1,540 cm^{-1}, surprisingly not indicative of aluminium stearate but rather zinc stearate. In cross sections, SR-μFTIR integrated absorption intensity maps confirmed a preferential accumulation of zinc stearate in the lower margins of paints. Characteristic peaks for aluminium stearate were less frequently discernible over more general broad carboxylate absorption, but integration of the narrow band 1,585–1,595 cm^{-1} (ν_a COO^- aluminium stearate) produced maps showing a correlation with high intensity areas in maps of broader range, typically dispersed 5–10 μm diameter masses.

Paints without aluminium stearate recorded more subtle differences between surfaces. The presence of aluminium stearate influenced zinc carboxylate distribution more strongly than pigment composition or oil type. Here, results are presented for an expanded range of ZnO containing paints including whites produced with soybean oil and several commercial Naples Yellow hue formulations with varied and more complex pigmentation than paints previously discussed. Previously reported paints and the current selection are detailed in Table 1.

Naples Yellow Hue: Linseed and Safflower Oil Based Paints

Winsor & Newton Naples yellow hue paints from 1978 to 1999 include lead white, ZnO and iron oxides, the older film additionally incorporates dolomite and neither contains aluminium stearate. FTIR spectra obtained from top and underside surfaces are similar although the generally broad carboxylate soap peak and oil carbonyl and triglyceride ester peaks appear smaller in underside spectra with a small side peak at 1,540 cm^{-1}. Both paints are dominated by lead carbonate and ATR-FTIR reveals little difference in carboxylate absorption irrespective of the presence of dolomite; spectra are comparable to those obtained from similarly dated Winsor & Newton lead white paints. Spectra are also similar for Gamblin Naples yellow hue which includes ZnO and iron oxides but is otherwise formulated differently with cadmium sulphide, chalk and no lead carbonate (Fig. 3). UVF cross section images (supported by SEM-EDX) suggest Gamblin paint incorporates more ZnO than

Table 1 Reference paints with composition as specified by manufacturers with additional pigmentation determined using optical microscopy, FTIR and SEM-EDX (italicised)

Paint manufacturer/product name/date of film casting	Composition: Pigment	Composition: Oil, additive	Underside FTIR 1,540 cm^{-1} peak
Naples Yellow hue			
Gamblin Naples yellow hue 1999	Cadmium sulphide, natural hydrated iron oxide, red iron oxide, zinc oxide; *aluminium silicates, chalk*	Linseed, alkali refined	Small
Grumbacher Naples yellow hue 1999	Zinc oxide, calcined natural iron oxide, cadmium selenium sulphide, cadmium zinc sulphide co-precipitated with barium sulfate (6:1)*	Linseed alkali refined + aluminium stearate	Strong
Winsor and Newton Naples yellow (London oil colour series 2) 1978	Basic lead carbonate, cadmium sulphide, iron oxide; *zinc oxide, dolomite*	Linseed	Negligible
Winsor and Newton Naples yellow 1999	*Basic lead carbonate, iron oxide, zinc oxide, chromium titanium yellow*	Not specified (linseed assumed)	Small
Titanium whites soybean oil based			
Speedball *Titanium white* 1981	Titanium dioxide, zinc oxide	Soybean + *aluminium stearate*	Strong
Custom control *Titanium white* 1998	Titanium dioxide, zinc oxide 12:1, aluminium silicate	Soybean modified + aluminium stearate	Medium
Custom control *Zinc white* 1998	Zinc oxide	Soybean modified + aluminium stearate	Strong
Titanium whites safflower oil based			
Winsor and Newton Titanium white 1978	Titanium dioxide, barium sulphate *(trace)*; *ZnO, dolomite*	Safflower (*+HCO?*)	Medium
Colorlab Titanium white 1981	Titanium dioxide, zinc oxide, barium sulphate 5:1:10	Safflower + aluminium stearate	Strong
Winsor and Newton Titanium dioxide A Permanent white 1980	*Titanium dioxide, zinc oxide, barium sulphate, magnesium carbonate, [Al, S]*	Safflower	Negligible
Winsor and Newton Titanium white 1999	*Titanium dioxide, barium sulphate, magnesium carbonate, zinc oxide (trace)*	Not specified (safflower assumed)	Negligible
Custom control *Titanium white* 1998	Titanium dioxide, zinc oxide 12:1, aluminium silicate	Safflower alkali refined + aluminium stearate	Small

Titanium whites linseed based			
Grumbacher Titanium white 1978	Titanium dioxide, zinc oxide 1:2; *trace aluminium silicate*	Linseed alkali refined + aluminium stearate	Strong
Grumbacher Titanium white 1999	Titanium dioxide, zinc oxide	Linseed alkali refined + *aluminium stearate*	Strong
Custom control *Titanium white* 1998	Titanium dioxide, zinc oxide 12:1, aluminium silicate	Linseed + aluminium stearate	Strong
Colorlab Titanium white 1981	Titanium dioxide, zinc oxide, barium sulfate 5:1:10; *[Al, S, Cl]*	Linseed + aluminium stearate	Strong
Gamblin Titanium white 1999	Titanium dioxide, zinc oxide; *chalk, trace aluminium silicate*	Linseed alkali refined	Very small
Lead whites safflower based			
Winsor and Newton Flake white 1978	Basic lead carbonate; *zinc oxide, barium sulphate, [Al, C]*	Safflower	Small
Winsor and Newton Flake white no.1 ASilver white 1980	Basic lead carbonate; *zinc oxide, aluminium silicate, chalk*	Safflower	Negligible
Winsor and Newton Flake white 1999	*Basic lead carbonate, zinc oxide*	Not specified (safflower assumed)	Small
Lead whites linseed based			
Grumbacher Flake white 1978	Basic lead carbonate, zinc oxide 20:1	Linseed alkali refined + aluminium stearate	Strong
Grumbacher Flake white 1999	Basic lead carbonate, zinc oxide	Linseed alkali refined + *aluminium stearate*	Strong
Zinc whites safflower based			
Winsor and Newton Zinc white 1978	Zinc oxide	Safflower	Small
Custom control *Zinc white* 1998	Zinc oxide	Safflower alkali refined + aluminium stearate	Strong
Zinc whites linseed based			
Grumbacher Zinc white 1978	Zinc oxide	Linseed alkali refined + aluminium stearate	Strong
Custom control *Zinc white* 1998	Zinc oxide	Linseed + aluminium stearate	Strong
Custom control *Zinc white* 1990	Zinc oxide	Linseed boiled	Negligible
Custom control *Zinc white* 1990	Zinc oxide	Linseed cold pressed	Negligible
Custom control *Zinc white* 1990	Zinc oxide	Linseed with litharge	Very small

*Ratios indicate relative proportion by weight of zinc oxide to other pigments combined where known

Winsor & Newton counterparts. Grumbacher paint similarly has a high proportion of ZnO which, with added aluminium stearate results in an intense underside peak at 1,540 cm^{-1} consistent with underside spectra from zinc oxide/aluminium stearate containing white paints.

UV fluorescent, low BSE density masses in Grumbacher paints have been previously shown to comprise aluminium stearate. Similar features in Gamblin paint are calcium based but have not been further characterised. In addition to large masses, calcium is also found in association with micron-sized carbon rich halos surrounding dispersed zinc based particles. Calcium driers function by competing for formation of carboxylate complexes, reducing mobility of fatty acid carboxylates with which they coordinate (Mallégol et al. 2000). Their alkalinity enables formation of stable salts with acid groups (Soucek et al. 2012). Although no drier is present[1], it is curious that calcium is detected in close association with zinc based particles in Gamblin paint with moderate and dispersed general carboxylate absorption despite a high concentration of ZnO.

Winsor & Newton Titanium white (1978) also has small (ca. 2 μm) dispersed low BSE intensity points but without the bright centres of Gamblin paint. These spots are carbon rich with trace contributions from elements generally present in the paint. The anomalous behaviour of this sample in recording strong underside zinc stearate FTIR absorption had been attributed to the likely inclusion of aluminium hydroxide in the formulation (Osmond et al. 2012). However there is less evidence for its presence in the 1978 paint than in later Winsor & Newton formulations which incorporate relatively large (10–20 μm) fluorescent inclusions high in carbon, aluminium and sulphur, in addition to similar magnesium based areas. A feature of the early Winsor & Newton paint not present in later titanium whites from Winsor & Newton or other manufacturers is a high proportion of dolomite. Dolomite has been used by Winsor & Newton in *Winton* Titanium white in conjunction with hydrogenated castor oil (HCO), an alternative rheology modifier to stearates.[2] HCO principally comprises glycerides of 12-hydroxystearic acid and stearic acid (Maskaev et al. 1971) and its presence could explain both the small carbon rich masses and the higher stearic acid concentration available for soap formation in this paint than comparable paints without HCO.

The strongest zinc stearate underside peak occurs in Grumbacher paints where similarities with control formulations with aluminium stearate indicate the additive contributes significantly to availability of stearic acid. Dissolution of ZnO to form zinc stearate is indicated in BSE images by irregular regions of reduced atomic density with less defined pigment particles. These regions are most prevalent in the lower margin of the film.

[1] R. Gamblin, personal communication. Email 21 September 2013.

[2] I. Garrett, personal communication. Email 20 September 2013.

Soybean Oil Paints

Previously reported results indicate reactivity of oil in paint formulations has some relationship to the rapidity and extent of zinc soap formation. Comparably pigmented paints prepared with safflower oil had lower carboxylate absorption than those produced with linseed oil, and the processing of linseed oil also influenced reactivity with ZnO. However, addition of aluminium stearate contributed generically to zinc stearate formation, diminishing the significance of oil type. Unfortunately it is not possible to compare soybean oil paints with and without aluminium stearate as only controls with the additive are represented, and analysis of Speedball Titanium white suggests it is similarly formulated. Control soybean paint surface spectra unusually have a small 1,540 cm^{-1} side peak adjacent to broad carboxylate absorption in addition to a stronger underside peak. Speedball paint has strong surface carboxyl doublets (oil plus acid) and negligible carboxylate absorption, in contrast with underside readings, consistent with high fatty acid concentration. Zinc stearate formation in the soybean + aluminium stearate range appears linked to ZnO concentration with the strongest 1,540 cm^{-1} peak at the protected underside of paint with ZnO as the primary pigment.

SR-μFTIR of three soybean based paints was undertaken to determine spectral variation through film thickness. The zinc and titanium/zinc controls displayed distribution patterns consistent with previously reported results for the same pigment combinations with linseed and safflower oils, with zinc stearate concentrated in the lower margin of films (Fig. 4). Aluminium stearate is localised within small spots although none is detected in the single pigment film; its absence coincides with the highest zinc stearate concentration. Integration of the ester carbonyl peak above 1,720 cm^{-1} (excluding that attributable to fatty acids) indicates levels of intact triglycerides are reduced where zinc stearate dominates.

SR-μFTIR of Speedball (Titanium/ZnO) soybean paint resolved more complex zinc carboxylate distributions including areas with split carboxylate vibrations. In these instances carboxylate peaks are comparable to doublets expected for zinc oleate (Robinet and Corbeil 2003) but other characteristic bands are not visible and spectra are otherwise indistinguishable from those assigned to zinc stearate/palmitate (Fig. 5). Similar spectra have been observed in paintings which GCMS analysis indicates are mixed zinc soaps with zinc palmitate predominant (Helwig et al. 2014). Another possible attribution includes hydrated stearates (Tachibana et al. 1970). Soybean oil has a higher combined percentage of palmitic, stearic and oleic acids than either linseed or safflower oils (typically 38 % versus 28–29 %) (Schilling et al. 2007). Chemical maps show the single and split peak ‘stearates’ broadly occupy the same region although are distinguished as adjacent when individual spectra (5×5 μm apertures) are interrogated. Split peak spectra are associated with stronger carbonyl and ester C—O absorptions through the upper half of the paint film, where hydration is greatest. CH_2 vibrations also occur at higher wavenumbers (2,926, 2,853 cm^{-1}) than single peak spectra (2,918, 2,849 cm^{-1}), consistent with a relatively polar molecular environment (Derrick et al. 1999). The low values applicable to single peak soap structures

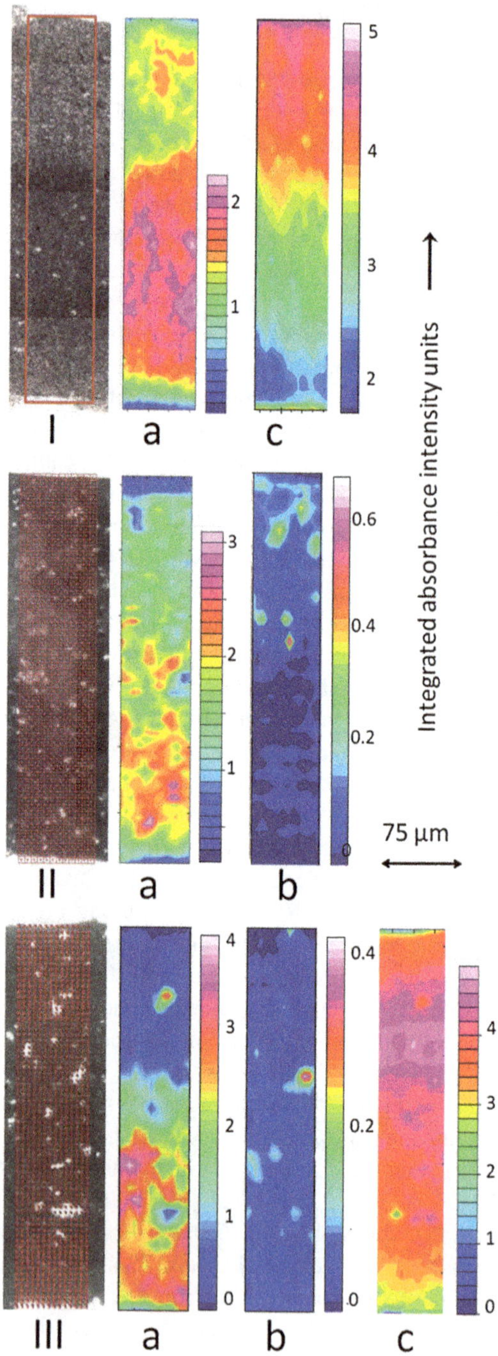

Fig. 4 Transmission SR-μFTIR integrated absorption intensity maps for wavenumber range (**a**) 1,525–1,555, (**b**) 1,580–1,600, (**c**) 1,720–1,760 cm^{-1} for soybean based paints with aluminium stearate (**I**) zinc white control; (**II**) titanium + zinc white and aluminium silicate control; (**III**) Speedball titanium + zinc white. Boxed optical image details of thin sections indicate regions mapped

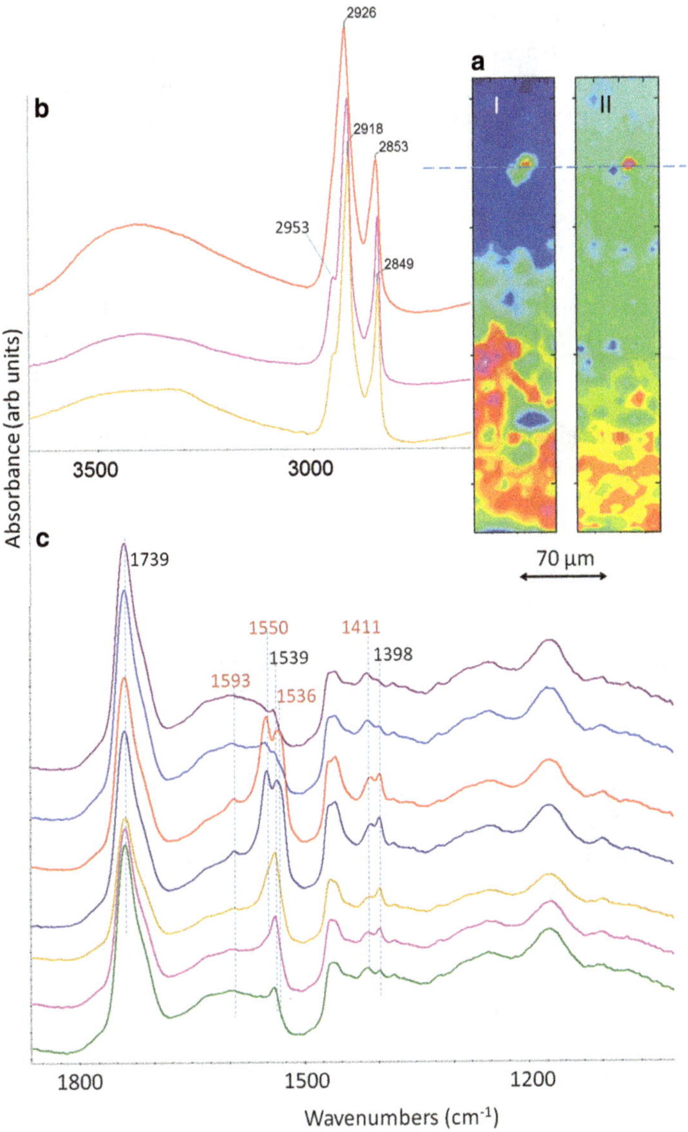

Fig. 5 Speedball titanium + ZnO soybean paint transmission SR-μFTIR (**a**) integrated absorption intensity maps for wavenumber range (**I**) 1,525–1,560 and (**II**) 2,947–2,960 cm^{-1}; (**b–c**) spectra extracted from a *line* though the hotspot indicated *centre top*. Hotspots in map I include spectra with both singlet v_a COO$^-$ (1,539) and doublet (1,536–1,550 cm^{-1}). C—H stretching region (**b**) shows singlets are distinguished by CH_2 peaks at lower wavenumbers and a v_a CH_3 2,953 cm^{-1} shoulder, localised in map II

are characteristic of highly compact trans-zigzag conformation of stearate chains with parallel packing (Sakai and Umemura 2002). Carboxylate doublets are also found in safflower Titanium/ZnO/aluminium silicate control paint with aluminium stearate, associated with spectra combining features of zinc and aluminium soaps where mixed structures may prevent compact packing.

Solvent Model

Over the experimental time frame of several months, the initial small volume of solids in individual vials was observed to increase significantly and form a partial suspension. This was most apparent with MKN ZnO and stearic acid in excess. Very limited changes were observed for Kremer models over the same period. For the purposes of accelerating timelines, the results discussed below refer to solvent models prepared with MKN ZnO.

Droplets sampled from models with ZnO in excess indicate that available stearic acid reacts quickly to form zinc stearate, even with minimal agitation. Microscopy shows stearic acid has greatly diminished after 5 h and is no longer present at 24 h. Fine, irregular waxy aggregates are the only solid evident in droplets sampled at 24 h. The FTIR spectrum of this substance is characteristic of zinc stearate, while FT-Raman and X-ray diffraction indicate ZnO is also present.

Zinc stearate formation is apparent in droplets sampled after 15 min from the aluminium stearate model. The characteristic 1,541 cm^{-1} COO^{-} peak accompanies disappearance of the small C=O acid peak in the spectrum (Fig. 6) suggesting zinc stearate initially forms by reaction with the most readily available free acid, a reaction producing water (Eq. 1).

There is little to visually distinguish aluminium and zinc soaps within dried droplets as both solids comprise irregularly sized white lumps. After 1 h, a fine perimeter ring dries separately to the bulk which FTIR characterises as stearic acid. Solids sampled from the centre of the droplet contain both zinc and aluminium soaps and spectra include peaks for Al-O/OH at ca. 986 and 3,694 cm^{-1} consistent with hydrolysis of aluminium stearate (Eq. 2):

$$AlSt_2OH + H_2O \rightarrow AlSt(OH)_2 + HSt \qquad (2)$$

By 24 h, there is little evidence of stearic acid; the purest zinc stearate phase dries at the centre of each droplet, while spectra from perimeter solids indicate a mixture of zinc and aluminium soaps. Insoluble gel lumps also form in the toluene and adhere to the vial walls following agitation. These comprise soaps with stronger Al-O contribution (Eq. 3):

$$2ZnO + 3AlSt_2OH + 2H_2O \rightarrow 2ZnSt_2 + 2AlSt(OH)_2 + Al(OH)_3 \qquad (3)$$

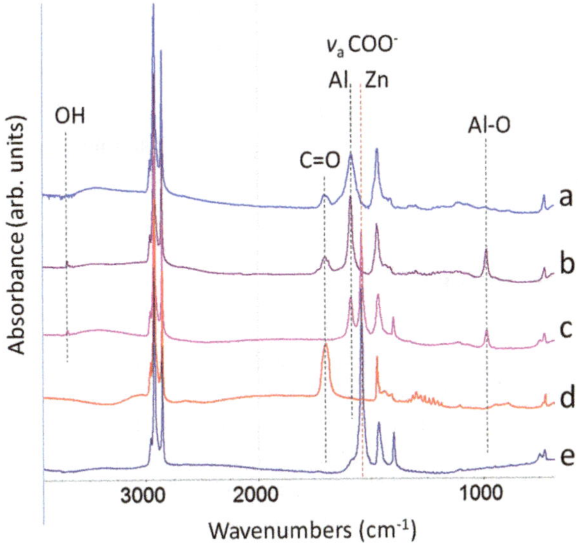

Fig. 6 Transmission FTIR spectral sequence obtained from ZnO and aluminium stearate in toluene. Aluminium stearate before mixing includes (**a**) free acid and (**b**) OH groups which have replaced stearate chains. Within minutes of combining zinc stearate peaks appear (**c**), and after several hours purer phases of stearic acid (**d**) and zinc stearate (**e**) are obtained

The 24 h aluminium stearate-model reaction sequence closely resembles what is observed in decade old reference paints. Incorporation of aluminium stearate increases concentration and availability of stearic acid typically released through hydrolysis. Additions of hydrogenated castor oil may function similarly. Hydroxyl groups replace stearate chains to form relatively polar molecules of aluminium mono- or di-stearate or aluminium hydroxide. In a paint film, increasing concentration of zinc stearate and contrasting polarity with Al-OH groups encourages phase separation. Zinc stearate is capable of highly efficient molecular conformation when associated with other similar molecules.

Beyond the presence of aluminium stearate, current and previously reported results indicate differences in ZnO reactivity are attributable to pigmentation and oil medium. Films pigmented solely with ZnO have higher surface area of zinc ions available to react with sources of stearic acid than comparable mixed pigment formulations. However only mixed-pigment control paints *with* aluminium stearate are available to compare with commercial formulations, where proprietary differences make it difficult to isolate possible influences of co-pigmentation on reactivity or propensity for phase separation.

Different oils have different fatty acid profiles influencing rate of reaction. Control paint prepared with ZnO and boiled linseed oil remains soft 20 years after casting with no discernible variation in FTIR absorbance between surfaces. Zinc stearate is less concentrated in decade-old Titanium/zinc/safflower oil control paint with aluminium stearate than comparable older films or films formulated with linseed oil. Paints with soybean oil and aluminium stearate are distinguished by accumulation of fatty acids and some zinc stearate at exposed surfaces.

Possible influences on carboxylate migration within reference paints cast onto Melinex have been discussed previously. Despite differences with typical painting supports, concentration of zinc soaps at interfaces between layers is implicated in instability, including a recent study of flaking paintings dated 1962–1980 by Vietnamese artist Nguyễn Trọng Kiệm (Osmond et al. 2014). Zinc oxide containing paintings are vulnerable to both intra- and interlayer failure.

Conclusion

Paint composition has implications for the reactivity of zinc oxide in oil based media. Rapid reactions in solvent models using nano-sized zinc oxide are comparable to those observed longer term in reference paints with larger artist grade pigment. Aluminium stearate is unequivocally shown to be a source of stearic acid which in the presence of zinc oxide will progressively be replaced by zinc stearate. Its inclusion in paints dominates more subtle differences in reactivity apparent between different oils. The availability of stearic acid from aluminium stearate or fatty acids released by oil hydrolysis facilitates formation of zinc soaps according to the amount of zinc oxide present. The influence of specific pigment combinations on carboxylate distributions has not been conclusively determined. Soaps with split carboxylate peaks revealed by high resolution SR-μFTIR may be indicative of mixed or hydrated soaps preventing compact packing configurations. Progressive reorganisation and aggregation of similar soap molecules is energetically favourable for C16 and C18 chain saturated zinc soaps as film chemistry increases in polarity with age. For zinc oxide containing paintings, specific pigment properties, fatty acid composition, additives and environmental history all influence component interactions with a range of aesthetic and structural consequences.

Acknowledgements Marion Mecklenburg, Smithsonian Museum Conservation Institute; Ljiljana Puskar Australian Synchrotron; and AMMRF Centre for Microscopy and Microanalysis staff, University of Queensland. Kate Helwig, Canadian Conservation Institute, is thanked for discussions initiated during ICOP. Professor John Drennan, The Australian Research Council and industry partners are acknowledged for their support through Linkage Project *The twentieth century in paint.*

References

Browne FL (1956) Swelling of paint films in water VIII: swelling of linseed oil paints in water and organic liquids. For Prod J 6:312–318

Corkery RW (2004) A variation on Luzzati's soap phases. Room temperature thermotropic liquid crystals. PCCP 6:1534–1546. doi:10.1039/B315595C

Derrick M, Stulik D, Landry J (1999) Infrared spectroscopy in conservation science. The Getty Conservation Institute, Los Angeles

Dredge P, Schilling MR, Gautier G, Mazurek J, Learner T, Wuhrer R (2013) Lifting the lids off Ripolin: a collection of paint from Sidney Nolan's Studio. J Am Inst Conserv 52(4):213–226

Elm AC (1957) Reevaluation of the function of pigments in paints. Official Dig (April):351–385

Gautier G, Bezur A, Muir K, Casadio F, Fiedler I (2009) Chemical fingerprinting of ready-mixed house paints of relevance to artistic production in the first half of the twentieth century. Part I: Inorganic and organic pigments. Appl Spectrosc 63(6):597–603. doi:10.1366/000370209788559584

Helwig K, Poulin J, Corbeil M-C, Moffat E, Duguay D (2014) Conservation issues in several 20th-century Canadian oil paintings: the role of zinc carboxylate reaction products. This proceedings, Chapter 11

International Zinc Association (2009) Commercial grades of zinc oxide. International Zinc Association. http://www.znoxide.org/commercial_grades.html. Accessed 7 June 2010

Jacobsen AE, Gardner WH (1941) Zinc soaps in paints: zinc oleates. Ind Eng Chem 33(10): 1254–1256

Mallégol J, Lemaire J, Gardette J-L (2000) Drier influence on the curing of linseed oil. Prog Org Coat 39:107–113. doi:10.1016/S0300-9440(00)00126-0

Maskaev AK, Man'kovskaya NK, Lend'el IV, Fedorovskii VT, Simurova EI, Terent'eva VN (1971) Preparation of 12-hydroxystearic acid, the raw material for plastic greases. Chem Technol Fuels Oils 7(2):109–112. doi:10.1007/bf00718698

Mecklenburg MF (2008) Meccanismi di cedimento nei dipinti su tela: approcci per lo sviluppo di protocolli di consolidamento – Failure Mechanisms in Canvas Supported Paintings: approaches for developing consolidation protocols. I Talenti. Metodologie, tecniche e formazione nel mondo del restauro Il Prato, Padova

Mehrotra KN, Shukla RK, Chauhan M (1995) Spectroscopic and conductometric studies of lanthanide soaps. Bull Chem Soc Jpn 68(7):1825–1831

Morley-Smith CT (1950) The development of anti-chalking French Process zinc oxides. J Oil Colour Chem Assoc 33:484–501

Osmond G (2012) Zinc white: a review of zinc oxide pigment properties and implications for stability in oil-based paintings. AICCM Bull 33:20–29

Osmond G, Boon JJ, Puskar L, Drennan J (2012) Metal stearate distributions in modern artists' oil paints: surface and cross-sectional investigation of reference paint films using conventional and synchrotron infrared microspectroscopy. Appl Spectrosc 66(10):1136–1144. doi:10.1366/12-06659

Osmond G, Ebert B, Drennan J (2014) Zinc oxide-centred deterioration in 20th century Vietnamese paintings by Nguyễn Trọng Kiệm. AICCM Bull 34:4–14

Phenix A (2002) Building models: comparative swelling powers of organic solvents on oil paint and the cleaning of paintings. V&A Conserv J 40 (Spring)

Robinet L, Corbeil MC (2003) The characterization of metal soaps. Stud Conserv 48(1):23–40

Sakai H, Umemura J (2002) Evaluation of molecular structure in Langmuir monolayers of zinc stearate and zinc 12-hydroxystearate by IR external reflection spectroscopy. Colloid Polym Sci 280(4):316–321. doi:10.1007/s003960100581

Schilling MR, Mazurek J, Learner TJS (2007) Studies of modern oil-based artists' paint media by gas chromatography/mass spectrometry. Modern paints uncovered. The Getty Conservation Institute, Los Angeles, pp 129–139

Soucek MD, Khattab T, Wu J (2012) Review of autoxidation and driers. Prog Org Coat 73(4):435–454, doi:http://dx.doi.org

Tachibana T, Kitazawa S, Takeno H (1970) Studies of helical aggregates of molecules. II. The sense of twist in the fibrous aggregates from the alkali metal soaps of optically active 12-hydroxystearic acid. Bull Chem Soc Jpn 43(8):2418–2421

van der Weerd J, van Loon A, Boon JJ (2005) FTIR studies of the effects of pigments on the aging of oil. Stud Conserv 50(1):3–22

Ware EE, Christman RE (1916) A study of the effect of storage on mixed paints. Ind Eng Chem 8(10):879–883

Winter G, Whittem RN (1950) Fluorescence and photo-chemical activity of zinc oxides. J Oil Colour Chem Assoc 33(November):477–483

Wöll C (2007) The chemistry and physics of zinc oxide surfaces. Prog Surf Sci 82:55–120. doi:10.1016/j.progsurf.2006.12.002

Its Surreal: Zinc-Oxide Degradation and Misperceptions in Salvador Dalí's *Couple with Clouds in Their Heads*, 1936

Katrien Keune and Gwendolyn Boevé-Jones

Abstract The painting *Couple with Clouds in their Heads* by Salvador Dalí (1936) from the Museum Boijmans van Beuningen in Rotterdam was seriously disfigured by ring formations on the edges of the painting and a matt, white-looking semi-transparent material on the surface. Paint samples, taken from affected and unaffected paint areas, investigated with light, scanning electron and ATR-FTIR microscopy and X-ray diffraction showed that the zinc white-containing oil paint was converted into zinc soaps, zinc formate dihydrate, zinc acetate dihydrate and zinc sulfide. The conversion of the zinc white started at the surface of the painting. Zinc soaps are formed by a reaction of zinc white and fatty acids from the oil. Formate and acetate, most likely emitted from the wooden frame, reacted with the zinc white pigment. Finally, the zinc sulfide is suggested to derive from zinc white reacting with hydrogen sulfide.

The high degree of degradation of the zinc white-containing paint and the ring formation is postulated to be triggered by the exposures of the panels to high temperatures which occurred during a 1-day photo shoot in 1936 shortly after completion of the paintings by Dali. Furthermore, it seems likely that the glazed frame created a (semi-)closed system and subsequently acid gasses diffused into the system via "leakages" due to a poor match between the frame and the panel, and may have initiated a reaction with the surface of the painting. As the glazed and framed panels were a non-ventilated system, the acid gasses did not distribute evenly over the surface, but rather were most intense at the edges of the painting. The ring formation caused by the gasses is compared with the phenomena of Liesegang rings, a reaction-diffusion process.

K. Keune (✉)
Van 't Hoff Institute for Molecular Sciences (HIMS), University of Amsterdam, Amsterdam, The Netherlands
e-mail: k.keune@uva.nl

G. Boevé-Jones
Redivivus Studio for the Conservation and Restoration of Paintings, The Hague, The Netherlands
e-mail: boeve@redivivus.nl

K.J. van den Berg et al. (eds.), *Issues in Contemporary Oil Paint*,
DOI 10.1007/978-3-319-10100-2_19

Keywords Dalí • Zinc white • Degradation • Zinc soap • Zinc acetate dihydrate • Zinc formate dihydrate • Zinc sulfide • Framing • Transparency • Concentric rings

Introduction

In 1936, Salvador Dalí painted the enigmatic *Couple with Clouds in their Heads,* a double portrait of himself and his wife Gala. Recognizing it as one of his seminal works, the two paintings were part of a photo shoot of Dalí and Gala taken by Cecil Beaton in the same year (Fig. 1). Photographic and archival evidence suggests that the large and highly illuminating photographic lights used for the shoot resulted in elevated temperatures in the studio.

The *Couple with Clouds in their Heads* became part of the collection of Edward James, who at the time was the most important patron and artistic collaborator with Dalí. Unfortunately, no photographs from the ensuing years could be found in the James Archives at Chichester to track the date at which the rings became visible. However, there are photographs of the painting taken in the late 1960s when the paintings came on loan to Museum Boijmans van Beuningen in Rotterdam,

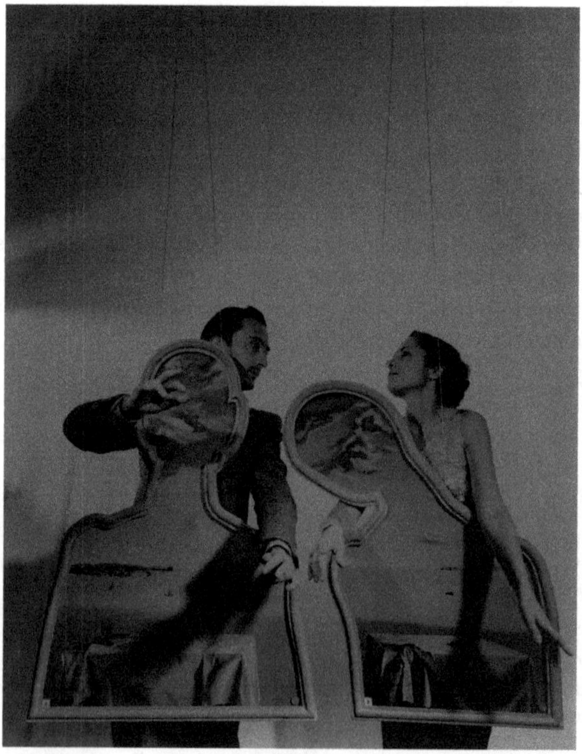

Fig. 1 *Couple with Clouds in their Heads* by Salvador Dalí (1936, oil on plywood), Museum Boijmans van Beuningen, Rotterdam, The Netherlands (Photo shoot of Dalí and Gala taken by Cecil Beaton (1936))

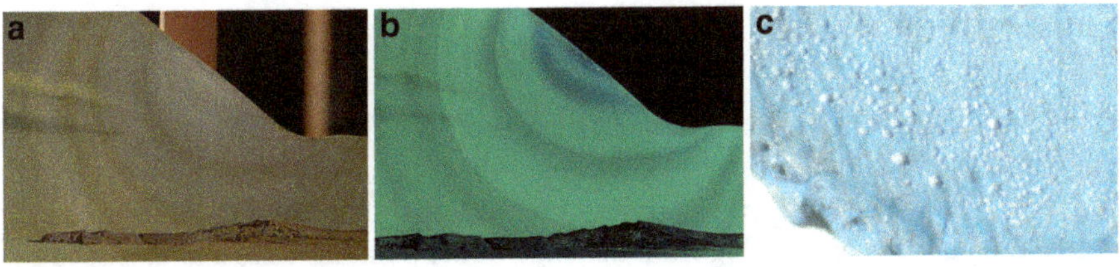

Fig. 2 Macroscopic image of the concentric rings on the right shoulder of the female under white light (**a**), UV light (**b**) and macroscopic detail of droplets of the soft whitish, transparent material in the blue sky of the male (**c**)

leaving a gap of approximately 32 years with no visual documentation. By this time, the paintings had significant visual alterations: ring formations on the paint surface had formed that were ambiguous in relationship to the original composition; subsequently there are publications which questioned the meaning of the rings with the assumption that they were intended by the artist (Fig. 2a, b). Recent close examination of the paintings clearly showed that the rings were not intentional, but rather related to the degradation of the paint. This paper describes the characterisation of the paint degradation, postulates the degradation mechanisms and the influences of external factors, and proposes a hypothesis on the ring formation mechanism.

Painting Technique

The *Couple with Clouds in their Heads* is painted on two bespoke plywood panels sized with animal glue with a subsequent white ground layer that is composed of lead white and quartz in drying oil and little paraffin. Dalí was sparing in his use of oil paint and he usually built up only a few relatively thin paint layers to achieve his final results. His mixtures were often composed of only a few pigments. For instance, the pale yellow-green paint used for the shoulders of the Gala figure are composed of zinc white oil paint with a little cobalt violet, cerulean blue and cadmium yellow. The paint of the blue sky is composed of zinc white with cerulean blue. What has become evident in the course of studying a group of nine paintings from the same period is that Dalí used different binding media to obtain certain effects of matt or gloss in his paints, leaving the paintings intentionally unvarnished (Boevé-Jones 2010). Very shortly after *Couple with Clouds in their Heads* was completed, both panels were framed each with a bespoke bentwood simple profile frame with glass glazing. This was completed for a gallery opening at Reid & Lefevre Gallery in London before the photo shoot and the wood had not yet even been painted, as was certainly always the intention. The unpainted wooden frame suggests an urgency to get the recently completed paintings framed with glazing as soon as possible.

Conservation History

In the preparation for the exhibition *Edward James, A Surreal Life* at the Museum Boijmans van Beuningen in 1998, the paintings were carefully studied for the first time. The concentric rings were observed on both paintings, the areas of degradation always originating from the edge of the painting were characterised by a matt, white-looking semi-transparent material often next to darker, altered looking paint zones (Fig. 2). The formations were found in areas of varying colors, but all mixed with white pigment. The whitish degradation material was observed in several forms such as minute droplets, worm-like tunnels or smooth ice-like layers at the painting's surface (Fig. 2c). Analysis on a few paint samples performed by the Cultural Heritage Agency of the Netherlands in 1997 indicated that zinc white degradation played a significant role in the problem (ICN 1997). At that moment, with limited time for the analysis and without a clear plan on how to proceed, the decision was made postpone any treatment. After a hiatus of 11 years and plenty of time to ponder the issue, the opportunity arose to further study the paintings and to propose a treatment plan. The availability of new techniques, such as imaging-ATR-FTIR, combined with novel insights into the degradation of oil paints, including the formation of metallic soap aggregates, made it possible to study the degradation phenomena in *Couple with Clouds in their Heads* in more detail. In 2009–2010, both panels and frames were conserved and restored for the first time. The whitish degradation products were mechanically removed with a Stabilo vinyl eraser as well as the use of an electric eraser with vinyl tips. The use of solvents was kept to a minimum to reduce the risk of initiating a new cycle of degradation. Retouching with Gamblin Conservation Colors and the application of a partial varnish (Regalrez) was carried out to integrate the permanently discolored rings into the surrounding areas.

Experimental

In order to gain a deeper understanding of the degradation phenomena observed on the *Couple with Clouds in their Heads* and the mechanisms that have taken place, paint samples were taken from both the Dali and Gala panels. Paint scrapings were analyzed with transmission FTIR, gas chromatography-mass spectrometry and X-ray diffraction (results not presented in this paper, see ICN (1997)). Three paint cross sections are presented here: two samples were taken from the pale yellow-green paint from the shoulders of Gala, one from the left-side in a degraded area in the center of the concentric rings, and one from the right-side where the paint seemed to have been unaffected. A sample from the Male was taken inside the ring deformation from the pale green sand, which revealed a transparent upper layer of paint. The samples were investigated with light, scanning electron and ATR-FTIR microscopy and X-ray diffraction (XRD).

Results

Comparison of Intact and Degraded Paint in Female's Shoulder

Light microscopy revealed that in the sample taken from an undamaged area, the so-called intact sample, shows a single thick light yellow paint layer was applied directly over the ground, while in the degraded sample the upper part of the yellow paint is completely transparent (Fig. 3a, b). Whereas the intact thick, light yellow paint shows characteristic UV fluorescence of zinc oxide, the transparent part is non-fluorescent. Furthermore, SEM backscatter images show a difference in morphology between the lower and upper part of the paint in the degraded sample (Fig. 3d). Zinc white particles are only visible in the lower part, and vertical and horizontal bands are observed in the transparent zone pointing to stress, expansion and/or crystallisation in the layer. In the intact paint, however, the zinc white particles are densely packed (Fig. 3c). Elemental analyses detected zinc and sulfur in the transparent layer, while sulfur is absent in the lower part of the degraded paint and the intact paint. Imaging-ATR-FTIR characterised zinc palmitate or stearate, so-called zinc soaps, with characteristic bands at: 1536 cm^{-1} ($\nu_{asym}(COO)$), 1395 cm^{-1} ($\nu_{sym}(COO)$) and 1449 cm^{-1} ($\nu_{bend}(CH_2)$). The zinc soaps were found in both the degraded and intact layers, and also in scrapings of the whitish surface material deriving from the concentric rings. Interestingly, spectra from analysis of the upper part of the degraded sample and the scraping of the whitish material show extra peaks, with peak maxima at 1573 cm^{-1} and an increase of the OH band vibration around 3357 cm^{-1} (Fig. 3e). These extra peaks are indicative of the presence of hydrate zinc formate (broad $\nu_{strech}(OH)$ around 3360 cm^{-1}, $\nu_{asym}(COO)$ at 1577 cm^{-1} and $\nu_{sym}(COO)$ at 1394 cm^{-1}) (Pienaar 2005). XRD analyses performed on the whitish transparent material identified zinc sulfide as the main component (ICN 1997).

Degraded Paint in Male Figure

Disturbing rings were also present at the edge of the lower right corner of the male figure. A sample was taken inside the ring deformation from the pale green sand, which revealed a transparent upper layer of paint (Fig. 4a). The coloured pigment particles are still visible within the transparent paint. The backscatter electron image of the sample shows a lack of pigment particles in the transparent part and the vertical and horizontal bands are also present (Fig. 4b). Elemental analysis detected zinc, sulfur and chlorine in the transparent part, while zinc alone was detected in a layer of 5 microns in thickness present at the surface. This difference between the bulk and the surface of the transparent layer is also visible in the imaging-ATR-FTIR images (Fig. 4c). In the bulk of the paint the characteristic peaks for zinc soaps at 1541, 1452 and 1395 cm^{-1} are present. In addition, the infrared spectrum revealed

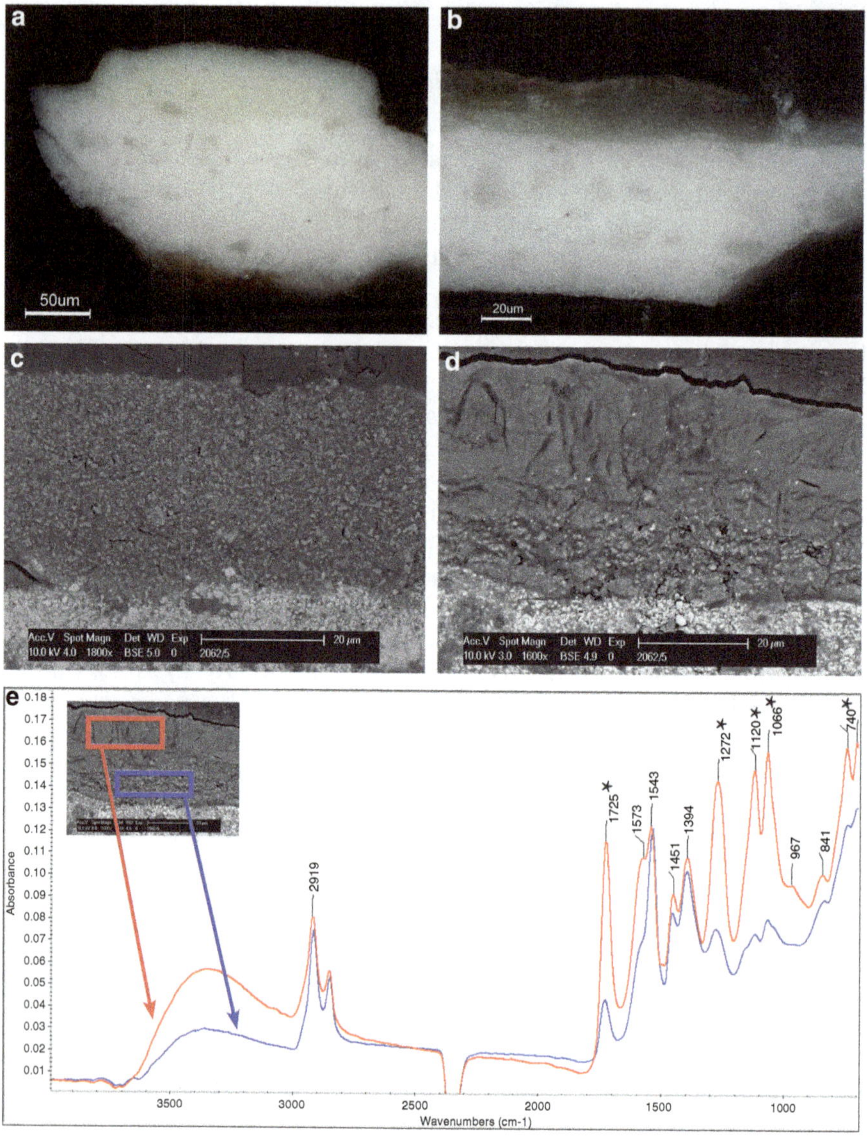

Fig. 3 Light microscopic image of the "intact" paint (**a**) and "degraded" paint (**b**) from the *left* and *right* female's shoulder at 200× and 500× magnification respectively, backscatter electron image of "intact" (**c**) and "degraded" (**d**) paint. Co-added FTIR spectrum derived from the *upper part* (in *red*) and *lower part* (in *blue*) of the "degraded paint" sample (**e**). The peaks indicated with a *star* are characteristic for the embedding medium

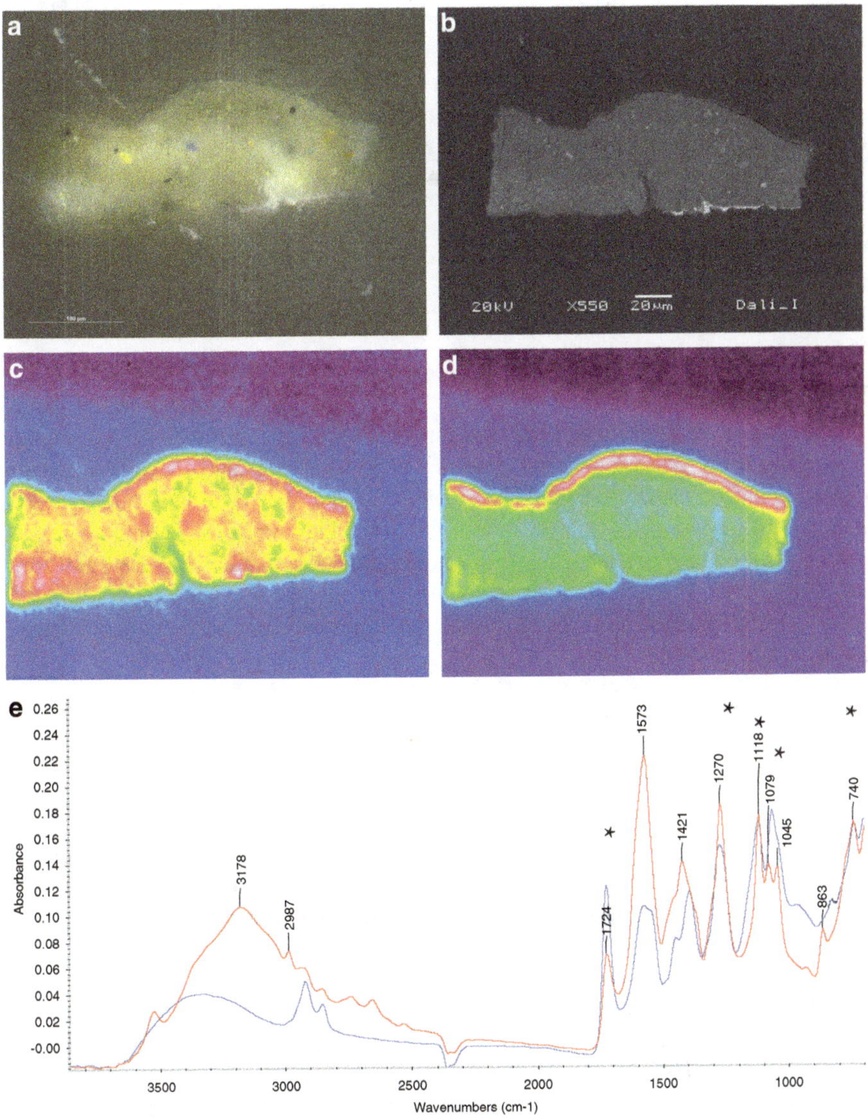

Fig. 4 Light microscopic image of paint sample taken from the *left lower* edge of the male (**a**), backscatter electron image (**b**), False colour-plots ATR-FTIR images at 1541 cm^{-1} (**c**) and 1573 cm^{-1} (**d**) (*red* is representative for high intensities). Co-added FTIR spectrum derived from the thin surface layer (in *red*) and bulk of the paint (in *blue*) (**e**). The peaks indicated with a star are characteristic for the embedding medium

a shoulder at 1573 cm^{-1} indicative of zinc formate (Fig. 4d). While the infrared spectrum of the thin surface layer shows an absence of zinc soaps, the presence of zinc carboxylate vibrations at 1573 and 1423 cm^{-1} plus a broad OH band around 3174 cm^{-1} are evident (Fig. 4e). This spectrum points to a mixture of zinc formate dihydrate and zinc acetate dihydrate (Johnson et al. 1981: broad ν_{strech}(OH) around 3130 cm^{-1}, ν_{asym}(COO) at 1558 cm^{-1} and ν_{sym}(COO) at 1440 and ν_{bend}(CH_3) at 1390 cm^{-1}).

Discussion

In *Couple with Clouds in their Heads,* the deterioration of zinc white paint is manifested as a whitish, semi-transparent film at the surface. No zinc soap aggregates were observed in the paint, as previously postulated as the cause of degradation in zinc white-containing paints (Van der Weerd et al. 2003; Osmond et al. 2005). In this study analyses showed that the zinc white-containing paint had been converted to zinc soap, zinc formate dihydrate, zinc acetate dihydrate and zinc sulfide. Zinc soaps are present in the so-called intact paint as well as in the degraded paint, while the others are only located in the degraded paint present at the surface of the painting.

Zinc-Containing Degradation Products

Zinc soaps are formed by a reaction between zinc white and free fatty acids, such as palmitic and stearic acid that are derived from the oil. The exposure of the just completed paintings to high temperatures and most likely high relative humidity as result of the photo shoot in combination with the (semi-)closed system of the glazed and framed panels created a "greenhouse" effect, which is likely to have promoted the saponification reaction on the surface of the paintings. Zinc soaps were also identified in the intact paint, since they are formed to a certain degree during ageing of the zinc white-containing oil paint.

Zinc sulfide can be formed by a direct and fast reaction between zinc oxide and hydrogen sulfide under normal conditions (Samokhvalov and Tatarchuk 2001). A reaction with sulfur dioxide is not likely here, as the zinc oxide would have absorbed sulfur dioxide resulting in zinc sulfite and zinc sulfate products (and these were not detected by FTIR). It is postulated that sulfur was originally present as hydrogen sulfide, and that the sulfur source must be external to the paint. Images from the photo shoot in 1936 show that Dalí and Gala physically embracing the suspended paintings, repeatedly. The position of their hands and arms in some of the photos corresponds to the location of the ring formations. Hydrogen sulfide produced by the body in low concentrations might have diffused into the system and thereby be an important source for the sulfur-compound in this case.

Zinc acetate and formate dihydrate are most likely the products of the reaction between zinc white and formic and acetic acid, emitted from the wooden frame although this remains unsubstantiated. The multiplex plywood supports are not a likely source, as these zinc products were found only on the surface of the painting. The wood from the frame emits significantly higher levels of acetic and formic acid, especially under higher temperatures (Gibson and Watt 2010).

Ring Formation

Currently, the formation and specific spatial distribution of the four zinc-containing products in the concentric rings is unclear. However, the position and shape of the rings suggest that there is a starting point that triggers the degradation process, possibly due to a leakage between the panels and the frame through which the gases could have diffused to the surface of the painting. The fact that the cut out panels do not match the frames exactly in several places makes this a likely scenario. In a non-ventilated system, as is the case here, the gas may not distribute evenly over the surface, but remains close to the edges (Di Pietro and Ligterink 2002). The pathway of the reaction of the acid gases at the paint surface may be comparable with the phenomena of Liesegang rings resulting in concentric rings. Liesegang rings are the banded patterns seen in some minerals and the formation of chemical striations of precipitating salt in a gel medium. However, the reaction-diffusion process by which Liesegang rings develop is not completely understood.

Related Paintings by Dalí

So far, the degradation phenomena of the concentric rings have only been found in *Couple with Clouds in their Heads* (1936). In 1937, another version of the *Couple with Clouds in their Heads* was made by Dalí, now in the collection of the Fondazione Isabella Scelesi in Italy. The second version is also framed with glazing and with a more perfectly fitting bentwood frame. This version has never undergone a photo shoot with extreme conditions and interestingly, although there was also a layer of white degradation, no ring formations were observed. Another painting from the Edward James collection, also dated 1936 is *Table Solaire*, which is in excellent condition; it also had a poorly fitted frame and a whitish material was observed in two localised areas on the edge of the panel. Further study is needed to investigate the possibility of a similar degradation product without the ring formation in this painting.

Conclusion

The present study showed that the combination of zinc white pigment in oil medium under glazed framing created a (semi-)closed system which, with the 1-day exposure to high temperatures, possibly high relative humidity and human sweat, resulted in the disfiguring rings and transparent and whitish surface material in *Couple with Clouds in their Heads.* That Dalí never intended these alterations is quite clear: the carefully constructed illusion and the modified variations in surface matt and gloss areas were damaged by the degradation process. The reaction of zinc white pigment with fatty acids from the oil, formate and acetate emitted from the wooden frame and hydrogen sulfide deriving from an external source resulted in conversion of the opaque paint to the transparent and whitish surface material. This case study illustrates that short exposure to extreme climate conditions, like high temperatures and presumably high relative humidity, and exposure to acid gasses can lead to dramatic alteration of the paint over time. Fortunately, through the restoration of the damaged areas, the original concept of the artist is again in tact and the artwork presentable in a manner consistent with the goals of the artist when he created *Couple with Clouds on Their Heads* in 1936.

Acknowledgements The authors would like to acknowledge the current owner of the *Couple with Clouds in their Heads*, Museum Boijmans van Beuningen, and Atelier Boersma, where the paintings were studied for the first time in 1997. We are grateful to the scientists from the Cultural Heritage Agency of the Netherlands (RCE) who contributed to this research: Karin Groen, Klaas Jan van den Berg, Peter Hallebeek, Suzan de Groot and Henk van Keulen. The authors thank Frank Ligterink (RCE) and Annelies van Loon (University of Amsterdam) for their discussion.

Appendix

Light Microscope

Paint cross-sections were examined under a Zeiss Axioplan 2 microscope both with incident polarised light and incident UV-light (from a Xenon-lamp and a mercury short arc photo optic lamp HBO, respectively). The filter set 'UV H365' used for examination in UV-light consists of the following filters: excitation BP 365/12, beam splitter FT 395 and emission LP 397.

Attenuated Total Reflection – Fourier Transform Infrared (ATR-FTIR) Imaging Microscope

The FTIR spectral data from paint cross-sections were collected on a Perkin Elmer Spectrum 100 FTIR spectrometer combined with a Spectrum Spotlight 400 FTIR microscope equipped with a 16 × 1 pixel linear Mercury Cadmium Telluride (MCT) array detector. A Perkin Elmer ATR imaging accessory consisting of a germanium crystal was used for ATR imaging.

Scanning Electron Microscopy – Energy Dispersive X-Ray Spectroscopy (SEM-EDX)

A sample from the male figure was analysed using a JEOL JSM 5910 LV SEM and Noran Vantage EDX system with Pioneer Norvar detector. Analysis of samples taken from the female figure were performed using a XL30 SFEG electron microscope (FEI, Eindhoven, The Netherlands) with an EDX system with spot analysis and elemental mapping facilities (EDAX, Tilburg, The Netherlands). Backscattered-electron images of the cross-sections were mostly taken at 20 kV accelerating voltage, at a 10 mm or 5 mm eucentric working distance, resp. Prior to SEM-EDX analysis, samples were gold coated to improve surface conductivity.

X-Ray Diffraction (XRD)

A scraping of the whitish semi-transparent degradation material was analysed by XRD, using a Discover D8 microdiffractometer with a General Area Detection Diffraction System (GADDS) two dimensional detector (Bruker AXS, Karlsruhe, Germany). The sample was applied in a little cedar oil on a Silicon zero diffraction plate. Diffractograms were acquired in reflection mode with CuKα radiation (40 kV, 30 mA). The GADDS software was used for integration and the Bruker AXS Eva software for phase identification using the PDF database.

References

Boevé-Jones GP (2010) Treatment report for Salvador Dali's "Couple with clouds in their heads." Unpublished manuscript, Museum Boijmans van Beuningen

Di Pietro G, Ligterink F (2002) Silver-mirroring edge patterns: diffusion-reaction models for the formation of silver mirroring on silver gelatin glass plates. JAIC 41(2):111–126

Gibson LT, Watt CM (2010) Acetic and formic acids emitted from wood samples and their effect on selected materials in museum environments. Corros Sci 52(1):172–178

ICN (1997) ICN rapport nr 1997-227, Cultural Heritage Agency of the Netherlands

Johnson MK, Powell DB, Cannon RD (1981) Vibrational spectra of carboxylato complexes—I. Infrared and Raman spectra of beryllium(II) acetate and formate and of zinc(II) acetate and zinc(II) acetate dihydrate. Spectrochim Acta Part A Mol Spectrosc 37(10):899–904

Osmond G, Keune K, Boon JJ (2005) A study of zinc soaps found in paintings at the Queensland Art Gallery. AICCM Bull 29:37–46

Pienaar A (2005) Metal carboxylate complexes relevant to the Fischer-Tropsch synthesis. Master thesis of Natural Science, University of Stellenbosch

Samokhvalov A, Tatarchuk BJ (2001) Characterization of active sites, determination of mechanisms of H_2S, COS and CS_2 sorption and regeneration of ZnO low-temperature sorbents: past, current and perspectives. Phys Chem Chem Phys 13:3197–3209

Van der Weerd J, Geldof M, Struik van der Loeff L et al (2003) Zinc soap aggregate formation in 'Falling Leaves (Les Alyscamps)' by Vincent van Gogh. Zeitschrift für Kunsttechnologie und Konservierung 17(2):407–416

Water Sensitive Oil Paints in the Twentieth Century: A Study of the Distribution of Water-Soluble Degradation Products in Modern Oil Paint Films

Anna Cooper, Aviva Burnstock, Klaas Jan van den Berg, and Bronwyn Ormsby

Abstract Water sensitivity has been observed on unvarnished, twentieth century oil paintings, which presents issues for surface cleaning. Previous studies have shown that manufacturers' formulations are contributing to one of the known causes of water sensitivity: magnesium sulphate heptahydrate formation. It has been proposed that the formation of this water soluble degradation product is a result of interactions between the paint additive magnesium carbonate with environmental sulphur dioxide at elevated relative humidity. The present study examined the presence of this water soluble degradation product on the surface and in the bulk paint films of naturally aged Winsor & Newton paint swatches made from the 1940s to the 1990s. Scanning electron microscopy with energy dispersive X-ray spectroscopy (SEM-EDX) was the primary method for detection. Heptahydrate crystalline entities on the surface were found to be below the detection limit for X-ray diffraction (XRD). Water immersion tests suggested that water sensitivity due to magnesium sulphate heptahydrate formation is largely a surface phenomenon. Painting case studies helped to contextualise the problem in light of future conservation treatments for water sensitive paintings.

A. Cooper
Painting Conservation Intern, Hamilton Kerr Institute, University of Cambridge, Mill Lane, Whittlesford CB22 4NE, UK
e-mail: annacooper87@hotmail.com

A. Burnstock (✉)
Department of Conservation and Technology, The Courtauld Institute of Art, Somerset House, The Strand, London WC2R 0RN, UK
e-mail: aviva.burnstock@courtauld.ac.uk

K.J. van den Berg (✉)
Cultural Heritage Agency of the Netherlands, Amsterdam, The Netherlands
e-mail: k.van.den.berg@cultureelerfgoed.nl

B. Ormsby
Tate, Millbank, London SW1P 4RG, UK
e-mail: bronwyn.ormsby@tate.org.uk

K.J. van den Berg et al. (eds.), *Issues in Contemporary Oil Paint*,
DOI 10.1007/978-3-319-10100-2_20

Keywords Water sensitive oil paint • Magnesium sulphate heptahydrate • Epsomite • Surface • Bulk • Winsor & Newton • Francis Bacon • Patrick Heron

Introduction

Water sensitivity of commercially prepared well-bound oil paints used on twentieth century, unvarnished paintings has been widely documented (Burnstock et al. 2006; Mills et al. 2008; Tempest et al. 2010; Silvester et al. 2014; Gaylor et al. 2008). Water sensitivity can affect discrete, single passages of oil paint or the entire surface of a painting, which may comprise multiple coloured oil paint films. It has been reported to occur in paint films and paintings from the 1900s to the present day, with the majority of those affected occurring between the 1950s and 1960s (Mills et al. 2008; Silvester et al. 2014). The degree to which paint films are affected can vary greatly, from acute sensitivity with only a single droplet of water, to those which can withstand longer exposure and a degree of mechanical action. Water sensitivity has been observed on oil paint films applied directly from the tube, indicating that manufacturers' formulations may be causal factors. Previous research has suggested that water sensitivity varies across manufacturing brands, with brand specific formulations displaying varying degrees of water sensitivity of differently coloured paints (Mills et al. 2008; Tempest et al. 2010). It was further noted that some pigments can exhibit water sensitivity in all paint brands, such as ultramarine, either due to the pigment itself or the additives used in relation to their formulations. In contrast, other pigments can range in sensitivity across paint brands, like raw sienna, exhibiting both high sensitivity and insensitivity depending on the manufacturer (Tempest et al. 2010; Silvester et al. 2014). The problem of water sensitivity is paramount due to the continued popularity of oil as a painting medium and a preference for unvarnished surfaces; this method of display necessitates the removal of surface dirt and is exacerbated when dry cleaning methods are not always effective.

Case studies of oil paintings exhibiting sensitivity to aqueous swabbing were found to have crystalline entities on the surfaces of afflicted films (Burnstock et al. 2006; Mills et al. 2008; Tempest et al. 2010; Silvester et al. 2014; Gaylor et al. 2008). Recent research undertaken by Silvester et al. characterised the crystalline products found on a water sensitive painting by Jasper Johns painted in 1964 and on a contemporary archival student-quality Winsor & Newton (W&N) paint swatch. Through elemental mapping, it was found that the crystals contained magnesium and sulphur (Silvester et al. 2014). XRD further characterised the crystalline materials as magnesium sulphate hexahydrate ($MgSO_4.6H_2O$) and magnesium sulphate heptahydrate ($MgSO_4.7H_2O$), the latter more commonly known as epsomite and more abundant in its occurrence on these affected paint films (Silvester et al. 2014). Silvester et al. hypothesised that the heptahydrate was formed by a reaction between the additive magnesium carbonate within the paint film and sulphuric acid (H_2SO_4), which forms from atmospheric sulphur dioxide (SO_2) in environments with an elevated relative humidity (Silvester et al. 2014; Thomson 1986). Evidence for this hypothesis derives from the exposure of experimental samples containing magnesium-based additives to sulphur dioxide at elevated relative humidity, which resulted in the formation of magnesium sulphate heptahydrate. Silvester et al. then drew links between peaks in environmental sulphur dioxide levels during the industrialised periods in Europe and North America of the 1950s–1970s, with a correlation between the peak in the number oil paintings known to exhibit sensitivity to aqueous swabbing (Silvester et al. 2014; Tempest 2009).

The water sensitivity of dried oil paint films of selected manufacturer's artists' oil paints which contain magnesium carbonate has therefore been found to be related to magnesium sulphate heptahydrate formation. The phenomenon is related in particular to paints manufactured by W&N until the middle of the 1970s. Magnesium carbonate was used by W&N in both their artists' quality and student quality ranges until this time (Garrett 2011a, b). W&N paints are a widely available international brand and are used by artists working across the globe.

The present study aimed to investigate the depth of sensitivity of paint films to water by assessing the presence of this degradation product in the bulk paint film. The goal was to inform conservation practice in relation to the treatment of water sensitive paints caused by magnesium sulphate heptahydrate formation.

Winsor & Newton Paint Swatches

International artists' colourmen and paint manufacturer Winsor & Newton produce a wide range of artist quality paints and materials. Keen to maintain standards and consistency, they were the first company to publish the constituents of their paints through their Composition and Permanence pamphlets from 1892 onwards (Winsor & Newton 1984), aiming to inform and reassure artists in the selection of their materials (Clarke 2009; Winsor & Newton 1984). As part of their quality control process, small batches of paint were painted onto commercially prepared canvas in order to record the touch dry times of paints (Fig. 1). This system allowed technicians to identify problematic batches and alter them accordingly (Garrett and Waldron 2010; Garrett 2011a). A range of swatches have been donated for research purposes and cover a selection of materials produced by W&N, including: Artists' Oil Colours (AOC), Student Oil Colours (WOC/Winton) and the Alkyd paint ranges from the mid-1940s to the early 1990s. The naturally aged AOC paints were selected for use in this study due to W&N's global and widespread artist's use, the use of the additive magnesium carbonate in W&N paints until 1975 (Garrett 2011a, b), as well as the unique availability of these samples for analysis and testing. Five swatches spanning a 45 year period throughout the 1940s–1990s were selected to investigate the presence of magnesium sulphate heptahydrate within paint films.

General trends in the water sensitivity of a wide range of oil colours were first established through empirical swab rolling tests with water. Some colours were not sensitive to water throughout the period studied. The majority of colours showed consistent water sensitivity in paints between the 1940s and 1970s, with a significant decline in solubility noted into the 1980s and early 1990s.

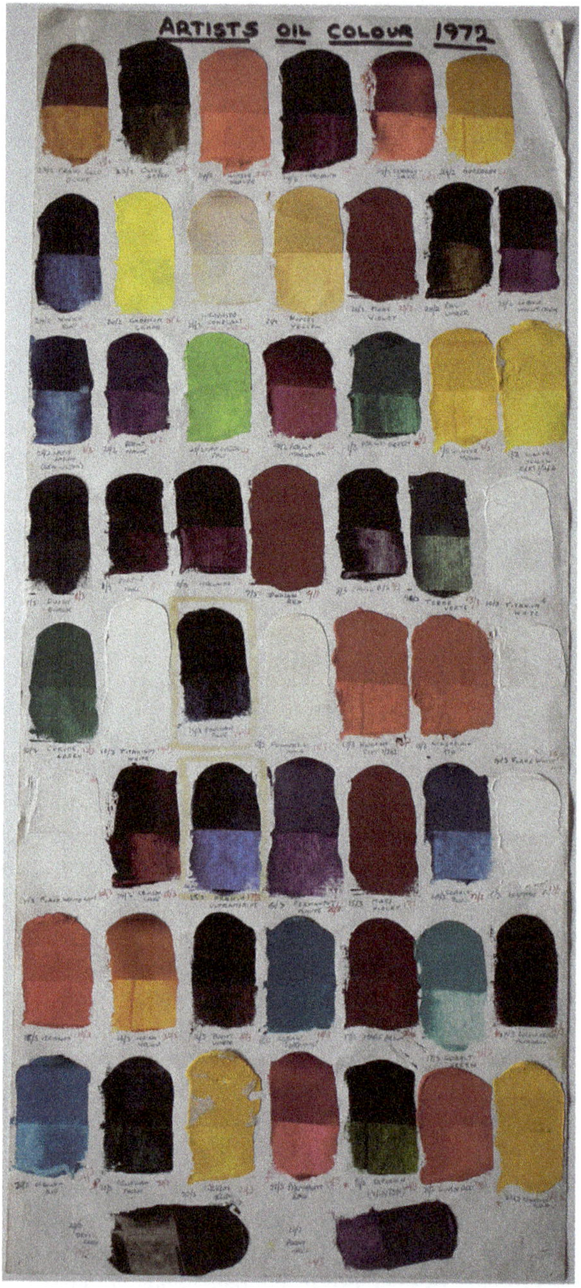

Fig. 1 Archival Winsor & Newton artists' oil colour paint swatch, dated 1972

A select group of pigments exhibited a degree of sensitivity into the later period. Two of the paints which were consistently highly sensitive across the period were cadmium lemon yellow and cobalt blue. Swatches of these paints were investigated in greater depth in relation to locating the presence of magnesium sulphate heptahydrate formation within paint films, and to see whether the declining sensitivity to aqueous solvents in the 1980s and 1990s is due to an absence of magnesium sulphate heptahydrate. These two pigments were also selected for study as the former contains sulphur inherently in its composition, whilst the latter does not, of interest in light of the connection established by Silvester et al. between water sensitivity and sulphur dioxide exposure at elevated relative humidity.

In the production of both cadmium lemon yellow and cobalt blue pigment, there is potential for excess free sulphur to be incorporated into the pigment during the manufacturing process. This is of importance as it could introduce a potential internal source of sulphur into the paint film and result in magnesium sulphate heptahydrate formation in the body of the paint film. In regards to cadmium yellows, the production derives from the precipitation of a soluble cadmium salt with a soluble sulphide (Remington and Francis 1954). W&N's manufacture of cadmium yellow required reacting cadmium with vinegar and carbon dioxide to produce cadmium carbonate, ($CdCO_3$) which is then reacted with molecular sulphur (S_8) to produce cadmium sulphide, a process deployed by W&N from the 1960s–1980s when they made this pigment in-house (Garrett and Waldron 2010; Garrett 2011a). For cobalt blue, internal sulphurous sources could derive from the high calcination temperatures required to combine cobalt oxide and an aluminium base to produce cobalt aluminate, although specific compounds vary depending on the manufacturer (Roy 2007; Remington and Francis 1954; Winsor and Newton 1982). Due to the high calcination temperatures, W&N did not make cobalt blue in-house, instead purchasing it from external sources (Garrett 2011b). After the manufacture of many pigments, including cadmium yellows, reference is made to a need for thorough washing before use, due to the by-products created as a result of manufacture (Remington and Francis 1954; Leone et al. 2005). W&N would grind and the wash the pigment after production, as instructed (Garrett 2011b), but the extent to which this removed residual sulphurous components is unclear. Therefore, identifying a definitive source of internal sulphur remains problematic, and whether this may have an impact on magnesium sulphate heptahydrate formation within paint films was investigated.

Identifying Magnesium Sulphate Heptahydrate

Magnesium sulphate heptahydrate, more commonly called epsomite, is a highly hygroscopic, water-soluble crystalline material. It has been characterised as rod-like crystalline entities from 2 to 60 μ in length on paint films using SEM, and by elemental analysis and mapping using SEM-EDX and XRD (Silvester et al. 2014). Secondary electron topographical imaging shows that epsomite generally sits flat, parallel to the surface making small depressions in the paint film where it occurs (Fig. 2a). When viewing the paint film at a slight angle, it is clear that the crystals are protruding from the surface as well as making small depressions within it. The epsomite shards are covered by a thin layer of organic material, which has

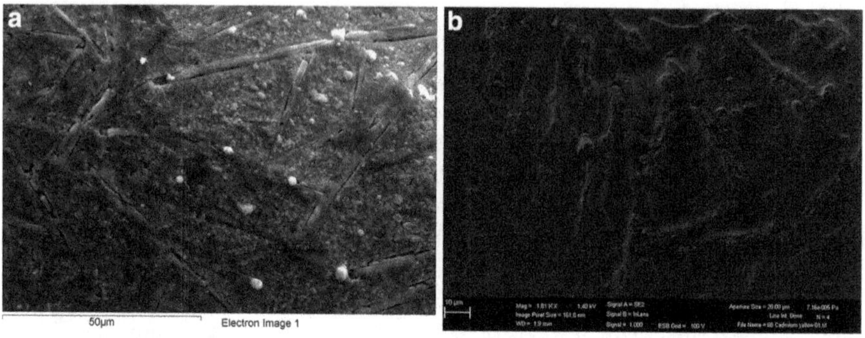

Fig. 2 (**a**) SEM secondary electron image of the surface of W&N cadmium lemon yellow, 1963. The epsomite crystals form indentations on the paint film (**b**) SEM secondary electron image of the surface of W&N cadmium lemon yellow 1968–1969. Crystals cover the surface underneath a skin of medium

cracked due to the disruption at the surface where epsomite has formed (Fig. 2b). The cracking of the surface "skin" of the paint may exacerbate the ingress of water by physically and mechanically weakening the paint film during swabbing.

Experimental

To identify epsomite within the bulk paint film, cross sections and SEM-EDX samples were taken from the W&N swatches and case studies. Separate surface and bulk samples of the W&N paints were taken for XRD. Surface and bulk samples were also taken from paint films which exhibited acute water sensitivity for water immersion tests.

Scanning Electron Microscopy

SEM imaging and elemental analysis of samples from the W&N swatches and case studies was carried out at the Natural History Museum, London using a LEO 1455 variable-pressure scanning electron microscope (VP-SEM). SEM-EDX was used to characterise the surfaces of samples. High resolution SEM images of the surface and bulk topography were captured with a Zeiss Ultra Plus field emission electron source, with low acceleration voltage and high vacuum (10–9 mBar). The use of low kV eliminated the need for the samples to be coated.

X-ray Diffraction

XRD was undertaken at RCE on samples from the W&N swatches using a Platform Discover D8 GADDS micro diffractometer (Siemens-Bruker) with a HI-STAR Area Detector. Two dimensional diffraction patterns were integrated over a range of chi, reducing the effect of non-random distribution of crystallinities. Phase identification was performed using a Diffrac Plus Eva software and the ICDD-PDF database. Exposure parameters were 40 kV and 30 mA for 300 s.

Water Immersion Tests

Water Immersion Tests from W&N swatch samples were undertaken. Surface and bulk samples were taken from cobalt blue and cadmium lemon yellow paints across the 45 year period studied. Each sample was placed surface up in a small vial and six droplets of deionised water were introduced. The results were recorded using a Leica M165C Microscope with Leica DFC 500 camera and LAS (Leica Application Suite) V3.7 software.

Results and Discussion

Magnesium Sulphate Heptahydrate Locations

Magnesium sulphate heptahydrate was visible on the surfaces of water sensitive W&N swatches from the 1940s to 1972 using SEM imaging. The size and concentration of crystals appeared to be influenced by the pigment present. Elemental analysis of W&N's cadmium lemon yellows across the periods studied confirmed peaks for cadmium, zinc, sulphur, magnesium, aluminium and barium. As a lemon yellow, the pigment is cadmium zinc sulphide (CdZnS) and the addition of zinc denotes the lighter shade of yellow. Therefore the additional EDX peaks correspond to the additives and extenders present. EDX analysis of the crystals on the surface of the cadmium lemon yellow samples from the 1940s to 1970s confirmed the presence of magnesium and sulphur. W&N's cobalt blues consistently contained cobalt and aluminium, from the pigment cobalt aluminate ($CoO.Al_2O_3$), as well as magnesium and sulphur, with zinc and calcium sometimes present. The crystals on the surface also confirmed peaks for magnesium and sulphur. As both samples exhibited acute water sensitivity, it was interesting to note that paints containing cadmium lemon yellow had a higher density of crystals of longer length, whereas the same compound found on cobalt blue swatches showed more widely dispersed shorter rods, despite empirical swab rolling tests being comparable (Fig. 3a, b).

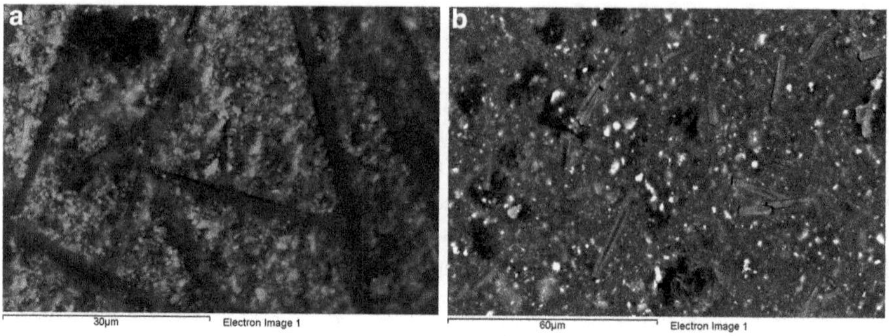

Fig. 3 (**a**) SEM Backscattered electron image of the surface of cadmium lemon yellow 1963 (**b**) SEM Backscattered electron image of the surface of cobalt blue 1963

Evidence for the formation of magnesium sulphate heptahydrate in the bulk of the paint films was not conclusive. Where crystalline deposits of the compound were evident in samples using imaging and elemental analysis, the concentration of crystals was below the 5 % detection limit for XRD. It was therefore not possible to characterise low concentrations of this compound in the bulk paint using this technique. Furthermore, the efficacy of using SEM-EDX to characterise crystals of epsomite in the bulk paint was compromised by the methods used for sampling the swatches. Using a scalpel to remove a cross section was particularly challenging for the soft, more medium rich paints and the possibility that surface material was moved into or deposited onto the bulk during this process cannot be ruled out. In some of the more brittle samples, it was possible to shear or shatter samples cleanly in order to avoid contamination; however the combination of low resolution XRD and sample artefacts highlighted the necessity to study the phenomena using a more sensitive method, such as synchrotron beamline analysis that provides sub micron resolution data.

A few epsomite crystals were identified in the bulk paint of a sample taken from cobalt blue 1968–1969 using SEM-EDX (Fig. 4a, b). The crystalline material is lighter in comparison to the darker surrounding organic matter, and the epsomite is smaller in size than those observed on the surface. Epsomite occurs in this instance at the very top and very bottom of the paint film, but not in the central body or 'bulk' of the paint. The cause of this is not clear, but may have been a result of sulphurous based material deposited on the pre-primed canvas prior to the paint application, or absorption of sulphurous material through the canvas and ground layer. Although this may show a potential for epsomite to form within the paint, it was not present within the other samples analysed. This single result where crystals were both smaller and less concentrated than those at the surface, suggests that this is an anomaly rather than a trend.

In relation to potential internal sulphurous sources during pigment manufacture, it appears that the lack of epsomite crystals in the bulk paint film show that any potential sulphur introduced into the body of the paint is not in sufficient quantity

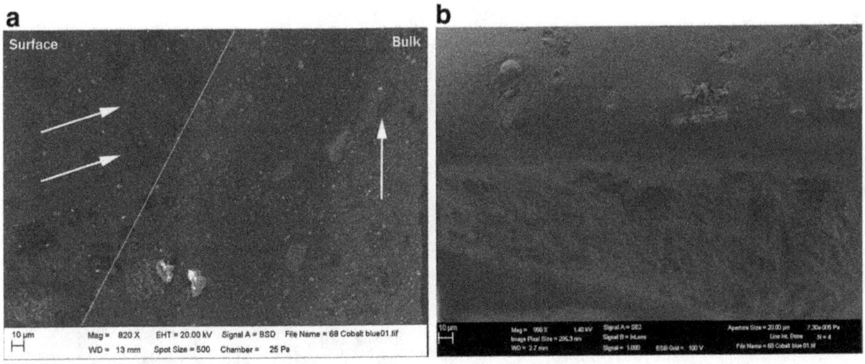

Fig. 4 (**a**) Backscattered electron image of surface and bulk W&N cobalt blue, 1968–1969. Epsomite crystals are evident on the surface and in the bulk paint (indicated with *white arrows*) (**b**) Secondary electron SEM image of the surface and bulk of the W&N cobalt blue, 1968–1969. A skin of medium is present on the surface with pockets of crystals emerging. Crystals are not visible throughout the paint strata and bulk in this image

to react with the magnesium carbonate additive and form epsomite, or that the sulphurous residues are not present in the correct configurations for the reaction to occur. It may be suggested that they been effectively removed through thorough washing of the pigment.

No epsomite was identified on the surface or in the bulk paint from swatches from the 1980s and 1990s. This result is consistent with empirical observations made during water swabbing tests in relation to the sensitivity of a wider range of pigmented W&N samples; the samples made from the 1980s onwards are significantly less water sensitive. This finding coincides with the dates of declining levels of atmospheric sulphur dioxide (Barry and Chorley 2003; Smith et al. 2001), when W&N stopped producing the majority of their in-house pigments (Garrett 2011b), W&N's replacement of the additive magnesium carbonate with micronised dolomite or 'microdol' which is calcium-magnesium carbonate in 1975 (Garrett 2011a, b), a reduction in the number of paintings noted to have sensitivity issues (Tempest 2009; Silvester et al. 2014) and the pattern of ancillary trends. Therefore the causes of sensitivity in these swatches are a result of other factors, which require further investigation, but could potentially include uneven pigment-medium distribution, underbound paint films, or incomplete polymerisation of the paint films.

Water Immersion Tests

Consistently, when water immersion tests were carried out on the bulk paint samples no reaction occurred, as expected based on the epsomite distribution seen in SEM (Fig. 5a). The samples remained sunken at the bottom of the vial, they may have

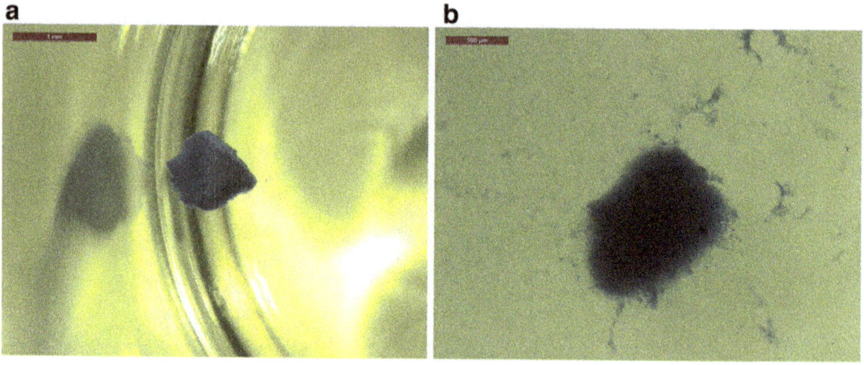

Fig. 5 (**a**) Water immersion test of a bulk film sample of cobalt blue 1968–1969 (**b**) Water immersion test of a surface sample of cobalt blue 1968–1969

swollen slightly, but the degree to which this occurred was difficult to detect with the low level magnification used. However, changes to the surface samples were more noticeable (Fig. 5b). Prior to immersion, cobalt blue 1968–1969 had a thick medium skin present, disrupted with epsomite, with some observations of epsomite in the lower bulk of the paint film (Fig. 4a, b). Upon immersion in water, the surface sample immediately solubilised. Long, oily chains carrying pigment particles were ejected from the sample and sat on top of the water's meniscus. The greasy material formed chains in all directions outwards from the sample. The reaction occurred for around 2 min, slowing after this with no visible change recorded after 10 min. When the reaction stopped, an oily network across the surface was apparent; the majority of the sample remained intact but swollen. It is suggested that the material leached out from the surface is what is being removed with aqueous swabbing.

The presence of an organic skin on cobalt blue 1968–9 was visible with the naked eye and could be seen clearly in cross section and SEM imaging (Fig. 4b). Phase separation and uneven pigment-medium distribution has been noted on a number of water sensitive paint films (Mills et al. 2008; Tempest et al. 2010; Silvester 2011). Although preliminary findings have suggested the skins are rich in diacids (Silvester 2011), further analysis is required to characterise exactly what these skins comprise. They are likely to be a compounding factor in the degree to which a paint film is water sensitive, as shown by the comparable solubility of the 1963 cadmium lemon yellow and cobalt blue paints, which show differing crystal densities (Fig. 3a, b). In the case of cobalt blue 1968–1969, the combination of the hygroscopic magnesium sulphate heptahydrate crystals present, the mechanical weakening of the surface structure, and the disruption of the organic skin increases the solubility of the surface. This leaves the bulk film, where the medium is more evenly distributed and only occasional epsomite crystals can be found, comparatively stable. This indicates that if epsomite crystals are found in the bulk or lower surface, they are not in sufficient quantity to affect the paint film, even on immersion in water.

Case Studies from Tate's Collection

Samples from twentieth century paintings identified with water sensitive paint passages from Tate's collection were examined to compare with W&N swatches of similar date. A number of paintings were examined and two will be presented in this study where water sensitivity is linked to magnesium sulphate heptahydrate formation: Francis Bacon's *A Figure in a Landscape*, 1945 (N05941) and Patrick Heron's *The Long Table with Fruit*, 1949 (T06850).

Francis Bacon Figure in a Landscape

Francis Bacon's *Figure in a Landscape* (Fig. 6) was painted in 1945, early in Bacon's career. It depicts a slumped figure, thought to be Bacon's lover Eric Hall, straddling a chair in Hyde Park which is adjacent to a microphone (Gale and Stephens 2008; Gowing 1983; Davies 1978). *Figure in a Landscape* is painted in oil, applied as thick impasto details over thin washes, with much of the ground still visible. Bacon is thought to have favoured Artists' Oil Colours by W&N, based on observations made on his studio contents (Russell 2009). He liked working in oil paint because it was able to retain the gestures of the artist during its application

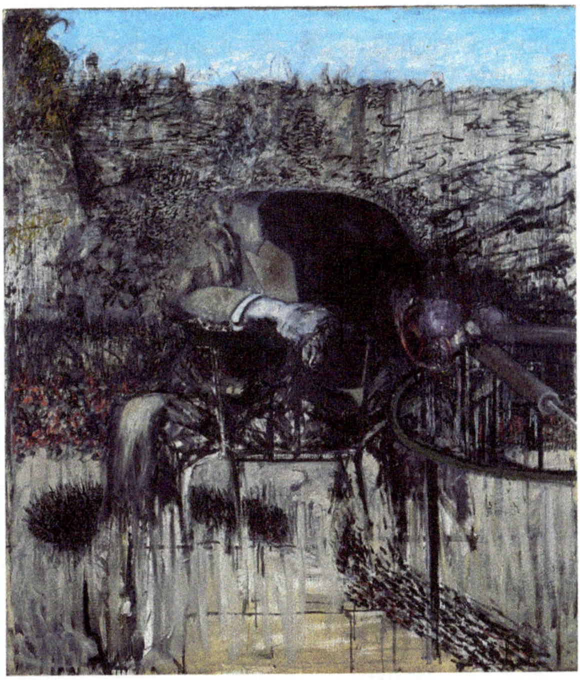

Fig. 6 Francis Bacon *Figure in a Landscape*, 1945. Oil on Canvas, 1,448 × 1,283 mm (Tate, London. © Tate 2014)

(Gale and Stephens 2008). Working straight from the tube, sometimes using the tube as an applicator, Bacon also experimented with varying the consistency of his paints by diluting them with turpentine (Hackney 1999). Other than his use of turpentine, Bacon is thought to have blotted some of his paints to leach out the medium. It is not likely that he would add additional oil or medium (Durham 1985).

A select few passages of paint in *Figure in a Landscape* are sensitive to aqueous solvents, whilst areas containing white are not sensitive, a finding comparable to the W&N swatches. The application of these sensitive paints varies: some are discrete strokes, with others worked with wet-in-wet mixing on the canvas. Some of the solubility noted could be a result of Bacon's paint manipulation, from the blotting of oil medium or the addition of excess turpentine resulting in an underbound paint film, a separate issue from water sensitivity.

Elemental analysis of the water sensitive paint passages confirmed the presence of magnesium carbonate as an extender, indicative of the use of W&N paint. Epsomite crystals were also found on the surfaces of these water sensitive paint passages. The density of the crystals present was quite sparse.

The formation of epsomite in this painting is of heightened interested due to the painting's physical history. Since its acquisition by Tate in 1950, the artwork has been displayed under Perspex (Tate File, N05941). Bacon did not varnish his works, instead favouring glass which he considered provided a unity to the painted surface whilst distancing the work from the viewer (Sylvester 1975; Hackney 1999). As a result, the surface is in excellent condition and has not required surface cleaning. The Perspex has provided a barrier to dust and dirt accumulation, but may also have protected the painting from significant sulphur dioxide exposure. As some epsomite crystals have formed, it could be suggested that the 5 years between the painting's completion, 1945, and acquisition, 1950, where it was presumably displayed without glazing, were sufficient for the fresh paint to react with sulphur dioxide from the atmosphere and form epsomite crystals. The low density of crystals observed might indicate that once contained in a controlled environment the formation slows and ceases once the sulphur dioxide has been 'used up'.

The density of crystals on the painting was significantly lower when compared to W&N swatches of the same date and pigment, confirming that the painting was not exposed to significant sulphur dioxide levels. Magnesium sulphate heptahydrate was not found in the bulk paint film of *Figure in a Landscape*, consistent with observations from the W&N swatches.

Patrick Heron The Long Table with Fruit

Patrick Heron's *The Long Table with Fruit* (Fig. 7) demonstrates his love of colour, applied in discrete passages in this still life painting. Heron applied the paint fluidly, demonstrating both wet-in-wet mixing and separate blocks of pure colour over a white ground. Heron sometimes used the tube to apply his paint directly to the

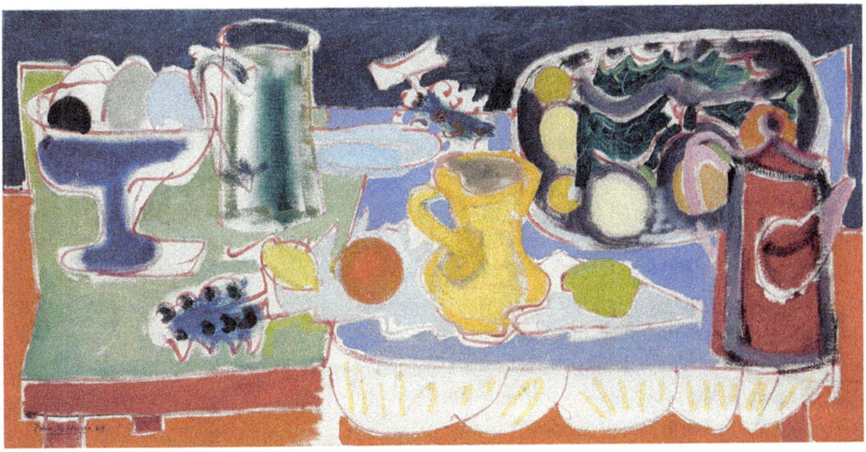

Fig. 7 Patrick Heron *The Long Table with Fruit*, 1949. Oil on Canvas, 457 × 914 mm (Tate, London. © Estate of Patrick Heron. All Rights Reserved, DACS 2014)

surface, such as the yellow detailing on the table cloth, and he also experimented with modifying the consistency of oil paint with turpentine (Gayford 1998; McNay 2002; Tempest 2009).

This painting has localised water sensitivity, occurring only in the orangey-red background and the yellow used for the central area of the jug. The vermilion, red paint film has a slight light scattering bloom on the surface and dirt particles have become embedded in the glossy skin. These two paint films contain magnesium carbonate, a marker for W&N paint indicating that water sensitivity may derive from magnesium sulphate heptahydrate formation.

The pale, water sensitive yellow shows small white crystalline material visible in the interstices of the brushwork. Small quantities of magnesium sulphate heptahydrate were observed on the surfaces of both of these paint films. The red paint passage shows small crystal rods observed beneath an uneven medium skin, as well as areas of uneven medium-distribution which may be compounding the water sensitivity in this case.

Water sensitivity in this case study is interesting in light of its conservation history and highlights an important observation. The painting was surface cleaned upon its acquisition in 1994, which is likely to have removed most of the epsomite present, explaining why only a few crystal shards are visible in SEM. Vermilion W&N swatches from the 1940s have a greater density of epsomite, in comparison to those crystals observed on the painting. This may be due to two reasons. Firstly, considering the date of cleaning and the lack of epsomite on swatches from the 1990s, it is unlikely new crystals have emerged since cleaning the painting, and so those identified are remnants of the earlier degradation product. The implication being that once removed and contained with a suitable environment, epsomite formation is likely to be arrested. Secondly, whilst the W&N swatches and painting are of comparable date and same pigment, the different environmental conditions they may have been exposed to since the 1940s may have resulted in marked differences in epsomite quantities originally, depending on their overall exposure to sulphur dioxide.

These case studies demonstrate that water sensitivity in paints containing magnesium carbonate, indicative of W&N paints, show the presence of magnesium sulphate heptahydrate crystals on the surfaces of these paint films. Where other case studies were examined and the water sensitive paint passages did not contain magnesium or show the presence of epsomite, this indicates that other casual factors, as yet unidentified, are contributing to sensitivity in other paint manufacturers. The case studies show the presence of epsomite on the surfaces of paintings and were unable to identify epsomite in the bulk films, consistent with findings from the W&N swatches. They also demonstrate that environmental conditions are critical for epsomite to form; once a painting is kept within a climate controlled, low sulphur dioxide environment, the probability of epsomite formation diminishes significantly. With overall declining levels of sulphur dioxide in the atmosphere and an absence of magnesium carbonate in W&N paint films post 1975, the probability is again reduced. In this way, if kept in suitable environments then epsomite formation is unlikely to continue.

Conclusion

The present study suggests that water sensitivity in artists' oil paints that can be attributed to the formation of epsomite is largely a surface phenomenon. The methods used in the study could not determine with precision whether epsomite as a degradation product is formed in the bulk paint film, as well as on the surface, although one single sample indicated that it is possible. Further studies using micro-analytical methods not available in the present study may resolve this question. However, circumstantial evidence from water swabbing and immersion tests showed that paints with surface epsomite exhibited extreme sensitivity to water, while the bulk paint showed less sensitivity, even on immersion in water. This suggests that the concentration of epsomite, if it is present in the bulk, is not sufficient or not accessible enough to water applied by swabbing to disrupt the integrity of the film.

The implication for conservation surface cleaning treatment is that there is likely to be a point at which paint films significantly reduce in water sensitivity just beneath the surface of the film. This sharp decline occurs when the epsomite crystals and medium skins have been removed. These findings may be critical in helping to define a treatment strategy that could limit the removal of original components of the paint film during cleaning, as it may be possible to thin or remove the water soluble skin which contains imbibed dirt particles without affecting the bulk of the paint film.

Acknowledgements This research was undertaken in collaboration between the Courtauld Institute of Art, London, TATE, London and the Cultural Heritage Agency of the Netherlands (RCE), Amsterdam. The authors would like to thank: Luc Megens, Jolanda van Iperen, and Marc Vermeulen (RCE), Annette King and Nelly von Aderkas (TATE), Alex Ball (National History Museum, London), Bill Luckhurst (King's College, London), Ian Garrett (Winsor & Newton), as well as Laura Mills, Hannah Tempest and Genevieve Silvester for their previous research which has helped to inform this study.

References

Barry RG, Chorley RJ (2003) Atmosphere, weather and climate, 8th edn. Routledge, London

Burnstock A, Van den Berg KJ, de Groot S, Wijnberg L (2006) An investigation of water-sensitive oil paints in twentieth century paintings. In: Learner T, Smithen P, Krueger J, Schilling M (eds) Modern paints uncovered: proceedings from the modern paints uncovered symposium, Tate Modern, London, 16–19 May 2006. Getty Conservation Institute, Los Angeles, pp 177–188

Clarke M (2009) A nineteenth-century Colourman's terminology. Studies in Conservation 54(3):160–169

Davies HM (1978) Francis Bacon: the early and middle years, 1928–1958. Garland Publishing Inc., New York/London

Durham A (1985) Note on technique. In: Francis Bacon: with essays by Dawn Ades & Andrew Forge, Exhibition at Tate Gallery, London, 22 May–18 August 1985. Thames & Hudson, London, pp 231–233

Gale M, Stephens C (eds) (2008) On the margin of the impossible. In: Francis Bacon, Exhibition at Tate Britain, London, 11 September 2008–4 January 2009. Tate Publishing, London, pp 14–27

Garrett I (2011a) 21st April. [Interview with Burnstock A, Learner T, Ormsby B, Van den Berg KJ]. Tate, London

Garrett I (2011b) 15th December. [Interview with Cooper A]. Winsor & Newton, London

Garrett I, Waldron P (2010) 2nd February (Interview with Burnstock A, Van den Berg KJ). Winsor & Newton, London

Gayford M (1998) "Looking is more interesting than doing anything else, ever": an interview with Patrick Heron. In: Sylvester D (ed) Patrick Heron, Tate Gallery, London, 25 June – 6 September 1998. Tate Gallery Publishing Ltd, London, pp 18–48

Gaylor S, Burnstock A, Vasconcelos A (2008) A technical study of seminal paintings from the 1960s by Robyn Denny in the modern British collection at the Gulbenkian Foundation, Lisbon. Zeitschrift für Kunsttechnolologie und Konservierung (ZKK) 22(1):63–72

Gowing L (1983) Francis Bacon. In: Francis Bacon: paintings 1945–1982 Exhibition Catalogue, National Museum of Modern Art, Tokyo 30 June–14 August 1983, pp 104–110

Hackney S (1999) Francis Bacon: three studies for figures at the base of a Crucifixion c. 1944. In: Hackney S, Jones R, Townsend J (eds) Paint and purpose: a study of technique in British art. Tate Gallery Publishing Ltd, London, pp 176–181

Leone B, Burnstock A, Jones C, Hallebeek P, Boon J, Keune K (2005) The deterioration of calcium sulphide yellow artists' pigments. In: ICOM Committee for Conservation (ICOM-CC) 14th triennial meeting The Hague, 12–16 September 2005: Preprints, vol II, James & James, London, pp 803–813

McNay M (2002) Patrick Heron: St Ives artist. Tate Publishing, London

Mills L, Burnstock A, Van Keulen H, Duarte F, Van den Berg KJ (2008) Water sensitivity of modern artists' oil paints. In: ICOM Committee for Conservation, 15th triennial meeting, New Delhi, 22–26 September 2008: Preprints, vol II. Allied Publishers, New Delhi, pp 651–659

Remington JS, Francis W (1954) Pigment: their manufacture, properties and use, 3rd rev edn. Leonard Hill Ltd, London

Roy A (2007) Cobalt blue. In: Berrie BH (ed) Artists' pigments: a handbook of their history and characteristics, vol 4. National Gallery of Art, Washington, pp 151–178

Russell J (2009) The mystery of the paint. In: Harrison M (ed) Francis Bacon: New studies (centenary essays). Steidl, Göttingen, pp 233–245

Silvester G (2011) An experimental investigation of the formation of sulphates in oil paint films exposed to gaseous sulphur dioxide with particular reference to the relationship between these sulphates and water sensitivity. Dissertation, Courtauld Institute of Art

Silvester G, Burnstock A, Megens L, Learner T, Chiari G, Van den Berg KJ (2014) A cause of water-sensitivity in modern oil paint films: the formation of magnesium sulphate. Studies in Conservation 59(1):38–51

Smith SJ, Pitcher H, Wisley TML (2001) Global and regional anthropogenic sulphur dioxide emission. Global Planet Change 29(1–2):99–199

Sylvester D (1975) Interviews with Francis Bacon. Thames & Hudson, London

Tempest H (2009) Water sensitive oil paint: an experimental investigation, with case studies, further characterising the causes of the phenomena. Dissertation, Courtauld Institute of Art

Tempest H, Burnstock A, Saltmarsh P, Van den Berg KJ (2010) Progress in the water sensitive oil project. In: Charola A, Mecklenburg M, Domenech-Carbo M-T (eds) New insights into the cleaning of paintings: proceedings from the cleaning 2010 international conference, 26–28 May 2010. Universidad Politecnica de Valencia, Valencia, pp 107–114

Thomson G (1986) The museum environment, 2nd edn. Butterworth-Heinemann, Oxford

Winsor & Newton (1982) Paint & painting: an exhibition and working studio sponsored by Winsor & Newton to celebrate their 150th anniversary, exhibition held at the Tate Gallery, London, 9 June–18 July 1982. Tate Gallery Publications Department, London

Winsor & Newton (1984) Artist's Colourmen's story. Reckitt & Coleman Leisure Ltd, Middlesex

An Investigation into the Viability of Removal of Lead Soap Efflorescence from Contemporary Oil Paintings

Alysia Sawicka, Aviva Burnstock, Francesca Caterina Izzo, Katrien Keune, Jaap J. Boon, Kathrin Kirsch, and Klaas Jan van den Berg

Abstract Metal soap efflorescent hazes and crusts are one of the most visually disturbing of all metal soap related alterations in paintings. The phenomenon presents particular challenges for conservators as the surface deposits that form are insoluble and intimately bound with the paint film. The superficial deposits accumulate following migration of metal carboxylates, a product formed within the paint film from reaction of free fatty acids with metal ions from pigments and/or driers. Upon deposition at the surface, the metal carboxylates undergo a process of remineralisation, the products of which are determined by the atmospheric conditions that the painting is exposed to. The material can appear optically similar to the whitish haze of organic efflorescence though manifestations also range from

A. Sawicka (✉) • A. Burnstock
Department of Conservation and Technology, The Courtauld Institute of Art, Somerset House, The Strand, WC2R 0RN London, UK
e-mail: alysia.sawicka@gmail.com

F.C. Izzo
Ca' Foscari University of Venice, Venice, Italy

Cultural Heritage Agency of the Netherlands, Amsterdam, The Netherlands
e-mail: fra.izzo@unive.it

K. Keune
Van 't Hoff Institute for Molecular Sciences (HIMS), University of Amsterdam, Amsterdam, The Netherlands
e-mail: k.keune@uva.nl

J.J. Boon
JAAP Enterprise for Art Scientific Studies, Nieuwendammerdijk 79, 1025LD Amsterdam, The Netherlands

K. Kirsch
Restauratieatelier Amsterdam, Asterweg 17 A13, 1031 HL Amsterdam, The Netherlands
e-mail: info@restauratieatelieramsterdam.nl

K.J. van den Berg
Cultural Heritage Agency of the Netherlands, Amsterdam, The Netherlands
e-mail: k.van.den.berg@cultureelerfgoed.nl

K.J. van den Berg et al. (eds.), *Issues in Contemporary Oil Paint*,
DOI 10.1007/978-3-319-10100-2_21

crystalline clusters to thick crusts, and in some cases the painting may consequently be rendered partially unreadable. This study aimed to investigate the removal of visually disturbing lead soap efflorescence with Ethylenediaminetetraacetic acid (EDTA), which sequesters the lead ions from the remineralised crust. Through a careful balance of variables such as pH, concentration and application, it is feasible to thin the superficial material significantly. By this means it is possible to saturate the surface of the paint film and render the previously disfiguring efflorescence transparent. Paintings previously treated with this method show no signs of recurrence and thus it appears to be a viable, long-term treatment option.

Keywords Efflorescence • Lead soaps • Free fatty acids • Remineralisation • Treatment • Ethylenediaminetetraacetic acid

Introduction

The Problem of Lead Soap Efflorescence

From the moment of execution, paintings undergo numerous chemical and physical processes, through which the paint film cures, ages and subsequently degrades. The nature of the surface of the painted picture is thus determined to a large extent by the chemical and physical processes that take place within the paint film. These processes within the three-dimensional network of a mature paint film can affect the structure's stability and may result in the migration of mobile components within the paint layers or to the surface of the painting.

This migration is the source for three distinct surface phenomena: metal soap aggregation, fatty acid efflorescence and metal soap efflorescence. Whilst the former have been the subject of extensive research,[1] and are thus commonly identified, metal soap efflorescence, namely lead soap efflorescence, has only recently been investigated (Van Loon et al. 2011; Van Loon 2008; Keune et al. 2007). Limited recognition of lead soap efflorescence as a distinctly different degradation phenomenon is due to the small number of analytical studies undertaken. The matter is further complicated by the visual similarity of the effects of some lead soap efflorescence to that caused by surface accretions of free fatty acids. The difference in solubility of the degradation products on empirical testing has provided criteria for identification of different causes of light scattering at the surface of paintings.

The phenomena of unknown cause can be described in general terms: *whitening*, *hazing*, or *white surface depositions*. However, clarification of specific terms may lead to greater enlightenment regarding the mechanism and appearance of lead soap

[1]Research carried out as part of the NWO priority programme MOLART (Molecular Aspects of Ageing in Painted Works of Art) (1995–2001) and the De Mayerne programme (2002–2006) at the FOM Institute, AMOLF.

efflorescence. The following terms clarified by Van Loon (2008), are considered appropriate, although meanings may differ from previous applications:

Chalking implies light-scattering of the surface film due to the breakdown of binding medium.

Blooming describes the migration of extractable components within the paint film and subsequent surface deposition, typically applicable to white organic precipitates.[2]

Efflorescence also refers to the process of migration and deposition, though is more specific for inorganic deposits.

In this paper, the phenomenon will thus be refered to as efflorescence, and quantified as inorganic/metal soap or organic/fatty acid to avoid confusion.

Hypotheses for the Formation of Lead Soap Efflorescence

The formation of metal carboxylates is a normal reaction product of mobile fatty acids with metal ions from pigments and/or driers. Lead and zinc carboxylates are most commonly identified, likely due to the fact that lead- and zinc-containing pigments are known to promote the de-esterification of the paint, although other forms may be encountered according to the pigments present.[3] When uniformly dispersed, the metal soaps are thought to confer flexibility to the paint film, though there appears a tendency towards aggregation, resulting in the large occluded masses of metal soaps which expand and finally break through the surface. Lead soaps may also migrate without aggregation and migrate towards paint-air interfaces, such as the surface of the paint film or cracking within a paint film. Upon deposition at the surface, the lead soaps also undergo a process of remineralisation, precipitating and crystallising in an energetically favourable orientation due to further reaction with atmospheric gases. The degree and nature of the remineralised salts are dependent on a variety of factors including the nature of the soap, the paint composition, treatment history and the environment to which the painting has been exposed (Van Loon 2008). It has been postulated that tendency towards migration without aggregation may be related to a low degree of metal coordination within the paint layer and susceptibility for acid hydrolysis (Noble and Boon 2007). This may account for the observation of lead soap efflorescence primarily on highly

[2]Though blooming is defined in this manner by both Van Loon (2008) and Akerlund (2012), the term has also been widely applied to visual changes in the varnish layer, most often due to penetration of moisture.

[3]For example potassium soaps have been demonstrated as a result of the leaching of potassium from smalt (Spring et al. 2005).

oil-absorbent porous paint layers containing pigments such as chalk, lakes, carbon black and earth pigments identified by Van Loon (2008).[4]

Efflorescence in Nineteenth- and Twentieth-Century Paintings

The propensity of paintings from the nineteenth and early twentieth centuries to metal soap efflorescence is likely to relate to the specificities of the materials employed. Following the invention of the collapsible paint tube and the opening of the first factory for pigment preparation, a number of pivotal developments were made in commercial paint production, including the incorporation of additives and manufacturing methods that affected the oxidation process, promoted hydrolysis and introduced or increased the proportion of fatty acid containing materials.

Developments in the composition of commercially prepared ground layers may be an important causal factor: the incidences of lead soap protrusions, efflorescence and associated variable translucency observed on unpainted, pre-primed canvases in the Olana archive are hypothesised to have been caused by migration of lead ions from ground or priming layers (Zucker 2007). The demand for a uniform, smooth surface that would absorb excess oil, supple enough to be rolled for storage and shipping, led to modification of recipes often at the expensive of quality. Given the profit motive that dominated the concerns of artists' colourmen, this modification included the addition of extenders, and unstable materials, which may have influenced the formation of degradation products. For example, the technical and economic difficulties in mass producing oil priming, which requires time to dry and mature, led to the use of power siccatives, and the brittleness which these imparted to the paint film was in turn counteracted through the addition of saponified fats and waxes.

Contemporary treatise, including Field's *Chromatography* (1862) and Church's *The Chemistry of Paints and Painting* (1890), discuss the use of sugar of lead (lead acetate drier), both incorporated in paints and strewn on pictures whilst wet, to promote quick drying. Notably, their warning of resulting 'efflorescence' is remarkably accurate; Church describes one of the results of the addition of lead acetate as:

> . . . the production of an immense number of small spots in the picture, sometimes appearing through the surface-varnish in the form of a white efflorescence. This efflorescence consists at first of lead acetate in crystals, but these soon attract carbonic acid from the air and become lead carbonate, which, in its turn, is changed into lead sulphide by the action of sulpheretted hydrogen.

---

[4]Ongoing research is being undertaken as part of the NWO Science4Arts PAinT project by Van Loon and Keune (http://www.s4a-paint.uva.nl/research-topics). Research will focus on the timescale of the phenomena and the identification of other key factors for soap formation migration, dissolution, recrystalisation and efflorescence in oil paint.

Paint formulations of the period are also critical to the formation of efflorescence as they offer a greater source of free fatty acids (Akerlund 2012). The inclusion of chemically inert bulking additives including barium sulphate, hydrated aluminium silicate and calcium carbonate, and the substitution of titanium dioxide for the carboxylate-forming zinc- and lead white, passively promote efflorescence through their exclusion from ionic reactions within the paint film.[5] Furthermore the absorbency of chalk in particular necessitates an oil-rich paint film, and the pigment-volume concentration of certain paint films illustrate that they effectively provide reservoirs of free fatty acids due to lack of metal coordination (Hinde et al. 2011; Ferreira et al. 2011). Alternatively components of efflorescence may themselves be included as additives, Tempest et al. (2013) observing a proportional relationship between the presence of stearic acid additives and of fatty acid bloom, and Mills et al. (2008) noting the presence of metal stearate additives as conducive to efflorescence and metal soap aggregation.

Treatment of Lead Soap Efflorescence

Suitable treatments for lead soap efflorescence have not been thoroughly investigated. The disfiguring material is insoluble in the range of organic solvents utilised for the removal of varnish, old restorations and indeed fatty acid efflorescence, and its intimate integration with the underlying paint precludes safe mechanical removal. The conservator thus faces the challenge of developing alternative methods. Effective removal of inorganic efflorescence from paint surfaces has not been considered feasible thus far, without compromising original paint beneath.

The most common approach is to reduce the degree of whitening by saturation with a low-viscosity resin, which fills the voids and micro-fissures of the surface film, reduces light-scattering and thus renders the efflorescence more transparent. Saturation however is not always feasible; it may be undesirable, as in unvarnished or locally-varnished paintings, or indeed impossible. In the latter scenario, surface whitening may be aesthetically integrated through retouching, though this could lead to large-scale covering of original paint, given the commonly widespread nature of the efflorescence. Van Loon (2008) concludes that in this case the only option is reconstruction by retouching. However, previous conservation treatments highlight an alternative. Chelating agents including triammonium citrate, diammonium hydrogen citrate and ethylenediaminetertraacetic acid have been deemed successful in a number of treatments aiming to remove inorganic salts from the surface of paintings (Morrison et al. 2007; Slavin 1990; Treatment reports CIA 1842 and CIA

[5]Hinde identified the inability of titanium dioxide to form stabilizing carboxylates as a key factor in the formation of fatty acid efflorescence in *Composition* by Serge Poliakoff (1956) (Hinde et al. 2011).

1939) though the action and the nature of the resultant surface remains unclear. This study investigates the application of the selected chelant for the removal of lead soap efflorescence in relation to these unresolved issues.

Introduced for the surface cleaning of paintings in the 1980s, chelating agents are now widely employed tools at the conservator's disposal, though concern has been raised over the use of chelants on painted surfaces. Pigmented films typically contain cations capable of sequestration, and thus it is critical to select an appropriate chelating agent based on an intimate knowledge of the material constituents of both the substance to be removed and the substrate it is to be removed from. For the removal of lead-containing material, the criteria proposed would thus be a chelant with a strong affinity for lead ions, a weak affinity for other metal ions potentially present in the underlying paint film and the ability to chelate in a pH range recommended for use on oil paint films, generally reported as a pH range from 5.5 to 8.5.

Based on these parameters the chelating agent ethylenediaminetetraacetic acid (EDTA) was selected for trial. Thorough knowledge of its chelation mechanism and parameters of use are published in the literature of the detergency, medical and soil industries (Hong and Jiang 2005; Kim et al. 2003; García et al. 2011). The chelant has a strong affinity for lead ions to enable efficient chelation of the efflorescent material within the pH range 5.5–8.5 and weaker affinity for most other metal ions potentially present in the underlying paint film to minimize risk of chelation of original material (Fig. 1). Its solubility in water facilitates appropriate methods of application, whilst familiarity within the conservation profession facilitates a critical judgment of the variable parameters of deployment beyond this study. There is both a history of use for the successful treatment of lead (and zinc) soap efflorescence, having been reported as empirically more effective for this purpose than other chelating agents such as TAC, and a potential for practical use in the selected case paintings for this study. Although EDTA's affinity for a select number of other cations (notably its particularly strong affinity for iron) warrants concern, literature suggests selectivity for lead might be increased by variation of parameters, namely pH. Conditional stability constants indicate preferential sequestration of Pb^{2+} ions over Fe^{3+} ions at pH 6.5–11. Given the prevalence of iron oxide pigments in paint films it was deemed critical to test this variable in order to provide guidelines for the safe use of EDTA on oil paint films.

Analytical Materials and Methods

Testing Materials and Approaches

Test surfaces were commercially-prepared, acrylic-primed canvases, to which a thin, even layer of Raw Sienna oil paint had been applied by brush in 2009. The paint films were completely cured, though remained flexible. With the exception of a

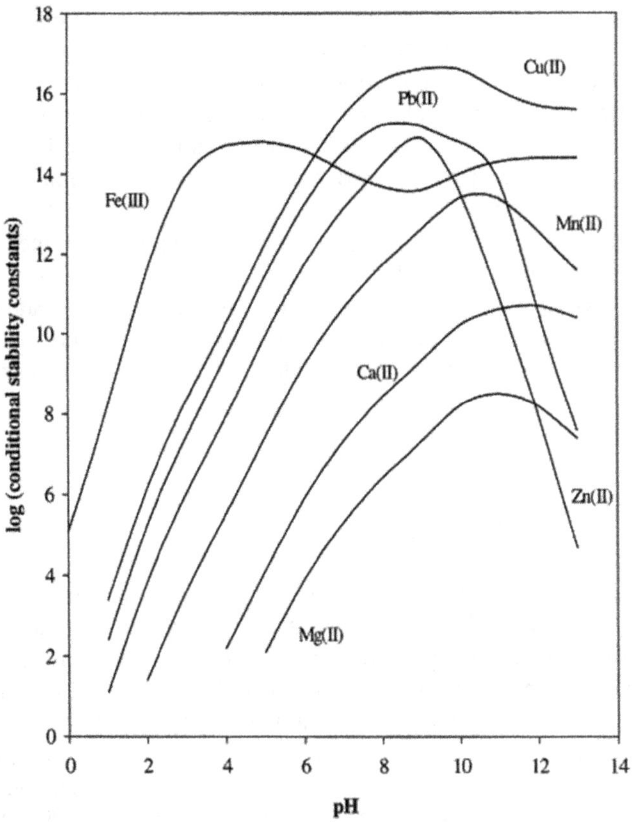

Fig. 1 Conditional stability constants of metal-EDTA complexes (Kim et al. 2003)

control surface, a layer of lead stearate was applied to the substrates, simulating the agglomeration of lead soaps on the surface of paintings. A total of six surfaces were prepared, four of which were judged as useful simulations of the intimate integration of the lead soaps with the paint film; (1) pressing the lead stearates onto the surface, (2) application of lead stearate onto reapplied wet raw sienna paint, (3) application of lead carbonate onto reapplied wet raw sienna paint and (4) application of a mixture of lead stearate and carbonate onto reapplied wet raw sienna paint.

The chelating ability of EDTA is dependent on a number of variables; EDTA is available both in its pure form and as several salts (differing in their respective solubilities and recommended for different pH ranges), the concentration of an EDTA solution determines the number of ligands available for the complexation of metal ions, the pH of a solution can optimize preference for different cations, and application methods affect the penetration of the solution, the mechanical action required and the evenness of removal. EDTA in its pure form was selected for testing, a pH range 5.5–8.5 (controlled by addition of ammonium hydroxide) was selected as appropriate for use on oil paint films, and concentrations of 2.5, 5, 7.5 and 10 % were chosen to cover the range of concentrations previously reported as effective. The solutions were applied as free solutions with a cotton swab, gelled with both low and high molecular weight methylcellulose, and gelled with Agar. Following treatment, surfaces were cleared with deionised water by swab rolling for 30 s, or in the case of areas tested with Agar gel, by a 1-min application of deionised water-Agar gel.

Efficacy of removal was assessed by a number of complimentary methods; visual observations and photomicrographs were recorded before and after treatment, SEM imaging was used to investigate the nature of the surface following treatment and the distribution of remaining lead compounds, and cross sections were compared to a reference sample to analyse the nature of the lead compound/paint layer relationship and the effect of the chelant on the iron oxide paint film.

Analytical Techniques

A HIROX digital microscope KH-7700 was used to capture full in-focus microphotographs and three-dimensional images, stacked by an algorithm determining in focus planes of a series of images (recorded at intervals) by contrast. The three-dimensional data enabled accurate measurements of surface profiles before and after cleaning. A dual illumination revolver zoom lens MXG-2500REZ with magnifications ranging from 35× to 2,500× was used and a FOBA stand used to compensate for the three-dimensionality of the objects studied. Additional photomicrographs were captured with a Leica M165C Microscope with Leica DFC 500 camera and Leica Application Suite software (Version 3.7).

SEM was undertaken at Rijksdienst voor Cultureel Erfgoed, Amsterdam (RCE) with Dr. Ineke Joosten – JEOL JSM-5910 LV Variable Pressure SEM used for SE and BSE imaging – The Natural History Museum, London (NHM) with Dr. Alex Ball – ZEISS Ultra Plus Gemini FEG used for SEI and LEO 1455 Variable Pressure SEM used for BSE – and King's College, London (KCL) – Hitachi S4000 SEM FEG used for BSE and EDX with INCA microanalysis suite.

ATR-FTIR on sample scrapings was performed using a Perkin Elmer Spectrum 1000 FT-IR combined with a GrasebySpecac Golden Gate Single Reflection Diamond ATR, from 4,000 to 400 cm^{-1} for 16 scans and a resolution of 4 cm^{-1}. ATR-FTIR on embedded cross-sections was performed using a Bio-Rad Stringray, combining the Bio-Rad FTS-6000 spectrometer equipped with a Bio-Rad UMA 500 infrared microscope with a 64×64 mercury-cadmium telluride (MCT) focal plane array camera. Analysis of the sample was carried out in reflection mode recorded with a 16 cm^{-1} spatial resolution, a step scan frequency of 1 Hz, and an UDR of 4. The reflection measurements were corrected by the Kramers-Krönig transformation.

Py-GC-MS analysis was performed using a Focus ISQ Thermo Quest mass spectrometer with a 20 m SLB5ms column, id 0.18 mm, ft 0.18 μm. One-shot

Pyrolysis at 550 °C. The inlet temperature was 300 °C, while the MS interface was at 280 °C. The temperature programme was set from 35 to 300 °C with a ramp of 10 °C/min, held isothermally for 3 min; total runtime 25 min. The MS was run in Full Scan mode (m/z 40–600) with a speed of 1.9 scans/s. Samples were derivatised with TMAH 2.5 %, using 2 internal standard (fatty acid C13 and C23) and data was processed using Xcalibur 1.4 software.

Test Results[6]

Concentration

An increase in concentration resulted in increased efficiency, with lead compounds affected faster and a greater degree of reduction achieved. Low concentrations (i.e. 2.5%w/v) resulted in a greater degree of damage to the underlying paint film, particularly when applied by swab. By reducing the quantity of EDTA available for chelation, the mechanism of removal was primarily mechanical, leading to redistribution of material over the test surface and damage of the paint film in test surfaces 2–4 where the lead compounds were more intimately bound with the substrate. The result of removal by mechanical action is evident in cross-section where craters remain as a result of dislodged material and by SEM imaging, parallel abrasion marks resulting from the rolling of the swab. A higher concentration did not consistently result in a greater degree of disruption to the paint film, the greater risk of damage to the paint substrate offset by greater efficiency, which consequently required less prolonged exposure of the paint film to the cleaning reagent, and a lesser degree of mechanical action.

pH

A higher pH led to increased efficiency both in the speed of action and the degree of reduction achieved, though it also commonly affected the paint layer beneath attested by colour removed on the swab. The resulting surface appearing rougher in SEM, the smooth organic surface layer of the medium-rich paint having been etched. However, again a more efficient removal reduced the length of application required, which could potentially offset the vulnerability of the paint film. Where the lead compounds were intimately bound with the paint surface, colour on the swab was more commonly noted for solutions of pH 5.5 than for pH 6.5, particularly in the case of lower concentrations of EDTA gelled with low molecular weight methylcellulose. Equivalent application in free solution did not consistently result in the affection of the paint substrate, and thus the detrimental affection appears due to

[6]Comprehensive results of tests carried out can be found in (Sawicka 2013).

the chelation of Fe ions rather than mechanical action. This correlates to conditional stability constants, which indicate a preference for Fe^{3+} ions below a pH of 6.5 (Fig. 1).

Application Method

Reduction in mechanical action necessitated by the application method largely minimized the damage caused to the paint substrate. Colour removed during testing with methylcellulose gels was partly due to the mechanical action of the swab during clearance, as reflected in the results of Agar gel testing, where (with one anomaly) no effect on the paint film was recorded. The lack of mechanical action due to clearance with Agar, also facilitated an even reduction of the lead compounds, SEM images illustrating the three-dimensional surface remaining in comparison to those smoothed and flattened by the rolling of a swab. Differing viscosities of methylcellulose gel were seen to be equally as effective. However, increased viscosity, which restricted interaction of the cleaning reagent and further reduced penetration due to the increased water retention of the gel, resulted in fewer cases of damage to the paint substrate.

Practical Application

Two case studies exhibiting lead soap efflorescence were selected to further investigate the potential of EDTA to remove the inorganic surface deposits beyond the limitations of a simulated system. A variety of manifestations of the inorganic efflorescence was visible across the two paintings, providing ample opportunity to assess the viability of EDTA as a treatment option for the range that a conservator may face. Testing was guided by the conclusions drawn from the simulated surfaces and the efficacy of removal was assessed by the same means.

Case Study 1: Portrait of Pieter Jacob Teding van Berkhout

Background

Portrait of Pieter Jacob Teding van Berkhout, executed in oil on a commercially-primed, basket-weave canvas (85.2 cm × 64.3 cm), was painted by the Dutch artist Sigisbert Chrétien Bosch Reitz in 1891/1894 (Fig. 2). The painting displays a white, unevenly-distributed surface material, py-GC-MS, ATR-FTIR and SEM-EDX analysis of which indicates a composition predominantly of lead stearate, lead

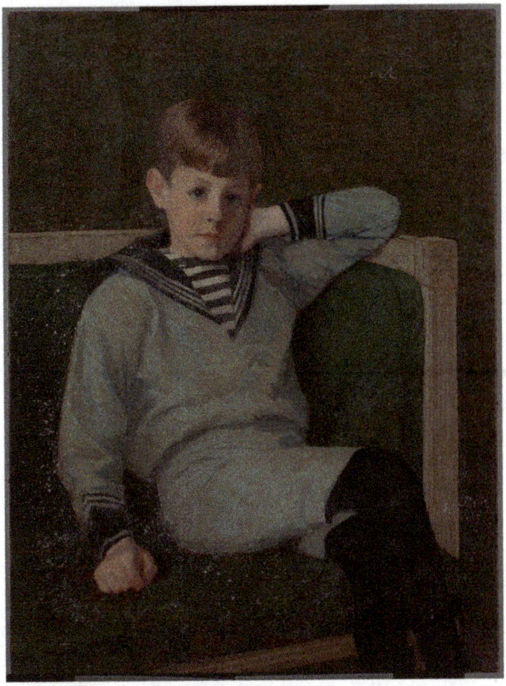

Fig. 2 *Portrait of Pieter Jacob Teding van Berkhout*, Sigisbert Chrétien Bosch Reitz, 1891/1894. Before treatment

palmitate and mineralized lead soaps.[7] The material appears in a variety of forms across the painting; as patches of densely packed crystals, as fine rod-shaped crystals in the valleys of brushstrokes and as a thin haze of particulate material. Distribution and form appears uncorrelated to the painting's composition although spots of the dense crystalline material are most prevalent in the green cushion (lower left quadrant), whilst the black stockings display a more homogenous distribution of fine crystals. The paler green background appears largely unaffected although crystalline material is visible in the upper-left and lower-right corners. Indeed, the horizontal band efflorescence across the lower right, suggests that the painting was partially protected by some means (the painting was part of a stack of paintings stored in a shed for some years) rather than a particular invulnerability to efflorescence in this area. Though the distribution of efflorescence cannot be rationalised by the presence of a specific material, it does appear influenced by material composition. Crystalline material is congregated around the edges of paint passages, in the organic material along the jaw, and around a series of glossy drips in the sitter's face. Crystal growth extends over cracks in the paint film though concurrently cracks pass through the efflorescence, indicating an extended period of formation.

Examination of the efflorescing surface with electron microscopy demonstrates the superficial growth of highly-scattering, plate-like structures, accumulating

[7] A previous technical study was carried out by Keune et al. (2007).

beneath a film of varnish. Aggregation of these lead rich shards has caused disturbance to the varnish film, undulations indicative of the three-dimensional structures beneath. Imaging of cross-sections with SEM illustrate the intimate integration of crystals within the paint surface, and the well-developed lamellar arrangements in the surface deposit, taken as an indication of remineralisation within lead soap aggregates (Keune et al. 2007).

Testing

Dense clusters of crystalline material located on the green cushion (lower left quadrant) were selected for testing of EDTA solutions, ample testing sites allowing as accurate a comparison of efficacy as possible. The surface was examined before and after cleaning with a Hirox 3D digital microscope.

Concentration

Free solutions displayed an increased efficiency with an increase in concentration. After c.50 swab rolls, solutions of 2.5 and 5 % (pH 6.5) did not result in any visible change, whilst thinning of the efflorescence was notable at higher concentrations. Despite this increased efficiency in the chelation of the lead-based efflorescence – including material lying beneath the surface of the varnish, particularly along lines of cracking – it did not appear to independently affect the metallic pigments of the paint film beneath.

pH

An increase in pH resulted in a greater degree of reduction. However, a higher pH corresponded to an incremental affection of the paint substrate. Given that no sensitivity was noted for any solutions of pH 6.5, it may be hypothesized that damage caused to the paint layer is partly due to affection of organic components at an alkaline pH, which also facilitated the increased reduction.

Application

Application of the cleaning reagent affected action and controllability. Whilst the swabbing of free solutions enabled better assessment of the surface during cleaning, an uneven removal was evident at higher magnification. Conversely, gelling solutions with methylcellulose limited mechanical action, resulting in a more even reduction (Fig. 3), and restricted penetration of the solution, which facilitated the use of solutions at a higher pH, without damage to the underlying paint film. Though Agar gel performed similarly in this respect, the efficiency of reduction

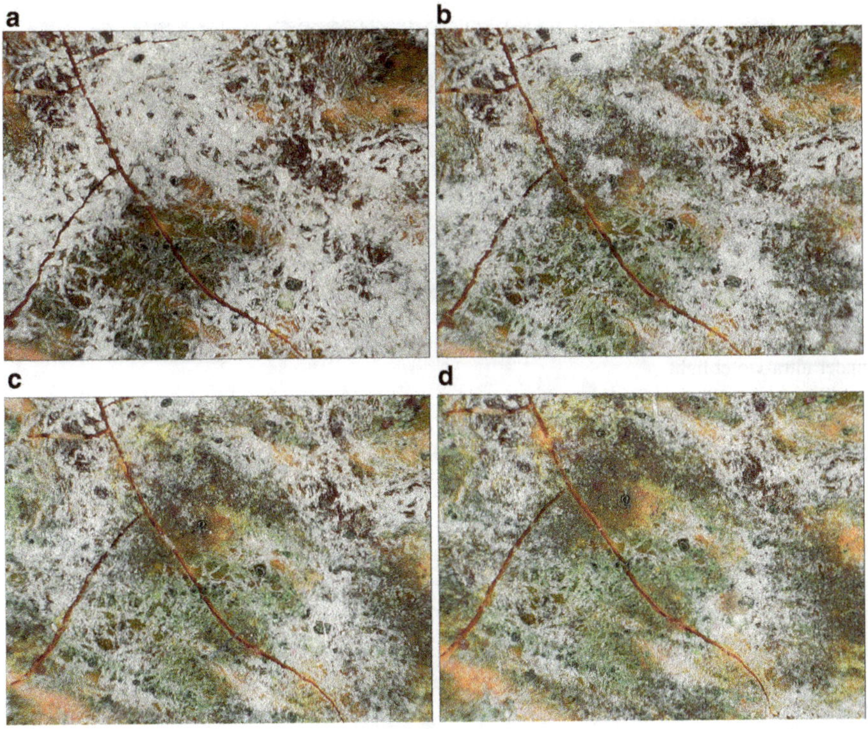

Fig. 3 Photomicrographs of lead soap efflorescence removal with MC gel, 5 % w/v EDTA, pH 6.5. (**a**) Before Cleaning; (**b**) After 30 s; (**c**) After 30 s (1 min total); (**d**) After 30 s (1.5 min total)

was reduced, requiring lengthier applications. Given the fine margin for error, it was difficult to estimate the critical point between chelation of the efflorescence and affection of the paint substrate in these prolonged applications. Shorter, repeated applications necessitated could be detrimental to the paint layers.

Evaluating the Success of Removal

Following testing, efflorescence from a site to the left of the sitter's proper right elbow was removed with a 5 % w/v solution of EDTA, pH 8.5, gelled with high molecular weight methylcellulose, applied for a total of 4 min. Visually, the surface appeared rougher than the smooth film of the surrounding varnish, though where the dense crystalline material had been located, nuances of the paint film beneath were now visible.

Examination of the treated surface in cross-section reveals a surface apparently clear of the translucent-whitish efflorescence evident in untreated samples, though ultra-violet light indicates a very thin film of the fluorescent material remains in the interstices of the rough paint surface (Fig. 4). This material is not highly scattering in

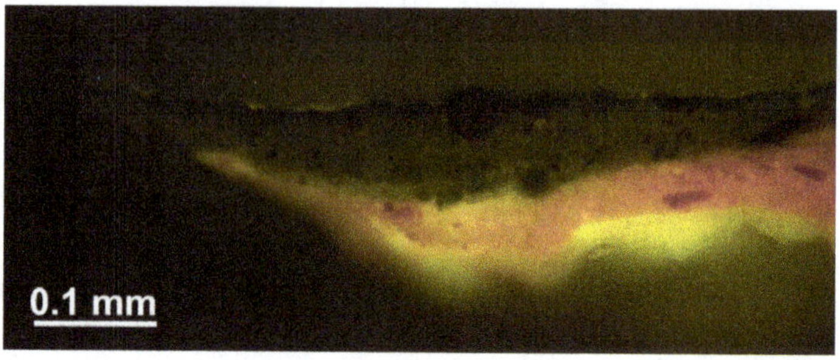

Fig. 4 Cross-section of sample taken after treatment with MC gel, 5 % w/v EDTA, pH 8.5, viewed under ultra-violet light

SEM backscatter mode, and thus it appears the lead has been successfully chelated, only the organic component remaining. Though the surface appears intact, a number of fractures are present. Whilst treatment does not seem to have initiated the cracks – present throughout the sample – it is possible the aqueous solution has exacerbated the issue. However, as the sample was divided it may be an artefact of the cleaving process.

Analysis of the treated surface with SEM reveals a fairly rough but even surface, with fine fracturing correlating to aforementioned observations. The pale surface of the sample in backscatter mode signifies the presence of a very thin film of lead-containing efflorescence with localised striations (relating to the parallel plate-like structures of the efflorescence) indicating an even and superficial thinning. Sharp-edged varnish residues suggest mechanical fracturing of the resin during treatment, though their surfaces appear unaffected, comparable to that of the intact varnish film.

Case Study 2: Return from the Front

Background

Return from the Front by the British artist Kate Elizabeth Olver c.1915, depicts a soldier returned from the front line during WW1 (Fig. 5). The painting, executed on a commercially-primed, basket-weave canvas (136.9 cm × 99.0 cm), exhibits a pale greyish surface material located predominantly over the lower third. The material is present in the interstices of the textured paint surface, as a coherent, flaky, particulate crust. Although cracks can be seen to pass cleanly through the crust, possibly indicating ageing of the paint film following its formation, the material can also be seen to accumulate around cracks, which have potentially provided transport lines for the migration of material to the surface.

Although efflorescence formation appears unattributable to distinct paint passages, some relation to the composition is highlighted by the isolated passage of

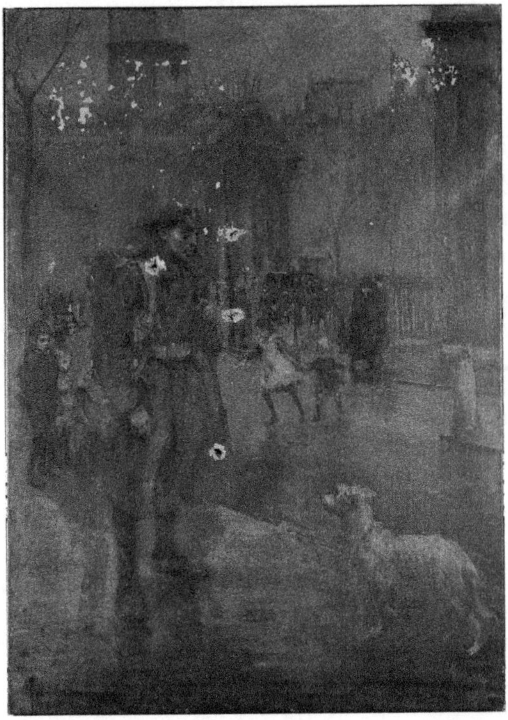

Fig. 5 *Return From the Front*, Kate Elizabeth Olver, c. 1915. Before treatment

efflorescence in the coat of the soldier, apparently demarcating the area between the soldier's underlying legs. It may thus be hypothesised that certain paint layers have facilitated or conversely inhibited the migration of efflorescence. Efflorescence appears less pronounced in areas thickly varnished, though it is unclear whether varnish has limited migration or is simply saturating efflorescence beneath. The efflorescence-free tips of the undulating paint texture also evidence a film of varnish in ultra-violet light, potentially supporting the latter theory, or alternatively suggesting that efflorescence is blocking the fluorescence of a consistent varnish film.

A combination of ATR-FTIR and py-GC-MS analysis of the superficial material suggests the presence of lead in the form of soaps. The sulphates and methane sulphonic acids detected, intimate the modification of lead soaps to an energetically-favourable mineral phase; lead sulphate. Given the presence of vermilion, the paint layers themselves present a potential source of sulphur, though given the imbibition of soot and dirt, the mineralised phase is likely formed superficially on reaction to atmospheric gases. SEM examination of an efflorescence-containing sample reveals a dense but friable crust covering the surface of the sample. It appears patchily integrated with organic medium, though largely resembles an accumulation of fine particles. The lead-containing material fluoresces with ultra-violet illumination and is strongly scattering in SEM backscatter mode, the latter illustrating the well-developed lamellar arrangements of the surface deposit and its intimate integration with the underlying paint film, with no clear interface visible.

Testing

The efflorescent crust located on the brown foreground was selected for testing of EDTA solutions, ample testing sites allowing as accurate a comparison as possible. The surface was examined before and after cleaning with a Leica M165C microscope.

Concentration

Testing free solutions of EDTA demonstrated the varying efficiency of lead chelation, increasing incrementally with an increase in concentration. Despite concern for the pigments of the underlying paint layer, concentration as an independent variable did not appear to increase the risk presented.

pH

Raising the pH of the solution resulted in increased efficiency of efflorescence removal. Although this did not necessarily affect the paint immediately beneath the crust, a pH of 7.5 and above did result in damage to the unaffected original paint film, affection exacerbated by the mechanical action of the swab. The effect was not observed in areas where a layer of varnish was present, which effectively provided a barrier against the solution.

Application

Application affected both action and controllability. Whilst swab-application of free solutions enabled a better assessment of the surface during cleaning, an uneven removal of efflorescence resulted and mechanical action of the swab caused abrasion to the vulnerable peaks of unvarnished paint. Gelling solutions with methylcellulose facilitated a more even reduction, although the mechanical action of clearance still abraded the peaks of the textured paint film when increasingly vulnerable at higher pHs. Repeated applications of gelled solutions proved more effective than a single, prolonged application and enabled better intermediary assessment of the surface, although they presented greater risk in light of the detrimental mechanical action of clearance. Agitation of the gel with a sable brush, helped to bring the efflorescent material into solution, thus providing a viable alternative.

Removal of the Varnish Layer

Removal of the efflorescence prior to the removal of vanish resulted in thin rings of the whitish surface deposit around the patches of vanish, where efflorescence and

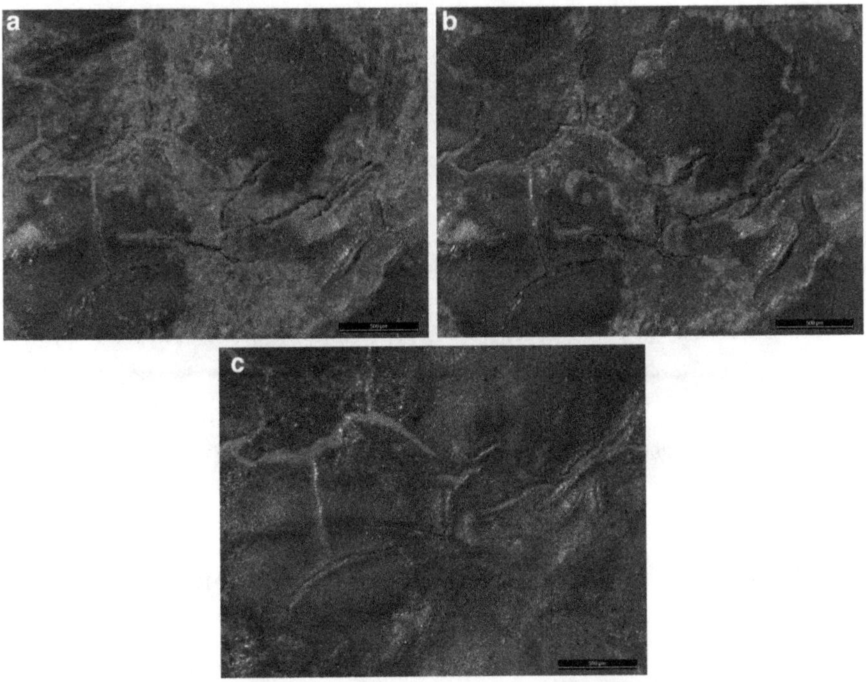

Fig. 6 Photomicrographs of lead soap efflorescence removal with MC gel, 5 % w/v EDTA, pH 6.5. (**a**) Before Cleaning; (**b**) After Cleaning (1.5 min application); (**c**) After removal of residual varnish

varnish were integrated, though these were seen to be eliminated with subsequent varnish removal (Fig. 6). Although initial varnish removal facilitated a more even reduction, residual varnish remained, necessitating further application of solvents. Given the varnish layer provided an effective barrier between paint film and cleaning solution, and that rings of residual efflorescence were cleared by the subsequent varnish removal, it was concluded that varnish should not be removed prior to treatment.

Evaluating the Success of Removal

Following testing conclusions, efflorescence from a selected site in the lower left foreground was removed with a 5 % w/v solution of EDTA, pH 6.5, gelled with high molecular weight methylcellulose, applied for a total of 2.5 min and agitated with a sable brush prior to clearance. Visually, the cleared surface appeared slightly rougher and more-light scattering than a saturated paint film, although reduction appeared very even and – with the exception of those integrated within the varnish film – very few visible residues remained.

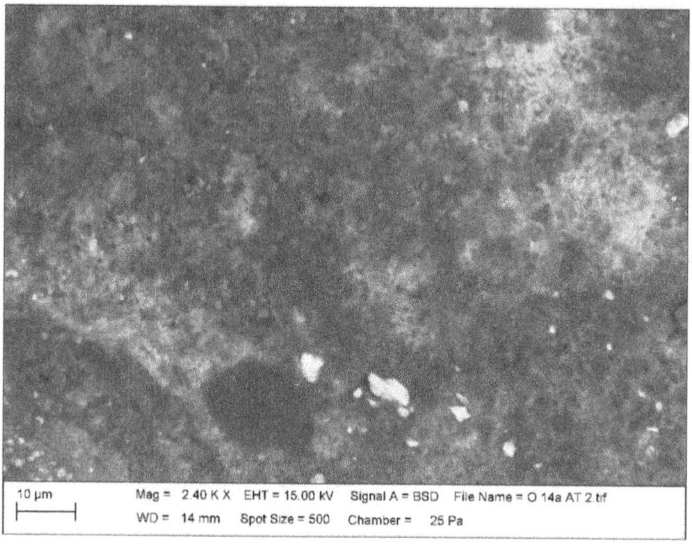

Fig. 7 Backscattered electron SEM image of sample after treatment with MC gel, 5 % w/v EDTA, pH 6.5

Microscopic examination of the treated surface in cross-section shows a thin layer of the pale greyish efflorescence, just discernable at the surface of the paint film. Ultra-violet illumination corroborates a relatively smooth and even layer, thinned just shy of the indistinct interface with the underlying paint film. Pigment particles appear partially bound within the remaining film of efflorescence and thus further thinning of the residual crust is inadvisable.

SEM analysis of the treated sample confirms the rough but fairly even surface topography. Given the intact film of efflorescence observed in cross-section, the topography likely relates to the surface of residual efflorescence as opposed disturbance of the paint film. This is corroborated by the backscatter image, which indicates thin residual efflorescence (Fig. 7). Brighter areas, indicating a greater quantity of residual efflorescence, appear partially linked to the presence of organic material remaining on the surface, which may have inhibited the access of the gel. The upper paint film appears in tact, with the underlying, pale purple layer only visible at the sheered edges. At high magnification, no residues of the gelled solution were distinguishable.

Post Treatment Recurrence of Lead Soap Efflorescence

The formation of lead soap efflorescence is influenced by both inherent material instability and environmental factors. Hypothetically therefore, the propensity for reformation of crusts after treatment may depend on both these factors, as indicated by studies that observed the common recurrence of associated phenomenon such as fatty acid bloom.[8]

It is possible that some conservation treatments may exacerbate the formation of efflorescence. Though difficult to assess, Van Loon (2008) suggested that reagents for varnish removal and surface cleaning may influence the formation of crusts and their composition, successive treatments providing the potential for a cumulative effect. Removal of the efflorescence with a chelating agent could thus impact future formation; the temporary moisture gradient set up by the aqueous cleaning method could potentially promote the migration of lead soaps. Moreover, the removal of the stable products of efflorescence may alter the equilibrium of the paint system, favoring migration to the cleaned surface.

Re-examination of two Arthur Walker paintings from the Royal Academy, which had previously undergone treatment with chelating agents to remove the visually disturbing lead soap efflorescence,[9] was carried out and the results compared to *Walker's Studio No. 1* by the same artist and *Portrait of Mrs Thorpe* by James Smetham, which had received no treatment of the superficial efflorescence. The propensity to recur was clearly demonstrated Walker's *Reclining Nude*, on which the efflorescence recurred within 4 years following its initial removal with a 2 % w/v solution of triammonium citrate. During this time the painting remained largely unvarnished, with only a thin and patchy local application over the figures and cushions (artist applied), and the painting underwent a series of flattening treatments involving exposure of the canvas to acute moisture. However following further treatment with a 2.5 % solution of ethylenediaminetetraacetic acid (pH 8.5) and the application of a synthetic varnish, no sign of recurrence was evident after a further 4 years of storage in the stable environment of the Royal Academy stores, kept at 20 °C at 50 % relative humidity. Similarly no indication of recurrence or further deterioration was identifiable on *Two Girls Reading a Book,* which was also varnished following treatment with EDTA, and *Walker's Studio No. 1* and *Portrait of Mrs Thorpe*, the efflorescence on which was dealt with through saturating with varnish and aesthetically integrated through retouching. Based on these case studies, preventive conservation measures might include considerations for environmental control and framing.

Conclusion

Identification of lead soap efflorescence is complicated by the wide range of forms it assumes. Case studies evidence manifestations varying from dense crystalline spots and thick layered crusts, to a fine haze, closely resembling that of organic

[8] A recurring cycle of organic fatty acid bloom was noted for the Stanley Spencer paintings in Sandham Chapel (Burnstock et al. (1993)).

[9] Treatment undertaken at the department of Conservation & Technology at the Courtauld Institute of Art from 2004 to 2009.

efflorescence. The visually indistinct nature of the degradation phenomenon has caused confusion, clarification further hindered by the use of ambiguous nonenclature. As relatively few paintings exhibiting this phenomenon have been thoroughly investigated, analysis of the efflorescence has proved critical in understanding the nature of the surface product and its formation. This in turn has enabled treatment methods to be more accurately tailored to the specifics of this phenomenon.

The two contrasting case studies explored, supported by testing with simulated efflorescing paint surfaces, substantiate the viability of reduction of disfiguring efflorescence with solutions of EDTA. Regardless of the contrasting forms of efflorescence, it was found to be possible in both cases to greatly reduce the material to a thin (and consequently far less disturbing) saturatable film, without damage to the underlying paint substrate. The aims of treatment must be tempered by the limitations posed by the indistinct interface between efflorescence and paint film. In this context, the reduction, though incomplete, should be viewed as a successful resolution. Though topographical analysis of treated samples indicated a fairly rough surface, the rough and uneven surface of both the untreated efflorescence and the underlying surface of the paint film, should be taken into account. In order to achieve a smooth and even surface, it would be necessary to unevenly thin the efflorescence mechanically, unviable given the propensity of mechanical action to damage the paint substrate.

The boundary between effective removal of the efflorescence and damage to the underlying paint film is relatively small, and hinges on a careful balance between the tested variables of concentration, pH and application. Test surfaces proved particularly useful despite the limitations of simulated testing, given the compositional similarity of the surface to the case studies, and consequently the conclusions drawn were borne true with practical testing. It was noted that mechanical action should be minimised, that concentration did not appear to independently affect the paint substrate and that a pH raised above 6.5 increased the efficiency of removal though simultaneously increased the vulnerability of the underlying paint. Gelling solutions of EDTA with methylcellulose proved effective in both limiting mechanical action and restricting penetration, whilst still facilitating critical judgement of when the desired level of reduction had been achieved. The mechanical action of clearance proved potentially damaging, though the risk was significantly reduced in comparison to the repetitive swab rolling of free solutions.

Practical testing highlighted the difficulties of judgement in the process of reduction. Due to the obscuring nature of some efflorescent crusts, the conservator must rely on inference as to the appearance of the original material they are attempting to reveal. However, comfort should be taken from the fact that analysis of the surface after treatment simply corroborated visual observations and thus reliance on good observational skills, due caution and the ability to make critical judgements should serve the conservator well. Following removal of the efflorescence, which does not appear to aggravate the inherent instabilities of the painting structure, recurrence of the phenomenon may be at least be stalled by maintenance of a stable environment, limiting exposure to moisture, and varnishing the painting where considered appropriate. Treatment with EDTA, with variables tailored to the specificities of the individual painting, thus offers a viable resolution to the issue of the visually disturbing aggregates, crusts and hazes of lead soap efflorescence.

References

Akerlund L (2012) Efflorescence: an investigation of selected paintings from the 19th to the 21st century with a preliminary experimental study of the role of moisture in the development of efflorescence. Courtauld Institute of Art Third Year Project, 2012

Burnstock A, Caldwell M, Odlyha M (1993) A technical examination of surface deterioration of Stanley Spencer's paintings at Sandham Memorial Chapel. In: ICOM Committee for Conservation, preprints of the 10th triennial meeting, Washington, DC, 22nd–27th August 1993. ICOM, Paris, pp 231–238

Church AH (1901) The chemistry of paints and painting, 3rd edn. Seeley and Co., London

Ferreira ESB, Boon JJ, Marone F, Stampanoni M (2011) Study of the mechanism of formation of calcium soaps in an early 20th century easel painting with correlative 2D and 3D microscopy; ICOM Committee for Conservation 16th triennial meeting, Lisbon 2011, Bridgland J (ed). Paper 1604

Field G (1891) Chromatography, Salter's edn. Winsor & Newton, London

García JMB, Torre SL, Serrano MR (2011) 'Utilización de ácido cítrico y EDTA en la limpleza de estructuras pictóricas', *Estudos de conservação e restauro*, vol 3. CITAR, Porto, pp 32–45

Hinde L, van den Berg KJ, de Groot S, Burnstock A (2011) Characterisation of surface whitening in twentieth century European paintings at Dudmaston Hall. In: ICOM Committee For Conservation 16th triennial meeting, Lissabon 2011, Bridgland J (ed), paper nr. 1310

Hong PKA, Jiang W (2005) Factors in the selection of chelating agents for extraction of lead from contaminated soil: effectiveness, selectivity and recoverability. In: Biogeochemistry of chelating agents. American Chemical Society, Washington, DC, pp 421–432

Keune K, Kirsch K, Boon J (2007) Lead soap efflorescence in a nineteenth-century painting; appearance, nature and sources of materials. In: AIC paintings specialty group postprints, Rhode Island, 16th–19th June 2006, vol 19. AIC, Washington, DC, pp 146–150

Kim C, Lee Y, Ong SK (2003) Factors affecting EDTA extraction of lead from lead-contaminated soil. Chemosphere 51:845–853, Elsevier

Mills L, Burnstock A, de Groot S, Megens L, Bisschoff M, van Keulen H, Duarte F, van den Berg KJ (2008) Water sensitivity of modern artists' oil paints. In: ICOM committee for conservation, preprints of the 15th triennial meeting, New Delhi, 2008. ICOM, London, pp 651–659

Morrison R, Bagley-Young A, Burnstock A, van den Berg KJ, van Keulen H (2007) An investigation of parameters for the use of citrate solutions for surface cleaning unvarnished paintings. In: Studies in conservation, vol 52, no 4, IIC, London, pp 255–270

Noble P, Boon JJ (2007) Metal soap degradation of oil paintings: aggregates, increased transparency and efflorescence. In: AIC paintings specialty group postprints, 16th–19th June 2006, vol 19. AIC, Washington, DC, pp 1–15

Sawicka A (2013) Clearing the 'Haze' of inorganic efflorescence: an investigation into the formation of lead soap efflorescence and the viability of its removal by means of the chelating agent Ethylenediaminetetraacetic Acid. Courtauld Institute of Art Third Year Project Dissertation, 2013

Slavin J (1990) The removal of salt deposits from decorative paintings on paper. In: Dirt and pictures separated: papers given at a conference held jointly by UKIC and the tate gallery, January 1990. UKIC, London, pp 49–50

Spring M, Higgitt C, Saunders D (2005) Investigation of pigment-medium processes in oil paint containing degraded smalt. In: National Gallery Technical Bulletin, vol 26, London, pp 56–70

Tempest H, Burnstock A, Saltmarsh P, van den Berg KJ (2013) Progress in the water sensitive oil project. In: Proceedings from of the cleaning 2010 conference, Valencia, 26–28 May 2010, Smithsonian Institute, pp 107–117

Van Loon A (2008) Colour changes and chemical reactivity in seventeenth-century oil paintings. PhD thesis, University of Amsterdam, MOLART Report 14, Archetype, London

Van Loon A, Noble P, Boon J (2011) White Hazes and surface crusts in Rembrandt's *Homer* and Related Paintings. In: ICOM Committee for Conservation, preprints of the 16th triennial meeting, Lisbon, 19th–23rd September 2011, CD-ROM, Criterio, pp 1–10

Zucker J (2007) Opaque to transparent: paint film defects in the Work of Frederic Church and the Hudson River School. In: AIC paintings specialty group postprints, Rhode Island, 16th–19th June 2006. AIC, Washington, DC, pp 33–41

Set Back the Race: Treatment Strategies for Running Oil Paint

Verena Franken, Gunnar Heydenreich, Elisabeth Jägers, Wolfgang Müller, Jenny Schulz, and Stefan Zumbühl

Abstract This study examines the phenomenon of semi-drying oil paint and liquefaction processes in contemporary oil painting. It aims to achieve a deeper understanding of the complex processes and to evaluate new treatment strategies for the subsequent curing of liquefied paint. Instrumental analysis of liquefied paint samples reveals that all samples contained predominantly semi-drying sunflower and/or safflower oil that have major impact on the liquefaction process. In thick paint layers low molecular weight polar fractions are created by decomposition of triglycerides that are mobile within the paint layer. Hardening of liquefied oil paint samples can be achieved by heating them to between 70 and 80 °C. The degree of hardness is dependent on the temperature, the length of exposure and the film thickness. A significant loss in weight suggests that low molecular weight components evaporate and that the hardening is primarily a physical process.

Keywords Sunflower oil • Safflower oil • Buff Titanium • Meese • Drying defects • Liquefaction • Heat treatment

V. Franken • G. Heydenreich (✉) • E. Jägers
Cologne Institute of Conservation Sciences, Cologne University of Applied Sciences, Ubierring 40, 50678 Köln, Germany
e-mail: verena.franken@smail.fh-koeln.de; gunnar.heydenreich@fh-koeln.de; elisabeth.jaegers@fh-koeln.de

W. Müller
H. Schmincke & Co.-GmbH & Co.KG, Otto-Hahn-Straße 2, 40699 Erkrath, Germany
e-mail: mueller@schmincke.de

J. Schulz
Private Conservator, Barthelstr.111, 50823 Cologne, Germany
e-mail: mail@jennyschulz.com

S. Zumbühl
Department Conservation HKB, Art Technological Laboratory, Bern University of Applied Sciences, Fellerstrasse 11, 3027 Bern, Switzerland
e-mail: stefan.zumbuehl@hkb.bfh.ch

K.J. van den Berg et al. (eds.), *Issues in Contemporary Oil Paint*,
DOI 10.1007/978-3-319-10100-2_22

Introduction

Non-drying oil paint is a phenomenon in contemporary art and paintings made using these materials may present challenging conservation problems. Case studies suggest that the dripping occurs often after about 5–7 years, in particular where the paint is thickly applied. The phenomenon occurs in paints from various manufacturers. Previous research has shown that the tacky and visco-elastic liquid paint occurs in superficial layers whereas deeper layers appear more stable (Pfandlbauer 1993; Schulz 2008). The affected paint layers are soft and tacky, and where thickly applied, a secretion of binding medium can occasionally be detected, and in an advanced phase the oil paint is viscose and may form drips. The aged and liquefied paint is unsightly and often requires treatment. It was also established that the phenomenon of liquefaction may stop once all the volatile components have leaked out. Case studies that exhibit dripping paint include paintings by Jonathan Meese, such as *Der Kotbart erwacht* (2001/2002) (Schulz 2008), *Portrait VII* (2003) and *Kleiner Hund* (2005) (Fig. 1).

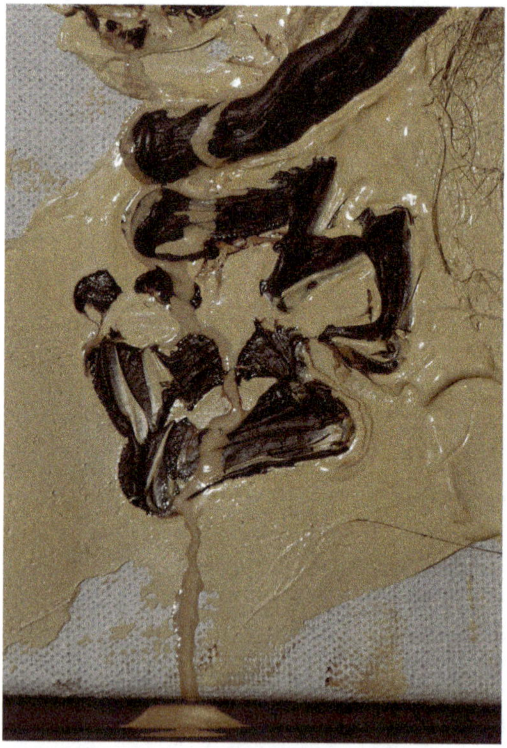

Fig. 1 Jonathan Meese, *Kleiner Hund*, 2005, Contemporary Fine Arts Berlin, detail with paint drips

Background and Current State of Research

The chemical alteration of oil paint has been investigated by a number of researchers (Boon and Ferreira 2006; Kamal-Eldin 2003; Van den Berg 2002; Mallégol et al. 2000; Frankel 1998; Chan 1987; Richardson and Korycka-Dahl 1984). Explanations for the cause of dripping paint suggest a physical rather than chemical liquefaction process dependant on rheological properties of the paint, together with the accumulation of low molecular weight species and unbound polar components within the paint caused by hydrolysis and oxidation (Götz 2003; Hoogland et al. 2007; Schulz 2011). The acidic products may be mobile within the film and influence on the liquefaction state.

Linseed, Poppy-Seed, Sunflower and Safflower Oil

To date linseed oil is the principal binding medium for oil paints and previous investigations on drying properties and drying problems have concentrated primarily on linseed oil. However, as paints containing linseed oil tend to yellow with age paint manufacturers also added other semi-drying oils like poppy-seed oil, particularly for white and blue paints.[1] In more recent decades poppy-seed oil was commonly replaced by semi-drying oils like safflower or sunflower oil. Sunflower oil and safflower oil exhibit different drying properties to linseed oil due to a relatively low content of poly-unsaturated fatty acids.[2] As a result, they form a less cross-linked insoluble network structure, which is weakly bonded. Preliminary studies showed that oil paints based on semi-drying oils achieve only a moderately stable paint matrix (Eibner 1922, pp. 186, 269). Following these early fundamental studies, it is possible to postulate that various processes may contribute to the liquefaction of oil paint. Paint composed of a high proportion of sunflower and other semi-drying oils may be one factor. Another factor may be the formation of mobile and volatile low molecular compounds as decomposition products from triglycerides in oil paints (cf. Zumbühl et al. 2014a). The mobile fractions may destabilise the paint matrix whereby a flowing phase can occur in the surface film. Drying of oil paint may be inhibited by the presence of selected pigments, fillers and organic additives such as aluminium and zinc stearate, clove oil or beeswax (Izzo 2011, p. 143). However, test films of pure sunflower and safflower oil have been observed to liquefy after 5–7 years and formed drips, suggesting that the process can occur in non drying oils alone.

---

[1]Related drying defects have been reported (Eibner 1930).

[2]Sunflower oil has a fatty acid composition of (by weight) 7 % C16:0 (palmitic acid), 5 % C18:0 (stearic acid), 23 % 18:1 cis-9 (oleic acid) and 65 % C18:2 (linoleic acid) (Schilling et al. 2007, p. 130).

Past Treatment of Affected Paintings

The increasing number of paintings affected by delayed dripping is of concern to artists, art collectors, paint manufacturers and conservators. There are so far no well-established methods for the treatment of paintings that exhibit liquefaction defects. Reworking of the painting by the artist or the application of a protective coating has caused ethical controversies (Althöfer 1980, p. 18). One approach has been to remove the drips and to rebuilt sunken impasto (Schulz 2008, p. 84). More recently conservators have attempted to search for possible means of stabilizing the paint and preventive treatment. Attempts were made to stabilise the softened paint by the injection of charged metal particles or to harden it with gamma rays (Schulz 2008, 2011). Whilst the first method appeared to be unsuccessful, the exposure to high energy radiation stabilised the paint. However it remains a theoretical approach as it is impossible to use this method on paintings for safety reasons.

Objective

The present study aimed to investigate the cause, the process and treatment of the phenomenon in paints containing semi-drying oils in order to expand our understanding of the complex reactions in oil paint. Practical tests explored the possibility of hardening liquefied paint using thermal treatment. Preliminary results suggested that heating changes the consistency of the liquefied oil paints but the mechanism and the potential for application of heat as part of a conservation treatment of a whole painting requires further study. This study aimed to clarify whether the curing attained by heating is primarily a physical process or produces chemical changes in the paint. The question whether a short exposure to high temperatures has the same effect as a longer exposure at a lower temperature was also investigated, together with the long term stability of heat treated paint in relation to fluctuations in relative humidity.

Experimental

Samples, Methodology and Analytical Instrumentation

Sources of sample material were

(a) Paintings that showed liquefaction defects
(b) Naturally aged paint samples that partly showed liquefaction defects and
(c) Fresh paint samples that were produced according to recipes of paints, which generated liquefaction defects in the past (Table 1)

Table 1 Description of the paint samples used for the experimental study

	Sample ID	Sample description	Composition	Wet film thickness/ amount	Support	Assigned for
(a)	a1 und a2 Buff Titanium, Daler Rowney®	Liquefied oil paint	Sunflower oil, titanium white, Iron-containing pigment, calcium-magnesium carbonate	Small droplet	Object slide	Heat treatment (80 °C, 70 °C, 65, 60 °C, 40 °C)
(b)	b1 Norma® Fleischfarbe Schmincke®	Liquefied oil paint	Linseed oil, sunflower oil, titanium white, calcium carbonate, additives	Small droplet	Object slide	Heat treatment (80 and 70 °C)
(b)	b2 Schmincke® recipe 11213	10 years naturally aged pink oil paint	80 % sunflower oil, 20 % linseed oil, safflower oil, calcium carbonate, cadmium pigments, additives	Pastose paint film	Cardboard	Heat treatment
(b)	b3 Schmincke® Rezept 11445	10 years naturally aged blue oil paint	25 % sunflower oil, 75 % linseed oil, safflower oil, calcium sulphate, blue pigments, additives	Pastose paint film	Cardboard	Heat treatment
(c)	c1 Schmincke®, recipe 11213, variation 3	Fresh, dark yellow oil paint	Sunflower oil, cadmium pigments, titanium white, calcium carbonate, additives	2,000 μm, size: ~ 2 × 4 cm	Polycarbonate film	Heat treatment (80 °C, 60 °C, 40 °C), various environmental conditions
(c)	c2 Schmincke®, recipe 11213/Variation 3	Fresh, dark yellow oil paint	Sunflower oil, cadmium pigments, titanium white, calcium carbonate, additives	Max. 7,000 μm	Object slide	Heat treatment (80 °C, 60 °C, 40 °C), various environmental conditions
(c)	c3 Schmincke®,	Fresh sunflower oil	Oil with cobalt siccative	0,10 g	Object slide	Heat treatment (80 °C, 60 °C, 40 °C), various environmental conditions

Samples (a) were provided by the studio of Jonathan Meese and by the conservation studio Brakebusch.[3] The affected paint is Buff Titanium from the series Georgian Oil Colours by Daler Rowney (Table 1). The paint contains mainly sunflower oil as binding medium.[4] The main pigment is titanium dioxide with traces of iron oxide and calcium-magnesium carbonate. The naturally aged paint samples (b) represent oil colour out of the series Norma® by H. Schmincke & Co. (b1). This paint contains linseed oil, sunflower oil, titanium dioxide, calcium carbonate, cadmium pigments and small additives such as beeswax. The samples were manufactured by H. Schmincke & Co. in the 1990s and in 2003 according to recipes from the 1990s. Some samples show liquefaction defects and these represent paint that was used by artists and where drying defects were observed in the past. Additionally, fresh paint samples (c) manufactured according to recipes from the 1990s and pure sunflower oil enabled the impact of external influences on the drying process to be monitored. Within this study we focus primarily on the aged and liquefied paint samples (a, b).

The paint samples were heated at different temperatures and for varying periods in a laboratory oven at 40–80 °C. The samples were exposed to increased temperatures for periods of 24 h up to a maximum of 25 days and examined at intervals of 1 day.[5] Temperature and humidity were monitored in the oven. With rising temperatures the RH dropped significantly to 3.5–14 %. After treatment sample material was exposed to a relative humidity of c. 45–55 % RH except one sample which was stored at 84 % RH.

All samples were examined by eye with a microscope (10–80× magnification) and mechanically with a needle and a scalpel in order to investigate stages of the drying process, changes of material characteristics and the impact of thermal treatment at different temperatures. Changes of the surface and inner paint bulk morphology were recorded photographically with a digital microscope, VIS and UV-light microscopy. Material properties such as hardness, stickiness and viscosity were evaluated continuously using a thin needle. After testing hardness with a needle and transecting the sample with a scalpel the degree to which the surface and the inner paint had dried could be evaluated.[6] Samples were

[3]The paint, which was supplied by the Restaurierungsatelier Brakebusch, was taken from a painting by Jonathan Meese from 2004. The painting was reworked in the artist's studio; the affected paint application was removed and replaced (personal communication, Börries Brakebusch, Düsseldorf).

[4]The P/S ratio is too high for linseed oil. NB: This brand is listed as using 'refined linseed and safflower oil' and 'sunflower oil for the whites' (Pearce 1992). An ESI spectra analysis of the tube colour (bought in 2006) shows a spectra with round about 90 % sunflower oil and no linseed oil (comp. Fig. 6).

[5]Every day one sample was removed from the oven for examination. Removed samples were not returned to the oven.

[6]For testing hardness with a fine needle a pressure of approximately 35 g (measured with analytical balance) was applied. To investigate the inner paint bulk morphology the samples were transected with a scalpel and the paint's cross-section was recorded photographically.

weighed every day with an analytical balance and thermogravimetric analysis was undertaken to qualify evaporating components under isotherm condition at 50 °C over 24 h.

Samples were analysed and examined using a Scanning Electron Microscope with back-scattered electron analysis (SEM-BSE) to localise the accumulations of organic compounds. Gas Chromatography–Mass Spectrometry (GC-MS) was applied to examine the chemical changes in samples (comp. appendix). As GC-MS cannot record all volatile components, as pyrolysis splitting occurs at *c*. 300 °C and components volatalise too rapidly, nano Electrospray Ionisation Mass Spectroscopy (ESI–MS) was used.

Fourier transform infrared spectroscopy (FTIR) was employed to characterise chemical alteration. Additionally, samples were prepared with a derivatisation technique using sulphur tetrafluoride SF_4 which facilitates the characterisation of functional groups. It helps to distinguish overlapping signals and to characterise acids in the samples (Zumbühl et al. 2011, 2014a). This standardised derivatisation technique facilitates the characterisation of the ageing and degradation conditions of the sample material. It is more specific, because a more precise breakdown in the carbonyl- and carboxyl-bands makes a distinction between ketones and acids possible (Zumbühl et al. 2014b).

Results

The experiments confirmed that heating cured the liquefied sample material in the paint films. The degree of hardness of the paint films was dependent on the temperature in the drying-cabinet, the length of exposure and the film thickness. A complete hardening was achieved at 80 °C after 5–6 days for the Buff Titanium samples (a1 and a2). These samples were soft and tacky before thermal treatment and initially developed a very thin, deformable surface skin before the paint beneath also became more solid (Fig. 2). Hardening of the H. Schmincke & Co. paint samples (Norma® Flesh colour) (b1) was achieved at the same temperatures in less time, probably as they exhibited a lower level of liquefaction before treatment.

The tests confirmed that thicker paint layers require a longer period of thermal treatment and apparently a minimum temperature was required to initiate the hardening process. Samples of Buff Titanium heated to 40 °C and 60 °C showed no obvious changes in hardness or viscosity after 6 days (a1 and a2). Even after more than 25 days at 60 °C these samples were soft enough to show plastic deformation, and threads of paint could be pulled from the samples. At 70 °C hardening was achieved after 10 days (Fig. 3). In the course of heating no changes in fluorescence were noticed. During exposure to heat a coalescence of the Buff Titanium sample could be observed, though this also occurred on reference samples at room temperature.

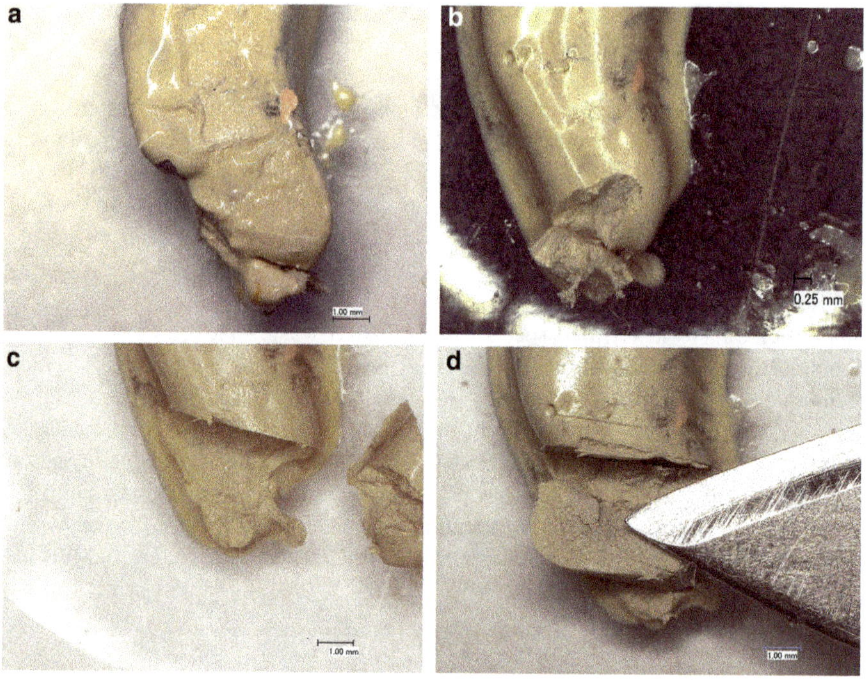

Fig. 2 Buff Titanium samples before (**a**) and after heat treatment at 80 °C for 2 days (**b**), 5 days (**c**) and 6 days (**d**)

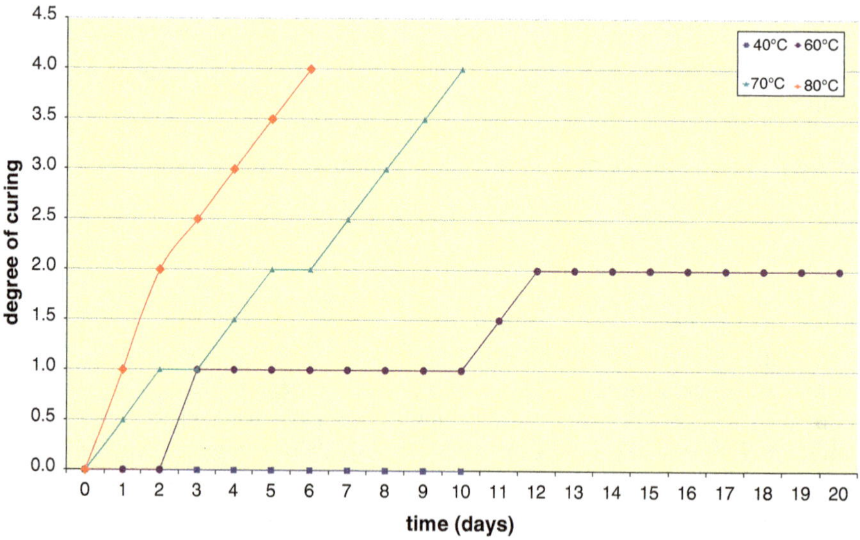

Fig. 3 Degree of curing for Buff Titanium samples after a heat treatment at different degrees (40, 60, 70 and 80 °C)

Thermogravimetric Analysis and Gravimetry

Thermogravimetric measurements showed that the hardening was accompanied by a considerable loss in weight. This also occurred in samples heated to below 50 °C. The loss in mass of the pink paint sample (b2) which contained a high proportion of sunflower oil (≈13 % linseed oil, ≈70 % sunflower oil, ≈17 % safflower oil) differed from the blue paint sample (≈66 % linseed oil, ≈23 % sunflower oil, ≈11 % safflower oil) (b3). The decrease for the pink paint heated at 50 °C was almost 20 % of the oil content within the first 6 h.[7] Compared to the blue paint (4 %), which contained more linseed oil, this is very significant. Although part of the initial loss in weight may be related to evaporation of water it is likely that unlinked low molecular weight volatile fragments of the oil film were also lost.

SEM – BSE Analysis

The SEM-BSE images of a cryo cutted cross-section of a paint sample exhibiting a liquefied surface show a darker zone in the upper section (Fig. 4). This zone represents the liquefied area of the paint bulk approximately 150 μm thick. In both zones the same chemical elements were detected in a same relative concentration

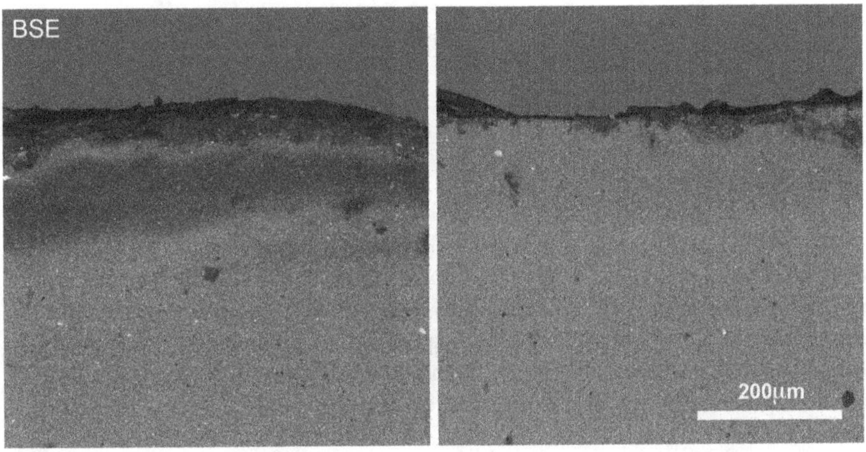

Fig. 4 SEM-BSE of a liquefied sample before (*left*) and after heat treatment at 80 °C for 24 h (*right*)

[7]Before analysis, samples were dried with silica gel for 7 days. The water content was determined for both samples and relates to the overall weight (blue: 0.18 %, pink: 0.23 %).

using energy-dispersive spectroscopy EDS, but in different amount (counts). The image contrast assessed from different elemental density using BSE indicates the presence of a higher concentration of organic material in the upper section. After thermal treatment this zone was hardened by evaporation of low molecular weight components to cause a change in pigment/volume concentration (Fig. 4).

FTIR Analysis

After thermal treatment of liquefied Buff Titanium paint Fourier transform infrared spectroscopy (FTIR) showed a slight shift of the $\nu(C=O)$ stretch vibration of the esters at 1,745 cm^{-1} to lower frequencies what indicates the presence of other carbonyl groups (Fig. 5).[8] These changes were already observed after 2 h of thermal treatment at 80 °C and increased after 24 h of exposure. The spectrum changes described occurred after 2 h of thermal treatment and no considerable changes were detected in spectra taken from samples heated for 2 days. Though the process of hardening was not complete after 24 h only a certain increase in viscosity was evident in the sample. After 1, 2 and 5 days of thermal treatment at 40 and 60 °C no changes could be detected in the FTIR spectra. The same spectral features were observed for the Buff Titanium sample (a1), which had already been hardened and was exposed for a month to a relative humidity of approximately 84 %. The sample exhibited a slight loss of hardness (needle test), which may partly relate to the formation of moisture equilibrium between the paint and the environmental conditions.

ESI–MS Analysis

The ESI – MS analysis spectrum of fresh sunflower oil (sample c3, high peak at m/z 903 of the triglyceride monomers) differs from the spectrum of running oil with drying defects (Fig. 6). The liquefied samples (a1 and a2) showed the presence of a variety of fractions with molecular masses below the triglyceride monomers (cf. Zumbühl et al. 2014a). This indicated the presence of a high amount of triglyceride fragments with a molecular mass between 200 and 900 Da. It reveals the relevance of the fragmentation reaction during the liquefying process.

[8]As a result of derivatisation with SF4 very low molecular acids evaporate, since in the reaction product acryl fluoride there is no hydrogen-bond possible. Low molecular weight components were detected with GC–MS.

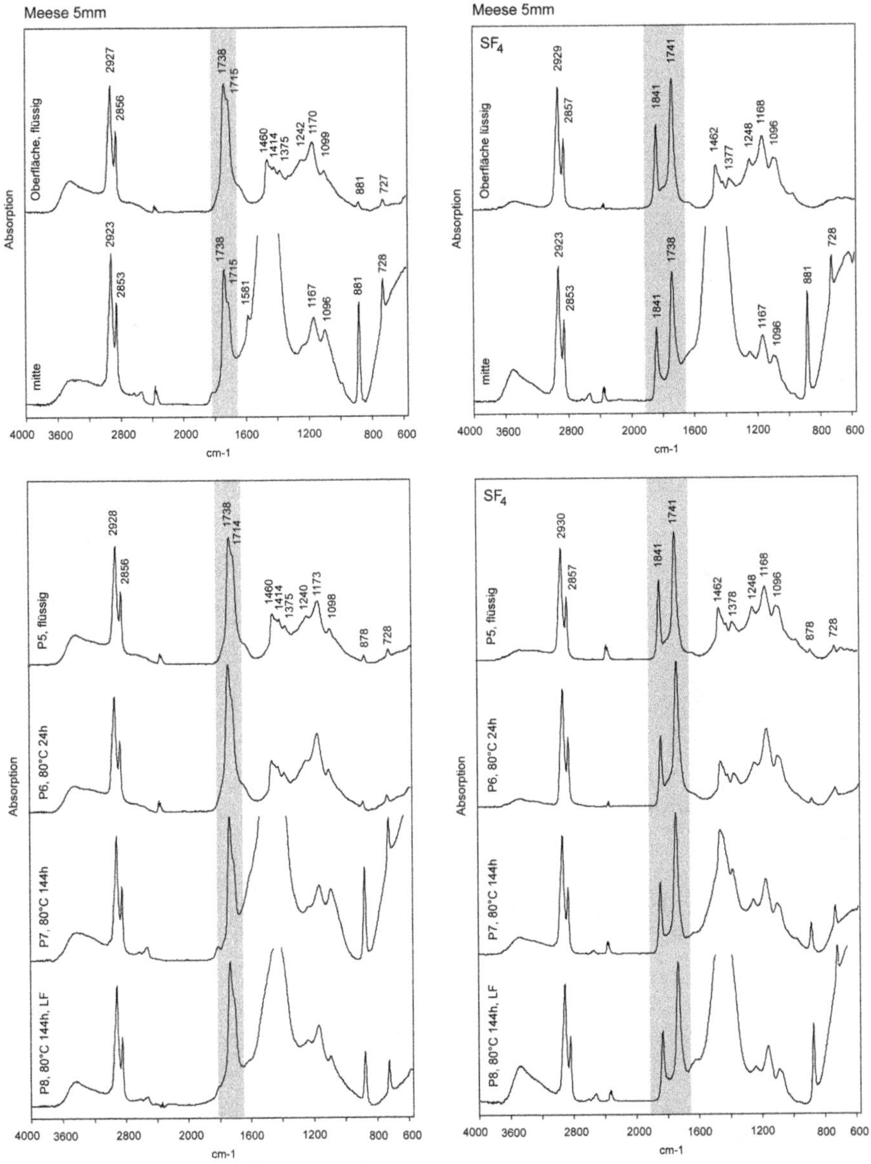

Fig. 5 FTIR spectra of a liquefied Buff Titanium sample before (*left*) and after (*right*) derivatisation using sulphur tetrafluoride before (*top*) and after (*bottom*) thermal treatment (80 °C/24 h)

GC–MS Analysis

For the liquefied Buff Titanium samples (a1 and a2) the following ratio was detected: The azelate/sebacate (A/Seb.) ratio is around 11, the azelate/suberate

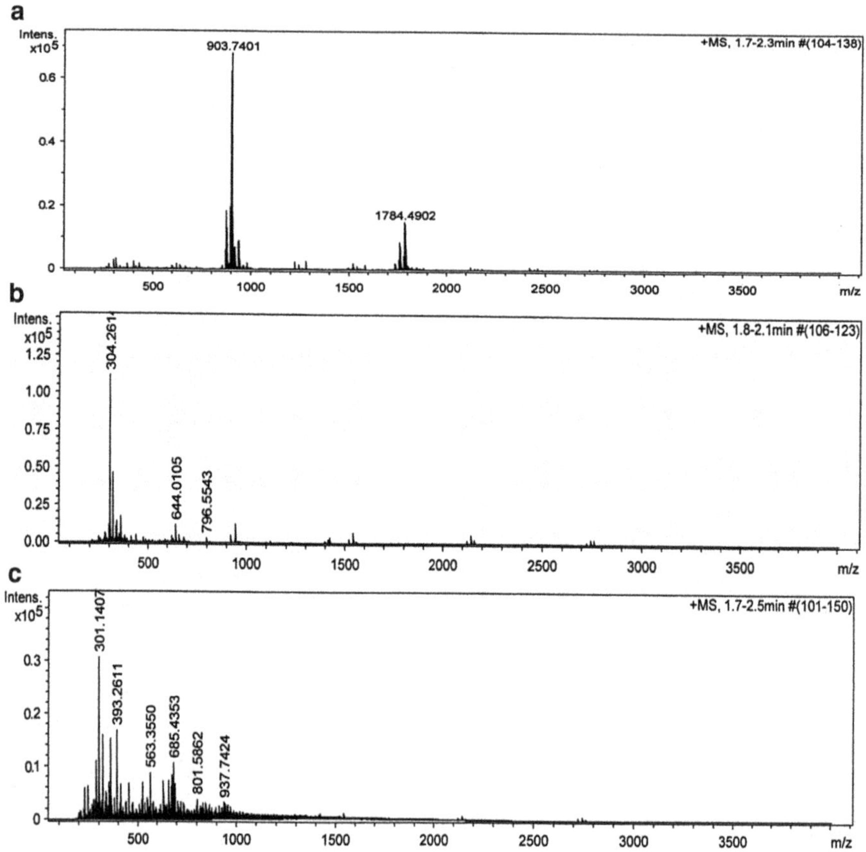

Fig. 6 ESI spectra positive mode, chloroform/methanol extract: (**a**) Fresh sunflower oil (sample c3, dilution 1:1,000) (**b**) Liquid phase Meese object (sample a1 + a2, dilution 1:100) (**c**) Liquid phase Meese object after 55 h by 80 °C (non-dilution)

(A/Sub.) ratio is around 3 and the palmitic acid/stearic acid (P/S) ratio is around 2.[9] After thermal treatment the ratios have not changed significantly.[10]

Discussion

Hardening of liquefied oil paint samples (a1 and a2, b1) was successful when samples were exposed to temperatures between 70 and 80 °C. The degree of hardness was dependent on the temperature, the length of exposure and the film

[9]Qualitative fatty acid ratios of Buff Titanium samples before heat treatment: A/P: 0.6, A/Sub.: 2.7, A/Seb 11.1, P/S: 2.1, O/S: 0.1, Dihydroxy C18/S: 1.6.

[10]Qualitative fatty acid ratios of Buff Titanium samples after heat treatment (6 h 80 °C): A/P: 0.4, A/Sub.: 2.7, A/Seb.: 9.6, P/S: 1.9, O/S: 0.1, Dihydroxy C18/S: None.

thickness. Heat treatment at 40 and 60 °C even after 25 days did not change the viscosity of the paint significantly. This confirms the hypothesis that a minimum temperature of more than 60 °C is required for activating the process of hardening. The coalescence of the sample is only partly attributable to the heat exposure and results from the fact that the sample material assumes the most energy-efficient state.

Analytical results demonstrate that hardening is related to a loss in weight due to the volatilisation of low molecular compounds from the paint. Analysis of liquefied samples before and after thermal treatment could contribute to a deeper understanding of the mechanisms that lead to the complex liquefying process. ESI measurements and FTIR results confirm a large fragmentation of the liquefied paint. Based on the distribution of mass it can be inferred that the liquefied products of decomposition are for the most part fragmented triglycerides. GC–MS analysis indicated that heat treatment caused a reduction in dihydroxystearate as a possible hydrolysis product. FTIR and ESI analysis suggest that the paint became more acidic and polar on heating. It may be concluded that liquefaction results from significant oxidative fragmentation reactions and it can be inferred that the liquefied products of decomposition are for the most part fragmented triglycerides. Thus, the oil paint would not polymerise further or dry in the future.

The phenomenon of liquefaction was more pronounced in thicker paint. A critical thickness is necessary to allow separation of the binding medium to initiate the flow process to form drips. Paint samples containing larger amounts of sunflower oil (samples b2) degrade faster than paints with higher linseed oil content (samples b3), and they exhibit differences in acidity. Paints made using semi-drying oils are likely to exhibit more significant fragmentation by beta-elimination. As the hardening process on drying is dependent on the equilibrium between the formation and decomposition of peroxides, the consequences of the decomposition reaction are more apparent in semi drying oils. Semi drying oils have therefore a considerably reduced viscosity through ageing, which is why binding-medium components are able to settle on the surface as a fluid medium-rich layer.

The FTIR-results together with the TGA-measurements and SEM investigations show a consistent pattern: the weight loss in paint samples increased with heating time and at higher temperatures. Significant loss in weight at temperatures around 50 °C (without accompanying changes in viscosity) may partly be related to the evaporation of water. It can be assumed that significant amounts of the liquid component of the paint evaporate at higher temperatures. The liquid medium content decreases during thermal treatment and the pigment-binding medium ratio appears to be altered, due to the loss of volatile low molecular components, illustrated in the SEM – BSE images before and after heat treatment (Fig. 4). This is indicated by a reduction in the binding medium content in the FTIR spectrum, characterised by the change of the signal intensity relative to the carbonate absorption of the filler.

Analysis of liquefied paint samples after 24 h of heat treatment showed no difference between the liquefied areas on the surface and the deeper firmer paint in terms of acidity. As the signal intensity of the acyl fluoride in the IR-spectrum after selective derivatisation of the carboxylic acids (using SF_4) around 1,841 cm^{-1} has not changed significantly the solid binding-medium components remain extremely acidic. Heating did not influence the polarity and degradation of the sample material[11]; instead the treatment primarily altered the pigment/binding media ratio. GC-MS analysis of samples before and after thermal treatment exhibited much the same chemical composition. Thus, the loss of a large single component cannot be confirmed.

Thermal treatment led to loss of volatile components but no significant changes in the polarity of the paint film. Thus, differences in acidity of the dicarbonatic acids and the fatty acids have apparently no effect on the liquefaction process. Because of the large percentage of photochemical inert oxidation products no significant increase of oxidative cross-linking can be expected and so far no cross-linking reaction could be detected in treated samples. Therefore the hardening would appear primarily to be a physical process.

Conclusions and Perspectives

Analysis of liquefied samples before and after thermal treatment contributes to a deeper understanding of the mechanisms that lead to the complex liquefying process. The liquefaction of oil paint is primarily dependent on the type of oil. Case studies suggest that semi-drying oils – sunflower and safflower oil are one of the main causes of liquefaction. Since unpigmented sunflower and safflower oil may liquefy after 5 years it seems that pigments and additives have a less significant influence on this process.

This study showed that hardening of liquefied oil paint can be achieved by heating. However, a minimum temperature is required to activate the process of hardening. During heating low molecular components migrate and evaporate, leading to weight loss in the films. Thermal hardening of the liquefied paint is primarily a physical process.

Further studies of the process of liquefaction and hardening of liquefied oil-paint have been initiated that examine a range of samples including different pigments and binding media to observe the drying and liquefaction processes under various environmental conditions. Thus we hope to gain a better understanding of the influence of paint compositions, additives, and layer thicknesses as well as changing relative humidity and light exposure. Also, the long-term behavior of the thermally treated paint samples and in particular at higher relative humidity will be examined in more detail. Furthermore, methods will be developed on how to apply heat locally to a painting. A painting by Jonathan Meese with significant liquefaction defects presently serves to develop a strategy and to evaluate an effective treatment.

---

[11]During SF4 treatment carbonates like lead carbonate or calcium carbonate are converted into fluorides, which are not IR active. Therefore no interference is show on the spectra after derivatisation. However dolomite is stable under dry derivatisation conditions. Therefore the Buff Titanium samples continue to show the signals from the bulking material.

Acknowledgements The authors would like to thank Bronwyn Ormsby (Tate, London), the studio of Jonathan Meese (Berlin), Peter Most (Berlin), Börries Brakebusch (Düsseldorf) and Helen Smith (Berlin).

Appendix

Instrumentation, materials and suppliers

Film application
Feeler gauge, 2 mm (Vogel Germany GmbH & Co. KG)

Support (paint and oil samples)
Microscope slides (Carl Roth GmbH + Co. KG)
Polycarbonate film, translucent, colourless (Modulor GmbH)

Heat treatment
Laboratory oven UM 100–800 (Memmert)

Data logger
Thermofox Universal with Multisensor and Thermoelement System, Hygrofox with Universal HUB (Scanntronik Mugrauer GmbH)
Results were processed using SoftFOX software
EL-USB-2, results were processed using EasyLog USB software (Lascar Electronics Ltd.)

Storage at high humidity
Potassium chloride (Carl Roth GmbH + Co. KG)

Microscopy
VHX-500FD digital microscope (Keyence)
Leica MZ Fluo microscope (visible light and BP 420-490), results were processed using Diskus software (Carl H. Hilgers)

Gravimetry
Analysis balance (Sartorius AG)

GS – MS: Sample preparation and conditions
Samples were derivatised using 20 μL Meth-Prep II (Alltech, UK) and heated for 1 h at 60 °C, then cooled to room temperature overnight.
GCMS analysis was carried out on a Varian Saturn CP-3800 GC coupled with a 1,200 L MS detector.
Column: Phenomenex DB5-MS; 30 m × 0.25 mm × 0.25 μm.
Oven program: held at 50 °C for 1 min then ramped to 320 °C at 10 °C/min then held for further 5 min at 320 °C.
Splitless injection volume of 1 μl.
Helium flow of 1.5 ml/min.

MS conditions: Source temp: 200 °C; transfer line: 300 °C, solvent delay of 9 min. EI mode (70 eV); scan group 1: 45–300 amu; Group 2: 45–700 amu at 16 min, every 1 s.

Results were processed using Varian Workstation software, version 6.8.

ESI-MS: Sample preparation and conditions

ESI mode: Positive; 50 bis 4,000 m/z; 2 GHz

Instrument: Bruker maXis 4G

Samples were extracted by chloroform/methanol (2:1) mixture and diluted with methanol (up to 1:1,000). Spectra are measured by institute for organic chemistry, University Duisburg-Essen.

References

Althöfer H (1980) Moderne Kunst: Handbuch der Konservierung. Schwann, Düsseldorf

Boon JJ, Ferreira ESB (2006) Processes inside paintings that affect the picture: chemical changes at, near and underneath the paint surface. In: Boon JJ, Ferreira ESB (eds) Reporting highlights of the De Mayerne Programme. Netherlands Organisation for Scientific Research (NWO), The Hague, pp 21–31

Chan HW-S (1987) Autooxidation of unsaturated lipids. Academic, London

Eibner A (1922) Über fette Öle. Leinölersatzmittel und Ölfarben. Verlag B. Heller, München

Eibner A (1930) Das Öltrocknen, ein kolloidaler Vorgang aus chemischen Gründen. Allgemeiner Industrie Verlag, Berlin

Frankel EN (1998) Lipid oxidation. The Oil Press LTD, Dundee Scottland

Götz E (2003) Zur Trocknungsproblematik pastoser Leinölfarben. Diploma-thesis, Cologne University of Applied Sciences

Hoogland FG, Van der Horst J, Boon JJ (2007) Liquefying oil paint in some late – twentieth-century paintings. In: Learner TJS, Smithen P, Krueger JW, Schilling MR (eds) Modern paints uncovered: proceedings from the modern paints uncovered symposium, Getty Conservation Institute symposium proceedings series. Getty Conservation Institute, Los Angeles, pp 282–283

Izzo F (2011) 20th century artists' oil paints: a chemical-physical survey. PhD dissertation, University Ca'Foscari of Venice

Kamal-Eldin A (2003) Lipid oxidation pathway. AOCS Press, Champaign

Mallégol J, Gardette J-L, Lemaire J (2000) Long-term behavior of oil-based varnishes and paints. Fate of hydroperoxides in drying oils. J Am Oil Chem Soc 77(3):249–255

Pearce E (1992) Artists' materials, which, why and how. A & C Black, London

Pfandlbauer S (1993) Leinöl in der modernen, zeitgenössischen Malerei. Diploma-thesis, Cologne University of Applied Sciences

Richardson T, Korycka-Dahl D (1984) Lipid oxidation. In: Fox FP (ed) Developments in dairy chemistry, vol 2, Developments series. Applied Science Publishing, London, pp 241–263

Schilling MR, Mazurek J, Learner TJS (2007) Studies of modern oil-based artists' paint media by gas chromatography/mass spectrometry. In: Learner TJS, Smithen P, Krueger JW, Schilling MR (eds) Modern paints uncovered: proceedings from the modern paints uncovered symposium, Getty Conservation Institute symposium proceedings series. Getty Conservation Institute, Los Angeles, pp 129–39

Schulz J (2008) "Flüssige Ölfarbschichten" in zeitgenössischen Gemälden. Untersuchung zu Ursachen – Möglichkeiten der Restaurierung am Beispiel des Werkes "Harvest" (1993) von Otto Piene. Diploma-thesis, Cologne University of Applied Sciences

Schulz J (2011) Jenny Schulz. Liquefying oil paint: contemporary paintings in danger? In *ICOM-CC 16th triennial conference Lisbon 19–23 September 2011: preprints*. ICOM Committee for Conservation. Critério–Produção Grafica, Lda

Van den Berg JDJ (2002) Analytical chemical studies on traditional linseed oil paints. PhD dissertation, University of Amsterdam

Zumbühl S, Scherrer N, Ferreira E, Hons S, Müller M, Kühnen R, Navi P (2011) Accelerated ageing of drying oil paint – an FTIR study on the chemical alteration. Problems of accelerated ageing under variable conditions of light, temperature and relative humidity. Zeitschrift für Kunsttechnologie und Konservierung 25(2):139–151

Zumbühl S, Scherrer N, Müller W (2014a) 'Derivatisation technique for infrared spectroscopy – characterisation of oxidative ageing products in modern oil paint' in this publication

Zumbühl S, Scherrer N, Eggenberger U (2014b) Derivatisation technique to increase the spectral selectivity of two-dimensional (2D) Fourier Transform Infrared Focal Plane Array (FT-IR FPA) imaging: analysis of binder composition in aged oil and tempera paint. Appl Spectrosc 68(2):458–465

Sensitivity of Modern Oil Paints to Solvents. Effects on Synthetic Organic Pigments

Diana Blumenroth, Stefan Zumbühl, Nadim C. Scherrer, and Wolfgang Müller

Abstract Synthetic organic pigments are widely used in contemporary artists- and house paints. They can be found in artworks since 1900. Because of their special particle properties and their solubility in solvents, however, synthetic organic pigments pose a special challenge in conservation treatments. Analyses have been carried out on 23 synthetic organic pigments in oil paint films with six representative solvents. Solubility of the pigments upon solvent exposure has been determined by UV-Vis-spectroscopy. Some pigments have shown high solubility and were even extracted out of the oil paint film. Detailed examination on the influence of pigment extractions from the oil paint film was carried out with light microscopy, Raman- and ATR-FTIR spectroscopy. Swab tests elucidated that all pigments are very sensitive to the combination of (mostly) polar solvents and mechanical stress. The study demonstrates that mechanically applying solvents to paint surfaces containing synthetic organic pigments is delicate due to (a) the solubility of the pigments themselves, (b) mechanical removal of pigment particles – likely supported by the temporary destabilization of the binder. These findings have important implications to conservation practice.

Keywords Synthetic organic pigments • Solvents • PY 3 • Oil paint • Solubility • Cleaning • Raman • UV-Vis spectroscopy • ATR-FTIR

D. Blumenroth (✉)
Cologne Institute of Conservation Sciences, Cologne University of Applied Sciences, Ubierring 40, D-50678 Köln, Germany
e-mail: diana@blumenroth.pro

S. Zumbühl • N.C. Scherrer
Bern University of Applied Sciences, Fellerstrasse 11, CH-3027 Bern, Switzerland
e-mail: stefan.zumbuehl@bfh.ch; nadim.scherrer@bfh.ch

W. Müller
H. Schmincke & Co. GmbH & Co. KG, Otto-Hahn-Str. 2, D-40699 Erkrath, Germany
e-mail: mueller@schmincke.de

K.J. van den Berg et al. (eds.), *Issues in Contemporary Oil Paint*,
DOI 10.1007/978-3-319-10100-2_23

Introduction

The conservation and restoration of modern art works show a new complexity of performance and sensitivity as a result of the huge variety of the materials used. Modern oil paints cannot simply be compared with "classic" oil paints, because of their different composition and thus different physical and chemical properties. An important part of this new characteristic is the wide use of synthetic organic pigments. The development of these pigments started already in the middle of the nineteenth century (Heumann and Friedänder 1898; Hübner 2006; Zerr and Rübenkamp 1921) and they can be present in works of art since the last quarter of the nineteenth century. Many new pigments have acquired a central role in paints due to their light fastness and high color intensity. The on-going development in this field has recently led to a huge variety of new products (Faulkner 2009; Smith 2002). Inorganic pigments are increasingly being displaced by synthetic organic pigments and they are now the most important group of pigments in the bright hues of modern artist's paints (Scherrer 2009). The interest in the properties of synthetic organic pigments has increased significantly in recent years, both in the historical (de Keijzer 1988, 1989, 1999, 2002; Frowein 2004; Fritsch 2006) and technological context (Lomax 2006; Lutzenberger 2009; Strauss 1984), as well as in relation to the specific requirements of the material analysis (Strauss 1984; Vandenabeele 2000; Kalsbeek 2005; Schäning 2005; Fritsch 2006; Lomax 2006; Ropret 2008; Schulte 2008; Frowein 2004; Scherrer 2009; Fremout 2012). As versatile as the applications of these pigments are, there also are many associated risks. New demands are therefore arising continuously with respect to the conservation of modern artists' paints, made with synthetic organic pigments. Some synthetic organic pigments found in "classic modernism" paintings of the early twentieth century, proved to be highly soluble (Zumbühl 2008). As a consequence, the specific particle properties as well as the sensitivity to solvents pose a particular challenge in conservation treatments of modern oil paints. In this context, the solvent sensitivity of 23 synthetic organic pigments in oil paints was tested under the various aspects (Blumenroth 2010).

Solvent Sensitivity of Organic Pigments and Their Modifications

Synthetic organic pigments consist of low molecular weight compounds (M <2,000 Da). In theory, a good solvatisation and solubility of these pigments in organic solvents can be predicted. The solvent sensitivity of these pigments is not evident in their chemical structure, as the properties are altered by selective modifications of the structure and encapsulation. Surface modifications are possible by physical or chemical adsorption of organic and metal-organic compounds or by in-situ polymerization of coating systems (Fu 2007; Lelu 2003; Sirikittikul 2004;

Viala 2002). Resins have long been used for modifying the pigment properties, e.g. through a process such as rosination (Schröder 1988). In this case, the resin acts as a growth inhibitor and is used to control the crystallization process. The cohesion of a pigment is reduced by coating the pigment surface and therefore improving its dispersibility (Schröder 1988; Sirikittikul 2004). An overview on the various derivatization types has not been attempted, as each class of pigments has another application system (Bugnon 1995). Next to organic additives, other types of coatings (encapsulation) are industrially possible. The stability of organic pigments can be achieved and optimized by the application of nano-layers of silicon dioxide (Yuan et al. 2005, 2008) or other metal oxides (Binkowski et al. 2000; Bugnon 1995; Tiarks 2001; Yuan et al. 2006). Through such inorganic coatings, the thermal stability, wettability, dispersibility, chemical stability and light and weather fastness are improved (Yuan et al. 2006, 2008). In order to ensure good processing properties, such encapsulated pigments – similar to inorganic pigments – can be polymerized with organic coupling agents (Tiarks 2001). Accordingly, the pigment properties will vary considerably depending on the surface modifications. Sometimes the modification of one property can lead to a decline of another. No general statement can be made, as the pigment processing can be material specific and vary from manufacturer to manufacturer. Accordingly, it is difficult to derive theoretical predictions for the resistance against solvents solely based on chemical composition. It is thus likely that early pigments may behave different to contemporary ones (Herbst and Hunger 1995). Properties of dissolution are thus influenced by the chemical nature, crystal structure, particle size, as well as surface modifications of the pigment products. The focus of this project thus was to determine solvent sensitivity of synthetic organic pigments and the paints produced with them experimentally (Blumenroth 2010).

Analysis

The sensitivity of artist's paints containing organic pigments to solvent action arises not only from the solubility of the pigment itself, but also from the characteristics of the surrounding binder matrix and the swelling behavior of the whole system. The solvent resistance of 23 organic pigments in oil paint was tested with the six solvents n-hexane, toluene, chloroform, diethyl ether, acetone and ethanol - comprising a spectrum of different interaction properties and cavitation energies (Marcus 1998, Zumbühl 2011). The solubility of the pigments was determined quantitatively on pigment/solvent mixtures as well as by immersion of artificially aged oil paint films.[1] The time-resolved extraction was determined by UV-Vis

[1]Mussini and Norma (H. Schmincke & Co. GmbH & Co. KG) paints were applied in a wet film thickness of 100 μm with a film applicator Model 360 (Erichsen GmbH & Co. KG)

spectroscopy. Following solvent treatment by immersion of oil paint films, the influence of pigment extraction along the surface was examined applying infrared spectroscopy ATR-FTIR. Raman spectroscopy was applied to study morphological effects of extraction on polished cross-sections of embedded oil paint films. Structural damage was examined under practical conditions with cotton swabs using the six solvents.

Results and Discussion

The Solubility of Synthetic Organic Pigments

The solubility of the organic pigments was defined according to the industry standard by immersion in solvent. Properties of dissolution are influenced by the chemical nature, crystal structure, particle size, particle size distribution, as well as surface modifications of the pigment products and the reaction temperature (Herbst and Hunger 1995). Almost all pigments of the series exhibited discoloration of the liquid. Suspensions were found especially in solvents with high density such as chloroform. In particular, pigments with very small particle sizes resulted in stable suspensions where the particle sedimentation was very slow. After filtration with a 0.2 μm syringe filter, the molecularly soluble pigments became apparent. In addition, the filtered solutions were evaporated to quantitatively detect the pigment residues. Pigments exhibiting solubility within the tested set were the yellow and orange azo pigments PY 3, PY 97, PY 153, PO 5, the orange pyrazolochinazolon PO 67, the red anthraquinone PR 83:1, phthalocyanine blue PB 15:6, dioxazine violet PV 23, and to a lesser extent the red azo pigments PR 188, PR 177, as well as the anthraquinone perylene PR 179, as summarized in Table 1. The solubility is similar to the sensitivities known from literature and industry. The main focus was thus set on the behavior of these pigments bound in oil. Time-resolved quantitative extractions were determined on artificially aged oil paint films with UV-Vis spectroscopy.[2] For eight pigments, the extraction of pigments from the oil paint film was determined. The azo PY 3, PY 97 and PO 5 exhibited particularly high extractability. This is of course partly due to the solubility of the pigments,

on glass (Slides Assistent Elka 2400) and silicon Hostaphanfoil® RNT 36 (Kremer GmbH & Co. KG). After a preliminary drying of 7 days under room conditions, the samples were artificially aged. The lighting was done with the fluorescent tubes True Lite® 5,500 K and Philips® UV–20 W/08 F20 T12 BLB with ≈5,800 lm/m^2 and 557 mW lm^{-1}/≈3,200 mW/m^2 at ≈35 °C/≈45–50 %rH for a minimum of 6 months.

[2]Perkin Elmer Lambda 650 UV/VIS/NIR spectrometer in quartz cuvettes (10 mm deep K282) over the spectral range of 200–1,000 nm at a resolution of 1 nm. Quantification was carried out in relation to the specific pigment absorption λ max (Blumenroth 2010). It was measured at a time interval of 1 min over a period of 16 min.

Table 1 Overview on the results comprising all pigment/solvent combinations tested in the series

Pigment				Paint samples						Cleaning samples					
C.I.	Pigment class	Sensitivity		Pigment solubility						Solvent sensitivity					
		Literature	Pigment pure	Hexane	Diethyl ether	Toluene	Chloroform	Acetone	Ethanol	Hexane	Diethyl ether	Toluene	Chloroform	Acetone	Ethanol
PY3	Monoazo	***	***	o	*	**	***	***	*	*	**	**	***	***	***
PY97	Monoazo		***	o	o	*	***	**	o	*	**	**	***	***	***
PO5	β-Naphthol	***	***	o	*	*	**	*	o	*	*	*	***	***	**
PR188	Naphthol AS		*	o	o	o	o	o	o	*	**	***	***	***	***
PY151	Benzimidazolone	o	o	o	o	o	o	o	o	o	*	*	**	***	***
PY155	Bisacetessigarylide	o	o	o	o	o	o	o	o	o	*	**	***	***	***
PR242	Disazo condensation	*	o	o	o	o	o	o	o	*	**	**	***	***	***
PY153	Metall complex	*	*	*	*	*	*	*	o	o	*	*	***	***	**
PY139	Isoindolinone	o	o	o	o	o	o	o	o	o	*	**	***	***	***
PB15:6	Phthalocyanine	***	***	*	*	***	**	**	*	o	*	*	**	**	**
PG7	Phthalocyanine	o	o	o	o	o	o	o	o	o	*	*	**	**	*
PG36	Phthalocyanine	o	o	o	o	o	o	o	o	o	o	**	***	***	**
PV19	Quinacridone	o	o	o	o	o	o	o	o	o	**	**	***	***	***
PR122	Quinacridone	o	o	o	o	o	o	o	o	*	**	**	***	***	***
PR209	Quinacridone	o	o	o	o	o	o	o	o	*	**	***	***	***	***
PR179	Perylene	o	*	o	o	o	o	o	o	*	**	**	***	***	***
PO43	Perinone	o	o	o	o	o	o	o	o	*	**	***	***	***	***
PR177	Antraquinone	o	*	o	o	o	o	o	o	o	**	***	***	***	***
PR83:1	Antraquinone CA	***	**	*	*	*	*	*	*	o	*	*	**	***	**
PB60	Indanthrene	o	o	o	o	o	o	o	o	*	**	**	***	***	***
PV23	Dioxazine	*	*	o	*	*	*	*	o	o	*	*	***	***	**
PR264	Diketopyrrolo-pyrrole		o	o	o	o	o	o	o	o	*	***	***	***	***
PO67	Pyrazolochinazolone	***	***	o	o	*	***	*	o	o	*	**	***	**	*
PW4	Zinc white		o	o	o	o	o	o	o	o	o	*	*	*	*

Solvent sensitivity: o insoluble, *low, **high, ***very high

but it is also influenced by the swelling behavior of the surrounding binder matrix (Zumbühl 2011). According to the strong dispersive force of interaction, chloroform often showed the highest extraction capacity. Good solubility usually is encountered along the polarity scale from the polarizable toluene up to the dipolar acetone. Towards the non-polar end of the polarity scale, the solubility decreases in general. The non-polar n-hexane is a poor solvent with synthetic organic pigments. The blue phthalocyanine PB 15:6 also exhibits high solvent sensitivity, with the highest extraction rate observed in toluene. In contrast to the well-soluble azo pigments, the solubility of the PV 23 and PY 153 proved modest and amounted to less than 0.005 mmol/l. The quantitative effect of extraction of the soluble PY 3 after immersion of the paint film in chloroform is presented in Fig. 1.

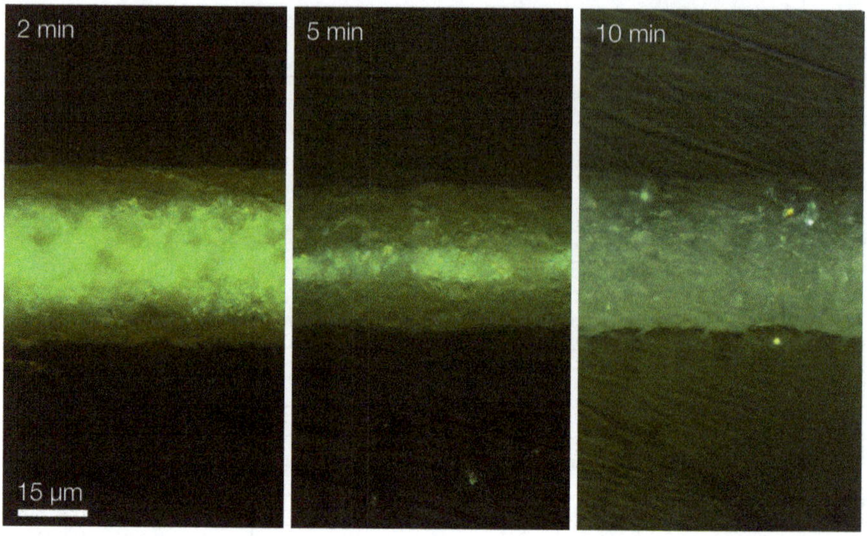

Fig. 1 Embedded PY 3 paint film cross-sections showing the time-dependent leaching front after immersion in chloroform. Bright field illumination with crossed polarization filters

The Influence of Solubility on the Oil Paint Film

The structural changes within the paint film were investigated using an artificially aged oil paint film containing the yellow azo pigment PY 3, which showed a strong solubility in the tests before. Applying infrared spectroscopy ATR-FTIR[3] to the surface of the leached films, the surface pigment response was tested upon solvent exposure of 1, 2, 5, 10 and 15 min (Fig. 2). After an immersion time of 1 min almost no pigment could be detected on the surface of the PY 3 oil paint film, even though no significant color changes can be observed visually.

With Raman spectroscopy,[4] the pigment concentration was examined along profiles across the embedded paint films on polished cross-sections.[5] All cases showed a steep concentration gradient (Fig. 3). The pigment was leached progressively from the exposed surfaces. Immediate leaching of the pigment occurs within the swollen parts of the film. This suggests very fast solubilization of the pigment with the solubility of the pigment being greater than the rate of diffusion of the solvent

[3]Avatar 360 FT-IR (Thermo Nicolet) ATR ZnSe-crystal Avatar Smart Miracle with 32 Scans and resolution of 4 cm^{-1}.

[4]Renishaw InVia Raman-spectrometer: 785 nm excitation, $<$12,1 mW on sample, 24 s per spot cumulative mode 6×4s, 1 µm step size, 100× objective, slit 65 µm standard in static mode, grating 1,200 l/mm, pinhole IN.

[5]Carbon evaporation coater Cressington Carbon Coater 108; embedding medium Technovit 2000LC.

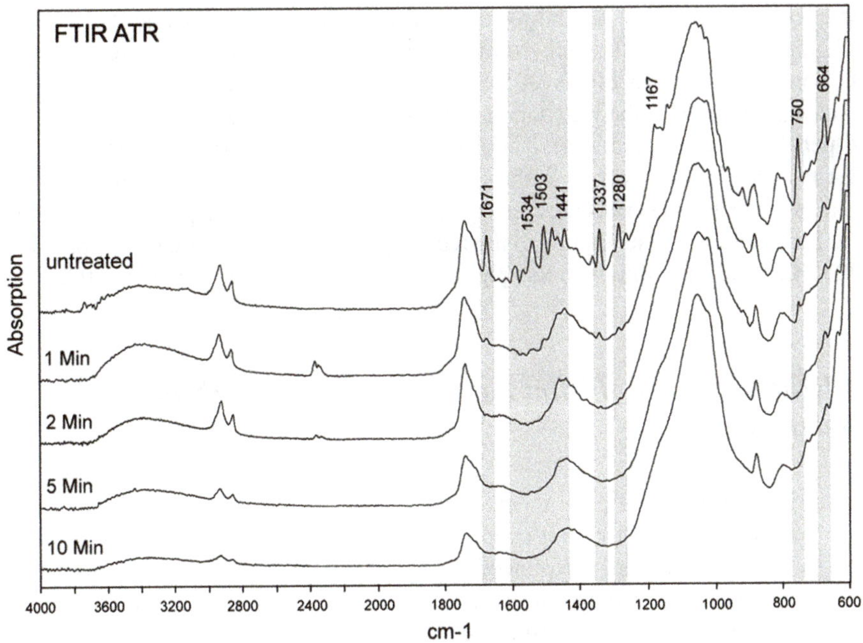

Fig. 2 Timed sequence of ATR-FTIR spectra on the PY 3 oil paint film after chloroform immersion: the pigment relevant response at the surface is lost within the first minute of solvent exposure

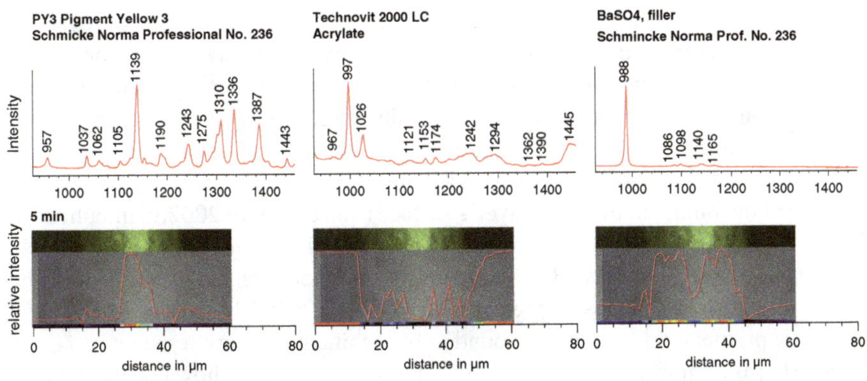

Fig. 3 The pigment loss, as observed in reflected light microscopy (Fig. 1), is also visible with Raman spectroscopy, documenting a sharp and clear delimitation of the solvent action horizon. Where the pigment and likely other paint additives are lost, the oil paint film becomes very porous and is infiltrated by the embedding medium

within the oil paint film. According to the signal response to ATR-FTIR, the pigment is being leached very efficiently from the surface (Figs. 2 and 3). Upon re-drying of the paint film, a porous film is left behind due to removal of pigment particles. This is also evident on embedded sample films, where the embedding resin was able to penetrate the leached zone. These results highlight the extremely delicate and fast response of paint layers with organic pigments to contact with solvents. The treatment of such materials thus sets a real problem to restoration practice. The same test procedure was also applied to the red 'insoluble' Disazo condensation pigment (PR 242). The paint film with this pigment, however, remained visibly unaffected.

Solvent Sensitivity of Paint Layers Containing Synthetic Organic Pigments

To test the sensitivity of paint layers containing synthetic organic pigments against solvents under real restoration conditions, cotton swab wipe tests with minimal pressure application were performed.[6] This allows the solvent to act in combination with light mechanical pressure on the surface. The tests were performed using commercially available solvent-moistened cotton swabs. They were wiped across the surface five times each as evenly as possible and with minimal contact pressure on the paint layer. While this test may not claim total objectivity, it nevertheless reproduces restoration practice and as such delivers useful information. The wiped cotton swab samples showed that all pigments within the series reacted very sensitively. Some exhibited pigment extraction at first contact with the cotton swab. This observation was independent from the pigment's classification with regard to the solubility examinations. In general, abrasion/extraction was lowest on cotton swabs soaked with non-polar solvents (Fig. 4). Solvents producing strong swelling of the oil binder caused an enhancement of this effect. Interestingly in contrast to the general solubility tests not only chloroform but also the polar solvents were most harmful. This may be explained by photochemical degradation and the change of the polarity of the binder in the paint layer's surface (Van den Berg 2002, Zumbühl et al. 2011). It also shows that the low wipe resistance is likely due to the small particle dimension of the pigments. By processing the pigments on a three-roll mill, the mean particle size of the tested pigments was between 8 and 12 μm. In contrast, the inorganic pigment zinc white was found to be significantly more resistant.

Shortly, as demonstrated in Fig. 4 and Table 1, these results show that the delicate behavior of paints with synthetic organic pigments against solvents is influenced by several factors. When considering the differences in solvent stability against the far less reactive classic oil paint made with inorganic pigments, one should bear in mind that the production of (metal) carboxylates as a factor of aging, has a stabilizing effect that is absent in modern paints with organic pigments.

[6] 100 μm oil paint films on slides, commercial cotton swabs.

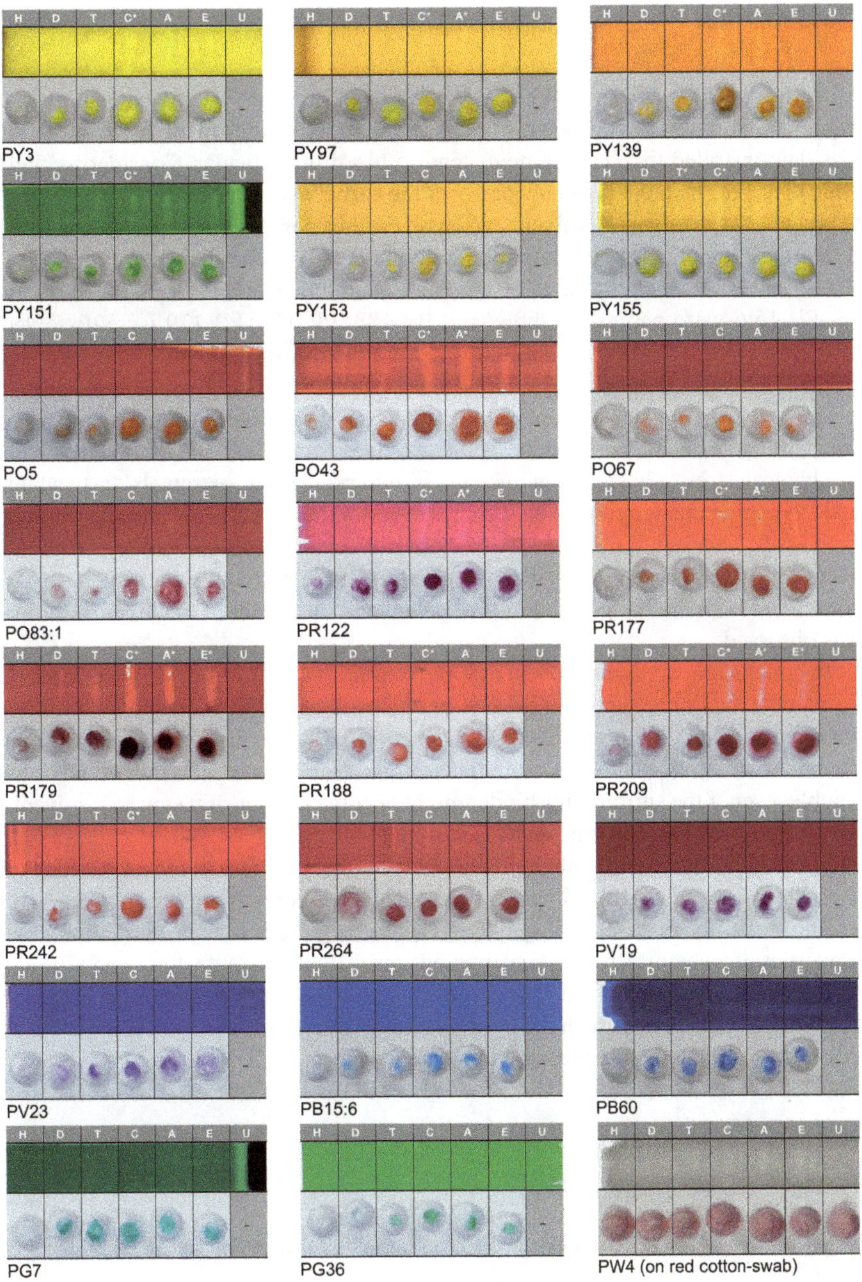

Fig. 4 The action of solvents in combination with mechanical stress to oil paint films with organic pigments can be regarded as dramatic. Due to the small particle size, even "insoluble" pigments will be extracted. Polar solvents seem to cause the most damage. (H = n-hexane; D = diethyl ether; T = toluene; C = chloroform; A = acetone; E = ethanol, U = untreated)

Conclusions

Modern oil paints are very different to classical oil paint systems with inorganic pigments. This sets new problems to conservation treatments involving solvents. It was demonstrated that azo pigments tend to bleed in particular. Considering their wide distribution, there are many paints of the early twentieth century that are likely to react very sensitively to solvent exposure. Furthermore, other pigments from very different pigment classes were also found to be soluble. The pigments PY 3, PO 5 PO 67 and PY 97 were found to be completely soluble, while PY 153, PR 83:1, PV 23, PB 15:6 were partially soluble, and PR 188, PR 177, PR 179 were dissolved to a lesser degree. The susceptibility of a paint film to extraction upon solvent contact, however, depends not solely on the solubility of the pigment molecule. It is also highly influenced by the swelling properties of the binder matrix, controlling the stability of such heterogeneous structures like paint films. In combination with mechanical stress, the reaction speed was observed to be extremely fast, which makes it rather challenging when it comes to restoration treatment handling. Factors such as the very small particle size of these pigments will further contribute to the overall rather poor wipe resistance – regardless of the pigment solubility. A strong correlation between the polarity of the solvent and the power of material extraction was observed.

In conclusion, applying solvents for the cleaning of surfaces made from modern oil paints containing synthetic organic pigments is extremely delicate with a high potential for inducing unintended detrimental effects. It is important to know that pigments that are soluble won't lose this feature, even in old age. It means they stay soluble, even if the surrounding binder might become less sensitive towards solvents during aging, as known from analysis.

Synthetic organic pigmented oil paints can create problems in the preparation of cross sections, because bleeding may make interpretation of samples difficult.

The difficulty of assessing synthetic-organic pigments is their variability. The properties of a chemically identical pigment can differ by manufacturing and further modifications, this aspect was not included. The wipe samples should be repeated at a more advanced time with the aged oil. If possible quantitative aspects, which at this time were just qualitative, should be included.

References

Binkowski S et al (2000) Preparation of pigments on modified precipitated silicas. Dyes Pigments 47:247–257

Blumenroth D (2010) Die Lösemittelempfindlichkeit synthetisch-organischer Pigmente in Ölfarben. MA-thesis, Berner Fachhochschule, Konservierung und Restaurierung HKB, Bern

Bugnon P (1995) Surface treatment of pigments. Treatment with inorganic materials. Prog Org Coat 29:39–43

de Keijzer M (1988) The blue, violet and green modern synthetic organic pigments of the twentieth century uses as artists' pigments. In: Modern organic materials. Preprints of the meeting, Edinburgh 1988. Scottish Society for Conservation and Restoration, Edinburgh, pp 97–103 de Keijzer M (1989) The colourful twentieth century. In: Modern art: the restoration and techniques of modern paper and paints – the proceedings of a conference. The United Kingdom Institute of Conservation, London, pp 13–20

de Keijzer M (1999) A survey of red and yellow modern synthetic organic artists' pigments discovered in the 20th century and used in oil colours. In: 12th Triennal meeting Lyon 29th August–3rd September, pp 369–374

de Keijzer M (2002) The history of modern synthetic inorganic and organic artists' pigments. In: Contributions to conservation: research in conservation at the Netherlands Institute for Cultural Heritage. James & James, London, pp 42–54

Faulkner EB (2009) High performance pigments. Wiley-VCH, Weinheim

Fremout W, Saverwyns S (2012) Identification of synthetic organic pigments: the role of a comprehensive digital Raman spectral library. J Raman Spectrosc 43:1536–1544

Fritsch A (2006) Hansa®-Pigmente: Werkstoffgeschichte, Werkstofftechnologie und Analytik. Diplomarbeit, Hochschule der Künste Bern, Konservierung und Restaurierung, Bern

Frowein S (2004) Synthetische organische Pigmente in zeitgenössischen Künstlerfarben – Angebot, neue Entwicklungen und Nachweismöglichkeiten. Diplomarbeit, Technische Universität, München

Fu SH, Fang KJ (2007) Preparation of styrene-maleic acid copolymers and its application in encapsulated pigment red 122 dispersion. J Appl Polym Sci 105(2):317–321

Herbst W, Hunger K (1995) Industrielle Organische Pigmente: Herstellung, Eigenschafen, Anwendung, 2. Auflage. VCH-Verlagsgesellschaft, Weinheim

Heumann K, Friedländer P (eds) (1898) Die Anilinfarben und ihre Fabrikation. Vieweg Verlag, Braunschweig

Hübner K (2006) 150 Jahre Mauvein. Chemie unserer Zeit 40:274–275

Kalbeek N (2005) Identification of synthetic organic pigments by characteristic color reactions. Stud Conserv 50(3):205–229

Lelu S (2003) Encapsulation of an organic phthalocyanine blue pigment into polystyrene latex particles using a mini-emulsion polymerization process. Polym Int 52(4):542–547

Lomax QS, Learner T (2006) A review of the classes, structures, and methods of analyses of synthetic organic pigments. J Am Inst Conserv 45(2):107–125

Lutzenberger K (2009) Künstlerfarben im Wandel – Synthetische organische Pigmente des 20. Jahrhunderts und Möglichkeiten ihrer zerstörungsarmen, analytischen Identifizierung. Herbert Utz Verlag, München

Marcus CM (1998) The properties of solvents. John Wiles & Sons, Chichester

Ropret P, Centeno SA, Bukovec P (2008) Raman identification of yellow synthetic organic pigments in modem and contemporary paintings: reference spectra and case studies. Spectrochim Acta A Mol Biomol Spectrosc 69:486–497

Schäning A, Schreiner M, Jembrih-Simbuerger D (2005) Identification and classification of synthetic organic pigments of a collection of the 19th and 20th century by FTIR. In: Piccolo M (ed) Proceedings of the sixth Infrared and Raman Users Group Conference (IRUG-6), Florence, 302–305, 2004

Scherrer NC, Zumbuehl S, Delavy F, Fritsch A, Kuehnen R (2009) Synthetic organic pigments of the 20th and 21th century relevant to artist's paint – Raman spectra reference collection. Spectrochim Acta A Mol Biomol Spectrosc 73:505–524

Schröder J (1988) Surface treatment of pigments. Prog Org Coat 16:3–17

Schulte F, Brezinka KW, Lutzenberger K, Stege H, Panne U (2008) Raman spectroscopy of synthetic organic pigments used in 20th century works of art. J Raman Spectrosc 29: 1455–1463

Sirikittikul D (2004) Modification of pigment surface with polymer monolayer. Cuvillier Verlag, Göttingen

Smith HM (2002) High performance pigments. Wiley-VCH, Weinheim

Strauss I (1984) Übersicht über synthetisch-organische Künstlerpigmente und Möglichkeiten ihrer Identifizierung. Maltechnik/Restauro Nr 4:29–44

Tiarks F, Landfester K, Antonietti M (2001) Encapsulation of carbon black by mini-emulsion polymerization. Macromol Chem Phys 202:51–60

Van den Berg JDJ (2002) Analytical chemical studies on traditional linseed oil paints, Academic Proefschrift. PhD thesis, Universiteit Amsterdam

Vandenabeele P, Moens L, Edwards HGM, Dams R (2000) Raman spectroscopic database of azo pigments and application to modern art studies. J Raman Spectrosc 31:509–517

Viala P (2002) Pigment encapsulation by emulsion polymerization, redispersible in water. Macromol Symp 187(1):651–662

Yuan J, Zhou S, Gu G (2005) Encapsulation of organic pigment particles with silica via sol-gel process. J Sol-Gel Sci Technol 36:265–274

Yuan J, Zhou S, Wu L, You B (2006) Organic pigment particles coated with titania via sol-gel process. J Phys Chem B 110:388–394

Yuan J, Xing W, Gu G, Wu L (2008) The properties of organic pigment encapsulated with assembly technique. Dye Pigment 76:463–469

Zerr G, Rübenkamp R (1921) Handbuch der Farbenfabrikation. Union Deutsche Verlagsgesellschaft, Berlin

Zumbühl S (2008) Material choice in the context of the artistic working phase – The use of synthetic organic colorants by Alexej von Jawlensky (1864–1941), Scientific workshop at the Doerner Institute München, Germany; Permanent Yellow, Irgazine Red, Heliogen Blue and Co.: Towards an improved micro-identification of synthetic organic pigments and dyestuffs in works of modern art, 12. Sept 2008

Zumbühl S (2011) Lösemittelempfindlichkeit von modernen Farbsystemen – Parametrisierung der Lösemittelsensitivität von Öl-, Alkyd- und Acryl-Künstlerfarben. Dissertation, Staatliche Akademie der Bildenden Künste Stuttgart, Fachbereich Kunsttechnologie, Stuttgart

Zumbühl S, Scherrer N, Ferreira E, Hons S, Müller M, Kühnen R, Navi P (2011) Accelerated ageing of drying oil paint – an FTIR study on the chemical alteration. Zeitschrift für Kunsttechnologie und Konservierung 25(2):139–151

Dry Cleaning: Research and Practice

Maude Daudin-Schotte and Henk van Keulen

Abstract The Cultural Heritage Agency of the Netherlands (RCE) 2006–2009 dry cleaning research project investigated a broad range of dry cleaning materials (latex sponges, make-up sponges, PVC erasers, Factis erasers, Gum powders, microfiber cloths, mouldable materials), each with very variable affinities to dirt and paint surfaces, from efficient and safe cleaning to abrasion and residues hazards. This paper presents a follow-up to the RCE 2006–2009 research, focusing on the long-term effect of potential chemical residues, facilitated by our practical experience acquired in workshops since 2010 together with further scientific analysis.

Keywords Mercaptobenzothiazole • Plasticizers • Residues • Long-term effect • Dry cleaning • Make-up sponges • Micropore sponge • Sofft Tools® • Absorene® • Electrostatic roll

Introduction

After gentle brushing and vacuuming of the surface, unvarnished paintings can typically be cleaned by swab rolling the surface with aqueous solvents, except in cases where surfaces are sensitive to water or other solvents (Burnstock et al. 2006; Ormsby et al. 2006; Mills et al. 2008; Tempest et al. 2013). In such instances alternatives mainly based on mechanical action, such as dry cleaning, are mandatory (Estabrook 1989; Paerlstein et al. 1982). The RCE 2006–2009 dry cleaning research project investigated a broad range of materials with very variable affinities to dirt and paint surfaces (Fig. 1; Daudin et al. 2013, 2014). The use of these materials might imply friction and subsequent elevation of surface temperature, potential risks of abrasion and polishing, micro cracks or flattening, and particulate or chemical residues.

M. Daudin-Schotte (✉)
Independent Painting Conservator, San Francisco, CA, USA
e-mail: DSConservation@free.fr

H. van Keulen
Cultural Heritage Agency of the Netherlands, Amsterdam, The Netherlands

K.J. van den Berg et al. (eds.), *Issues in Contemporary Oil Paint*,
DOI 10.1007/978-3-319-10100-2_24

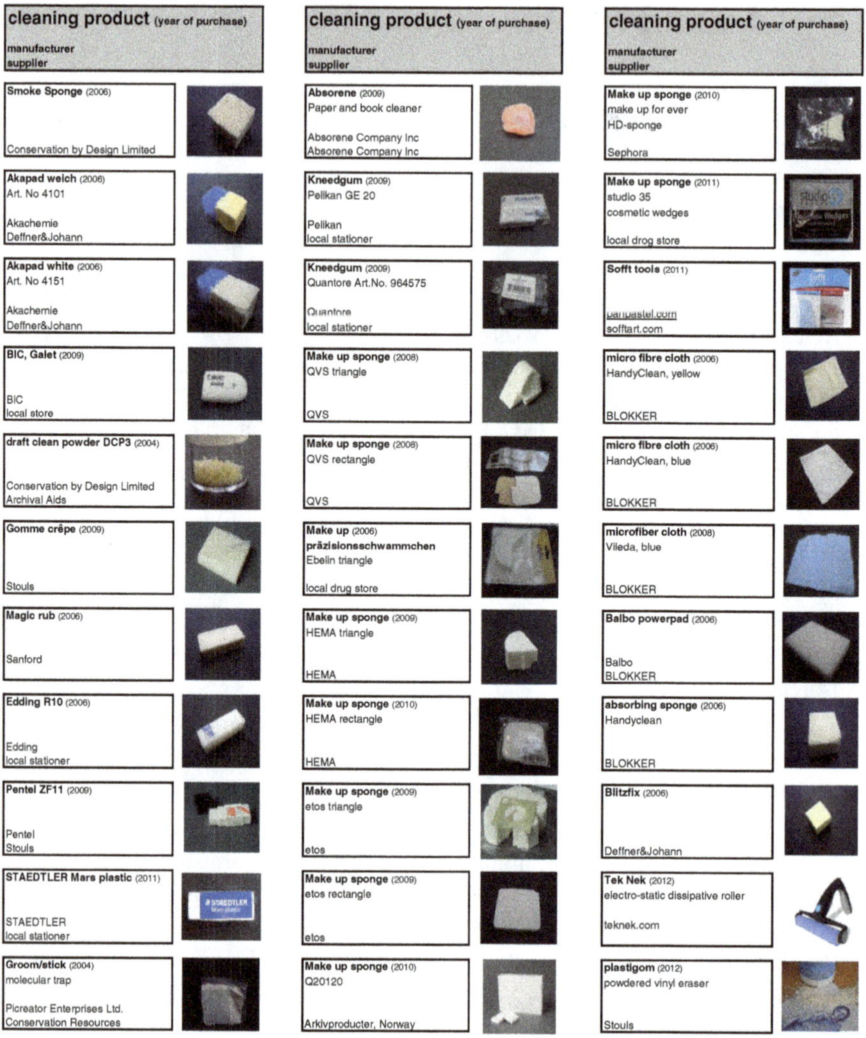

Fig. 1 Table of dry cleaning materials tested since 2006, shown are the product name, year of purchase, supplier information and a photograph. A table with a complete summary of the results of the investigations can be downloaded from the RCE website (http://www.cultureelerfgoed.nl/sites/default/files/documenten/drycleaning%20table.pdf)

Research and practice showed for instance that Absorene® or Groom/Stick® have tendencies to leave particulate residues behind; while smoke sponges or make-up sponges proved successful and suitable for dirt removal from paint surfaces (Daudin et al. 2013; Froment 2009). This paper presents a follow-up to the RCE 2006–2009 research, facilitated by our experience acquired in workshops since 2010 together with further scientific analysis.

Research into Long-Term Effects of Dry Cleaning

For the RCE 2006–2009 project, the composition of the dry cleaning materials was determined using (py-) GCMS, FTIR and SEM-EDX. Part of the materials containing organic solvent extractable compounds, such as plasticizers and antioxidants (chemical residues), have been selected for further testing. The selection was based on the type and the effect that the extractable compound might have on the treated paint surface. The dry cleaning materials selected were Groom/Stick®, Akapad® weich and white, Smoke Sponge® and Magic Rub®.

Zinc white pigmented oil paint swatches were cleaned using the selected dry cleaning materials, and the presence of extractables on or in the surface of the paint layer was examined. After that, the swatches were naturally aged, in daylight behind glass, for 5 years. The swatches were examined as part of the follow-up project.

For the follow-up project, the surface of an oil painting, aged naturally for 40 years, was treated with organic solvent extractable compounds present in dry cleaning materials (Oruç-Iddi 2013).[1]

The compounds applied were plasticizers di-iso-octyl phthalate (DIOP) and tributyl phosphate (TBP), UV absorber Tinuvin 292®, moisturizer polyethylene glycol 400 (PEG-400), anti-oxidant butylated hydroxyl toluene (BHT), and curing accelerator mercaptobenzothiazole (MBT).

Although not present as an extractable compound, MBT is present in material analysis of some of the make-up sponges. It was added to the series of extractables tested, as it is a potential danger for paint surfaces.

The compounds were applied in a range of concentrations, from 10 times below to 100 times above the concentration that could reasonably be expected for a normal cleaning treatment (0.04–40 $\mu g/cm^2$, based on the residue of 0.75 $\mu g/cm^2$ of plasticizer present after vigorous cleaning with a PVC eraser). The paint surface was artificially light-aged.[2]

Results of Long-Term Effects of Extractables

The zinc white in oil paint swatches were examined after aging using microscopy and a preparation needle. No changes such as yellowing, softening, and stickiness were observed due to the presence of the extractables.

After artificially aging, the surface of the organic compounds treated painting was visually, optically inspected and mechanically tested with a spring balance (pressure tests).

[1] Thanks to Zuhura Oruç-Iddi. Oruç-Iddi, Z. Droge reiniging: Het effect van zes soorten residuen op ongeverniste olieverflagen, master thesis, University of Amsterdam 2013.

[2] Aged for 5.8 years under museum conditions assuming reciprocity.

The paint surface treated with MBT (all concentrations) showed cracks and reduced flexibility; no effects could be observed for the other organic compounds.

In general, except the tested PVC erasers, the tested dry cleaning materials show no or very low amounts of chemical residues present on the surface of an oil paint after cleaning.

Although results indicate that the long-term effects of the tested chemical residues are very low/negligible, materials with low amounts of additives are preferable.

Some manufacturers have produced new materials such as 'Phthalate free PVC erasers' (Staedtler Mars® plastic). In this PVC eraser, which is commonly plasticized with phthalates, the plasticizers have been replaced by a mixture of carbonic acid long chain alkane, phenyl esters. This confirms the fact that, even if manufacturers indicate an eraser as phthalate-free, any PVC eraser manufacturing process will require a plasticizer.

Make-up sponges are popular for cleaning due to their softness. Polyurethane make-up sponges e.g. HD forever® and Studio 35® were found to contain Tinuvin® or vitamin-E, as (natural) antioxidant. In these sponges, vulcanizers or accelerators, as in their rubber counterparts, could not be detected.

The Follow-Up Project in Practice

Workshops practical sessions followed guidelines from 2006 to 2009 project conclusion (Daudin et al. 2013) and provided an opportunity to test the recommended dry cleaning materials on a variety of paint surfaces.

Review of the RCE 2006–2009 Project Conclusion

The gradual test approach proposed in 2009 remains up to date.[3] Even though our former concerns about plasticizers found in PVC erasers has been revised from the above research results on long-term effects, PVC and factis based erasers are still considered sensitive materials with potential abrasion hazards.

We advise to choose TDI urethane based make-up sponges over other components to avoid potential MBT concern.[4] Also, practice validated so far the swelling

---

[3]Standard test approach recommends to test (first to last): make-up sponges (TDI), microfiber clothes (thin and thick ones), Akapad® white (brushed off three times after cleaning), smoke sponge (only for semi glossy to glossy surfaces).

[4]Make-up sponges' tests, no matter their brands, proved successful both regarding preservation of the topography and efficiency in the cleaning, but only SBR or isoprene rubber-based make up sponges (ex. ETOS -isoprene rubber-, HEMA -SBR-) potentially contain MBT.

of TDI sponges when immersed in water, confirming this is a simple yet reliable method to determine the type of a make-up sponge.

Moreover, exchanges with workshop participants allowed a review of the cleaning materials in the published study in particular those found to be associated with residue hazards, such as Absorene® and other malleable materials. This plasticine-type material is popular because pleasant to use and efficient, but proved to leave sticky, unstable residues with ageing. Our practical sessions provided strategies for minimizing residual deposits, as Absorene® pink particulate residues are easily identifiable under microscopy (as opposed to the Groom/Stick® ones), as well as strongly fluorescent in UV light. Therefore, in some cleaning cases with complex topography patterns requiring using a malleable material, Absorene® characteristics facilitate the identification and immediate removal of its particulate residues along the cleaning, which considerably limits the damage caused by their interaction with the paint layer.

Additional Guidelines for the Application Method

Additionally to cleaning materials recommended in the 2009 conclusion,[5] workshops aimed to provide with more specific guidelines while extending the range of dry cleaning options.

The uneven nature and components of deposited dirt (Perry 1990; Phoenix and Burnstock 1990) required a more defined and systematic cleaning method. A strategy in two approaches has been determined: the first focuses on the surface, combining the use of different materials depending on the paint colour and topography to evenly remove embedded dirt. The second approach targets the dirt thickness, aiming to remove superposed layers of grease or dust.

These approaches, used alternately during cleaning, provide a better understanding of the nature of dirt (as for instance Absorene or smoke sponge show an affinity with grease when gum powder or microfiber cloth are more efficient on dry dirt) and aim to provide guidelines for even cleaning.

The shape of cleaning materials always proved crucial, as for example a same solvent applied with a cotton swab or with a make-up sponge will not equally affect the paint surface. With this in mind we successfully tested the principle of combining shredded materials with compact ones, which in some cases meant to alternatively apply the very same material in different shapes (e.g. smoke sponge). This versatile and subtle application method has been included to our cleaning options.

[5]Respectively Make-up sponges (TDI), microfiber Clothes, Akapad® white, and smoke sponge.

Additional Tests and Results

Multiple new materials have been tested (results observed with naked eye, light microscopy, SEM EDX[6] and under UV fluorescence), from erasers to jelly tabs, micropore sponges to electrostatic roll,[7] sometimes with promising results.

Vinyl-Based Gum Powder

Plastigom®[8] is already popular in conservation practice, and has been tested here as an alternative to hard PVC erasers, comparing with gum powder based on vulcanized vegetable oil,[9] only with a more even particles sizing.

Despite a slight harsher contact with the paint layer than the factis-based gum powders, Plastigom® results have been encouraging, showing so far a satisfying cleaning efficiency for a correct preservation of the topography and confirming it as an alternative to regular PVC erasers.

Micropore Sponges

Investigating the range of micropore sponges, Studio 35®[10] make up sponges and Sofft Tools® painter art covers have been introduced in the 2011 workshops. These tools complement the texture and absorbing capacities of a make-up sponge with a very flexible handling, allowing for much more precision. They are manufactured for artists, for paint applications, and come in multiple shapes that could be an asset for conservation treatments. Sofft Tools®, analyzed to contain a mixture of isoprene and styrene butadiene rubber, proved efficient for dirt removal purposes, especially on impasted areas[11] (Fig. 2).

---

[6] SEM imaging by the Universitat Politècnica de València, Spain; Studio 20I21 Conservation and restauration, Portugal; Cesmar7 and University of Pavia, Italy.

[7] Natus Jelly Tab® sensors, pen erasers, Art gum® eraser, Staedtler Mars® plastic 526 50, Plastigom®, Studio 35® Make up sponges, Sofft Tools®, and Teknek® roll. (Cf. table).

[8] Vinyl-based gum powder (Cf. table) http://www.stouls.com/pages/conservation_en/produits_conservation.php?gamme=17 (Accessed April 8th 2014).

[9] DCP3, Cf. table.

[10] TDI based make up sponge, vitamin E added (Cf. table).

[11] http://www.sofftart.com/products.html#knives (Accessed April 8th 2014).

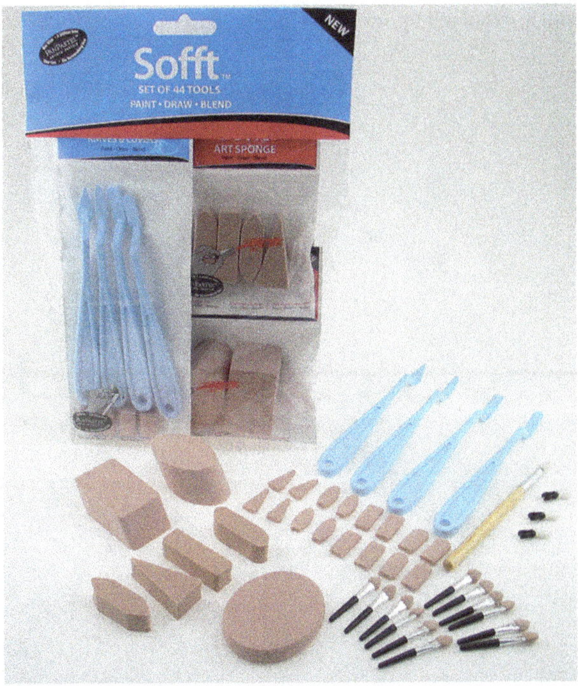

Fig. 2 Sofft Tools®

Non-contact Hand Roll

Another novel material introduced in the 2012 workshops is the Teknek® DCR hand roll, which uses electrostatic charging to attract loose dirt, which is then deposed on an accessory paper sheet.[12] The Teknek® Company, specialized in cleaning sensitive substrates such as electronics, presents the hand roll as an elastomer rubber roller with and engineered adhesive pad. Supposedly this pad will not leave any adhesive residues on the roll (Fig. 3).

Cleaning tests on both oil paint and gouache surfaces achieved a satisfying yet superficial dirt removal, comparable to results achieved with microfiber clothes.

Teknek® application is limited to relatively loose dirt deposited on flat surfaces. One of its interesting assets could be its ability for selective removal, as it proved to precisely remove pencil marks from over the dirt layer.

It is yet to determine if this interesting method could be developed to fully suit conservation cleaning purposes.

[12]http://www.teknek.com/index.php/products/hand-rollers/dcr (Accessed April 8th 2014).

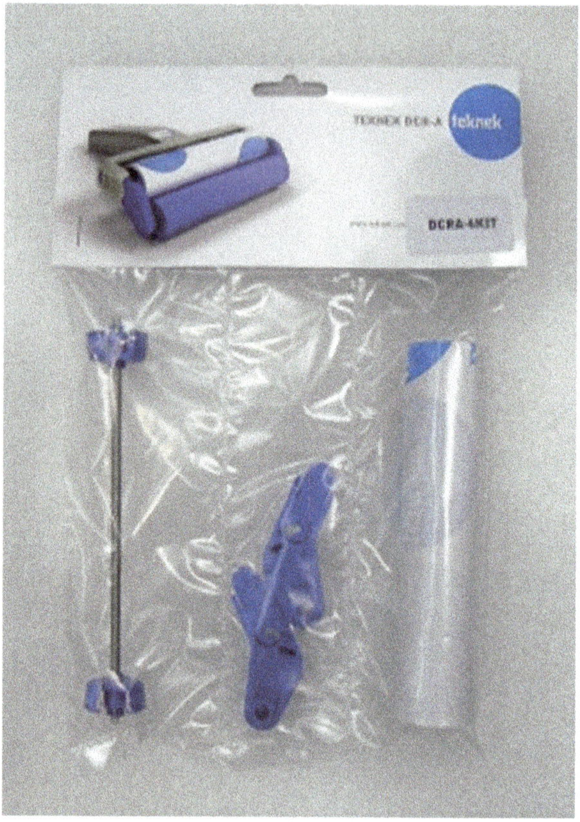

Fig. 3 Hand roll by Teknek®

Discussion and Conclusion

The present study indicated that routine use of most of the dry cleaning materials tested by conservators carries a low risk of creating long-term changes in paint. An exception was the vulcanizing agent mercaptobenzothiazole; an amount of 0.04 $\mu g/cm^2$ is enough to alter the flexibility of the paint surface. Although plasticizers are expected to soften the paint, our experiments indicated that on application on a well-cured, 30 year old oil paint surfaces, no discernible changes occur.

The materials are sold for their physical properties, but their chemical composition is not guaranteed to be permanent.

The range of materials that is available keeps growing, for purchase either at a local store or in specialized conservation stores. Each of these materials can be used shredded and/or compact to complement another, to allow better affinities with the nature of dirt, achieving subtler cleaning results.

Our 2009 dry cleaning standard test approach is augmented with Sofft Tools®, and does not exclude Absorene® paper and book cleaner. Practical experience indeed showed that, when Absorene® residues are cautiously tracked under UV-light fluorescence and removed during the cleaning, this material is safely and efficiently able to remove dirt from specific paint surfaces.

In addition to fine-tuning the method of application of dry cleaning materials since 2010, new materials are finding their way into dry cleaning conservation practice, including a promising method for dirt removal based on the principle of electrostatic charging.

Acknowledgements Klaas Jan van den Berg and the RCE; Madeleine Bisschoff; Polly Saltmarsh; Beatriz Verissímo Mendes, the RCE 2006–2009 Project 'klankbordgroep'; Aviva Burnstock; Emilie Froment and the University of Amsterdam; Zuhura Oruç-Iddi, master student University of Amsterdam; Laura Fuster-López and the Universitat Politècnica de València; Paolo Cremonesi, Ambra Giordano, Ilaria Saccani, Cesmar7 and University of Pavia; Pedro Pardinhas, Joanna Correia, Marta Palmeira, Nancy Fonseca, Conservação restauro 20I21; Nathalie Le Dantec and the Institut National du Patrimoine.

References

Burnstock A, van den Berg KJ, de Groot S, Wijnberg L (2006) An investigation of water-sensitive oil paints in 20th century paintings. In: Learner T, Smithen P, Krueger JW, Schilling MR (eds) Modern paints uncovered conference, proceedings from the Tate Modern symposium. The Getty Conservation Institute, Los Angeles, pp 177–188

Daudin-Schotte M, Bisschoff M, Joosten I, van Keulen H, van den Berg KJ (2013) "Dry cleaning approaches for unvarnished paint surfaces", post prints Valencia international conference "New Insights into the cleaning of paintings, (Cleaning 2010)". Smithsonian Institution Scholarly Press

Daudin-Schotte M, van Keulen H, van den Berg KJ (2014) For the cultural heritage agency of the Netherlands, and Giordano, A.; Saccani, I. for the Cesmar 7, collaborative paper, "Analysis and application of dry cleaning materials on unvarnished paint surfaces, RCE project 2006–2009", in Quaderno/Cesmar 7 publication, to be published in 2014

Estabrook E (1989) Consideration of the effect of erasers on cotton fabric. J Am Inst Conserv 28:79–96

Froment E (2009) Dry cleaning workshop. A collaboration between Amsterdam University and ICN. ICN Internal report, Instituut Collectie Nederland, Amsterdam

Mills L, Burnstock A, de Groot S, Megens L, Bisschoff M, van Keulen H, Duarte F, van den Berg KJ (2008) Water sensitivity of modern artists' oil paints. In: Bridgland J (ed) Preprints ICOM-CC 15th triennial conference New Delhi, vol II. Allied Publishers, New Delhi, pp 651–659

Ormsby B, Learner T, Schilling M, Druzik J, Khanjian H, Carson D, Foster G, Sloan M (2006) The effects of surface cleaning on acrylic emulsion paintings – a preliminary investigation. In: Weyer C (ed) Oberflächenreinigung; materialien und methoden. VDR schriftenreihe, 2nd edn. Theiss Verlag, Stuttgart, pp 135–149

Oruç-Iddi Z (2013) Droge reiniging: Het effect van zes soorten residuen op ongeverniste olieverflagen. Master thesis, University of Amsterdam

Paerlstein EJ, Cabelli D, King A, Indictor N (1982) Effects of eraser treatment on paper, J Am Inst Conserv 22(1), Article 1, pp 1–12

Perry R (1990) Problems of dirt accumulation and its removal from unvarnished paintings: a practical review, in dirt and pictures separated. United Kingdom Institute for Conservation and the Tate Gallery, London, pp 3–6

Phoenix A, Burnstock A (1990) The deposition of dirt: a review of the literature, with scanning electron microscope studies of dirt on selected paintings, in dirt and pictures separated. United Kingdom Institute for Conservation and the Tate Gallery, London, p 14

Tempest et al (2013) Progress in the water sensitive oil project, post prints Valencia international conference "New Insights into the cleaning of paintings, (Cleaning 2010)", Smithsonian Institution Scholarly Press, Washington, DC, pp 107–114

New Approaches to Surface Cleaning of Unvarnished Contemporary Oil Paintings – Moist Sponges and Cloths

Beatriz Veríssimo Mendes, Klaas Jan van den Berg, Luc Megens, Ineke Joosten, and Maude Daudin

Abstract The properties and cleaning efficacy of five different sponges and two microfibre moistened cloths were studied on both naturally and artificially aged and soiled oil paint surfaces. The cleaning efficacy of the materials were compared to that of conventional cotton swab cleaning with water and saliva. All materials tested showed good performance in terms of cleaning compared to swab rolling with water or saliva. The tested moist isoprene make up sponge performed best of all materials on the artificeally aged and soiled surfaces.

Keywords Surface cleaning • Aqueous cleaning • Contemporary art • Microfibre cloths • Makeup sponges

Introduction

Dirt deposition on painted surfaces is a common problem for artworks resulting in the increase of light scattering on the surface and causing effects such as de-saturation of hue and luminosity decrease for both light and dark colours (Green 1990).

In the case of unvarnished objects the dirt will adhere directly to the painted surface and in time it becomes embedded making its safe removal a common and complex problem for conservators (Perry 1990).

The nature of dirt depends greatly on the environment. Size and distribution rate of airborne contaminants in a museum environment varies with factors like internal and external environment, ventilation systems time of year and visitor's

B. Veríssimo Mendes (✉)
Faculdade de Ciências e Tecnologia, Universidade Nova de Lisboa, Lisbon, Portugal

Cultural Heritage Agency of the Netherlands, Amsterdam, The Netherlands
e-mail: beatrizverissimo@gmail.com

K.J. van den Berg (✉) • L. Megens • I. Joosten • M. Daudin
Cultural Heritage Agency of the Netherlands, Amsterdam, The Netherlands
e-mail: k.van.den.berg@cultureelerfgoed.nl

K.J. van den Berg et al. (eds.), *Issues in Contemporary Oil Paint*,
DOI 10.1007/978-3-319-10100-2_25

presence (Nazaroff et al. 1990). Deposition processes depend mainly on the nature of the dirt and the painted surface, particle size, as well climate parameters such as temperature and relative humidity (Phenix and Burnstock 1990) (Nielsen et al. 1978). In addition, physical characteristics such as texture, porosity, roughness, hardness and tack have a role in deposition of dirt (Nielsen et al. 1978).

The removal of dirt from an artwork is a complicated process and damages to the surface such as abrasion and polishing, depend on the approach and cleaning materials used, rather than strictly professional skills (Hackney 2010). A common surface cleaning technique is swab rolling with water, water and additives or solvent combinations (Wolbers 2000). However, conservators increasingly use new materials such as dry or moistened sponges and cloths.[1] Due to smoothness and ease of handling, these materials are pursued as a useful addition to common practice cleaning materials. There is concern, however, about their durability, performance when used on unvarnished painted surfaces and potential extractable components.

Studies have shown the utility and limitations of some of these materials when applied dry to e.g. water sensitive painted surfaces (Daudin-Schotte et al. 2011). The aim of the present study was to observe the cleaning effect and surface changes caused by sponges and cloths used moist –combined with demineralised water – on different unvarnished oil paint surfaces. The results are evaluated in terms of material characteristics, ease of handling and cleaning efficacy using visual, microscopy and electron microscopy techniques (Morrison et al. 2007; Daudin-Schotte et al. 2011).

Materials Used in This Study

Cleaning Materials

The cleaning materials chosen are two sponges, three makeup sponges of different composition and two microfiber cloths of different roughness listed in Table 1. The composition and additives of these materials as well as their cleaning performance while used dry were previously studied.[2]

The polyvinyl alcohol – Blitzfix® and semi-vulcanised isoprene rubber Smoke Sponge® as well as the microfire polyethylene therephthalate (PET) cloths are sold by conservation suppliers – the three makeup sponges are not sold as such.[3]

[1]Verbal communication with colleagues and conservators participating in the *Dry Cleaning Workshops* of January and June 2010, Instituut Collectie Nederland, Hobbemastraat 22, Amsterdam.

[2]http://www.cultureelerfgoed.nl/sites/default/files/documenten/drycleaning%20table.pdf.

[3]Hema is a department store and Etos is a drugstore. Both are chain stores in the Netherlands, Germany and Belgium. There are new sponges on the market. ALthough most claim to be PU sponges testing is needed to determine any additives present.

Table 1 Results of characterisation of the cleaning materials, elemental analysis, composition, observations with SEM, water retention and handling abilities Water retention was studied for each material by placing a moist weighed material on blotting paper and weighing it again after 10 min

Cleaning materials (comercial name)	Elemental analysis	Chemical composition[a]	Observation (SEM)	Water retention	Handling
Blitzfix	C, O, Si	Polyvinylalcohol (PVAL)	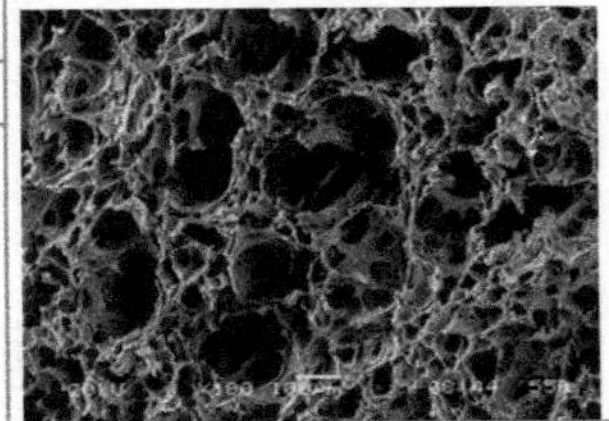	++	Easy
		Starch			Smooth
		Filler			Soft
					Dense
Smoke Sponge	C, O, Ca, S	Isoprene rubber	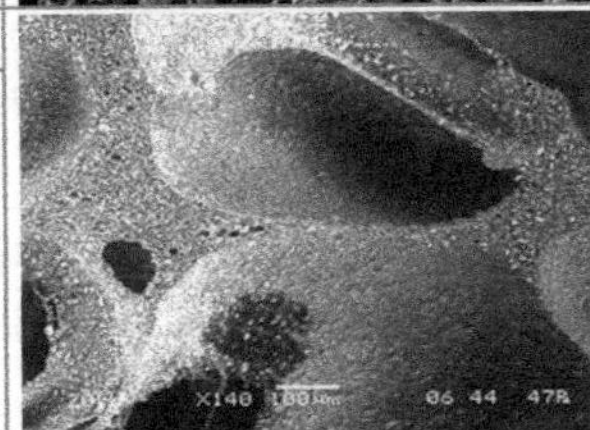	–	Easy
		Chalk			Dense
					Tacky
HD Makeup Sponge	C, O, Zn	Polyurethane Ether (PU)	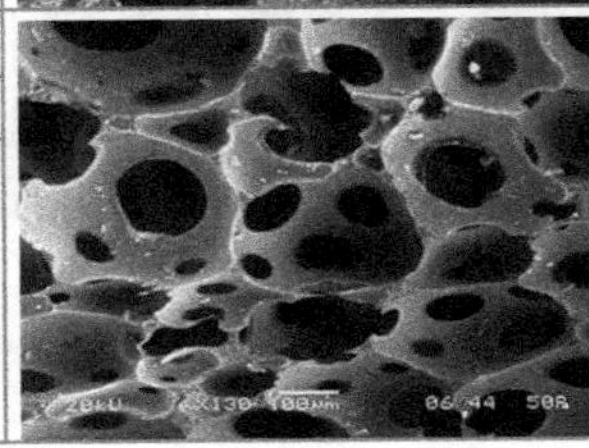	–	Easy
		Tinuvin			Smooth
		Zinc Oxide?			Soft
					Flexible
					Semi Dense

(continued)

Table 1 (continued)

Cleaning materials (comercial name)	Elemental analysis	Chemical composition[a]	Observation (SEM)	Water retention	Handling
Hema Makeup sponge	C, O, Zn, K, S, Si	Styrene butadiene rubber (SBR) Butylated Hydroxy Toluene (BHT) Diethyldithiocarbamate Mercaptobenzothiazole Zinc Oxide	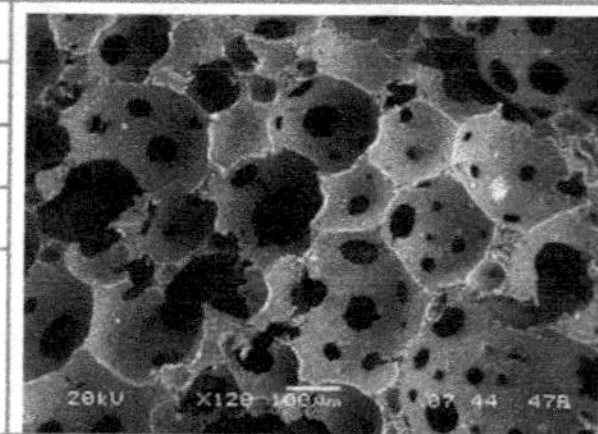	+/−	Easy Smooth Soft Flexible
Etos Makeup sponge	C, O, S, Zn, Si	Isoprene rubber Butylated Hydroxy Toluene (BHT) Zinc Oxide	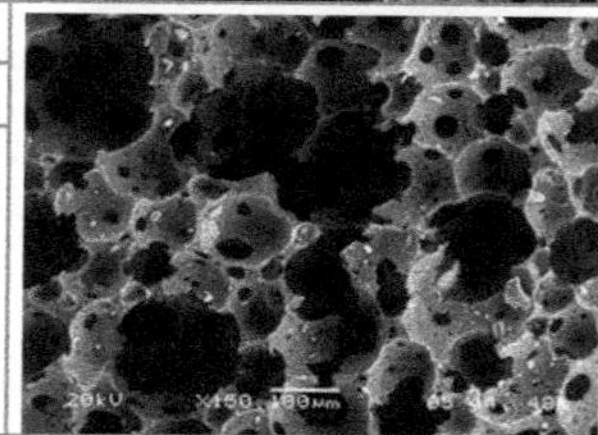	+	Easy Smooth Soft Flexible
Yellow cloth (more rough)	C, O, Al	Polyethylene therephthalate (PET) (polyester) – Nylon 6 (polyamide)	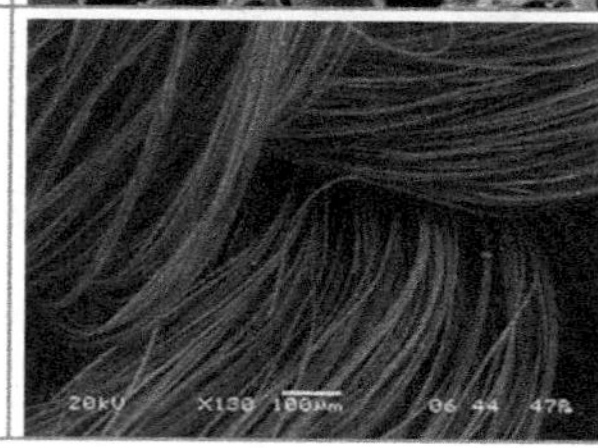	+/−	Difficult Smooth Not flexible

Blue cloth	C, O, Al	Polyethylene therephthalate (PET) (polyester) – Nylon 6 (polyamide)	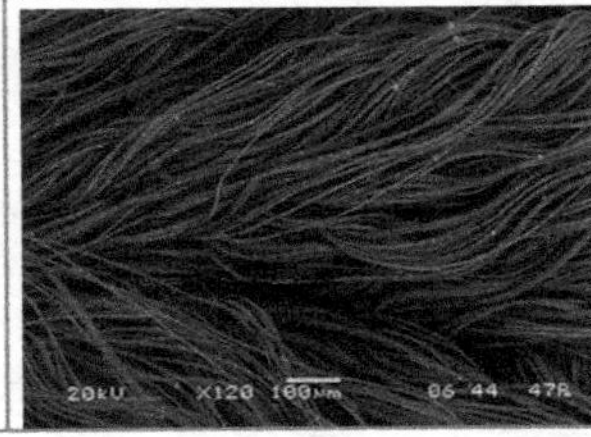	+/–	Difficult Smooth Not flexible

[a]http://www.cultureelerfgoed.nl/sites/default/files/documenten/drycleaning%20table.pdf
A percentage of water loss was calculated. Water retention was categorized very good ++ (less than 50 %), good + (from 50 to 60 %), average +/– (60 to 70 %), low – (70 to 80 %) and finally very low – (higher than 80 %)

Ease of handling, structure, softness, smoothness and flexibility were assessed during cleaning procedure, by touching and by observations with the microscope and Scanning Electron microscopy (SEM).

The cleaning performance of the new materials was compared with that of cotton swabs, applied with water and saliva.

Paint Surfaces

For this study two types of paint surfaces were used. The first set exists of artificially aged and soiled paint surfaces. The second set consists of a deaccessioned unvarnished monochromatic oil painting dating from 1977.

Set (I) consisted of Old Holland tube oil paints, painted out with flat and impasto areas on canvas boards.[4] The paints consisted of cadmium yellow, cadmium red and ultramarine blue in oil (Mills et al. 2008). Surfaces were cured and artificially light aged at 10,000 lux at room temperature to the equivalent of 46 years of exhibition in recommended conditions, 8 h a day 200 lux, assuming reciprocity. The paints and their ageing regime were also used in the previous dry cleaning study (Daudin-Schotte et al. 2011).

The red and yellow paints were artificially soiled while the panels of cobalt blue were left unsoiled for pilot tests. In order to maximise the variety of dirt, soiling procedures included the application of particulate dirt collected from an old cellar[5] with a microfiber cloth; followed by spraying on grimy dirt, prepared after a recipe for artificial dirt (Ormsby and Phenix 2010). The soiled surfaces were further aged, at a constant temperature of 50 °C and relative humidity variations from 30 to 80 % every 6 h, for 21 days, to increase the dirt retention of the painting.

Each reference panel was divided in four. The remaining panels were divided into two, separated by a reference area that remained unsoiled throughout the project (Fig. 1).

The monochromatic green painting surface dated from 1977 and had been kept in standard storage conditions at RCE Collections in Rijswijk. The surface was smooth and dirt was scarce and relatively loosely bound, except for fingerprints on the margins. Infrared Absorption Spectrometry – Fourier Transform Attenuated Total Reflectance Infrared Spectrometry (FT-ATR-IR), X-ray Fluorescence Spectroscopy (XRF) analyses and light microscopy showed on the canvas a zinc white, barium sulphate and chalk containing ground layer followed by a grey oil paint layer

[4]By Polly Saltmarsh in 2008.

[5]At the former ICN, Gabriël Metsustraat 8. Light microscopy and SEM showed that the dirt was mainly particulate inorganic matter conaining silica, calcium, sulphur and aluminium, indicating that much of the dust was related to cement and building material.

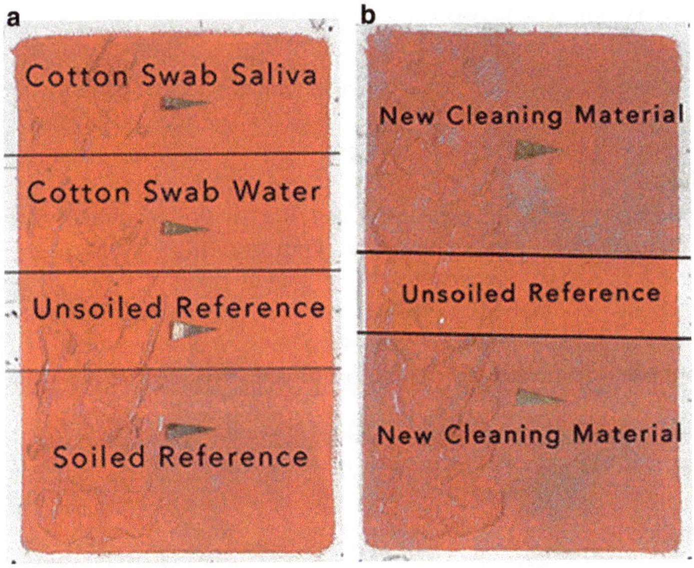

Fig. 1 Layout prepared panels where areas remained soiled and unsoiled throughout the study and remaining areas were cleaned with cotton swab in the case of the reference panel (**a**) or new materials in case of remaining panels (**b**). In this image it is also visible the markers that allowed observations in the exact same spots at different stages of the project

containing titanium white and carbon black. The top paint layer contained an unidentified possibly organic green pigment in oil.

The dirt on the panels and dirt applied on carbon tape was characterised using SEM, XRF and light microscopy. Cross sections from each kind of painted surface were made and studied by microscopy.

Analytical Instruments

The light microscope used was a Zeiss Axioplan 2 with Zeiss neofluar lenses. Photomicrographs were taken with a Zeiss Axiocam MR with an Axiovision 4.8 program. Raking light was provided by a Leica cold light source.

25× and 100× magnifications were used.

A Scanning Electron Microscope with Energy Dispersive X-ray Detector (SEM-EDX, JEOL 5910LV) was used in the secondary electron mode to examine the paint surface without further preparation. Due to charging of the painted surface, low ranges of accelerating voltages between 1.7 and 5 kV were used. Magnifications ranged from 100× to 750× up to 1,000× or 3,000×. A copper tape marker was glued on the surface of each area, in order to facilitate finding the same spot during the various stages of the project (Morrison et al. 2007).

Cleaning Tests

Cleaning tests were performed under ambient temperature and humidity. In order to better understand the mechanical action of the cleaning materials, without abrasion effects caused by the dirt, pilot tests were made on unsoiled prepared panels of cobalt blue.

All sponges were cut into pieces of approximately 2 cm^2 and were applied onto the surface using very light pressure and then slightly turned, minimizing mechanical action. The cloths were cut into small portions, wrapped around the index finger and then applied on the surface in the same manner as described for the sponges. Cotton swabs on bamboo sticks were rolled on the surface.

To minimise the possibility of leaving organic residues on the paint surface (Daudin-Schotte et al. 2011), all sponges and cloths were rinsed with water and pressed in between sheets of tissue paper, leaving them hardly wet to the touch. While cleaning, the materials were frequently renewed. Clearance procedures were not performed as only water was used. Results were assessed by visual observations with the naked eye as well as light microscopy and SEM.

Evaluation Parameters

Surfaces were compared with the reference area of each panel and recorded in evaluation tables. The deaccessioned surface has no unsoiled original area; therefore the cleanest area on the painting is used as a reference.

The evaluation is given semi-quantitatively, assigning marks ranging from 0 to 10, where 10 represents the best result possible.

The overall cleaning effectiveness is determined based on previously established parameters (Daudin-Schotte et al. 2011), particularly visual changes to the paint surface, particulate residues and cleaning results. Visual change is the most significant of the three as it has the highest impact on the appreciation of an artwork, and is irreversible.

Visual changes are indicative of surface microscopic topography changes such as flatenning, abrasion or polishing. This can cause an increase in gloss or create a matte surface. A patched surface is usually indicative of a combination of both abrasion and polishing. Visual changes in this study refer to gloss changes due to polishing and abrasion and are assessed with the naked eye complemented by observations under the light microscope when needed and measurments with a glossimeter. Scratches to the pained surface can only be observed under the light microscope or SEM.

Particulate residues, only observed under the light microscope, takes into account the number of residues found in 1 cm^2 of the surface. Sometimes the residues could be brushed of unless they were pressed into the surface. This parameter is to determine if there are residues left behind that cannot be seen with the naked eye.

Finally cleaning level refers to how clean and also how even a surface was. It was assessed by the observation of the surface with the naked eye, light microscopy and SEM.

Cleaning materials were also assessed in terms of their handling and water retention properties. Water retention was studied for each material by placing a moist weighed material on blotting paper. After 10 min the cleaning materials were removed from the blotting paper and weighed once more. Results for water retention and handling are presented in Table 1.

Results & Discussion

Solvent Extractable Components

The amounts of solvent extractable components in the cleaning materials were determined for five different solvents: tap water, demineralized water, an aqueous solution of 1 % triammonium citrate, 2-propanol and iso-octane. The cleaning materials were extracted with the solvents, and the weight of the extracted material was determined after evaporation of the solvent.

Results for the percentage of extractable components show that tap and demineralised water produced the lowest amounts of extractable residue compared to triammonium citrate (TAC), and organic solvents. The highest percentage of extracts is observed for the TAC solution. After the solvent evaporation, the TAC extractions leave behind a translucid gel-like substance.

The exact nature and composition of the extracts will be subject of further study.

Observation and Characterisation of the Artificial and Cellar Dirt

The combination of artificial and cellar dirt forms a whitish film over the surface. Observations with the light microscope and SEM on the cadmium yellow reference area show a superficial, dry, grainy and heterogeneous layer, which is mainly concentrated in the impastos and ridges of the brush strokes. Analyses with EDX identify silica calcium, sulphur and aluminium. Minor peaks of sodium and potassium are also observed. High contents of alumina-silicates are to be expected as both types of dirt contain a considerable amount of cement and building dust.

Natural dirt on the monochromatic green surface and cross section was limited and was not identified.

Pilot Tests

Pilot tests were performed on the unsoiled cobalt blue paint surface.

The smoke sponge had shown high percentages of extractable components (Table 1). In the tests it showed poor water retention and released crumbles onto the surface. Therefore the material was concluded to be unsuitable and expelled from further tests.

The PVA blitzfix sponge indicated its potential for polishing a surface whereas the PU, SBR and isoprene makeup sponges showed no noticeable changes to the surface. Finally pilot tests for both cloths show minor signs of abrasion and polishing.

Cleaning Tests

Degree of visual change observed mainly with the naked eye, shown in Fig. 2 was assessed by comparison to reference surfaces.

The particulate residues were caused by the cutting of cleaning materials into smaller pieces. These particulate residues were observed only under the light microscope or with the SEM. Results for particulate residues can be seen in Fig. 3.

The cleaning level was assessed mainly with the naked eye and SEM. General observations with microscope showed that dirt in crevices usually remains on the painted surface after cleaning tests suggesting the materials cannot reach them.

Results for the cleaning level obtained are observed in the bar graph of Fig. 4.

Cotton swabs with water and with saliva produced similar results on all surfaces Figs. 2, 3, and 4. The degree of visual change is comparable, but slightly better for the cotton swab with water Fig. 2. These materials left medium amounts of fibres on the painted surface, and they only produced medium cleaning level. Observations with the naked eye and microscope showed areas of dirt mainly concentrated on the impasto areas, scratches (Fig. 5) and broken impastos (Fig. 6).

Cotton swabs are easy to use but the texture of this material is sometimes too rough and open, occasionally causing the fibres to grab impasto tops and break them off (Fig. 6).

The following images show a surface observed with SEM, before and after cleaning the same area with cotton swab and saliva. These ilustrate the type of surface damage observed.

It is noted that all the cleaning materials produce varied degrees of visual change depending on the type of surface (Fig. 2). All cleaning materials caused lower degrees of change on the deaccessioned surface (Fig. 2).

The blitzfix, SBR and the PU makeup sponges caused significant amount of visual change to the surfaces (Fig. 2). The blitzfix and the PU makeup sponge caused gloss increase whereas the SBR sponge had results ranging from gloss increase to a patched surface. The extent of visual change is possibly related to the materials density lack of flexibility and water retention capabilities, as this might influence the mechanical interaction between material and surface (see Table 1).

The isoprene makeup sponge produced good cleaning levels and created less visual change than any other material (Figs. 2 and 4). This makeup sponge also produced better results when compared to the cotton swab with water and saliva.

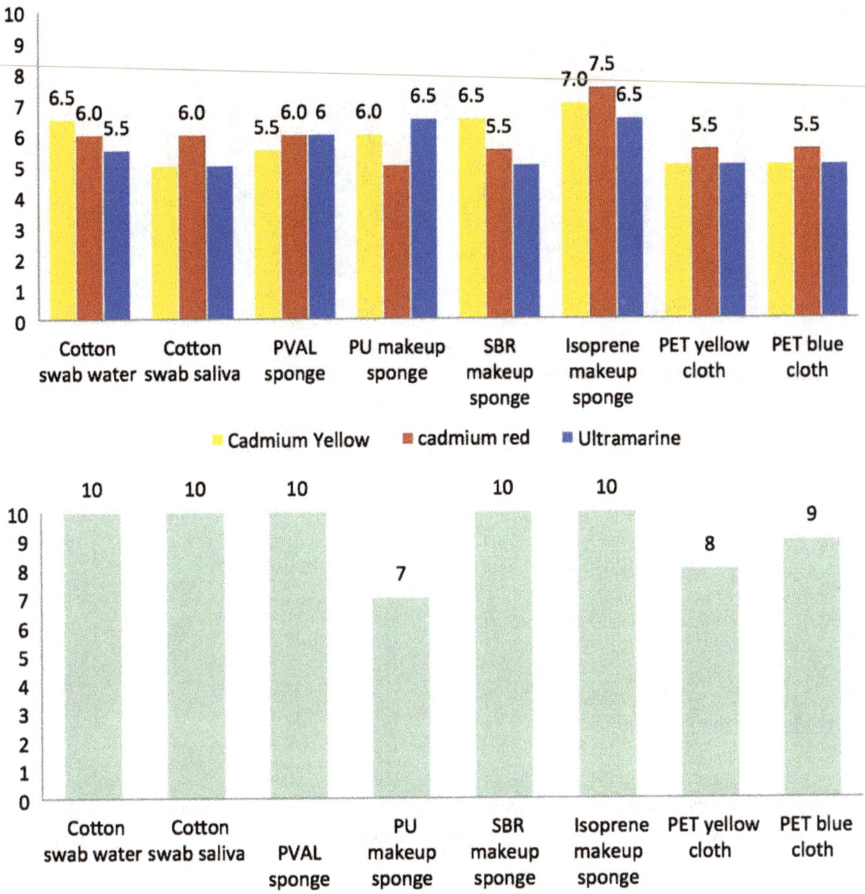

Fig. 2 Test results showing the degree of visual change observed on the painted surface for all materials tested on prepared (*above*) and deaccessioned (*below*) surfaces. Results were rated from 0 to 10, 10 being the best result possible, in this case, the least change to the painted surface. These results were obtained mainly by observations with the naked eye and glossimeter in terms of gloss changes, and with the light microscopy in the impasto areas to observe possible topography damages

Microfiber cloths produced low results on both the prepared and the deaccessioned surfaces (Figs. 2 and 4). Observations under the microscope showed scratches and flattened impastos.

For all prepared surfaces it was observed that the flat portion of the surfaces was easier to clean than the impastos. Most materials had difficulty in reaching the crevices where dirt is deposited. Sponges and cotton swabs caused damages to the impasto areas such as flattening or breaking of the impastos (Fig.6).

Material residues were to be found either loose on the surface or attached on flattened or broken impastos (Fig. 6). Of the new cleaning materials, the cloths left

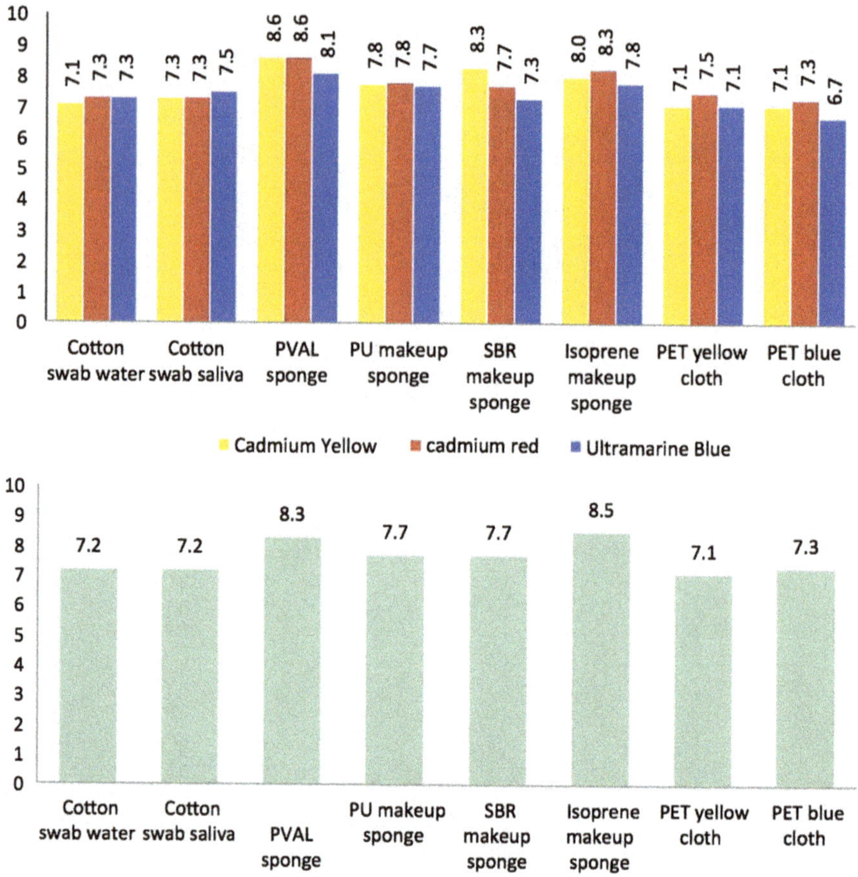

Fig. 3 Test results of particulate residues observed for all materials tested on prepared (*above*) and deaccessioned (*below*) surfaces. Results were rated from 0 to 10, 10 being the best result possible, in this case, the least amount of particulate residues. These results were obtained mainly by observations with light microscopy in a 1 cm^2 area

the highest amounts of residues behind (Fig. 3). These result mostly from loose threads after cutting the cloths; leaving the cloths intact should reduce the amount of residues.

It is important to note that the prepared mock-up surfaces used in this study were relatively young, artificially aged and that the dirt on them was gritty. The mature monochromatic green paint was a well-cured and aged surface. All cleaning materials presented better cleaning levels on the mature monochromatic surface than on the prepared surfaces (Fig. 4).

Dry cleaning tests performed for the makeup sponges and cloths on the mature green surface showed that all these materials produced better results when used moist than when used dry. Observations revealed that using the cleaning materials used dry the surface was polished or abraded and dirt was not successfully removed.

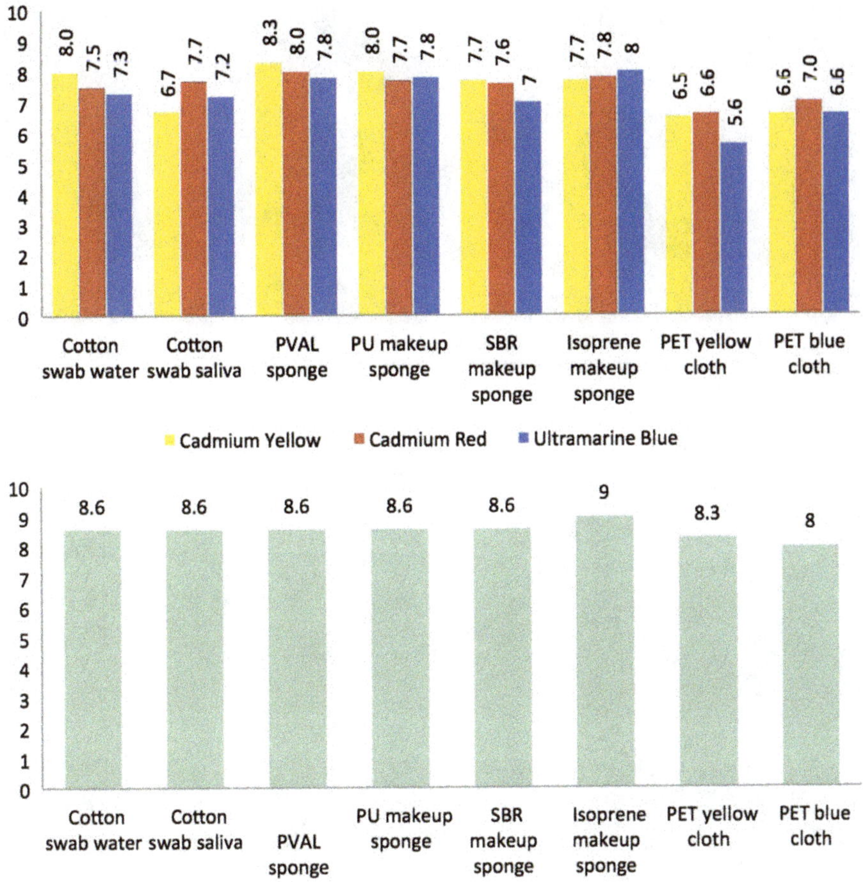

Fig. 4 Cleaning level results on the prepared (*above*) and deaccessioned (*below*) surfaces. Results were rated from 0 to 10, 10 being the best result possible, in this case, the least change to the painted surface. Results were obtained mainly by observations with naked eye, SEM and light microscopy

In terms of surface damage, particulate residues and cleaning level, the good results obtained with the moist isoprene makeup sponge moist, were comparable to results obtained while used dry (Daudin-Schotte et al. 2011).

Of all materials, the cloths produce the most abrasivon and polishing results. Test results for moist microfiber cloths show a higher amount of surface damage and lower cleaning levels, than dry cleaning results (Daudin-Schotte et al. 2011). Such a difference might be caused by an alteration in the action of these cloths when they are moist, by the combination dirt/painted surface used, or by differences of handling of the material by the assessors.

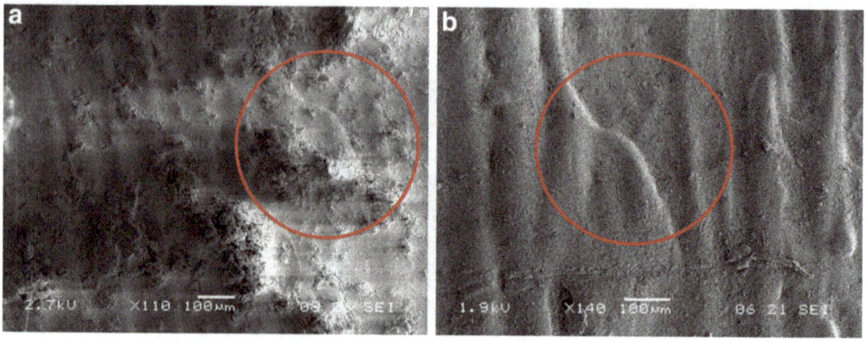

Fig. 5 Cadmium yellow reference sample before (a) and (b) after cleaning with cotton swab and saliva. The circle indicates recognizable surface topography that helps determine that both observations are in the same area of the painted surface. SEM images (SE mode) ×110 (a) and ×140 (b) magnification

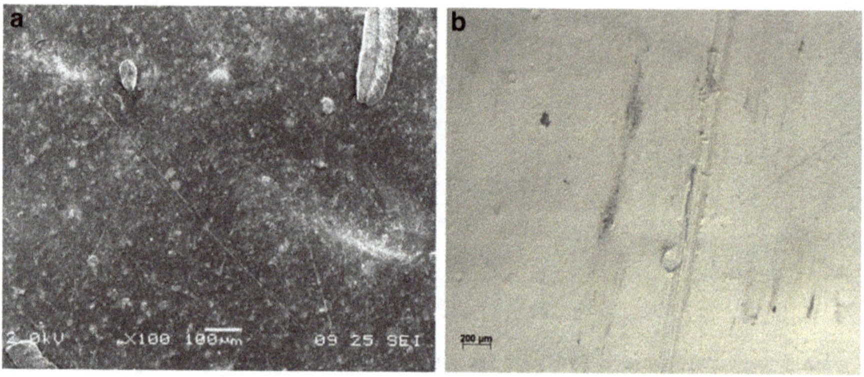

Fig. 6 (a) Example of scratches on the mature monochrome surface cleaned with the PET blue cloth. Image taken with (raking) light microscopy, ×25× magnification. (b) Scratches on areas tested with cotton swab and saliva on Ultramarine Blue. SEM images (SE mode) ×100 magnification

Conclusions

The present study shows the potential of sponges used moist for the removal of dirt in unvarnished painted surfaces. From all cleaning materials tested, the isoprene makeup sponge shows the best performance. This makeup sponge is easy to handle, has good water retention and is flexible, adapting well to the painted surface. Results from cleaning tests demonstrate that for the prepared panels used in this study, moist make up sponges perform better than cotton swabs with water or saliva.

PET cloths are the least successful, as they have a tendency to affect young and mature painted surfaces due to their abrasive action. The extent of polishing of a surface cleaned with a sponge is related to the density of this material.

The research of the cleaning materials has highlighted the issue of the composition and additives present in the best performing materials. The problem with sponges from drugstores such as the ones used in study, is that the composition of these materials is not guaranteed by the producer. Although not tested in the present study, the potential residues containing antioxidants and the extractable component of the sponges are of concern. The present study indicates a high percentage of extractable components for organic solvents. Further studies must be done in order to determine the long term effect of extractable components on a painted surface.

The possibility of the extraction of potentially harmful additives of sponges by rinsing them with solvents; or the feasibility of the production of a sponge without harmful additives should be considered.

The authors are aware that the combination of dirt and artificially aged painted surfaces used in this study are not necessarily representative when compared to the variability of "real, natural" dirt and naturally aged painted surfaces. Tests performed with the same cleaning material show that the well cured and aged deaccessioned surface, suffered less abrasion and polishing than the relatively young albeit light aged surfaces of the prepared panels. However, cleaning test results on these surfaces give a good insight on the potential use of the cleaning materials on surfaces which are young, have some surface textured and a grittier dirt layer.

Practical guidelines
Based on the results of this study, a few practical guidelines could be suggested. When facing an array of sponges, the softer and least dense isoprene sponge with the lowest amount of additives in its composition should be chosen. To minimize potentially harmful extractable component on the surface always rinse the cleaning material. For all sponges it is advised to use pre cut pieces since cut pieces tend to leave more particulate residues on the surface. Smaller pieces should be cut out only when the surface has some texture and cleaning has to be more localised. Abrasion can be minimised by avoiding dragging the materials on the painted surface, instead it is advised to gently dab the surface.

References

Daudin-Schotte M, Bisschoff M, Joosten I, van Keulen H, van den berg KJ (2011) Dry cleaning approaches for unvarnished paint surfaces, Post prints cleaning 2010. Smithsonian Institution Scholarly Press, Valencia

Green T (1990) Surface dirt removal from unvarnished paint films. In: Hackney S, Townsend J, Eastaugh (eds) Dirt and pictures separated. The United Kingdom Institute of Conservation, London, pp 51–55

Hackney S (2010) The art and science of cleaning paintings. In: Laura F-L, Elena Charola A, Marion M, Teresa Doménech-Carbó M (eds) Cleaning 2010 new insights into the cleaning of paintings. Universidad Politecnica de Valencia, Valência, pp 15–16

Mills L et al (2008) Water sensitivity of modern artists' oil paints. ICOM preprints of the 15th triennial meeting New Delhi, 22–26 Sept 2008

Morrison R, Bagley-Young A, Burnstock A, van den Berg K, van Keulen H (2007) An investigation of parameters for the use of citrate solutions for surface cleaning unvarnished paintings. Stud Conserv 52:255–270

Nazaroff W, Ligocki M, Ma T, Cass G (1990) Particledeposition in musuems:comparison of modeling and measurments results. Aerosol Sci Tech 13:332–348

Nielsen H, Kornum L, Säberg O (1978) Dirt retention on painted surfaces. J Coat Technol

Ormsby B, Phenix A (2010) Artificial dirt recipe

Perry R (1990) Problems of dirt accumulation and its removal from unvarnished paintings: a practical review. In: Hackney S, Townsend T, Eastaugh N. United Kingdom Institue for Conservation of historic and Artistic Works, London, pp 3–10

Phenix A, Burnstock A (1990) The deposition of dirt: a review of the literature, with scanning electron microscope studies of dirt on selected paintings. In: Hackney S, Townsend J, Eastaugh N (eds) Dirt and pictures separated. The United Kingdom Institute for Conservation of Historic and Artistic Works, London, pp 11–18

Wolbers R (2000) Cleaning painted surfaces aqueous methods. Archtype Publications, London

Agar – A New Tool for the Surface Cleaning of Water Sensitive Oil Paint?

Annegret Volk and Klaas Jan van den Berg

Abstract Water sensitivity of well-bound twentieth century oil paints is a well known phenomenon among paintings conservators. It complicates conservation processes involving the use of water, particularly surface cleaning treatment.

This article introduces agar gels as an approach to the challenge of surface cleaning of water sensitive oil paint.

An experimental cleaning study using test paint samples was conducted and evaluated with scanning electron microscopy, digital 3D microscopy and optical microscopy of thin cross sections.

Agar gels prepared with deionized water and brushed on the paint surface before gelation exhibited great potential. An agar gel loaded with an o/w microemulsion in low concentration showed best cleaning results on the artificial surface dirt used in this study.

Keywords Surface cleaning • Unvarnished • Water sensitive • Oil paint • Modern paint • Agar gel

Introduction

Surface cleaning is among the most frequently executed tasks in paintings conservation, requiring conservators to make complex decisions. They need to consider the integrity of the artwork, the cleaning systems and their properties, as well as to reflect on their own expertise in order to evaluate the effect and impact of treatment. At the same time conservators are committed to work within a tight ethical framework, demanding minimum intervention. This requires the conservator to carry out the cleaning treatment with the least possible impact on the artwork. The selection of conservation materials should therefore not impact on the re-treatability

---

A. Volk (✉)
Paintings Conservator, Frankfurt, Germany
e-mail: Annegret.volk@gmail.com

K.J. van den Berg
Cultural Heritage Agency of the Netherlands, Amsterdam, The Netherlands

K.J. van den Berg et al. (eds.), *Issues in Contemporary Oil Paint*,
DOI 10.1007/978-3-319-10100-2_26

of the painting as well as not leaving residual cleaning material on the paint that may alter the surface through degradation processes. Scientific research that investigates the causes of cleaning issues and offers new cleaning materials increasingly allows for tailoring unique solutions in conservation practice (Mecklenburg et al. 2013; Ormsby et al. 2013).

The decision to undertake surface cleaning is mostly based on visual observation. Action is required when surface soiling negatively affects the appearance of the artwork and with it its meaning. Soiling also has the potential to participate in degradation reactions at the paint surface and to become embedded into paint films. Therefore cleaning is not only understood as a restoration that attempts to approach the artwork's assumed aesthetic intention; but also as a preventive treatment to reduce the deposition of superficial dirt on the unvarnished paint surface.

The deliberate lack of a varnish layer on modern and contemporary oil paintings attributes a new significance to their surface features. Differences in gloss and surface structure generated by a particular choice of materials can critically contribute to the intended appearance of the artwork (Learner 2012, Von Saint-George 2013). Preserving these aesthetic characteristics should be a major aim during cleaning.

Where dry cleaning methods are not effective, aqueous methods are frequently used for the removal of well attached surface dirt. Unvarnished oil paints that are sensitive to water or protic solvents present an exceptional challenge to conservators because "standard" cleaning procedures potentially lead to paint removal. Water sensitivity in modern well bound oil paints is a well known and studied phenomenon affecting artworks, for example, by Jasper Johns, Karel Appel and Robyn Denny (Wijnberg et al. 2007; 2008; Burnstock et al. 2007; Cooper et al. 2014).

This study presents a novel approach for the removal of superficial dirt from water sensitive oil paint surfaces using agar gels. The results of the agar gel cleaning tests were compared to the commonly used method of aqueous swab rolling. An evaluation of both methods, including their application with different additives – a surfactant, a chelating agent and the addition of an oil-in-water (o/w) microemulsion – has been conducted, with regards to maximized cleaning effects and minimized alterations to the paint surface. All cleaning tests were performed on experimental oil paint samples and evaluated with scanning electron microscopy, digital 3D microscopy and optical microscopy carried out on thin cross sections. The main aim of the study was to provide conservators with an initial assessment on the impact of agar gel cleaning on water sensitive oil paints, in preparation for their application to actual paintings.

Current State of Research

Epsomite – A Source for Water Sensitivity

The phenomenon of water sensitivity in modern well bound oil paints has been described comprehensively and different causes have been investigated (Burnstock et al. 2007; Wijnberg et al. 2008; Mills et al. 2008; Tempest et al. 2013). A phase

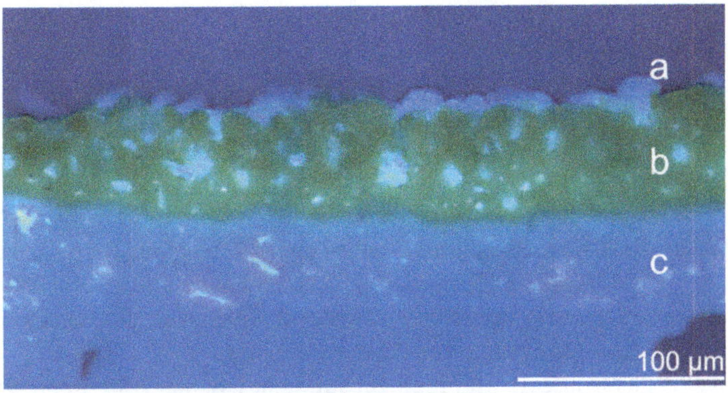

Fig. 1 Thin cross section of an oil paint sample in UV excitation; after exposure to SO_2 crystals are protruding through the paint surface, the crystalline material was identified by Silvester et al. (2014) as epsomite using XRD analysis on the same paint samples after incubation with SO_2, (**a**) epsomite (**b**) oil paint layer containing magnesium carbonate (**c**) ground layer

separation of hydrophilic pigment and binder, mainly observed in ultramarine paints (Burnstock et al. 2007), as well as the formation of water soluble organic degradation products in the binding medium[1] have been linked to water sensitivity.

This study focuses on the effects of cleaning of oil paints which had become water sensitive due to the formation of water soluble magnesium sulphate salts. Recent investigations by Silvester et al. (2014) have explained the underlying mechanisms behind the formation of this material: magnesium carbonate, a constituent of some modern tube oil paints, can react upon exposure to an airborne sulphurous/sulphuric source, resulting to form water soluble magnesium sulphate heptahydrate (epsomite). The crystalline entities disrupt the paint surface (Fig. 1), leading to a physically weakened paint film, especially when dissolved upon exposure to water. The combination of dissolving epsomite with mechanical impact during surface cleaning, promotes the removal of paint, causing associated changes in surface topography, gloss and UV fluorescence. It has been demonstrated that this degradation process mainly affects the medium rich surface of an oil paint (Cooper et al. 2014). This pigmented medium rich layer determines the gloss and saturation of the paint film and is therefore significant for the visual perception of the paint.

Surface Cleaning with Agar Gels

To minimize the risk of removing paint from water sensitive oil paint surfaces during surface cleaning, two requirements need to be met: decreased diffusion of water into

[1]Research into the formation of water soluble degradation products during aging of oily binding media is still ongoing at the Cultural Heritage Agency of the Netherlands, Amsterdam.

the paint, thereby preventing the dissolution of the epsomite crystals; and minimized mechanical action on the surface. Agar gels might satisfy both demands by reduced penetration of water into the substrate, and causing minimal mechanical action on the paint surface upon both, application and removal of the gel.

Agar agar, a polysaccharide derived from seaweed, is widely known as a gelling agent for biological culture media.[2] It has been studied and evaluated positively as a material for the cleaning of delicate surfaces, which are for instance porous or sensitive towards solvents and mechanical action (Campani 2006; Anzani et al. 2008; Gorel 2010; Tortajada Hernando and Blanco Domíngez 2013; Finozzi et al. 2013). The gels offer controlled release of water or aqueous reagents to the substrate surface, and the removal of superficial material from it.

Agar forms physical gels with high gel strength. Upon heating to 90–95 °C the agar powder dissolves in water, building a random coil polymer colloid solution or sol. When cooling down to 30–40 °C -depending on the type of agar- a network of double helixes develops through hydrogen bonding, resulting in a stiff, rigid gel (FAO 1990). Rigid gels are formed starting with a concentration of 1.5 % agar powder in water (weight/volume).

Two modes of application meet the demands of various objects and their surfaces. On flat objects the gels can be used cold, in an already gelled state, and in the case of paintings with impasto, warm agar solutions can be brushed on as a sol. Gelation then takes place within few seconds. After application, water is slowly released into the substrate surface depending on dwell time; a process called syneresis (Anzani et al. 2008). Ideally the surface soiling is dissolved by the water and subsequently absorbed into the gel. To date no comprehensive model of how agar works as a surface cleaning agent has been proposed. It has been described as a "molecular sponge" that takes up dissolved material with the help of osmosis (Anzani et al. 2008). It is also possible that surface dirt is taken up through capillary action within the large cavities of the molecular gel structure (Anzani et al. 2008).

The gel is removed with minimal mechanical action by lifting it off the surface. Therefore, agar gels should never be used on matte, underbound paints where low cohesion and poor adhesion of the paint layer onto the ground are evident. Removal of pigment and whole paint flakes is to be expected in these areas.

No further clearance step is required after gel removal. There are no concerns regarding residues of pure agar gels on porous surfaces, assuming that an appropriate application technique is used (see Section "Agar Gel Cleaning Tests"). An examination using GC/MS has evaluated the amount of residues for short applications of microbiological grade agar as very low (Anzani et al. 2008).

It is possible to modify the water in agar gels by adjusting the pH, adding chelating agents or surfactants (Campani 2006). Enhancing the cleaning effect by adding these materials, however, leaves residues behind that require appropriate

[2] Agar agar must not be mistaken as the actual nutrients in culture media, like yeast or malt extract, which are usually added to the gel.

clearance. A study by Sánchez-Ledesma et al. (2013) indicates that it is possible to rinse additives in agar gels with the application of pure water agar gels.

Experimental Setup

Experimental Paint Samples

An experimental study was designed and conducted on standardized, artificial test paint films. The composition of these films is known accurately, which facilitated reproducible experiments.

Model oil paint films containing cadmium zinc sulphide yellow, magnesium carbonate, alkali refined linseed oil and free fatty acids were prepared via a previous study in 2010. The paints were applied to a Melinex® support in a thickness of 90 μm with a draw down bar.[3] In 2011 these samples were exposed to gaseous SO_2 and raised levels of relative humidity to induce the formation of epsomite crystals that protruded from the paint surfaces (Fig. 1).[4]

Swab rolls moistened with deionized water were performed to compare the water sensitivity of the samples before and after incubation with SO_2. Before incubation the test oil paints were not sensitive to water at all. A dramatic increase in water sensitivity occurred after crystalline epsomite formation; whereby paint removal was detected after two swab rolls.

The water sensitive samples were artificially soiled. The artificial dirt recipe used in this study was adapted from Ormsby et al. (2013). It comprises 2 g ivory black, 0.5 g ochre, 1.75 g silica, 20 g kaolin, 10 g gelatin powder, 10 g soluble starch, 17.5 g cement, 10 mls olive oil and 20 mls mineral oil. The particulate ingredients were sieved with a mesh size of 75 μm. The mixture was diluted in petrol ether and carefully brushed on the paint surfaces in one stroke. Some ingredients of the artificial dirt recipe slightly fluoresce under UV excitation. Therefore a remote loss in fluorescence indicates dirt removal.

Accelerated thermal ageing was carried out with the aim of increasing the bond between the applied dirt and the paint surface. It was performed in a Vötsch VC 0200 climate chamber for 3 weeks at 50 °C and humidity cycles changing from 30 to 80 % every 6 h. It was noted that after accelerated ageing the visible epsomite crystals had vanished from the sample surfaces (See Results and Discussion) (Volk 2012).

[3]Part of the test paints were artificially light-aged with fluorescent light tubes to the equivalent of 24 years ageing in museum lighting conditions (Silvester et al. 2014), assuming reciprocity. Interestingly, no influence on the formation of crystalline epsomite and concomitant water sensitivity was found compared to the naturally aged samples.

[4]A detailed explanation of the experimental paint sample preparation and the incubation with SO_2 is published in Silvester et al. 2014.

Testing Procedure

Aqueous Swab Rolling

Swab rolling was performed with deionized water, deionized water and the addition of 1 % tri-ammonium citrate (TAC), 0.2 % sodium dodecyl sulphate (SDS) and swab rolling with pure o/w microemulsion described in Section "Agar Gel Cleaning".[5] These systems were also used in the agar gels for comparison.

Both application procedures – gel application and swab rolling – are not directly comparable. Since swab rolling is still frequently used for surface cleaning operations, it illustrates the "standard" conservation approach and its consequences for water sensitive oil paints.

Agar Gel Cleaning

Preparation of Agar Gels

All agar gels were prepared au bain marie with deionized water in a beaker. Following recommendations by Anzani et al. (2008) homogenous gels were obtained by heating the agar-water-mixture twice above 95 °C for several minutes and allowing it to cool down beneath gelation point -under 30 °C- in between. The evaporated water was replenished after every cooking phase.

Preliminary Agar Gel Application Tests

Preliminary tests were performed to determine the best working properties for dirt removal, impact on the paint surface and residues. Gels containing 2–5 % (w/v) agar powder in pure deionized water were applied cold as rigid gels as well as brushed onto the surface of water sensitive test oil paint samples right before gelation. All brushed on gels had working temperatures around 37 °C. Upon contact to the cold paint surface they gelled instantly. Dwell times ranged from 5 min to 24 h.

Agar Gel Cleaning Tests

For the main series of agar cleaning tests, gels with 3 % (w/v) agar powder in deionized water were prepared. All gels were brushed onto the experimental paint sample surfaces before gelation and removed after 5 min. The treatment was repeated for a second time after leaving the surface to dry for 2 h.

---

[5]The action of swab rolling is highly dependent on the executor. Only accumulated experience allows conservators to develop a sense of the impact their treatments have while performing them. This leads to permanent adaption of movements during swab rolling to increase the cleaning effect while minimizing harm. While this is not a scientific approach it reflects the everyday work of conservators.

Cleaning tests were performed with agar in pure deionized water, agar in deionized water with the addition of 1 % TAC, at pH 7, and the addition of 0.2 % SDS. For the last set of tests 10 %, 20 % and 30 % of the deionized water was replaced with an oil-in-water (o/w) microemulsion, consisting of 85 % deionized water, 4 % SDS, 6 % Pentan-1-ol and 5 % petrol ether (Gorel 2010).

The additives were incorporated in the gels in order to enhance their cleaning efficacy by metal ion complexation (TAC) and increasing their affinity to dissolve the hydrophobic portions of the dirt (SDS and o/w microemulsion).

Evaluation of the Cleaning Tests

The performance of agar gels for removing artificial surface dirt from water sensitive paint sample surfaces was studied in comparison to aqueous swab rolling. The evaluation was performed visually in visible light and in UV excitation, as well as using scanning electron microscopy (SEM), digital 3D light microscopy and light microscopy of thin cross sections (for technical details see Appendix). Criteria for the evaluation of the cleaning tests were the removal of surface soiling, the removal of paint, as well as changes in surface topography, gloss and UV fluorescence.[6]

Results and Discussion

Aqueous Swab Rolling

Cleaning tests with damp cotton swabs produced very uneven results. While abrasion occurred on elevated areas of the surface topography, the interstices retained residual agglomerations of dirt (Fig. 2). Moreover, it was not possible to obtain adequate control over the dirt removal process and any impact on the underlying paint layer. Dirt removal was always accompanied by paint removal (in the form of pigment transfer to the swabs), with associated changes in topography, UV-fluorescence and gloss.

By comparison the best cleaning result was achieved while swab rolling with pure deionized water. Enhancing the cleaning effect of water with the additives increased the negative effects on the surface. In these cases, more paint was removed and the surface topography was more disturbed.

Swab rolling with the o/w microemulsion led to almost immediate and substantial paint loss. The cleaning system acted extremely quick and changed the paint topography, UV-fluorescence and gloss to a significant degree.

---

[6]In this study a remote loss of fluorescence points towards a removal of fluorescent artificial dirt; whereas progressing loss of fluorescence is a good indicator for the disruption of the skin of medium (Van den Berg and Burnstock 2014).

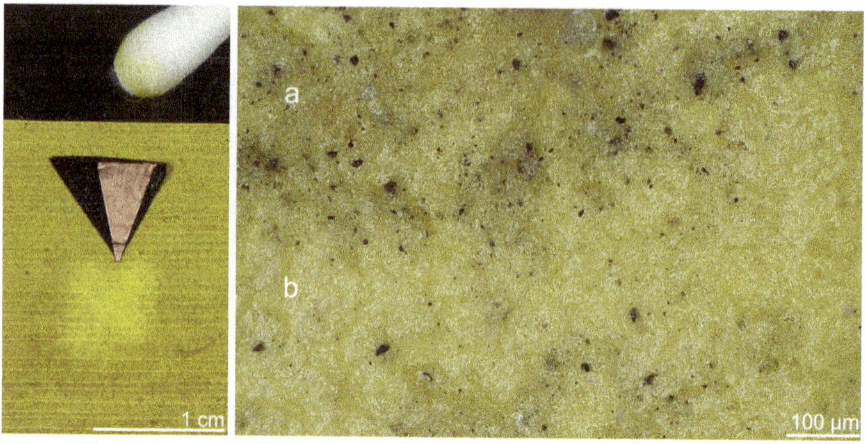

Fig. 2 Water sensitive oil paint sample after cleaning with 0.2 % SDS in deionized water; (*left*) area of cleaning is inhomogeneous with visible paint removal on the cotton swab, (*right*) digital 3D microphotograph illustrating (**b**) paint removal in elevated areas whereas (**a**) depths are still soiled

The lack of control during swab roll cleaning can be explained with the increased vulnerability of the paint film towards the water. The artificial soiling, which was more hydrophobic than the paint layer acted as a protective layer upon water exposure for a short amount of time. As soon as the water penetrated the paint layer and mechanical force was applied in horizontal direction; the dirt was undercut and removed together with paint (Fig. 3).

Uneven soiling removal can be explained through insufficient contact between the cotton swab and the paint surface topography. Only the elevated areas of the samples were exposed to the solvent and mechanical action (Fig. 2).

Swab rolling with an o/w microemulsion illustrated the potential aggressiveness of these systems on water sensitive oil paint surfaces due to their broad solubility spectrum and enhanced cleaning activity.

Agar Gel Cleaning

Preliminary Agar Gel Application Tests

Preliminary tests were carried out to determine optimal conditions for gel preparation and application.

As expected, brushed-on agar gels generally attached much better, even to flat surfaces, and produced more homogenous cleaning results than the rigid gels also tested.

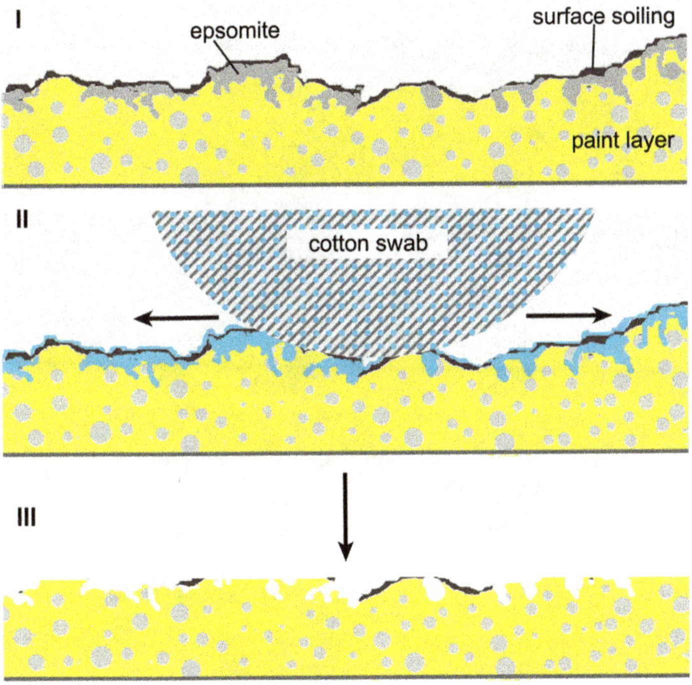

Fig. 3 Schematic cross section of a water sensitive oil paint layer during aqueous swab rolling; (**I**) before cleaning, (**II**) during cleaning with a moist cotton swab epsomite crystals dissolve, leaving a physically weak paint surface that is sensitive towards mechanical impact, (**III**) after cleaning the paint surface is left with abraded peaks and residues of dirt in depths

2 % (w/v) gels, using agar powder in deionized water, exhibited poor water retention. An area of free water developed on the paint surface around the gels after applications as short as 2 min. On the other hand, the 5 % (w/v) gels were highly viscous and difficult to handle.

Repeated short exposure times of 5 min were found to be less damaging for water sensitive oil paint surfaces, than one longer application. Leaving the gels on the surface for 4 h or longer resulted in complete penetration of water into the paint, due to the effect of syneresis. After long term treatment the surface topography of the water sensitive paint samples had changed completely, developing small blisters over the treated area.

The removal of the gels was facilitated by applying relatively thick films of at least 0.5 cm on the samples, with no notable difference in cleaning efficacy.

Gels left to dry for 24 h performed poorly on water sensitive paint samples. The thin films of pure agar adhered extremely well to the paint surfaces, resulting in the deformation of samples and even delamination of the paint layer from its support.

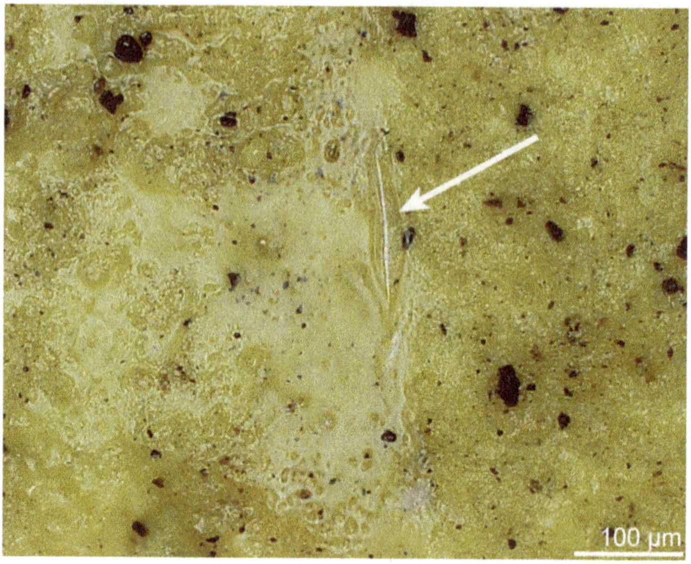

Fig. 4 Digital 3D microphotograph of a surface where agar gel cleaning had been performed; the glossy area is covered with a thin residue of agar agar

Issues concerning residues arose after the detection of glossy areas surrounding the borderline of the gel applications left for longer than 30 min (Fig. 4). Microscopic observations confirmed that these were thin residues of agar agar. This indicates that small portions of low-viscous gel were secreted to the margin areas, forming a thin film that dried quickly and attached very well to the surface. The glossy residues, however, could be avoided by applying the gels at temperatures below 40 °C, thereby increasing viscosity, and removing them after a short dwell time of less than 30 min.

Agar Gel Cleaning Tests

All agar gel cleaning tests, where the gels had been applied twice for 5 min and leaving the paint surface to dry for 2 h in between, produced varied but more homogenous cleaning results when compared to the aqueous swabbing tests.

Applications of agar in deionized water showed a poor cleaning performance by visual observation in visible light. However, SEM and digital 3D microscopy revealed an even reduction of surface soiling (Fig. 5). In this case, larger, transparent particles had been predominantly removed, leaving a film of oily material and smaller particles. The small particles mainly consisted of the pigments used in the artificial soil, which explains the visually dirty surface. Microscopy of thin sections taken from these areas confirmed that no paint had been removed and revealed that

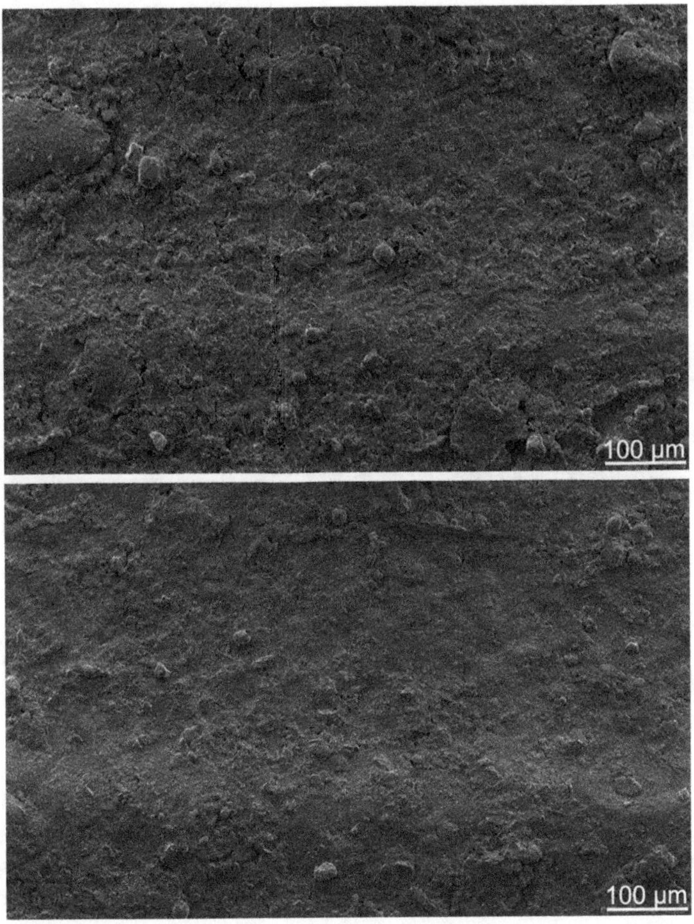

Fig. 5 SEM image of test paint surface area; (*above*) before cleaning and (*below*) after cleaning with an agar gel prepared with deionized water

the paint surface had retained its topography. In addition, no changes in gloss and UV fluorescence were detected.

The agar gel with 0.2 % SDS slightly increased the cleaning efficacy in comparison to pure deionized water agar gels; however, artificial soil was still present on the paint surface. No paint was removed during cleaning; therefore the gloss and surface topography were intact. A slight change in UV fluorescence was detected. This indicates removal of fluorescent dirt.

Cleaning with an agar gel formulated with 1 % TAC in deionized water resulted in significant paint removal. Around 35 % of the original layer thickness was lost in the thin cross sections. SEM images showed exposed magnesium carbonate particles in high magnification.

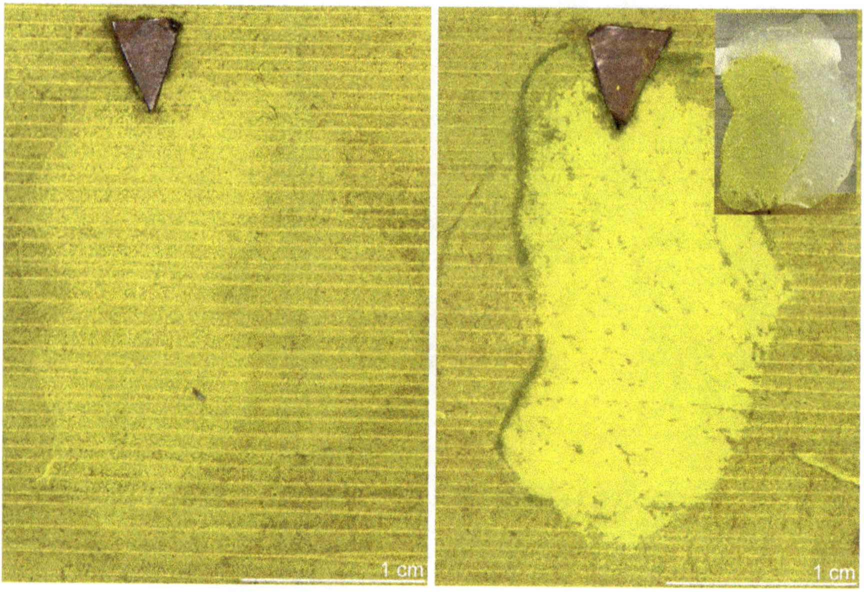

Fig. 6 Test paint samples after cleaning with an agar gel, prepared with 10 % o/w microemulsion and 90 % deionized water (*left*) and prepared with 30 % o/w microemulsion and 70 % deionized water (*right*); agar gel after removal from the paint surface (*upper right corner*)

Agar gels with the addition of 10 % microemulsion produced an overall even cleaning result (Fig. 6). Visual observations showed a clear reduction of dirt. Thin cross sections verified that no paint had been removed during cleaning; the integrity of the surface was preserved, with no changes in topography and gloss. Yet a small change in fluorescence occurred; again indicating the removal of fluorescent dirt.

Increasing the relative amount of o/w microemulsion resulted in more aggressive cleaning. At a proportion of 20 % microemulsion, the gel caused a visible removal of paint after the second application. Equally, the first application of an agar gel stocked with 30 % microemulsion removed the uppermost paint surface together with the dirt layer (Fig. 6). The overall impact of these more active gels was slightly less severe than the cleaning with the TAC-agar gel.

The more even cleaning results obtained with agar gels can be attributed to the absence of mechanical action parallel to the paint surface, and the fact that the brushed on sols adapted to the surface topography before gelation. The subsequent release of solvent onto the paint surface and removal of dirt therefore occurred at the same rate in all areas (Fig. 7).

Gels prepared with deionized water showed no damaging effects. The lower impact of this method can be either attributed to the strong water retention of the gel, which decreased the dissolution of the crystalline epsomite, or the reduced

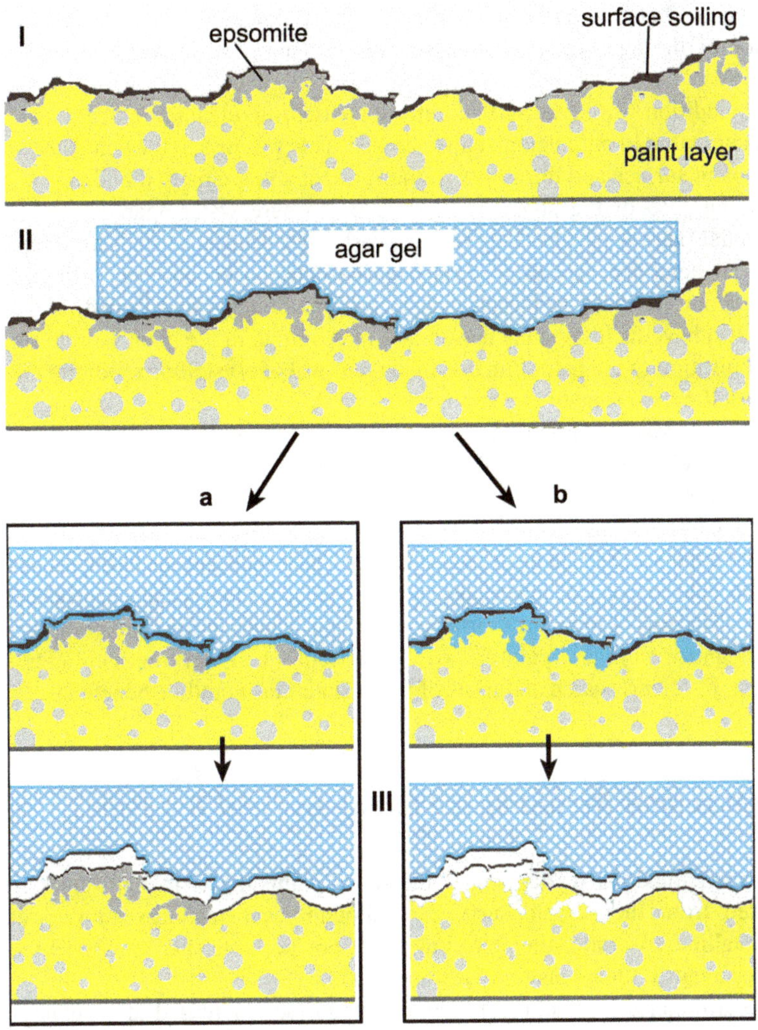

Fig. 7 Schematic cross section of a water sensitive oil paint layer during surface cleaning with a deionized water agar gel, illustrating two possible processes; (**I**) before cleaning, (**II**) during cleaning, (**III**) after cleaning; (**a**) the strong water retention reduces the dissolution of epsomite crystals, the paint surface is completely intact after cleaning; (**b**) although the epsomite crystals are partially dissolved the reduced mechanical action leads to a preservation of the original surface topography, after cleaning the epsomite crystals are removed

mechanical action on the surface (Fig. 7). It is however well known that water will not remove ionic and fatty soiling efficiently without the use of additives. The cleaning efficacy of pure water agar gels was therefore limited to the removal of particulate dirt and water soluble entities of the dirt layer. The addition of surfactants, chelators and o/w microemulsions predictably enhanced the cleaning efficacy of the agar gels, which also increased the risk of damage to the paint films.

The addition of 0.2 % SDS slightly increased the cleaning efficacy of the gel when compared to pure deionized water gels. This was due to the enhanced wetting of the paint surface and the ability of the surfactant to emulsify oily dirt.

Adding 1 % of the chelating agent TAC to pure water agar gels resulted in substantial paint loss. The TAC led to a fast sequestration of the salt components of the artificial dirt, resulting in its fast dissolution and removal. Although TAC does not preferentially sequester magnesium ions (Phenix and Burnstock 1992), it could also have facilitated the dissolution of epsomite in the paint sample surfaces. Destabilisation of the paint film due to the removal of crystalline epsomite made the paint itself more prone to chelation, which might have affected the organic binding medium (Morrison et al. 2007), and resulted in paint loss.

Optimal cleaning in terms of dirt removal and impact on the set of paints evaluated was obtained, replacing 10 % of deionized water with the o/w microemulsion in the gel. Clearly, dissolution of the oily surface dirt was enhanced by the addition of petrol ether and SDS in the microemulsion. This highlights the ability of agar gels to control the delivery of potentially harmful materials to a water sensitive oil paint surface. Because of the many variables involved, selecting the right composition is not straightforward, and paint removal can occur when working with additives in agar gel. Further research and trials of these materials would be beneficial.

Considerations Regarding the Use of Artificial Test Paints

After accelerated thermal ageing of the soiled paints the visible epsomite crystals vanished from the sample surfaces. Although clear readings on the removal of crystalline epsomite after cleaning were not possible, the experimental paint samples retained their sensitivity towards water.[7]

The artificial dirt attachment to the paint surface increased after ageing. Thin cross sections revealed that in some areas small amounts of paint were partially migrating into the dirt layer. A softening of the experimental paint samples must have taken place due to a combination high temperature conditions and paint samples that had not fully cured. Although the possibility to work with well defined systems is an advantage in general, this demonstrates the limitations of working with relatively young test paint films.

[7] We speculate that superficial epsomite, being extremely sensitive to moisture, has dissolved as a result of high relative humidity in the artificial ageing procedure.

Conclusion

This study aimed for a risk assessment of two different aqueous cleaning methods – swab rolling and agar gel cleaning – on experimental water sensitive oil paint samples in order to control the removal of dirt and the preservation of the paint surface. The uppermost surface of the paint layer was studied as this area affects the visual perception of the gloss and saturation of the paint.

The use of alleged "standard" conservation approaches, such as swab rolling with deionized water, implicated substantial risks for water sensitive oil paint surfaces. Risks included uncontrolled paint removal with associated changes in surface topography, gloss and UV fluorescence. By removing a surface layer from water sensitive oil paints the material integrity of the painting is disturbed and the aesthetic appearance of the artwork is affected due to induced changes in surface features. This strong impact is therefore ethically questionable.

On the whole, agar gels produced promising results and proved to be an interesting alternative method for the aqueous cleaning of water sensitive oil paints. Agar gels prepared with pure deionized water enabled the even removal of superficial surface soiling without a visible impact on the test paint surface. Whether this positive result can be attributed to the strong water retention of the gel or the reduced mechanical action on the paint surface or a combination of both was not evident. Further investigation with regards to the leaching of water soluble epsomite during agar gel cleaning is needed.

The most efficient properties of pure water agar gels on water sensitive substrates were found to occur with brushed on agar sols before gelation, with high agar powder concentrations (3 % and higher), short application times and with repetitions when required. It was also observed that the potential for leaving agar gel residues was dependent on application practice.

Incorporating additives into agar gels produced differing cleaning results on water sensitive paint samples ranging from improved cleaning effects to destructive impact on the paint. However, concerns regarding residues of cleaning agents and their clearing remain and require further study.

Risk assessments facilitate cleaning tests for practising conservators but can never replace their thorough evaluations prior to conservation operations on originals. This study was executed using experimental test paints that were incubated with high levels of SO_2 as well as artificially soiled and aged. Naturally aged and soiled water sensitive oil paintings could present diverse challenges to which agar gels do not offer a solution. This research focussed on oil paints that had become water sensitive due to crystalline epsomite formation. Other sources for water sensitivity might require different cleaning approaches, which needs further examination and research. In conservation practice a conductivity meter is a useful diagnostic tool to help determine whether epsomite dissolution is causing water sensitivity (Volk 2012; Soldano and Van den Berg 2014).

Ultimately, the complex challenges involved in the cleaning of water sensitive oil paintings emphasize the ongoing necessity to protect these artworks from progressing dirt deposition wherever possible.

Acknowledgements This research was conducted at the Cultural Heritage Agency of the Netherlands (RCE) in Amsterdam as part of a diploma thesis at the Stuttgart State Academy of Art and Design.

The authors wish to thank the colleagues at RCE who have been particularly supportive: Suzan de Groot, Ineke Joosten and Luc Megens. We would like to thank Genevieve Silvester for her paint samples as well as other colleagues who highly informed this study: Aviva Burnstock, Courtauld Institute of Art; Meta Chavannes and Louise Wijnberg, Stedelijk Museum Amsterdam; Emilie Froment, University of Amsterdam; Jolanda van Iperen, Rijksmuseum Amsterdam; Christoph Krekel and Volker Schaible, Stuttgart State Academy of Art and Design; Bronwyn Ormsby, Tate and Karolina Soppa, Bern University of the Arts.

Appendix

Technical Details

SEM

A JEOL JSM 5910 LV variable pressure scanning electron microscope (VPSEM) was used for secondary electron imaging of the paint surfaces before and after cleaning. Imaging was performed, without coating, in high vacuum with a low accelerating voltage of 1.7 kV before and 1.3 kV after cleaning.

Digital 3D-Microscopy

A Hirox KH 7700 digital 3D-microscope was used with the dual illumination revolver zoom lens MXG-2500REZ for digital 3D microscopy.

Optical Microscopy of Thin Cross Sections

Optical microscopy was done with a ZEISS Axioplan 2 stereomicroscope with Epiplan objectives and polarized incident light, transmitted light and UV light. The source for visible light was a halogen HAL 100 lamp.

UV light was produced by a mercury short arc photo optic lamp HBO 50/ac. The filter set “UV H 365”, consisting of excitation BP 365/12, beam splitter FT 395 and emission LP 397, was used for examination in UV-light.

Images were taken with the help of a Canon SLR camera and processed with axiovision software.

Samples for thin cross sections were embedded in Technovit 7100 and cut with a fully motorized Leica rotary microtome RM 2165 with a thickness of 16–18 μm according to a methodology that was developed at the Stuttgart State Academy of Art and Design (Soppa et al. 2013).

Materials and Distributors

Agar (Type 05040)	Fluka
Cement (Type I)- Schnellzement	Knauf
Gelatine powder	Kremer Pigmente
Ivory Black	Lukas-Nerchau
Kaolin	Kremer Pigmente
Mineral oil – Smeerolie	Hema
Ochre	Kremer Pigmente
Olive oil	Albert Heijn
Pentan-1-ol	Sigma Aldrich
Petrol ether 100–140°	Kremer Pigmente
SDS	Interchema
Silica	Kremer Pigmente
Soluble potato starch	Honig
TAC	Sigma Aldrich

References

Anzani M et al (2008) Gel rigidi di Agar per il trattamento di pulitura di manufatti in gesso. Use of rigid agar gels for cleaning plaster objects. CESMAR7 (Quaderno 6)

Burnstock A et al (2007) An investigation of water-sensitive oil paints in twentieth-century paintings. In: Learner TJS et al (eds) Modern paints uncovered. Proceedings from the modern paints uncovered symposium May 2006, J.Paul Getty Trust, Los Angeles, pp 177–188

Campani E (2006) L'uso di agarosio e agar per la preparazione di "gel rigidi". Use of agarose and agar for preparing "rigid gels". CESMAR7 (Quaderno 4)

Cooper A et al (2014) Water sensitive oil paints in the 20th century. These proceedings, Chapter 20

FAO (Food and Agriculture Organization of the United Nations) (1990) Training manual on Gracilaria culture and seaweed processing in China. In: FAO corporate document repository. http://www.fao.org/docrep/field/003/AB730E/AB730E00.htm. Accessed 11 Sept 2012

Finozzi A et al (2013) Use of agar cyclododecane for cleaning tests on a frail painting. In: Mecklenburg MF et al (eds) New insights into the cleaning of paintings. Proceedings from the cleaning 2010 international conference, Valencia, May, 2010. Smithsonian contributions to museum conservation, number 3. Smithsonian Institution Scholarly Press, Washington, DC, pp 205–207

Gorel F (2010) Assessment of agar gel loaded with micro-emulsion for the cleaning of porous surfaces. In: CeROArt. http://ceroart.revues.org/1827?lang=fr. Accessed 3 May 2011

Learner T (2012) Modern paints. In: Hill S, Rushfield A (eds) Conservation of easel paintings. Routledge, New York, pp 242–251

Mecklenburg M et al (2013) New insights into the cleaning of paintings. Proceedings from the cleaning 2010 international conference, Valencia, May, 2010. Smithsonian contributions to museum conservation, number 3. Smithsonian Institution Scholarly Press, Washington, DC

Mills L et al (2008) Water sensitivity of modern artists' oil paints. In: Janet Bridgland (ed) ICOM committee for conservation, 15th triennial meeting, New Delhi, Sept 2008. Preprints vol II. Allied Publishers, New Delhi, pp 651–659

Morrison R et al. (2007) An investigation of parameters for the use of citrate solutions for surface cleaning unvarnished paintings. Stud Conserv 52(4):255–270

Ormsby BA (2013) An empirical evaluation of a range of cleaning agents for removing dirt from artists acrylic emulsion paints. In: AIC specialty group postprints, 23rd edn., pp 77–87
Phenix A, Burnstock A (1992) The removal of surface dirt on paintings with chelating agents. Conservator 16:28–38
Sánchez-Ledesma A et al (2013) Effects of commercial soaps on unvarnished painted surfaces: A pilot study for their assessment. In: Mecklenburg MF et al (eds) New insights into the cleaning of paintings. proceedings from the cleaning 2010 international conference, Valencia, May, 2010. Smithsonian contributions to museum conservation, number 3. Smithsonian Institution Scholarly Press, Washington, DC, pp 185–196
Silvester G et al (2014) A cause of water-sensitivity in modern oil paint films: the formation of magnesium sulphate. Stud Conserv 59(1):38–51
Soldano A, Van den Berg KJ (2014) Investigation into the surface conductivity of water-sensitive modern oil paints. This proceedings, Chapter 27
Soppa K et al (2013) Lokalisierung von Konsolidierungsmitteln in Gemälden durch Fluoreszenzmarkierung. Teil I: Einführung in die Verfahrenstechnik und Anwendungsbeispiele bei aufstehender Malschicht auf textilem Bildträger. Zeitschrift für Kunsttechnologie und Konservierung (ZKK) 2/2013, pp 195–217
Tempest H et al (2013) Sensitivity of oil paint surfaces to aqueous and other solvents. In: Mecklenburg M F et al (eds) New insights into the cleaning of paintings. Proceedings from the cleaning 2010 International Conference, Valencia, May, 2010. Smithsonian contributions to museum Conservation, number 3. Smithsonian Institution Scholarly Press, Washington, DC, pp 107–114
Tortajada Hernando S, Blanco Domíngez M (2013) Cleaning plaster surfaces with agar-agar gels: evaluation of the technique. Ge-Conservación 4:111–126
Van den Berg KJ, Burnstock A (2014) Twentieth century oil paint. The interface between science and conservation and the challenges for modern oil paint research. These proceedings, Chapter 1
Volk A (2012) Approaches towards the surface cleaning of water sensitive oil paint with aqueous methods. Diploma thesis Staatliche Akademie der Bildenden Künste Stuttgart, unpublished
Von Saint-George C (2013) Die Ästhetik der Oberfläche-Vom Umgang mit Firnis in der deutschen Malerei um 1900 mit abschließendem Blick auf die "Brücke". Zeitschrift für Kunsttechnologie und Konservierung (ZKK) 1/2013, pp 105–124
Wijnberg L et al (2007) Jasper Johns untitled 1964–65. Art Matters 4:68–80
Wijnberg L et al (2008) Watergevoelige olieverfschilderijen: werkwijze van Jasper Johns en Karel Appel. kM 66, pp 23–26

Investigation into the Surface Conductivity of Water-Sensitive Modern Oil Paints

Alexia Soldano and Klaas Jan van den Berg

Abstract This study investigated the practical use of conductivity measurements on modern unvarnished oil paint surfaces. The aim was to correlate recent developments in the measurement of paint conductivity with observations during surface cleaning of samples of unvarnished oil paint. Richard Wolbers has demonstrated the use of conductivity measurements for adapting aqueous cleaning solutions to remove soils from particular acrylic surfaces so as to minimise the ionic disruption at the surface of paintings. We have observed the differences in response from oil-based paint films through an experimental investigation of sample paint films as well as a painting by Jasper Johns *Untitled 1964–1965*. The conductivity measurements recorded on these surfaces were compared to the observed water-sensitivity of the different paint surfaces in the aim of finding possible correlations. It was evident from this study that the link between sensitivity and swelling of the paint film in the case of oil films is not as clear as it is with acrylics. The different additives, in the form of salts, can influence the measured conductivity.

Keywords Conductivity • Aqueous cleaning • Water sensitivity • Ionic solutions • Jasper Johns • Unvarnished oil paint • Swelling

Introduction

Recent developments in surface cleaning science have included measuring the conductivity of a paint surface as a guide to developing a method for wet-surface cleaning works of art (Wolbers 2000). Conductivity measurements have proven useful in the provision of guidelines for cleaning acrylic paintings with respect to minimizing paint swelling potential during wet surface cleaning using aqueous ionic

A. Soldano (✉)
Private Conservator, Paris, France
e-mail: alex_soldano@hotmail.com

K.J. van den Berg
Cultural Heritage Agency of the Netherlands, Amsterdam, The Netherlands
e-mail: K.van.den.Berg@cultureelerfgoed.nl

K.J. van den Berg et al. (eds.), *Issues in Contemporary Oil Paint*,
DOI 10.1007/978-3-319-10100-2_27

solutions (Ormsby 2009, 2010). However, the surface conductivity of unvarnished modern oil paintings, especially those that are sensitive to aqueous cleaning, has not been fully explored. Paint sensitivity, in the case of oil paints can be caused by a number of different reasons (Mills et al. 2008; Silvester et al. 2014), however, we must also consider swelling as a response to aqueous swabbing.

The aim of this study is to explore links between surface conductivity and sensitivity to aqueous swabbing during surface cleaning on a set of specified samples.[1] The theory of matching the conductivity of the paint surface and the cleaning solution predicts the prevention of unnecessary swelling (Wolbers et al. 2013; Osmond and Carter 2013). This study will explore the potential for using data derived from surface conductivity measurements for designing appropriate aqueous cleaning solutions to minimize the sensitivity of some modern oil paint films. This was experimented on a range of water sensitive paint surfaces. Data was also obtained from surface conductivity measurements of a painting *Untitled 1964–1965* by Jasper Johns (Burnstock et al. 2006; Silvester et al. 2014).[2]

Definition of Conductivity and Its Application to Paintings

The conductivity of a solution is the measure of its ability to conduct an electric current. It is influenced by its ionic concentration, the type of ions present[3] as well as temperature and time. The conductivity of a paint surface is measured by bringing some of the ions present in the paint film to diffuse into deionised water (zero conductivity) and then measuring the conductivity of that solution using a conductivity meter. The conductivity of a paint film can be affected by pigment type, ionic additives in the paint formulations as well as superficial deposits and even cleaning system residues.

The unit for measuring conductivity is the milliSiemens per centimetre (mS/cm) or microSiemens per centimeter (μS/cm). It is most commonly measured using a well-plate conductivity meter.[4] A drop of the material to be measured is placed on the paint surface for a given time (1 min has become standard)[5] using a pipette. The drop is then retrieved and placed in the reservoir of the conductivity meter

[1]The tests done in this study were carried out in 2010 (Soldano 2010).

[2]Painting inventory n° A24222 Stedelijk Museum, Amsterdam. It was investigated previously for its unusual sensitivity for aqueous treatment. Studies revealed that the upper paint layer contains high concentrations of soluble salts.

[3]Ions are electrolytes which are solutions of acids, bases or salts.

[4]Commonly used is the Horiba B-173 Twin Conductivity Meter (not manufactured anymore) or HORIBA LAQUAtwin B-771 Compact Conductivity Meter.

[5]Preliminary investigations have shown that shorter measuring times demonstrate more inaccuracies. However, in an effort to minimize risk of using water on sensitive surfaces, 1 min has been suggested (Ormsby et al. 2013). This enables values to be compared among conservators for one given surface type.

which contains two metal plates. The strength of the current that passes through is measured. The minimum volume to cover both plates is approximately 50 μL.

Recent developments have included using filter paper discs and agarose gel pellets (Larochette 2012) as in the case of sensitive surfaces to aqueous products, there is an elevated risk of leaving a permanent mark on the surface after the retrieval of the water droplet.

Swelling of Paint Films

Recent studies have demonstrated the reduction of paint disruption through the swelling of acrylic paint by adjusting the conductivity and pH of the aqueous solution used as the cleaning solution (Stavroudis and Doherty 2013; Wolbers et al. 2013; Dillon et al. 2012; Kampasakali et al. 2011). Other studies have focused on the swelling of the oil component in paint films through solvents and its effect on loss of pigment binding and vulnerability to mechanical pick-up during cleaning (Phenix 2002, 2013).

The model for paint swelling behaviour operates according to osmosis, as presented in Fig. 1 (Wolbers 2009).

For hypotonic solutions, the cleaning solution contains fewer ions than the material's surface, for example deionised water. When in contact, some of the ions from the paint will move into the solution, causing swelling and/or shrinkage. With a hypertonic solution, the solution contains a higher ionic content than the paint surface (for example a TAC solution). The ions from the solution will diffuse into the paint film to reach an equilibrium state, causing the paint to swell and/or shrink. In the case of an isotonic solution, the ion content is relatively similar to the paint film. Therefore the exchange will have a net zero effect. As the isotonic system causes the least disruption to the surface of the paint film (translated as 'paint sensitivity'), these conditions are ideally the ones to be used in an aqueous treatment.

This study explores the links between oil paint sensitivity and conductivity of oil paint surfaces and cleaning solutions based on the theories described above, on a set of experimental samples. A range of solutions of different ionic character were tested at different adjusted conductivities on paint surfaces which demonstrated different degrees of sensitivity swabbing using water.

Test Samples and Painting

The samples selected for study, as described in Table 1, are unvarnished oil paint surfaces. Two paint samples (A and B) originate from discarded paintings made in the 1970s which have been kept in storage.[6] The three other set of samples

[6]Cultural Heritage Agency of the Netherlands (RCE) collection depots, formerly known as the Rijksdienst voor Beeldende Kunst (RBK).

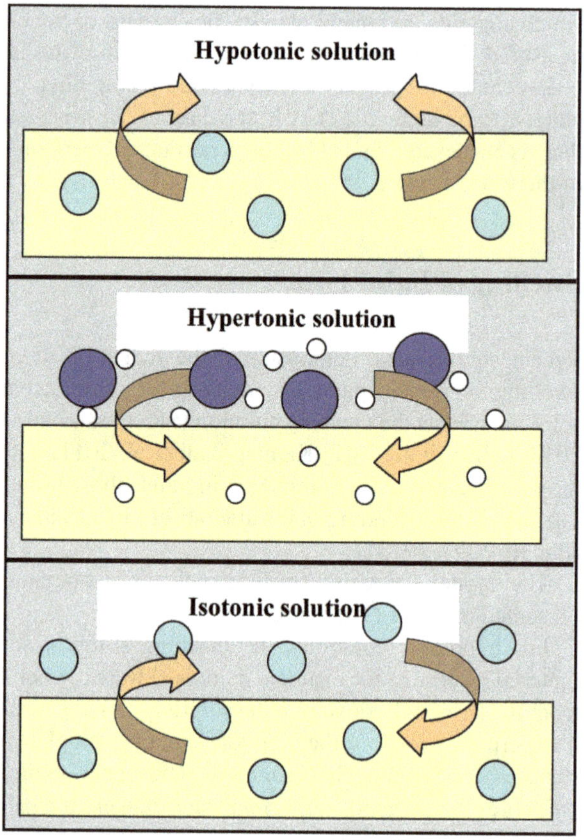

Fig. 1 Ionic exchange when a solution comes in contact with a paint film (Wolbers 2009)

were cast from paint tubes onto small canvas boards[7] and were artificially aged in 2008.[8] These were stored in relatively protected conditions in the RCE Amsterdam laboratory.

The conductivity of each sample was measured using a 50 μL droplet of deionised (DI) water left on the surface for 1 min. Three measurements were taken in total and an average conductivity value was calculated.

Sensitivity was determined by counting the number of damp cotton swab rolls rolled on the surface of each paint until colour was observed on the white swab

[7]Previous student research at RCE by Hannah Tempest; paint outs were made by Polly Saltmarsh.

[8]After 3 months of curing, artificial aging with fluorescent light tubes (36 W Philips Colour 96, in combination with Lexan® polycarbonate sheets to filter out ultraviolet) of test paint films was carried out at c. 44 % RH and 25 °C. Equivalent of 25•2 years of museum lighting (200 lux, 8 h a day, assuming reciprocity).

Table 1 Sample description

Sample	Paint/Pigment type	Paint brand	Average conductivity	Average water sensitivity (from 1 to 5)	Date of creation	Comments
A	Mixture of blue and red paint (Ultramarine)	Unknown	195 μS/cm	3	1976	Thin layer of dust visible on the surface. Paint appears brittle. The surface is smooth
B	Blue paint (Ultramarine)	Unknown	137 μS/cm	3	1976	Thin layer of dust visible on the surface. The surface is slightly powdery
C	Raw Sienna[a]	Old Holland	12 μS/cm	3	2008	No surface dust. Surface is very smooth
D	Raw Sienna[a]	Talens	138 μS/cm	5	2008	No surface dust. Surface is very smooth
E	Chromium oxide green[a]	Talens	8 μS/cm	4	2008	No surface dust. Surface is very smooth

[a]Paints composition described in Mills et al. (2008); sensitivities described in Tempest et al. (2013)

(the sensitivity point). The number of swab rolls to the sensitivity point on all paint surfaces tested ranged from 4 to 48. Sensitivity was then rated on a scale from 1 to 5, one being a very low sensitivity with no disruption of the film surface before over 40 swab rolls. Five represents a very high sensitivity with colour visible on the cotton swabs after 4–12 swab rolls.

The solutions selected for testing were: DI water (conductivity of ~0 μS/cm); sodium chloride and sodium phosphate dibasic (both salts) and triammonium citrate (TAC, chelating agent).[9]

Each solution was chosen for its different inherent ionic character and because they are already used as ionic buffers in conservation (Wolbers 2000). Each solution was made in a range of conductivities, from 50 μS/cm to 10,000 μS/cm. All solutions were adjusted to a neutral pH of 7.5.[10,11]

A cotton swab was dipped in the solution, slightly blotted prior to application and then rolled on a section of the paint surface until colour/pigment was visible. The number of swab rolls before this sensitivity point was noted and rated according to the scale described above. This was repeated with all solutions on all the different paint surfaces at least five times to obtain sufficient data.

Results

Table 2 presents the sensitivity ratings attributed to each section after swabbing with the different aqueous solutions adjusted to the range of conductivities presented above. These are averaged over the five different readings that were obtained on each sample. The results obtained were fairly consistent with low standard deviation. In green are highlighted the solutions that caused the least paint sensitivity on the samples. Both are occurring on sample E. As visible in Table 1, sample E is the paint surface with the lowest conductivity value (8 μS/cm).

In red are highlighted the solutions which caused most sensitivity when swabbing. It is interesting to note that on sample D, the salt adjusted solutions at high conductivities caused high sensitivity of the paint surface. Sample D has an average conductivity of 138 μS/cm and was the most sensitive of our paint samples at the beginning of the study.

[9]Carbonated water was first included in the list of cleaning solutions. Because of its inherent conductivity of approximately 1,000 μS/cm it could not be made in the same ranges and was later excluded.

[10]Wolbers recommends cleaning oil-bound paint surfaces at a pH between 5.5 and 8.5 to minimise potential hydrolysis of oil paints (Wolbers 2000). We have chosen a pH of 7.5.

[11]NaCl adjusted to pH 7.5 with sodium hydroxide. Sodium phosphate dibasic adjusted to pH 7.5 with hydrochloric acid.

Table 2 Sensitivity ratings when swabbing with adjusted conductivity solutions

Solution and adjusted conductivity values (µS/cm)		Sample number and sensitivity rating[a]				
		A	B	C	D	E
		Discarded paintings		Prepared 2008 samples		
DI water	0	3	3	3	5	4
NaCl	50	4	4	4	4	3
	500	4	4	3	4	3
	1,000	5	5	3	4	4
	2,500	4	4	3	5	3
	5,000	3	3	4	5	2
	10,000	4	3	4	5	3
Sodium phosphate dibasic	50	4	3	3	5	3
	500	4	3	3	4	4
	1,000	5	4	3	5	3
	2,500	5	4	4	5	4
	5,000	4	4	4	5	4
	10,000	4	3	5	5	4
Triammonium citrate	50	4	4	3	4	2
	500	4	4	4	4	4
	1,000	4	4	4	4	4
	2,500	4	4	3	4	4
	5,000	4	4	4	4	4
	10,000	5	5	5	5	4

[a]Solvent sensitivity scale paint surface: 1 -very unsensitive, 5 -highly sensitive, See text.

Case Study: A Water-Sensitive Painting

The surface of most colours except black and white in Jasper Johns' *Untitled 1964–1965* was found to be sensitive to water (Burnstock et al. 2006). Recently the presence of water soluble magnesium sulphate heptahydrate (epsomite) has been given as an explanation for its water sensitivity. The salt is present in the uppermost

Table 3 Conductivity and sensitivity values on different colours of *Untitled 1964–1965*

Colour	Conductivity (μS/cm)	Sensitivity rating from 1 to 5 (number of swab rolls)	Pigments/compounds revealed in paint analysis[a]
Yellow	600	4	Cadmium (zinc) yellow, zinc white and barium sulphate (and/or lithopone), magnesium carbonate and aluminium compounds in oil
Blue	1,170	5	Cobalt blue, sodalite and/or french ultramarine, lead white, titanium dioxide (anatase), barium sulphate, magnesium carbonate, quartz, calcite in oil
Light blue	400	1	NA
Red (background)	360	3	Cadmium red, red iron oxide, barium sulphate, magnesium carbonate, unidentified aluminium compounds in oil
Red (rainbow)	810	4	NA
Purple	500	3	Cobalt violet, red lake containing Br, barium sulphate, magnesium carbonate and aluminium compounds
Orange	590	2	Cadmium sulphide orange, quartz, barium sulphate, magnesium carbonate and aluminium compounds
Orange (blue panel)	840	5	NA
Green	225	1	Viridian, quartz, barium sulphate, zinc and aluminium compounds
Black	187	3	NA
Grey	135	1	NA
White	200	NA	NA
Canvas priming	280	NA	Lead white, chalk, barium sulphate in oil

[a]Paint analysis results from Burnstock et al. (2006). Magnesium sulphate most probably present on surfaces with paint containing magnesium sulphate (Silvester et al. 2014, Cooper et al., Chapter 20, these Proceedings)

layer in the paint and was formed from magnesium carbonate present as filler after reaction with atmospheric sulphur dioxide (Silvester 2011; Silvester et al. 2014; Cooper et al. 2014).

Conductivity measurements were made parallel to sensitivity tests on the painting. The tests were carried out on different colours and results are presented in Table 3.

The conductivity values measured on the different sample oil paint surfaces range between 8 and approximately 200 μS/cm. The Jasper Johns painting presented much higher conductivity values compared with the other oil paint surfaces that were measured previously while also presenting surprisingly high water-sensitivity. The most sensitive paint surfaces (orange and blue) showed the highest conductivity values.

Observations and Conclusions

While reported surface conductivity measurements on acrylic paints range between 0 and 1,200 μS/cm (Ormsby et al. 2013), the measurements recorded from most oil paint surfaces in this study were significantly lower, from 5 to 200 μS/cm. However, the water sensitive paints in *Untitled 1964–1965* presented readings much higher than this (135–1,170 μS/cm, with water sensitive surfaces above 300 μS/cm). Whereas the low readings on experimental paint samples surfaces is likely to be related to salts and soluble components in surface dirt,[12] the high readings in *Untitled 1964–1965* may be explained by the dissolution of epsomite crystals embedded in the medium at the paint surface during measurement.

The potential for using conductivity readings for indicating the presence of water soluble salts is highlighted in this study. This may become useful for suggesting the presence of epsomite or other water soluble salts during a diagnostic examination.[13] This observation is in accordance with a recent study by Volk on a painting by Karel Appel, where high conductivities of over 600 μS/cm are linked with water sensitive surfaces whereas conductivities below 300 μS/cm are found on non-sensitive paints (Volk 2012).[14]

It is clear that the swelling effect in acrylic paints is not the same as in oil systems. In acrylics, swelling is affected by the coalescent phase during drying, the additives (surfactants migrating to the surface, thickening agents, extenders), and free acrylic acids in the film (Ormsby et al. 2007; Ormsby and Learner 2009). In oil paint systems, soluble components such as free fatty acids will also be present, however swelling will depend very much on the type of oil paint. Some oil paints which contain high amounts of polyvalent ions (such as lead) form dense ionomeric structures which are not prone to swelling with aqueous solvents (van den Berg et al. 1999). Swelling induced by ion exchange and leeching of components from the paint film will more readily occur where relatively high contents of organic degradation products are present which will have quite low conductivity readings. Swelling could also occur in films with high conductivity readings caused by high contents of soluble salts such as epsomite.

Evidence suggests that oil paint films tend to demonstrate lower surface conductivity measurements. Following Richard Wolbers' observations that aqueous systems should be used at a conductivity approximately ten times the one measured on the surface, appropriate solutions for surface cleaning might be in a range of 50–2,000 μS/cm (Wolbers 2000). However, the present study did not show a linear relationship between ionic strength, conductivity and paint sensitivity, especially with the salt-based systems. The action of the aqueous system is therefore

[12]An extensions to this study will notably investigate the influence of specific dirt deposits on conductivity.

[13]Identification of epsomite otherwise has to be carried out using instrumental analysis.

[14]An article on the technical research and surface cleaning of this painting is in preparation for ZKK.

more complex involving the adjustment of ionic content and pH. Non-mechanical cleaning or clearing methods such agar gels may then become useful (Volk and Van den Berg, this volume).

In conclusion, this study suggests that measuring conductivity of oil paint surfaces showed potential for empirically characterising different broad classes of water sensitive paint. Methodology can influence results and it is therefore essential that conservators agree on a standard method for recording data.[15] Difficulties during this research were posed by the lack of sufficient data from which to draw conclusions. To follow up this research, a set of wider data set will be compiled by initiating and gathering conservator-led conductivity measurements on a broader range of paint surfaces designed to provide a closer understanding between the paint type, degree of paint sensitivity, and choice of treatment. A database will be created from case studies that will allow compilation of more information.[16] In addition, future research will aim to test more conductivity-adjusted solutions as well as their possible adaptation to oil paint surfaces.

Acknowledgements Bronwyn Ormsby, Senior Conservation Scientist, Tate, London. Louise Wijnberg, Painting conservator, Stedelijk Museum, Amsterdam

References

Burnstock A, van den Berg KJ, de Groot S, Wijnberg L (2006) An investigation of water-sensitive oil paints in 20th century paintings. In: Learner T (ed) Reprints of the modern paints uncovered conference, London 2006. Getty, Los Angeles, pp 177–188

Cooper A, Burnstock A, van den Berg KJ, Ormsby B (2014) Water sensitive oil paints in the twentieth century: a study of the distribution of water-soluble degradation products in modern oil paint films. These proceedings, Chapter 20

Dillon C, Lagalante F, Wolbers R (2012) Acrylic emulsion paint films: the effect of solution pH, conductivity, and ionic strength on film swelling and surfactant removal. Stud Conserv 59(1):52–62

Kampasakali E, Ormsby B, Phenix A, Schilling M, Learner T (2011) A preliminary study into the swelling behaviour of artists' acrylic emulsion paint films. In: Bridgland J (ed) ICOM-CC 16th triennial conference preprints, Portugal, International Council of Museums, Italy, paper 1008

Larochette Y (2012) Wolber's world: a review of a textile wet-cleaning workshop held in Oaxaca, Mexico. WAAC Newsl 34(1):24–26

Mills L, Burnstock A, de Groot S, Megens L, Bisschoff M, van Keulen H, Duarte F, van den Berg KJ (2008) Water sensitivity of modern artists' oil paints. In: ICOM committee for conservation 15th triennial meeting, New Delhi, pp 651–659

Ormsby B (2009) Cleaning of acrylic paint surfaces (CAPS): research into practice Colloquium, July 7–11. Getty Center, Los Angeles

Ormsby B (2010) Personal communication

[15] For example: 100 μL droplet left on the surface for 1 min (or agarose pellets for sensitive surfaces). If measurements are taken in a different way, the result obtained may be influenced.

[16] In the aim of collecting as many elements as possible for each case study while using the data with a critical perspective.

Ormsby B, Learner T (2009) The effects of wet surface cleaning treatments on acrylic emulsion artists' paints – a review of recent scientific research. Rev Conserv 10:29–41

Ormsby B, Learner T, Foster G, Druzik J, Schilling M (2007) Wet-cleaning acrylic emulsion paint films: an evaluation of physical, chemical and optical changes. In: Learner T, Smithen P, Kruger JW, Schilling MR (eds) Modern paints uncovered. Los Angeles, Getty Conservation Institute, pp 187–198

Ormsby B, Barker R, Hellen R, Smithen P (2013) Cleaning acrylic emulsion paintings: case study treatments, evaluation and reflections. In: What's changing: theories and practices in the restoration of contemporary art. Castello di Rivoli

Osmond G, Carter A (2013) The effect of conductivity on water solubility: cleaning a modern Chinese oil painting. In: New insights into the cleaning of paintings: proceedings from the cleaning 2010 International Conference Universidad Politecnica de Valencia and Museum Conservation Institute, Smithsonian Institution Scholarly Press, pp 115–118

Phenix A (2002) The swelling of artists' paints in organic solvents. Part 1, a simple method for measuring the in-plane swelling of unsupported paint films. J Am Inst Conserv 41(1):43–60

Phenix A (2013) Effects of organic solvents on artists' oil paint films: swelling. In: New insights into the cleaning of paintings: proceedings from the cleaning 240 2010 international conference Universidad Politecnica de Valencia and Museum Conservation Institute, Smithsonian Institution Scholarly Press, pp 69–76

Silvester G (2011) An experimental investigation of the formation of sulphates in oil paint films exposed to gaseous sulphur dioxide with particular reference to the relationship between these sulphates and water sensitivity. Courtauld Final Year Project, London

Silvester G, Burnstock A, Megens L, Learner T, Chiari G, van den Berg KJ (2014) A cause of water-sensitivity in modern oil paint films: the formation of magnesium sulphate. Stud Conserv 59(1):38–51

Soldano A (2010) Investigation of the practical use of conductivity measurements on water-sensitive modern oil paints. (unpublished student research)

Stavroudis C, Doherty T (2013) The modular cleaning program in practice: application to acrylic paintings. In: New insights into the cleaning of paintings: proceedings from the cleaning 2010 International Conference Universidad Politecnica de Valencia and Museum Conservation Institute, Smithsonian Institution Scholarly Press, pp 139–145

Tempest H (2013) Sensitivity of oil paint surfaces to aqueous and other solvents. In: Mecklenburg MF (ed) New insights into the cleaning of paintings. Proceedings from the cleaning 2010 International Conference, Valencia, May, 2010, vol 3, Smithsonian contributions to museum conservation. Smithsonian Institution Scholarly Press, Washington, DC, pp 107–114

van den Berg JDJ, van den Berg KJ, Boon JJ (1999) Chemical changes in curing and ageing oil paints. In: Bridgland J (ed) ICOM committee for conservation 12th triennial meeting, Lyon, France, 29 August–3 September 1999. James & James, London, pp 248–253

Volk A (2012) MA Dissertation. Staatliche Akademie der Bildenden Künste Stuttgart

Wolbers R (2000) Cleaning painted surfaces: aqueous methods. Archetype, London

Wolbers R (2009) Cleaning of acrylic paint surfaces (CAPS): research into practice Colloquium, July 7–11. Getty Center, Los Angeles

Wolbers R, Norbutus A, Lagalante A (2013) Cleaning of acrylic emulsion paints: preliminary extractive studies with two commercial paint systems. In: New insights into the cleaning of paintings: proceedings from the cleaning 2010 International Conference Universidad Politecnica de Valencia and Museum Conservation Institute, Smithsonian Institution Scholarly Press

Index

K.J. van den Berg et al. (eds.), *Issues in Contemporary Oil Paint*,
DOI 10.1007/978-3-319-10100-2

GPSR Compliance
The European Union's (EU) General Product Safety Regulation (GPSR) is a set of rules that requires consumer products to be safe and our obligations to ensure this.

If you have any concerns about our products, you can contact us on

ProductSafety@springernature.com

In case Publisher is established outside the EU, the EU authorized representative is:

Springer Nature Customer Service Center GmbH
Europaplatz 3
69115 Heidelberg, Germany

www.ingramcontent.com/pod-product-compliance
Lightning Source LLC
Chambersburg PA
CBHW060744150926
51501CB00003B/83
* 9 7 8 3 3 1 9 1 0 0 9 9 9 *